AF606470

VOLUME SIXTY THREE

ADVANCES IN BOTANICAL RESEARCH

ADVANCES IN BOTANICAL RESEARCH

Series Editors

Jean-Pierre Jacquot
Professeur, Membre de L'Institut Universitaire, de France, Unité Mixte de Recherche INRA, UHP 1136 "Interaction Arbres Microorganismes", Université de Lorraine, Faculté des Sciences, Vandoeuvre, France

Pierre Gadal
Professor honoraire, Université Paris-Sud XI, Institut Biologie des Plantes, Orsay, France

VOLUME SIXTY THREE

MITOCHONDRIAL GENOME EVOLUTION

Volume Editor

Laurence Maréchal-Drouard
Institut de Biologie Moléculaire des Plantes,
UPR 2357 du CNRS, Université de Strasbourg,
Strasbourg, France

Amsterdam • Boston • Heidelberg • London
New York • Oxford Paris • San Diego
San Francisco • Singapore • Sydney • Tokyo
Academic Press is an imprint of Elsevier

Academic Press is an imprint of Elsevier
The Boulevard, Langford Lane, Kidlington, Oxford, OX51GB, UK
32, Jamestown Road, London NW1 7BY, UK
Radarweg 29, PO Box 211, 1000 AE Amsterdam, The Netherlands
225 Wyman Street, Waltham, MA 02451, USA
525 B Street, Suite 1900, San Diego, CA 92101-4495, USA

First edition 2012

ISBN: 978-0-12-394279-1
ISSN: 0065-2296

Printed and bound in USA
12 13 14 10 9 8 7 6 5 4 3 2 1

CONTENTS

CONTRIBUTORS

Linda Bonen
Biology Department, University of Ottawa, Ottawa, Canada

Françoise Budar
INRA, UMR1318, Institut Jean-Pierre Bourgin, Versailles, France; AgroParisTech, Institut Jean-Pierre Bourgin, Versailles, France

Gertraud Burger
Robert Cedergren Centre of Bioinformatics and Genomics, Department of Biochemistry, Université de Montréal, Montréal, QC, Canada

Anne-Marie Duchêne
Institut de Biologie Moléculaire des Plantes, UPR 2357 du CNRS, Université de Strasbourg, Strasbourg, France

Owen Duncan
Australian Research Council Centre of Excellence in Plant Energy Biology, University of Western Australia, Crawley, WA, Australia

Pavel Flegontov
Biology Centre, Institute of Parasitology, Czech Academy of Sciences and Faculty of Sciences, University of South Bohemia, České Budějovice (Budweis), Czech Republic

Sota Fujii
University of Kyoto, Oiwakecho, Kitashirakawa, Sakyo-ku, Kyoto, Japan

Philippe Giegé
Institut de Biologie Moléculaire des Plantes du CNRS, University of Strasbourg, Strasbourg, France

Elzbieta Glaser
Department of Biochemistry and Biophysics, Arrhenius Laboratories for Natural Science, Stockholm University, Stockholm, Sweden

Anthony Gobert
Institut de Biologie Moléculaire des Plantes du CNRS, University of Strasbourg, Strasbourg, France

Diego González-Halphen
Instituto de Fisiología Celular, Universidad Nacional Autónoma de México, México D.F., Mexico

José M. Gualberto
Institute de Biologie Moléculaire des Plantes-CNRS-UPR2357, Université de Strasbourg, Strasbourg, France

Bernard Gutmann
Institut de Biologie Moléculaire des Plantes du CNRS, University of Strasbourg, Strasbourg, France

Nancy J. Hepburn
Center for Plant Science Innovation, University of Nebraska, Lincoln, NE, USA; Department of Agronomy and Horticulture, University of Nebraska, Lincoln, NE, USA

Koh-ichi Kadowaki
National Agriculture and Food Research Organization (NARO), Institute of Crop Sciences, Tsukuba, Ibaraki, Japan

Kanika Jain
Center for Plant Science Innovation, University of Nebraska, Lincoln, NE, USA; School of Biological Sciences, University of Nebraska, Lincoln, NE, USA

Beata Kmiec
Department of Biochemistry and Biophysics, Arrhenius Laboratories for Natural Science, Stockholm University, Stockholm, Sweden

Kristina Kühn
Institut für Biologie, Humboldt-Universität zu Berlin, Germany

B. Franz Lang
Robert Cedergren Centre of Bioinformatics and Genomics, Department of Biochemistry, Université de Montréal, Montréal, QC, Canada

Julius Lukeš
Biology Centre, Institute of Parasitology, Czech Academy of Sciences, and Faculty of Sciences, University of South Bohemia, České Budějovice (Budweis), Czech Republic

Morgane Michaud
Institut de Biologie Moléculaire des Plantes, UPR 2357 du CNRS, Université de Strasbourg, Strasbourg, France

Jeffrey P. Mower
Center for Plant Science Innovation, University of Nebraska, Lincoln, NE, USA; Department of Agronomy and Horticulture, University of Nebraska, Lincoln, NE, USA

Monika W. Murcha
Australian Research Council Centre of Excellence in Plant Energy Biology, University of Western Australia, Crawley, WA, Australia

Claire Remacle
Genetics of Microorganisms, Institute of Plant Biology, University of Liège, Liège, Belgium

Elizabeth Rodríguez-Salinas
Instituto de Fisiología Celular, Universidad Nacional Autónoma de México, México D.F., Mexico

Pascal Touzet
Laboratoire de Génétique et Evolution des Populations Végétales, UMR CNRS 8198, Université des Sciences et Technologies de Lille - Lille1, Villeneuve d'Ascq cedex, France

Minoru Ueda
Department of Botany, Graduate School of Science, Kyoto University, Kyoto, Japan

James Whelan
Australian Research Council Centre of Excellence in Plant Energy Biology, University of Western Australia, Crawley, WA, Australia

PREFACE

The present volume is part of a comprehensive series of books of the «*Advances in Botanical Research*» that will describe the genome organization of photosynthetic bacteria up to land plants. This volume focuses on mitochondrial genome evolution in photosynthetic organisms. It includes 12 chapters written by internationally recognized authors.

The opening chapter by B. Franz Lang is logically dedicated to "Mitochondrial and eukaryotic origins". The symbiotic introduction of mitochondria that occurred some one billion years ago represents a key event in eukaryotic evolution. The origin of the eukaryotic cell and of mitochondria has intrigued life scientists for many decades. However, remaining traces are difficult to find and speculative evolutionary theories for the origin of eukaryotes and mitochondria have been developed. Several lines of evidence support that mitochondria originate from an α-proteobacterial symbiont. In this chapter, B. Franz Lang is going to discuss how the close α-proeobacterial outgroup not only improves the phylogenetic resolution of the mitochondrial heritage but also allows adressing question on eukaryotic origins. Exciting questions on the first emergence of mitochondria in eukaryotes will be presented and the two major hypotheses on the origin of eukaryotes, the "serial endosymbionts" and the symbiogenesis" hypotheses will be deeply discussed.

Present-day mitochondrial genomes have mostly nothing in common in term of form organization, expression and gene content with the ancestral bacterial genome from which they derive. Among the different forces responsible for the evolution of plant mitochondrial genomes, the loss and transfer of genes to the nucleus or the acquisition of foreign DNA play a crucial role. This is the topic of the two following chapters. In their contribution entitled "Gene content and gene transfer from mitochondria to the nucleus during evolution", Minoru Ueda and Koh-ichi Kadowaki discuss insights into gene transfer and substitution revealed by comparative genomic analysis. During the process of gene transfer from the mitochondria to the nucleus, transferred mitochondrial genes must undergo several changes to be expressed in the nuclear DNA. They present interesting data showing that studying the mitochondrial genome of angiosperms is a powerful way to understand the steps leading to a complete gene transfer. In the contrary, plant mitochondrial DNA can acquire foreign DNA either from the plastidial or the nuclear genomes *via* intracellular transfer or from other species *via* horizontal transfer. In the third chapter, Kanika Jain, Nancy J.G. Hepburn and Jeffrey P. Mower explain the "Role of horizontal transfer in shaping plant mitochondrial genomes". The role played by parasitic plants, the importance of recombination events and the fixation of the transferred gene by drift in the mechanisms of transfer will be tackled. Beyond the evolutionary significance of such process, it is obvious that understanding mechanisms of transfer into mitochondria could offer clues for a still missing plant mitochondrial transformation system.

As emphasized above, mitochondrial function depends on the coordinate action of nuclear and mitochondrial genomes. One of the best examples to illustrate the importance of mitochondrial-nuclear interaction is the trait cytoplasmic male sterility (CMS). CMS is determined by plant mitochondrial genomes and many mitochondrial genes that determine CMS can be suppressed or counteracted by the products of one or more nuclear genes known as restorer-of-fertility genes. This CMS trait has been found and is used in numerous crop species for hybrid production. In the fourth chapter entitled " Mitochondrial genome evolution and gynodioecy", Pascal Touzet explains how the coexistence of female and hermaphrodite plants, a process called gynodioecy, is under the nuclear-cytoplasmic control implicating CMS genes and unclear restorers. From an evolutionary point of view, the effect of the dynamics of gynodioecy on mitochondrial diversity and the question whether gynodioecy contributes to the fast evolution of CMS mitochondrial genomes is tackled. Our knowledge on the importance of mitochondrial-nuclear interactions in evolutionary issues is now expanded rapidly. Françoise Budar and Sota Fujii in their contribution "Cyto-nuclear adaptation in plants" also provide several lines of evidence demonstrating how the co-evolution of organellar (mitochondrial and also plastidial) and nuclear genomes led to the co-adaptation of the different genetic compartments at the species level. In a first part, they present the mechanisms underlying this co-adaptation and illustrate them by experimental data. In a second part, the question of the contribution of organelle genomes to the adaptation of plants to their environment in frame of the cyto-nuclear co-adaptation process is tackled. Improving our knowledge in this domain of research will undoubtedly be of crucial importance for crop improvement and breeding strategies in a near future.

In chapter 6, intriguing features of the "Mitochondrial genomes of photosynthetic euglenids and alveolates" are presented by Pavel Flegontov and Julius Lukes. The world of protists illustrates well the huge eukaryotic diversity. Euglenida belong to the eukaryotic supergroup Excavata, the members of which possess the most varied mitochondrial genomes in terms of their structure and gene content. Another interesting feature of the Euglenida is the presence of a green plastid acquired by secondary endosymbiosis. The major peculiarities of the mitochondrial genome of the most studied member of Euglenida, Euglena gracillis, will be detailed here. On the other hand, photosynthetic dinoflagellates and parasitic apicomplexans with a relic plastid constitute a large and diverse group within alveolates. All species of this group share the most reduced mitochondrial genome found. Very "bizarre" non-canonical features in structure and expression will be presented. From the contribution of P. Flegontov and J. Lukes, an evolutionary picture that emerges is that of convergent evolution of radically reduced mitochondrial genomes.

Another very exciting example of the evolutionary history of plant mitochondrial genomes is linked to the existence of introns. In the following chapter, Linda Bonen summarizes the "Evolution of mitochondrial introns in plants and photosynthetic microbes". Two types of introns, categorized as group I or group II according to their distinctive structural properties, play a crucial role in mitochondrial gene expression. Thanks to the huge amount of mitochondrial genome sequences obtained over recent years, the importance of introns in mitochondrial gene expression has been strengthened. The behavior of introns, the machineries required for their splicing, the relationships between splicing and other expression events in mitochondria and their potential roles are highlighted here.

Among eukaryotes, the group Archaeplastidia contains the photosynthetic organisms. They can be classified into three lineages, namely Glaucophyta, Rhodophyta and Chloroplastidia. Among the latter lineage belong higher plants and green algae. Chapter 8 written by Elizabeth Rodriguez-Salinas, Claire Remacle and Diego Gonzales-Halphen and entitled "Green algae genomics: a mitochondrial perspective" is entirely devoted to green algae mitochondrial genomes. Quite surprisingly, the increasing number of fully sequenced mitochondrial genomes of green algae revealed an extraordinary diversity in their genome organization, structure and gene content. This information, in synergy with the sequences of the plastidial and nuclear genomes of green algae, helps us to understand the phylogenetic relationships among green algae, the interconnection between their mitochondria and chloroplasts, their mitochondrial metabolic plasticity and their adaptative advantages. Green algae represent a biomass that appears to be more and more important for energy and food production. Deciphering the fundamental processes of the evolution and function of their mitochondria as well as their interaction with the other compartments of the cells represent major challenges for the next following years.

Dysfunction of mitochondrial DNA inheritance often results in detrimental phenotypes such as CMS. During evolution, plant mitochondria (and chloroplasts) have developed several strategies to maintain the integrity of the organellar genomes. In chapter 9 entitled «Recombination in the stability, repair and evolution of the mitochondrial genome», Kristina Kühn and José Gualberto describe our present knowledge on these processes. In particular, recombination processes seem to play central roles in mitochondrial DNA repair and replication. Deregulation of recombination in organelles results in genomic instability. The different recombination pathways (e.g. frequent recombination via large repeated sequences, ectopic recombination involving medium size repeats or illegitimate recombination involving sequence microhomologies), the occurrence of alternative mitochondrial DNA configurations and its dynamic nature will be discussed here as well as the multiple factors that act in these pathways to maintain plant mitochondrial DNA.

The following chapter "Mitochondrial genome evolution and the emergence of PPR proteins" deals with the evolutionary history and diversity of the pentatricopeptide repeat (PPR) proteins. In this contribution Bernard Gutmann, Anthony Gobert and Philippe Giegé highlight the most exciting recent developments related to the huge PPR family. Three main features of higher plant mitochondria are the occurrence of very large genomes, the existence of specific gene expression processes such as splicing and RNA editing and the vey high numbers of PPR proteins as compared to other eukaryotes (80 times more PPR genes in higher plants than in animal genomes). As emphasized by the authors, the incidence of these features should not be considered as independent but should rather be regarded as evolutionary connected phenomena. The huge PPR protein family was discovered only ten years ago in plants, but is clearly one of the most fascinating families of proteins working in plant organelles and this timely review relates their evolution in relation with the occurrence of the specific gene expression processes existing in plant mitochondria.

Present-day mitochondrial genomes only contain a very limited set of genes (e.g. about 35 protein-coding and 20 structural RNA-coding genes in Arabidopsis mitochondrial DNA). As plant mitochondria contain more than 1000 proteins, the biogenesis and the function of mitochondria mainly rely on the expression of nuclear-encoded genes. Thus, the majority of mitochondrial proteins and a small number of transfer RNAs (tRNAs) are imported from the cytosol to the organelle. Therefore a crucial evolutionary step was the establishment of specific intracellular macromolecule trafficking, processing and assembly machinery to allow the come back of proteins and RNAs into the organelle. Furthermore, in plants, the presence of another organelle of cynaobacterial origin, the chloroplast that also imports most of its proteins, leads to an even more complex situation. The two last chapters of this volume are devoted to protein and RNA mitochondrial import in plants. Monika Murcha, Elzbieta Glaser and James Whelan in their section " Evolution of protein import pathways" present a very detailed analysis on the targeting peptides, the signal recognition and the processing steps of mitochondrial proteins. They also offer a wide overview on the different components and multiple machineries that emerged during evolution to translocate mitochondrial proteins across the double mitochondrial membrane. The second contribution in this research field written by Morgane Michaud and Anne-Marie Duchêne and entitled "Macromolecules trafficking to plant mitochondria" gave more emphasis on three other important aspects of macromolecule trafficking. First, they focus on the phenomenon of dual targeting of proteins to mitochondria and chloroplasts, which raises interesting questions for inter-organelle communication. Second, they present the latest developments in an emerging area of research that is the subcellular localization of cytosolic mRNAs to the vicinity of mitochondria. This offers new perspectives in the biogenesis of mitochondria in the plant cells, but the implicated processes are still largely unknown. Finally, the importance of tRNA mitochondrial import for the plant mitochondrial translation machinery, and our present knowledge on the mechanisms implicated are described.

As a whole, this up-to–date book covers a wide range of topics, and must therefore form an authoritative reference in the field for years to come. It will be useful for a wide range of students and for scientists wishing to follow progress not only inside but also outside their own specialized field.

CONTENTS OF VOLUMES 35–62

Series Editor (Volumes 35–44)

J.A. CALLOW
School of Biosciences, University of Birmingham, Birmingham, United Kingdom

Contents of Volume 35

Contents of Volume 36

Contents of Volume 37

Contents of Volume 38

Contents of Volume 39

Contents of Volume 40

Contents of Volume 41

Contents of Volume 42

Contents of Volume 43

Contents of Volume 44

Series Editors (Volume 45–60)

JEAN-CLAUDE KADER
Laboratoire Physiologie Cellulaire et Moléculaire des Plantes, CNRS, Université de Paris, Paris, France

MICHEL DELSENY
Laboratoire Génome et Développement des Plantes, CNRS IRD UP, Université de Perpignan, Perpignan, France

Contents of Volume 45

Contents of Volume 46

Contents of Volume 47

Contents of Volume 48

Contents of Volume 49

Contents of Volume 50

Contents of Volume 51

Contents of Volume 52

Contents of Volume 53

Contents of Volume 54

Contents of Volume 55

Contents of Volume 56

Contents of Volume 57

Contents of Volume 58

Contents of Volume 59

Contents of Volume 60

Contents of Volume 61

Contents of Volume 62

CHAPTER ONE

Mitochondrial and Eukaryotic Origins: A Critical Review

B. Franz Lang[1] and Gertraud Burger

Robert Cedergren Centre of Bioinformatics and Genomics, Department of Biochemistry, Université de Montréal, Montréal, QC, Canada

[1]Corresponding author: E-mail: franz.lang@umontreal.ca

Contents

Abstract

It is now common knowledge that mitochondria originated about one billion years ago from a symbiotic α-Proteobacterium that resided inside a host cell. Yet, two fundamental questions remain unanswered: (1) what is the exact nature of this host organism? and (2) when did distinctive eukaryotic features emerge – simultaneously with or well before mitochondrial symbiosis? These questions have been the subject of extensive speculation due to limited phylogenetic evidence for evolutionary events

Advances in Botanical Research, Volume 63
ISSN 0065-2296,
http://dx.doi.org/10.1016/B978-0-12-394279-1.00001-6

that long ago. The two most influential theories on the origin of eukaryotes are the Serial Endosymbiont theory and the symbiogenesis hypotheses, both with multiple variants. According to the first, eukaryotes evolved gradually from an ancient lineage devoid of mitochondria (termed Archezoa), and eventually acquired an endosymbiotic α-Proteobacterium that was transformed into a subcellular organelle. In contrast, the more recent symbiogenesis theory posits that eukaryotes originated by metabolic syntrophy of an archaeal cell and a bacterial cell. The merger of two simple prokaryotic organisms would have given rise to the highly complex eukaryotic cell with its subcellular structures such as nucleus and cytoskeleton, and, according to one variant of the symbiogenesis theory, the hydrogen hypothesis, simultaneously to mitochondria. The unquestioned tenet of this hypothesis is that evolution proceeds from simple (primitive) to complex (higher in an Aristotelian sense) life forms. This review confronts these theories with known evolutionary principles and results from rigorous molecular phylogenies. Although all theories on eukaryotic origins lack evidence for one or more of the predicted evolutionary intermediates (in the form of modern descendants), symbiogenesis scenarios have an additional shortcoming: they do not account for the evolutionary time span required for creating thousands of genes that encode eukaryotic subcellular structures.

1. INTRODUCTION

The origin of the eukaryotic cell and of mitochondria has intrigued life scientists for many decades. However, with the corresponding evolutionary events dating back a billion or more years, traces are difficult to find in genome sequences of extant species. Furthermore, there have been repeated periods of massive species extinction. Due to a limited supply of reliable phylogenetic inferences, current theories for the origin of eukaryotes and mitochondria are quite speculative, and some explicitly disregard phylogenetic results.

With a rapidly growing collection of genome sequences, improved analytical methods, and reports of sophisticated phylogenetic analyses, a reality check is now in order to test for potential incoherencies in current theories on mitochondrial and eukaryotic origins. More specifically, we here revisit the question whether mitochondria emerged indeed only once, and how traditional theories on the origin of eukaryotes and mitochondria hold up against more recent symbiogenesis theories. In addition, we discuss recent phylogenetic analyses on the rooting of the eukaryotic tree and on the origin of Archaea-related eukaryotic genes (see Glossary), and the impact of these analyses on our understanding of early eukaryotic evolution.

2. THE EVOLUTION OF EUKARYOTES IS ONE OF SYMBIOTIC RELATIONSHIPS

Among the three domains of cellular organisms, Bacteria, Archaea, and Eukarya, the latter stands out by a phagocytotic mode of nutrition (most protists are microbe-eating predators) as well as numerous symbiotic relationships. For instance, animals require an elaborate population of diverse, intestinal microbes, most plants rely on mycorrhizal fungi and soil bacteria, and many unicellular eukaryotes, in particular amoeba, often carry (proteo-) bacterial intracellular symbionts.

The interplay between eukaryotes and bacterial endosymbionts can be ephemeral and casual, but also highly complex, intimate, and long lasting. An example of an orchestrated relationship is flies that depend on bacteria (e.g. *Buchnera* and *Wigglesworthia*) for viability and fecundity (Wernegreen, 2002). An intriguing instance for more occasional relationships is the symbiosis between the ciliate *Paramecium* and *Holospora* bacteria (a distant relative of rickettsial pathogens). *Holospora* bacteria multiply and mature within the nucleus, and eventually depart to start a new infection cycle, surprisingly without apparent damage to the host cell (Fujishima, Dohra, & Kawai, 1997; Fujishima & Fujita, 1985; Görtz, Lellig, Miosga, &Wiemann, 1990; Wiemann & Görtz, 1991). These examples demonstrate that eukaryotes are receptive or at least tolerant to accommodating intracellular guests, dedicating various complex subcellular structures to predation and to hosting of endosymbionts. In addition to bacterial partners, eukaryotes also team up with other eukaryotes, such as certain non-photosynthetic protists that unite with algae, which has led in several instances to secondary endosymbiosis, and when repeated, even tertiary and higher-order endosymbioses.

The above examples illustrate that the evolution of eukaryotes is characterized by symbiotic relationships, which implies that the genetic material of extant eukaryotes originates from several, if not a wide variety of sources. This fact makes the infernce of mitochondrial and eukaryotic origins a challenging task and requires most thorough phylogenetic analysis. Another important point is the outcome of symbiotic relationships. In the case of mitochondria and plastids, which equip the host with oxidative phosphorylation and photosynthesis, the benefit for the host seems evident. In other cases, an advantage is uncertain (e.g. *Holospora* symbionts), or the relationship is detrimental for one of the partners (e.g. pathogenic Rickettsiales symbionts; phylogenetic relatives of *Holospora* and mitochondria). While ancient symbiotic events are often viewed as driven by selective advantage for both partners, it is equally possible that they have started out either by pure accident, or as a consequence of biological warfare. With the strong propensity of eukaryotes to engage over and over again in new symbiotic relationships, the question arises if, as widely assumed, mitochondria have indeed emerged only once. Do we have phylogenetic evidence making multiple bacterial acquisitions improbable?

3. A SINGLE ORIGIN OF MITOCHONDRIA?

Mitochondria have become a crucial element in the debate on eukaryotic origins, because apparently all extant eukaryotic lineages either have or once had such an organelle. The assumption is that mitochondria, hydrogenosomes and mitosomes originate from the very same symbiotic event. But with an estimated emergence of mitochondria about a billion years ago (Brinkmann & Philippe, 2007), what is the accuracy of this statement? Are we able to distinguish between one or several ancient symbiotic events from within a given bacterial group? The somewhat sobering realization is that single- and multi-gene phylogenies alike have difficulty pinpointing a precise mitochondrial origin, which some analyses place within the Rickettsia lineage, others as a sister group to Rickettsia-related bacteria, and again others within free-living α-Proteobacteria (e.g. (Andersson, Karlberg, Canback, & Kurland, 2003; Andersson *et al.*, 1998; Esser *et al.*, 2004; Gray & Spencer, 1996; Lang *et al.*, 2005; Sicheritz-Ponten, Kurland, & Andersson, 1998)).

The adhesion of mitochondria and *Rickettsia*-like bacteria may indeed be an artefact of phylogenetic inference caused by A + T-rich gene sequences and high evolutionary rates. A similar issue arose recently regarding a group of marine bacteria (SAR11 clade) that was claimed to be related to mitochondria (Thrash *et al.*, 2012). It now appears that SAR11 bacteria, Rickettsiales, and mitochondria are grouped artefactually (they all share a high A + T content). When using more sophisticated phylogenetic methods and more realistic evolutionary models, SAR11 bacteria no longer associate with mitochondria (Brindefalk, Ettema, Viklund, Thollesson, & Andersson, 2011; Rodriguez-Ezpeleta & Embley, 2012). It is possible that the same is true for Rickettsiales and mitochondria, and that available data are still insufficient to solve this question.

Under these circumstances, it is most difficult to demonstrate a single ancestry of mitochondria beyond reasonable doubt, i.e. to distinguish between one or more symbiotic events. Certain gene acquisitions may result not from symbiosis but rather from predation (following the provocative 'you are what you eat' postulate; Doolittle, 1998). Transient bacteria–eukaryote associations may also have left a genetic footprint, further complicating phylogenetic inference.

In conclusion, eukaryotic history has offered ample opportunity for repeated mitochondriogeneses involving different α-Proteobacteria. Therefore, the view of a single mitochondrial origin clearly needs to be re-evaluated by rigorous and comprehensive phylogenomics with much broader datasets.

4. TRADITIONAL HYPOTHESES ON THE ORIGIN OF EUKARYOTES AND MITOCHONDRIA

The serial endosymbiosis theory (SET) posits that eukaryotes evolved gradually from a primitive stage to highly complex cells (the amitochondriate Archezoa), which at some time point acquired mitochondria through endosymbiosis (Doolittle, 1980, 1981; Margulis, 1981; Taylor, 1987). That extant eukaryotes are indeed genetic chimera was confirmed early on by phylogenetic studies, in which mitochondria and plastids group with two different bacterial lineages (Gray, Sankoff, & Cedergren, 1984). The finding of unicellular phagocytotic eukaryotes without mitochondria (initially termed Archezoa; including Microsporidia, Metamonada such as *Giardia*, and Parabasalia such as *Trichomonas*) became a corner stone of SET (Cavalier-Smith, 1983). These organisms live under anaerobic conditions and contain a regular nucleus, a cytoskeleton and a more or less rudimentary endomembrane system. They were thought to be primitively without mitochondria, implying that the hallmarks of eukaryotes were already present prior to endosymbiosis. This view was reinforced by some (but not all) ribosomal RNA (rRNA)-based phylogenies, placing these protists as the earliest diverging eukaryotic lineages (Leipe, Gunderson, Nerad, & Sogin, 1993; Sogin, Gunderson, Elwood, Alonso, & Peattie, 1989; Vossbrinck, Maddox, Friedman, Debrunner-Vossbrinck, & Woese, 1987; Woese & Fox, 1977).

The notion of Archezoa came into discredit with two realizations. First, the basal tree position of the above anaerobes turned out to be a phylogenetic artefact attracting fast-evolving species to the outgroup (Budin & Philippe, 1998; Kumar & Rzhetsky, 1996; Philippe & Adoutte, 1998; Stiller & Hall, 1999). Second, genes were discovered in these taxa that are involved in mitochondrial biogenesis (such as heat shock proteins) and group with mitochondrial and α-proteobacterial counterparts in phylogenetic analyses. Furthermore, several of these proteins were shown to be targeted to hydrogenosomes or to previously unrecognized cryptic organelles (also termed mitosomes or cryptons). This suggested that these latter organelles all derive from mitochondria (Bui, Bradley, & Johnson, 1996; Embley & Hirt, 1998; Germot, Philippe, & Le Guyader, 1996; Germot, Philippe, & Le Guyader, 1997; Mai *et al.*, 1999; Regoes *et al.*, 2005; Roger *et al.*, 1998; Tovar, Fischer, & Clark, 1999; Williams & Keeling, 2003). In conclusion, there is strong evidence that organisms previously thought to be primitively without mitochondria have secondarily lost this organelle.

These findings raise the question why all extant eukaryotes have or once had a mitochondrion, and why they all possess the distinctive characteristics of eukaryotes such as a nucleus and a cytoskeleton. It seems that the host of the mitochondrial endosymbiont was already a complex eukaryotic cell. But why is there no trace of truly amitochondriate eukaryotes? The reason might be that either biologists have failed to detect them, or that these organisms became distinct, out-competed by derived, mitochondriate, secondarily anaerobic eukaryotes. Curiously, lack of evidence for primitively amitochondriate eukaryotes was proclaimed by some as the downfall of the archezoan hypothesis, although it is illogical to turn a lack of evidence for a given theory right around as evidence against it.

5. SYMBIOGENESIS HYPOTHESES ON THE ORIGIN OF EUKARYOTES

The analysis of complete eukaryotic genomes confirms that eukaryotes are most complex genetic mosaics, with a more Archaea-related information processing machinery and bacteria-related operational genes involved in metabolism (Lopez-Garcia & Moreira, 1999; Ribeiro & Golding, 1998; Rivera, Jain, Moore, & Lake, 1998; Zillig *et al.*, 1989a; Zillig *et al.*, 1989b). As discussed above, SET explains the chimeric eukaryotic nature by endosymbiosis of an amitochondriate eukaryote with a bacterium. The other diametrically opposite interpretation is that eukaryotes only emerged by symbiogenesis (Koonin, 2010), involving metabolic symbiosis (syntrophy)

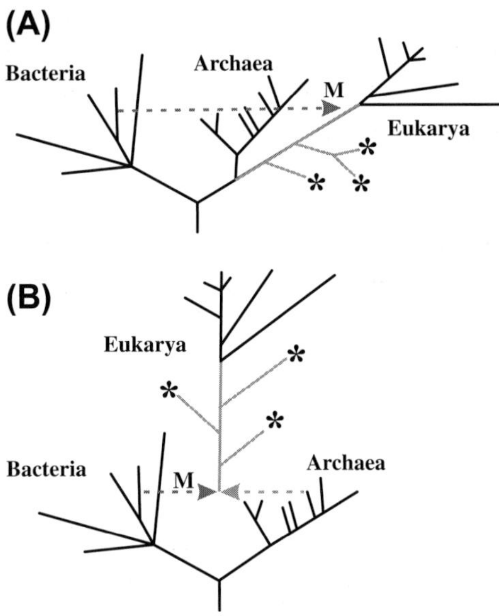

Figure 1.1 Alternative hypotheses on the origin of the eukaryotic cell. (A) The iconic three-domain tree of life based on rRNA sequence data (tree redrawn from Woese *et al.*, 1990), assuming an origin of eukaryotes by stepwise evolution from an amitochondriate eukaryote (archezoan; blue line, extinct lineages marked by dotted lines and asterisks), and subsequent acquisition of the mitochondrion (M) through endosymbiosis with an α-Proteobacterium. Note that the rooting of the tree is chosen arbitrarily within Bacteria (see also text). (B) Symbiogenesis scenario (hydrogen hypothesis), drawn as in (A) with the same phylogenetic distances. Note that eukaryotic evolution has to be extremely rapid compared with (A). This interpretation implies a long period of extinctions prior to the divergence of extant eukaryotes. The tree shape is inconsistent with the mitochondrial big bang-shaped phylogenies that fit scenario (A). For interpretation of the references to colour in this figure legend, the reader is referred to the online version of this book.

between an archaeal and a bacterial partner (Fig. 1.1). Probably the best known theory of this genre is the hydrogen hypothesis (Martin & Müller, 1998), which posits that the endosymbiotic α-Proteobacterium equipped with oxidative phosphorylation was hosted by a methanogen archaeal cell. Under the early anaerobic terrestrial atmosphere, the bacterial partner produced hydrogen, which was utilized by the hydrogen-consuming archaean. The gradual integration of both metabolic systems would have ultimately given rise to the typical eukaryotic metabolism, a portion of which localized in mitochondria, and the distinct eukaryotic cell structure. Variations of this hypothesis assume different and additional bacterial partners and another time point when mitochondria came into play (Forterre, 2010;

Moreira & Lopez-Garcia, 1998), although phylogenetic support for these more complicated scenarios is lacking. In the following, we focus on the theoretical implications of the symbiogenesis concept (Cavalier-Smith, 2009; Embley & Martin, 2006; Forterre, 2010; Gray & Arch bald, in press; Koonin, 2010; Poole & Penny, 2007).

6. IMPLICATIONS OF THE SYMBIOGENESIS HYPOTHESES

According to the hydrogen hypothesis on the origin of eukaryotes, uptake of the α-Proteobacterium (that gave rise to mitochondria) triggered an accelerated evolution of the archaeal host, leading to the nucleus, cytoskeleton, subcellular compartmentalization, and the machinery for microbial predation. This postulate, when placed into an evolutionary context, implies several inconsistencies, most of which also apply to other variants of symbiogenesis theory.

6.1. Mitochondriate Cells without Typical Eukaryotic Features

According to the hydrogen hypothesis, cell forms existed temporarily that had a mitochondrion, but neither a nucleus, a cytoskeleton nor an endomembrane system. However, descendents of these postulated organisms are not known, and it is difficult to rationalize why mitochondriate precursors of eukaryotes disappeared, despite an alleged advantage through metabolic syntrophy and tolerance to oxygen in an increasingly oxygen-rich environment. While the classic archezoan scenario struggles with lack of evidence for truly amitochondriate eukaryotes, the hydrogen hypothesis struggles with lack of evidence for intermediate forms of mitochondriate eukaryotes.

6.2. Why are Similarities between Eukaryotic and Archaeal Genes Weak...

If one of the two eukaryote-producing fusion partners was indeed an archaeal cell, then a given set of eukaryotic and archaeal gene sequences should align well in multiple sequence alignments. Yet this is not the case; more than half of the sequence position is not aligned with confidence, even for the most conserved genes (Cox, Foster, Hirt, Harris, & Embley, 2008). Furthermore, based on metabolic considerations, most symbiogenesis theories postulate an archaeal methanogen as a host. However, molecular

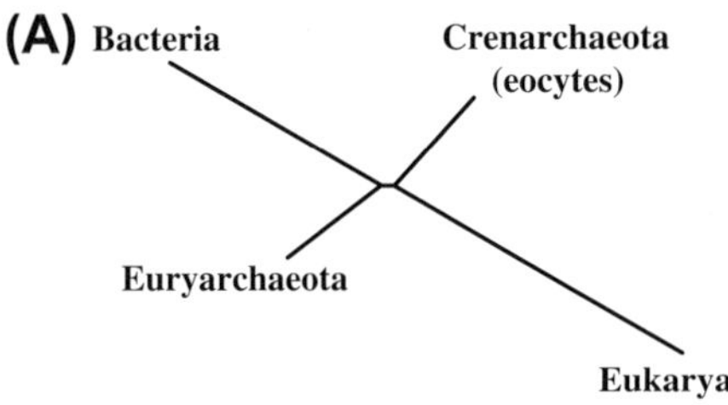

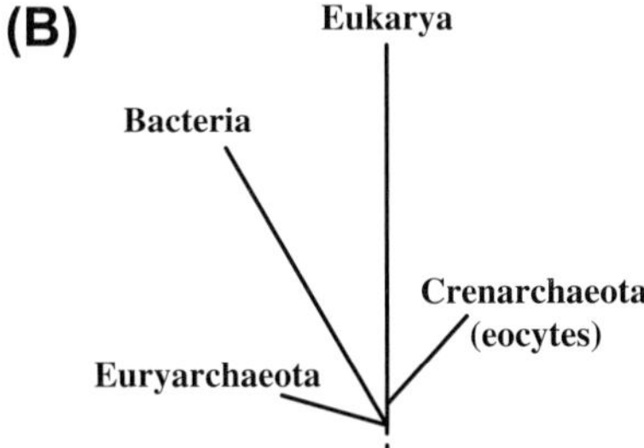

Figure 1.2 Phylogenetic tree with 49 Archaea-related eukaryotic proteins. Schematically redrawn after Cox *et al.* (2008). (A) Unrooted tree with four rather than three separate phylogenetic domains. (B) The same tree rooted arbitrarily at the divergence of Euryarchaeota and Bacteria.

phylogenies do not associate Archaea-related eukaryotic genes within methanogens, but instead with Crenarchaeota (eocytes; Fig. 1.2), and this with high statistical confidence (Cox *et al.*, 2008; Foster, Cox, & Embley, 2009). The corresponding phylogenetic analyses were conducted by using most sophisticated (realistic) composition-heterogeneous evolutionary models (such as CAT) on a carefully chosen, combined set of ~50 protein-coding genes. A long-branch-attraction (LBA) artefact is most unlikely, as it would attract the two fast-evolving groups, Eukarya and Bacteria, rather than Eukarya and Archaea. The only way to defend symbiogenesis is that the particular lineage of methanogens from which the host is supposedly derived is unrelated to extant methanogens, and thus became extinct.

6.3. ...while α-Proteobacterial Signatures are so Conspicuously Strong?

There is clear evidence for an origin of mitochondrial genes (whether encoded in mtDNA or transferred into the nucleus) from a source close to α-Proteobacteria (Andersson *et al.*, 1998, 2003; Derelle & Lang, 2011; Esser *et al.*, 2004; Gray & Spencer, 1996; Lang *et al.*, 2005; Sicheritz-Ponten *et al.*,

1998). Sequence alignments used for these inferences had to be trimmed only marginally, indicative of a relatively short phylogenetic distance to α-Proteobacteria. That eukaryotic genes display such unequal degrees of relatedness to genes from Archaea and Bacteria contradicts the supposition that Archaea-related and Bacteria-related genes of eukaryotes arose simultaneously, in a fusion event.

6.4. Genes of the Endosymbiont should Evolve Even Faster than Those of the Host

In endosymbiotic relationships, the genes of the symbiont are under much higher adaptive pressure than the genes of the host. Genes of endosymbiont origin that are transferred to the host's nucleus have to be adapted to new rules of expression and regulation. In addition, for gene products that act inside the mitochondrion, targeting signals have to be invented for import into the organelle. Present-time gene transfer from plant mtDNA to the nucleus testifies to substantial sequence changes (Adams *et al.*, 1999; Archibald & Richards, 2010). Therefore, when comparing phylogenies of nuclear genes that are mitochondria-like with trees of Archaea-related genes, the branch lengths should be shorter and the resolution higher in the latter trees, if the two gene classes originated by a merger as postulated by the hydrogen hypothesis. Yet quite the opposite is observed. Mitochondrial phylogenies (whether based on mtDNA-encoded or nucleus-transferred genes) point clearly to a mitochondrial origin close to α-Proteobacteria. In addition, mitochondrial genes allow rather confident rooting of the eukaryotic tree (Derelle & Lang, 2011) (Fig. 1.3).

6.5. Nuclear Genes Fall into Four Categories

Phylogenomic analyses distinguish four major categories of eukaryotic genes. The largest fraction contains a variety of poorly conserved genes that are of little value for phylogenetic inference (also including laterally transferred genes and potential sources of δ-proteobacterial (Moreira & Lopez-Garcia, 1998) Chlamydia-like genes that are somehow connected to photosynthesis (Huang & Gogarten, 2007), and genes of potential viral sources (Forterre, 2010; Richards & Archibald, 2011)). Genes that are well-conserved across eukaryotes further divide into three categories: Archaea-related, Proteobacteria-related, and eukaryote-specific genes (coding for eukaryote-specific functions). For the question on the origin of eukaryotes, this latter group is central. According to symbiogenesis hypotheses, these

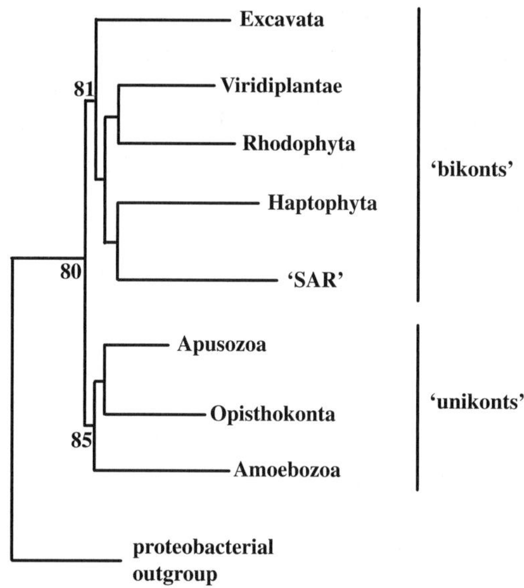

Figure 1.3 Eukaryotic tree rooted with α-Proteobacteria. Excavata: jakobids, euglenids, heterolobosean amoeba. Viridiplantae: plants and green algae. Rhodophyta: red algae. Haptophyta: *Emiliania, Pavlova*, etc. SAR: stramenopiles, alveolates, rhizarians. Apusozoa: *Thecamonas*. Opisthokonta: animals, fungi, choanoflagellates, nucleariids, etc. Amoebozoa: slime moulds. The support values represent Bayesian jackknife values (CAT model). *Adapted from Derelle and Lang (2011).*

well-conserved genes emerged shortly after the posited fusion event, but why then is there no trace of their origin (as seen with Archaea-related and Proteobacteria-related eukaryotic genes)?

One might argue that symbiogenesis occurred at a most ancient point in time, leaving ample time for eukaryote-specific genes to emerge and diverge. Yet this scenario is in conflict with phylogenies as stated earlier. It is also inconsistent as to the unexplained retention of useless genes. A mitochondrion-early scenario implies that oxidative phosphorylation was introduced when atmospheric oxygen concentrations were low, and that the corresponding gene set was preserved (although not required) over a very long evolutionary period. This seems unlikely, given what we know from present-day anoxic habitats. These environments are populated with eukaryotes that have highly degenerate mitochondria (hydrogenosomes or mitosomes), lack oxidative phosphorylation, usually tolerate low levels of oxygen, and none of them resembles a phylogenetically ancient eukaryote (Blankenship & Hartman, 1998; Forterre & Gribaldo, 2007).

6.6. Not only Distinctive Eukaryotic Genes but also Distinctive Molecular Machineries

The three-domain division of organismal life is based on phylogenetic inference using rRNA sequence data with eukaryotes well separated from Archaea and Bacteria (Vossbrinck *et al.*, 1987). In addition, as correctly pointed out (Forterre, 2010), the eukaryotic cytosolic ribosome is so different from the bacterial and archaeal ribosome that it is difficult to imagine that Eukarya arose through a fusion of an Archaea with a Bacteria, and that an Archaea-type ribosome transformed into the eukaryotic one during a period of accelerated evolution. There is no good reason that justifies re-engineering of the host's well-tuned translation system. Mito-ribosomes in contrast, particularly those from jakobid protists (e.g. *Reclinomonas americana*) (Lang *et al.*, 1997), are strikingly bacterial, although mitochondria undergo accelerated reductive evolution. They are evidently of relative recent bacterial origin. The same case can be made for the bacteria-related RNA polymerases in jakobids (Lang *et al.*, 1997) and for RNase P RNAs (Seif, Cadieux, & Lang, 2006). The best explanation for cytoplasmic ribosomes to be that distinct is that they are the product of a long evolutionary past that occurred well before the acquisition of mitochondria.

6.7. Endosymbiosis without Phagocytosis

From a cell biology point of view, extant eukaryotes stand out because of their complex subcellular structures that allow phagocytosis (for food), as well as acquisition and hosting of endosymbionts, features that are virtually non-existent in Bacteria and Archaea. To engulf the α-proteobacterial precursor of mitochondria, the ancestral host cell should have possessed these subcellular structures – a description that fits an archezoan host but not an archaeal host. A similar idea was promulgated earlier (Cavalier-Smith, 2009; Poole & Neumann, 2010; Poole & Penny, 2007), although without considering that the propensity for endosymbiosis needs the invention of a large set of additional specialized genes.

The viewpoint that the host already possessed phagocytosis and the prerequisites for endosymbiosis is consistent with the apparent big bang–like radiation of mitochondriate eukaryotes, as seen in phylogenetic trees (Philippe *et al.*, 2000b). Since each of the diverse lineages possess all the distinctive features of eukaryotic cells, emergence of the corresponding eukaryote-specific genes must have occurred prior to this radiation (Brinkmann & Philippe, 2007).

Note that the finding of a few eukaryote-related genes in Archaea but not in Bacteria does not question this view. For instance, the histone-like genes in both major groups of Archaea (Cubonova, Sandman, Hallam, Delong, & Reeve, 2005; Sandman & Reeve, 2005), which are viewed by some as genuine precursors of eukaryotic histones, could also be simplified gene versions that stem from ancient Eukarya (Philippe & Forterre, 1999). In other words, with no reliable rooting of the tree of life, a root within eukaryotes cannot be excluded. Also bacteria might once have had such genes, but lost them secondarily in the course of reductive evolution (which is consistent with the relatively fast evolution of bacterial genomes). In the end, contemplating individual genes, out of phylogenetic context, is of little value for inferring deep evolutionary scenarios.

6.8. Evolution not only Proceeds from Simple to Complex

Symbiogenesis theories presume a priori that complex eukaryotic cells can only derive from simple prokaryotic ancestors. Yet there is no evidence that evolution always follows a path from simple to complex, which is a popular concept borrowed from homocentric philosophies. Simplification and streamlining are equally important evolutionary principles, as testified by genome reduction seen in endosymbionts and organelles. Living examples of reductive evolution are eukaryotes adapted to anaerobic environments such as *Giardia* (McArthur *et al.*, 2000), and yeasts and fission yeasts that possess only a fraction of genes usually found in other fungi.

In summary, symbiogenesis theories tend to overlook the time requirements for the evolution of complex cellular features. In addition, they are at odds with phylogenetic data as discussed in the following section.

7. THE QUEST FOR THE ROOT OF THE EUKARYOTIC TREE FROM A MITOCHONDRIAL PERSPECTIVE

Due to the large phylogenetic distance between Archaea-related nuclear genes and genes from extant archaeal species, attempts to root the eukaryotic tree are plagued by LBA artefacts (Brinkmann, van der Giezen, Zhou, Poncelin de Raucourt, & Philippe, 2005; Philippe *et al.*, 2000b). Current rooting attempts rather rely on the taxonomic distribution of rare genetic events (proposing a position between unikonts and bikonts; Stechmann &

Cavalier-Smith, 2002) and gene counting across eukaryotic lineages (Katz, Grant, Parfrey, & Burleigh, 2012). Both approaches are problematic because they rely on only a handful of unweighted characters selected by subjective criteria. The longevity of the controversy is therefore no surprise; nor is the fact that apologists are switching camps.

An alternative to rooting the eukaryotic tree with an archaeal outgroup is the use of mitochondrial proteins (mtDNA- and nucleus-encoded) and a proteobacterial outgroup. For this approach to be legitimate, several assumptions have to be made. (1) All extant eukaryotic lineages have or once had mitochondria; (2) there is a single mitochondrial origin, i.e. trees built with eukaryote-specific (see above) and Archaea-related genes are consistent in topology (which they are within the limits of statistical confidence); and (3) nucleus-encoded genes of mitochondrial origin can be distinguished from genes that originate from transient symbiotic events and single gene transfers. Furthermore, it is of critical importance that the sampling of bacterial and eukaryotic species is as broad as possible, because systematic phylogenetic error introduced by both data may obstruct the positioning of the root. Obviously, the evolutionary model for inferences must be chosen in a way that systematic error is minimized (e.g. see the controversy over the position of SAR11 bacteria mentioned above).

To select genes that are suitable for such phylogenetic analyses, mitochondrial proteome data provide a lead (Karlberg, Canback, Kurland, & Andersson, 2000; Kumar *et al.*, 2002) because gene products of mitochondrial origin are usually located in this organelle (although with notable exceptions, e.g. RNA and DNA polymerase in a number of taxa) (Shutt & Gray, 2006). Yet, with a large fraction of mitochondrial proteins not well-conserved and rapidly evolving, their evolutionary origin often remains obscure. Among the ~800 proteins with high-scoring Blast hits to α-Proteobacteria (Gabaldon & Huynen, 2007; Szklarczyk & Huynen, 2010), many are either paralogs, lack universal distribution, or do not carry sufficient signal for phylogenetic analysis. In the end, only a few hundred well-conserved protein-coding orthologues may ever be used for rooting the eukaryotic tree.

In a recent publication (Derelle & Lang, 2011), a dataset of 42 mitochondrial proteins (most encoded by the nucleus) provides reasonably high support for a eukaryotic root between unikonts (animals + fungi + Amoebozoa and certain protists) and bikonts (plants and all other protists), a topology proposed earlier by others (Stechmann & Cavalier-Smith, 2002) (Fig. 3). An important challenge for this analysis is the amino acid bias (introducing systematic error) due to the unequal base composition across taxa: a high A + T content in mtDNA-encoded and rickettsial sequences; a more balanced nucleotide ratio in nuclear genes; and a very low A + T content in free-living α-Proteobacteria. In this analysis, mitochondria branch as a sister group to Rickettsia-related bacteria, which might be due to an LBA-type systematic error (Felsenstein, 1978). Although the authors did not observe changes in tree topology when removing mtDNA-encoded data or species whose genes are A + T rich, this issue has to be revisited with a much more diverse bacterial outgroup, more genes and better eukaryotic taxon sampling.

8. DID EUKARYOTES EMERGE FROM WITHIN ARCHAEA OR *VICE VERSA*?

As explained above, rooting eukaryotes with Archaea-related eukaryotic genes is currently unfeasible, due to the large phylogenetic distance to genes from extant Archaea, which makes trees susceptible to strong LBA artefacts (Brinkmann *et al.*, 2005). An additional complication is the rapid divergence of eukaryotic lineages in a big bang–like evolutionary event with extremely short internal vertices, whose branching order is therefore difficult to resolve (Philippe, Germot, & Moreira, 2000a).

On the other hand, it is feasible to identify the divergence point of Archaea-related eukaryotic genes and archaeal genes (Cox *et al.*, 2008; Foster *et al.*, 2009), because Archaea are diverse and evolve relatively slowly. As mentioned earlier, the corresponding phylogenetic analyses were conducted on a carefully chosen combined set of ~50 highly conserved proteins, yielding robust support for eukaryotes forming a sister group to (or potentially intermingling with) Crenarchaeota. The set of sequences used includes more than ten operational genes, calling for an amendment of the common view that this gene category comes mostly from the bacterial partner (Lopez-Garcia & Moreira, 1999; Ribeiro & Golding, 1998; Rivera *et al.*, 1998; Zillig *et al.*, 1989a,b).

The most important implication of the phylogenetic results from Embley's group (Cox *et al.*, 2008; Foster *et al.*, 2009) is that they defy the three-domain concept of life. Monophyly, as suggested by the influential three-domains hypothesis (Woese, Kandler, & Wheelis, 1990), was already criticized earlier as lacking significant support (Brown, Douady, Italia, Marshall, & Stanhope, 2001; Daubin, Gouy, & Perriere, 2002; Lake, 1988;

Tourasse & Gouy, 1999). The tree of Cox *et al.* (2008), in its unrooted form, suggests a total of four domains of life, Euryarchaeota, Bacteria, Crenarchaeota and Eukarya (Fig. 2A). Note that a rooting of this tree would not change this view, only redefine sister group relationships.

9. CONCLUSIONS

In recent years, considerable progress has been made in the understanding of eukaryotic evolution. We now have solid phylogenetic evidence for a relatively recent origin of mitochondria, allowing not only rooting of extant mitochondriate eukaryotes but also a more precise identification of the proteobacterial predecessor of mitochondria. Furthermore, we have strong phylogenetic evidence that an important portion of nuclear genes is related to Crenarchaeota. This newly recognized genealogy points to four instead of three domains of life: Euryarchaeota, Bacteria, Crenarchaeota and Eukarya. What remains unresolved is the root of the tree of life, since all approaches, including phylogenies with gene paralogs, remain unreliable (due to too few informative characters).

Virtually all permutations of possible eukaryotic origins have been postulated, and all current hypotheses have to assume extinction of transitory lineages, notably one event for SET and several for the symbiogenesis scenarios. Now it is time to go back and:

- search for missing descendants of predicted eukaryotic precursors;
- sequence and analyse genomes from a much broader collection of protists and bacteria close to the mitochondrial origin;
- refine phylogenomic analyses, and
- improve evolutionary models and inferences in order to avoid systematic error.

GLOSSARY

Archaea A group of organisms constituting one domain in the iconic three-domains-of-life hypothesis; the group is subdivided into Crenarchaeota (eocytes) and Euryarchaeota (including methanogens).

Archezoa Hypothetical eukaryotes that never had a mitochondrion; not to be confused with extant amitochondriate species that secondarily lost mitochondria and carry vestiges of this organelle.

Protists A non-taxonomic term for eukaryotes other than fungi, animals, and plants; predominantly unicellular. Protists represent more biological diversity than the three former groups together. Fungi, animals, and plants emerged from within protists.

Eocytes Synonym of Crenarchaeota. One of the two large archaeal groups; according to phylogenetic analyses, Archaea-related genes (see below) in extant eukaryotes are most closely related to Eocytes.

Archaea-related genes Genes in eukaryotic genomes with significant sequence similarity to archaeal, but not to bacterial genes. Note that this relatedness does not specify their origin (which varies with the rooting of the tree of life).

Phagocytosis Uptake of a cell by another cell, usually by sequestration into a food vacuole where digestion takes place.

Informational genes Genes involved in genetic information transfer and processing; principal components of replication, transcription, and translation.

Operational genes Genes involved in biosynthesis and metabolism.

LBA Long branch attraction (Felsenstein, 1978); a phylogenetic artefact leading to the incorrect grouping of fast-evolving species or their attraction to a distant outgroup, due to evolutionary model violations and underestimation of repeated sequence change. LBA can also be seen as a concomitant attraction of short branches (due to a true phylogenetic signal), and attraction of long branches (due to model violations).

CAT One of the evolutionary models used in phylogenetic reconstruction based on protein sequence; CAT uses categories of distinct, site-wise amino acid profiles (inferred from the data, i.e. multiple sequence alignments). Inferences with this model have been shown to be least prone to LBA.

Phylogenomics Phylogenetic inferences based on genome-wide sets of well-conserved (minimizing noise and sequence bias for an optimized phylogenetic signal) gene orthologues (only trees built with orthologues reflect the species tree). Laterally transferred genes have to be strictly excluded from the dataset.

SET Serial Endosymbiosis Theory positing an origin of mitochondria by endosymbiosis with an amitochondriate host (archezoan) that had the distinctive features of extant eukaryotes.

Symbiogenesis A concept postulating endosymbiosis itself, rather than progressive (serial) evolution, as the cause of the emergence of eukaryotic features. Symbiogenesis theories have also been termed fusion theories, in which fusion stands for endosymbiosis in a wide sense.

Unikonts Controversial name for a large subdivision of eukaryotes including opisthokonts (Metazoa, Fungi, and related unicellular lineages), Amoebozoa, and Apusozoa. The second large subdivision, bikonts, includes all remaining eukaryotic groups combined (including plants, all algal groups, jakobids, euglenids, trypanosomatids, alveolates, rhizarians, etc.).

ACKNOWLEDGEMENTS

The authors thank Hervé Philippe for comments on the manuscript. Financial support was provided from NSERC (grant 194650, BFL), the Canadian Research Chair Program, CIHR (grant MOP-79309, GB), and Genome Quebec/Genome Canada.

REFERENCES

Adams, K. L., Song, K., Roessler, P. G., Nugent, J. M., Doyle, J. L., Doyle, J. J., et al. (1999). *Proceedings of the National Academy of Sciences of the United States of America, 96*, 13863–13868.

Andersson, S. G., Karlberg, O., Canback, B., & Kurland, C. G. (2003). *Philosophical Transactions of the Royal Society B: Biological Sciences, 358*, 165–177, discussion 177–9.

Andersson, S. G., Zomorodipour, A., Andersson, J. O., Sicheritz-Ponten, T., Alsmark, U. C., Podowski, R. M., et al. (1998). *Nature, 396*, 133–140.

Archibald, J. M., & Richards, T. A. (2010). *BMC Biology, 8*, 147.

Blankenship, R. E., & Hartman, H. (1998). *Trends in Biochemical Sciences, 23*, 94–97.

Brindefalk, B., Ettema, T. J., Viklund, J., Thollesson, M., & Andersson, S. G. (2011). *PLoS One, 6*. e24457.

Brinkmann, H., & Philippe, H. (2007). *Advances in Experimental Medicine and Biology, 607*, 20–37.

Brinkmann, H., van der Giezen, M., Zhou, Y., Poncelin de Raucourt, G., & Philippe, H. (2005). *Systematic Biology, 54*, 743–757.

Brown, J. R., Douady, C. J., Italia, M. J., Marshall, W. E., & Stanhope, M. J. (2001). *Nature Genetics, 28*, 281–285.

Budin, K., & Philippe, H. (1998). *Molecular Biology and Evolution, 15*, 943–956.

Bui, E. T., Bradley, P. J., & Johnson, P. J. (1996). *Proceedings of the National Academy of Sciences of the United States of America, 93*, 9651–9656.

Cavalier-Smith, T. (1983). In W. Schwemmler, & H. E. A. Schenk (Eds.), *Endocytobiology II* (pp. 1027–1034). Berlin: De Gruyter.

Cavalier-Smith, T. (2009). *International Journal of Biochemistry & Cell Biology, 41*, 307–322.

Cox, C. J., Foster, P. G., Hirt, R. P., Harris, S. R., & Embley, T. M. (2008). *Proceedings of the National Academy of Sciences of the United States of America, 105*, 20356–20361.

Cubonova, L., Sandman, K., Hallam, S. J., Delong, E. F., & Reeve, J. N. (2005). *Journal of Bacteriology, 187*, 5482–5485.

Daubin, V., Gouy, M., & Perriere, G. (2002). *Genome Research, 12*, 1080–1090.

Derelle, R., & Lang, B. F. (2011). *Molecular Biology and Evolution, 29*, 1277–1289.

Doolittle, W. F. (1980). *Trends in Biochemical Sciences, 5*, 146–149.

Doolittle, W. F. (1981). *Science, 213*, 640–641.

Doolittle, W. F. (1998). *Trends in Genetics, 14*, 307–311.

Embley, T. M., & Hirt, R. P. (1998). *Current Opinion in Genetics & Development, 8*, 624–629.

Embley, T. M., & Martin, W. (2006). *Nature, 440*, 623–630.

Esser, C., Ahmadinejad, N., Wiegand, C., Rotte, C., Sebastiani, F., Gelius-Dietrich, G., et al. (2004). *Molecular Biology and Evolution, 21*, 1643–1660.

Felsenstein, J. (1978). *Systematic Zoology, 27*, 27–33.

Forterre, P. (2010). *Research in Microbiology, 162*, 77–91.

Forterre, P., & Gribaldo, S. (2007). *HFSP Journal, 1*, 156–168.

Foster, P. G., Cox, C. J., & Embley, T. M. (2009). *Philosophical Transactions of the Royal Society B: Biological Sciences, 364*, 2197–2207.

Fujishima, M., Dohra, H., & Kawai, M. (1997). *Journal of Eukaryotic Microbiology, 44*, 636–642.

Fujishima, M., & Fujita, M. (1985). *Journal of Cell Science, 76*, 179–187.

Gabaldon, T., & Huynen, M. A. (2007). *PLoS Computational Biology, 3*. e219.

Germot, A., Philippe, H., & Le Guyader, H. (1996). *Proceedings of the National Academy of Sciences of the United States of America, 93*, 14614–14617.

Germot, A., Philippe, H., & Le Guyader, H. (1997). *Molecular and Biochemical Parasitology, 87*, 159–168.

Görtz, H. D., Lellig, S., Miosga, O., & Wiemann, M. (1990). *Journal of Bacteriology, 172*, 5664–5669.

Gray, M., & Spencer, D. (1996). In D. Roberts, P. Sharp, G. Alderson, & M. Collins (Eds.), *Evolution of microbial life* (pp. 109–126). Cambridge University Press.

Gray, M. W. & Archibald, J. M. In R. Bock, V. Knoop (Eds), *Origins of Mitochondria and Plastids*, Dordrecht, in press.

Gray, M. W., Sankoff, D., & Cedergren, R. J. (1984). *Nucleic Acids Research, 12*, 5837–5852.

Huang, J., & Gogarten, J. P. (2007). *Genome Biology, 8*, R99.

Karlberg, O., Canback, B., Kurland, C. G., & Andersson, S. G. (2000). *Yeast, 17*, 170–187.

Katz, L. A., Grant, J. R., Parfrey, L. W., & Burleigh, J. G. (2012). *Systematic Biology*.

Koonin, E. V. (2010). *Genome Biology, 11*, 209.

Kumar, A., Agarwal, S., Heyman, J. A., Matson, S., Heidtman, M., Piccirillo, S., et al. (2002). *Genes & Development, 16*, 707–719.

Kumar, S., & Rzhetsky, A. (1996). *Journal of Molecular Evolution, 42*, 183–193.

Lake, J. A. (1988). *Nature, 331*, 184–186.

Lang, B. F., Brinkmann, H., Koski, L., Fujishima, M., Goertz, H. D., & Burger, G. (2005). *Japanese Journal of Protozoology, 38*, 171–183.

Lang, B. F., Burger, G., O'Kelly, C. J., Cedergren, R., Golding, G. B., Lemieux, C., et al. (1997). *Nature, 387*, 493–497.

Leipe, D. D., Gunderson, J. H., Nerad, T. A., & Sogin, M. L. (1993). *Molecular and Biochemical Parasitology, 59*, 41–48.

Lopez-Garcia, P., & Moreira, D. (1999). *Trends in Biochemical Sciences, 24*, 88–93.

Mai, Z., Ghosh, S., Frisardi, M., Rosenthal, B., Rogers, R., & Samuelson, J. (1999). *Molecular and Cellular Biology, 19*, 2198–2205.

Margulis, L. (1981). *Symbiosis in cell evolution*. San Francisco, CA: Freeman.

Martin, W., & Müller, M. (1998). *Nature, 392*, 37–41.

McArthur, A. G., Morrison, H. G., Nixon, J. E., Passamaneck, N. Q., Kim, U., Hinkle, G., et al. (2000). *FEMS Microbiology Letters, 189*, 271–273.

Moreira, D., & Lopez-Garcia, P. (1998). *Journal of Molecular Evolution, 47*, 517–530.

Philippe, H., & Adoutte, A. (1998). In G. Coombs, K. Vickerman, M. Sleigh, & A. Warren (Eds.), *Evolutionary relationships among protozoa* (pp. 25–56). Dordrecht: Kluwer.

Philippe, H., & Forterre, P. (1999). *Journal of Molecular Evolution, 49*, 509–523.

Philippe, H., Germot, A., & Moreira, D. (2000a). *Current Opinion in Genetics & Development, 10*, 596–601.

Philippe, H., Lopez, P., Brinkmann, H., Budin, K., Germot, A., Laurent, J., et al. (2000b). *Proceedings of the Royal Society B: Biological Sciences, 267*, 1213–1221.

Poole, A. M., & Neumann, N. (2010). *Research in Microbiology, 162*, 71–76.

Poole, A. M., & Penny, D. (2007). *Bioessays, 29*, 74–84.

Regoes, A., Zourmpanou, D., Leon-Avila, G., van der Giezen, M., Tovar, J., & Hehl, A. B. (2005). *Journal of Biological Chemistry, 280*, 30557–30563.

Ribeiro, S., & Golding, G. B. (1998). *Molecular Biology and Evolution, 15*, 779–788.

Richards, T. A., & Archibald, J. M. (2011). *Current Biology, 21*, R112–R114.

Rivera, M. C., Jain, R., Moore, J. E., & Lake, J. A. (1998). *Proceedings of the National Academy of Sciences of the United States of America, 95*, 6239–6244.

Rodriguez-Ezpeleta, N., & Embley, T. M. (2012). *PLoS One, 7*. e30520.

Roger, A. J., Svard, S. G., Tovar, J., Clark, C. G., Smith, M. W., Gillin, F. D., et al. (1998). *Proceedings of the National Academy of Sciences of the United States of America, 95*, 229–234.

Sandman, K., & Reeve, J. N. (2005). *Current Opinion in Microbiology, 8*, 656–661.

Seif, E., Cadieux, A., & Lang, B. F. (2006). *RNA, 12*, 1661–1670.

Shutt, T. E., & Gray, M. W. (2006). *Trends in Genetics, 22*, 90–95.

Sicheritz-Ponten, T., Kurland, C. G., & Andersson, S. G. (1998). *Biochimica et Biophysica Acta, 1365*, 545–551.

Sogin, M. L., Gunderson, J. H., Elwood, H. J., Alonso, R. A., & Peattie, D. A. (1989). *Science, 243*, 75–77.
Stechmann, A., & Cavalier-Smith, T. (2002). *Science, 297*, 89–91.
Stiller, J. W., & Hall, B. D. (1999). *Molecular Biology and Evolution, 16*, 1270–1279.
Szklarczyk, R., & Huynen, M. A. (2010). *Proteomics, 10*, 4012–4024.
Taylor, F. J. (1987). *Annals of the New York Academy of Sciences, 503*, 1–16.
Thrash, J. C., Boyd, A., Huggett, M. J., Grote, J., Carini, P., Yoder, R. J., et al. (2012). *Scientific Reports, 1*, 13.
Tourasse, N. J., & Gouy, M. (1999). *Molecular Phylogenetics and Evolution, 13*, 159–168.
Tovar, J., Fischer, A., & Clark, C. G. (1999). *Molecular Microbiology, 32*, 1013–1021.
Vossbrinck, C. R., Maddox, J. V., Friedman, S., Debrunner-Vossbrinck, B. A., & Woese, C. R. (1987). *Nature, 326*, 411–414.
Wernegreen, J. J. (2002). *Nature Reviews Genetics, 3*, 850–861.
Wiemann, M., & Görtz, H. D. (1991). *Journal of Bacteriology, 173*, 4842–4850.
Williams, B. A., & Keeling, P. J. (2003). *Advances in Parasitology, 54*, 9–68.
Woese, C. R., & Fox, G. E. (1977). *Proceedings of the National Academy of Sciences of the United States of America, 74*, 5088–5090.
Woese, C. R., Kandler, O., & Wheelis, M. L. (1990). *Proceedings of the National Academy of Sciences of the United States of America, 87*, 4576–4579.
Zillig, W., Klenk, H.-P., Palm, P., Leffers, H., Pühler, G., Gropp, F., et al. (1989a). *Endocytobiosis and Cell Research, 6*, 1–25.
Zillig, W., Klenk, H. P., Palm, P., Puhler, G., Gropp, F., Garrett, R. A., et al. (1989b). *Canadian Journal of Microbiology, 35*, 73–80.

CHAPTER TWO

Gene Content and Gene Transfer from Mitochondria to the Nucleus During Evolution

Minoru Ueda[†] **and Koh-ichi Kadowaki**[*,1]

[*]National Agriculture and Food Research Organization (NARO), Institute of Crop Sciences, Tsukuba, Ibaraki, Japan

[†]Department of Botany, Graduate School of Science, Kyoto University, Kyoto, Japan

[1]Corresponding author: E-mail: kadowaki@affrc.go.jp

Contents

Abstract

Organelles such as mitochondria and chloroplasts are derived from endosymbionts. Gene transfer events from organelles to the nucleus, in which organelle genes are translocated and activated in the nucleus, have been occurring over evolutionary time. Complete gene transfer requires several steps because of the differences in transcriptional and translational machinery between the organelles and the nucleus (cytoplasm). In addition, protein signals appropriate for sorting need to be acquired. In angiosperms, more genes (in particular, ribosomal protein genes) are encoded in the mitochondrial genome than in vertebrates and fungi. Furthermore, the number of genes in the mitochondrial genome varies among plant species. These clues suggest that mitochondrial gene transfer to the nucleus is still ongoing in angiosperms. Thus, the mitochondrial genome in angiosperms is a good tool for the study of gene transfer events from the mitochondria to the nucleus and provides a way of understanding the steps of symbiosis in angiosperms. In this review, we discuss insights into gene transfer and substitution revealed by comparative genomic analysis in angiosperms.

Advances in Botanical Research, Volume 63
ISSN 0065-2296,
http://dx.doi.org/10.1016/B978-0-12-394279-1.00002-8

1. INTRODUCTION

It is generally accepted that mitochondria and chloroplasts are descendants of α-proteobacteria and cyanobacteria, respectively. Most of the genes in the ancestral endosymbiont have either been translocated to the nuclear genome of the host cell or have been lost during evolution after the initial endosymbiotic event (Gray, 1992; Martin, 2003).

The complete mitochondrial genome has been sequenced for about 70 animal species including such phyla as the Chordata, Arthropoda, Mollusca, and Nematoda. With few exceptions, a typical animal mitochondrial genome contains 13 protein-coding genes: seven subunits of the NADH ubiquinone oxidoreductase complex (*nad1–6* and *nad4L*), three subunits of the cytochrome *c* oxidase complex (*cox1–3*), a single subunit of the ubiquinol cytochrome *c* oxidoreductase complex (*cob*), and two subunits of ATPase (*atp6* and *atp9*) (Boore & Brown, 1998; Wolstenholme, 1992). On the other hand, more than 14 complete mitochondrial genomes have been sequenced to date in angiosperms (Sloan, Alverson, Storchova, Palmer, & Taylor, 2010), and angiosperm mitochondrial genomes show great divergence regarding protein-coding gene content among species. These variations in gene content between plant species strongly suggest that gene transfer from the mitochondrial to the nuclear genome is an ongoing process and that angiosperm genomes retain processes through which mitochondrial genes have been activated during gene transfer events (Adams & Palmer, 2003; Brennicke, Grohmann, Hiesel, Knoop, & Schuster, 1993), whereas gene transfer is almost completed in animal mitochondrial genomes.

Many gene transfer events from the mitochondrial to the nuclear genome have been identified in land plants, particularly in angiosperms. During the process of gene transfer from an organelle to the nucleus, transferred genes must undergo several changes. These include gene translocation, acquisition of a promoter, a poly(A) signal, a targeting signal, and subsequent elimination of the original sequence from the organelle genome (Brennicke *et al.*, 1993). Although symbiosis is an important process in biology, little is known about the mechanisms of any of the above steps. In this review, we discuss insights about the mechanisms of gene transfer from mitochondria to the nucleus gained from analyses using angiosperm genomes.

2. GENE CONTENT OF LAND PLANT MITOCHONDRIA

The complete mitochondrial genome sequences of various species have been determined. A limited number of genes are encoded, and their relative positions are largely conserved among vertebrate mitochondrial genomes. They are typically ~16 kb in size. With few exceptions, such as gender-specific genes (M-*orf*, F-*orf*, and H-*orf*) in freshwater mussels (Breton *et al.*, 2011), all animal mitochondrial genomes contain 13 genes for proteins (Boore, 1999). This is in marked contrast to the land plant mitochondrial genome sequences such as liverwort *Marchantia polymorpha* (Oda *et al.*, 1992), *Arabidopsis thaliana* (Unseld, Marienfeld, Brandt, & Brennicke, 1997), *Beta vulgaris* (Kubo *et al.*, 2000b), and *Oryza sativa* (Notsu *et al.*, 2002), among which gene order and gene content are highly variable (Bullerwell & Gray, 2004; Sloan *et al.*, 2010). For details of mitochondrial genome organization, evolution and recombination, see Chapter 9.

Liverworts are considered to represent the basal group of terrestrial embryophytes (Steemans *et al.*, 2009). *M. polymorpha* mitochondrial genome encodes 17 kinds of ribosomal protein genes that have already been lost from vertebrate mitochondrial genomes. In angiosperm mitochondrial genomes, the situation is more complex. The number of ribosomal protein genes in mitochondrial genomes varies among species, which suggests that transfer from mitochondria to the nucleus of the 17 mitochondrial ribosomal protein genes has occurred independently and frequently during the evolution of land plants (Adams & Palmer, 2003; Sloan *et al.*, 2010). A mitochondrial ribosomal protein gene missing from one plant species but encoded by another plant species is likely to be encoded by the nuclear genome in the former species because ribosomal proteins are essential for protein synthesis. These discrepancies in gene content strongly support the idea that gene transfer from the mitochondrial genome to the nuclear genome is an ongoing process in angiosperms. Several transfer events from the mitochondrial to the nuclear genome have been identified, and they deepen our understanding of the steps of gene transfer including gene translocation, acquisition of regulatory elements for expression and a targeting signal, and elimination of the original sequence from the organelle genome (Fig. 2.1A).

3. GENE TRANSLOCATION AND INTEGRATION

During evolution, most genes originally encoded by the endosymbiont have been translocated to the nuclear genome. The details of these observations are described in Chapter 3. Other unidirectional DNA translocations between organelles, from chloroplasts to mitochondria and from the nucleus to mitochondria, have also been observed; for example, the ribulose bisphosphate carboxylase large subunit gene (*rbcL*) in the *Zea mays*

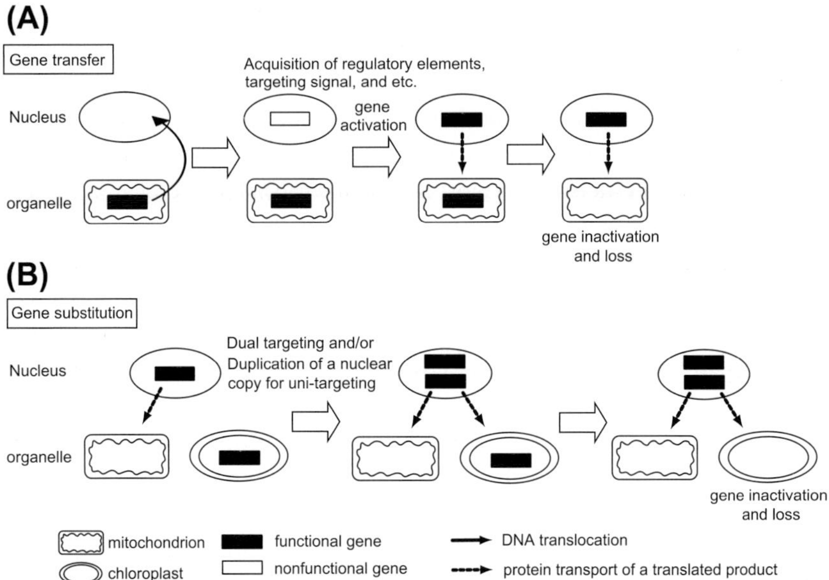

Figure 2.1 Schematic representation of the processes of gene transfer and gene substitution. Different steps in gene transfer (A) and gene substitution (B) are shown and described. Gene substitution is achieved not only by transfer of a gene from an organelle but also via neofunctionalization of a gene that was originally irrelevant to the substituted gene. In this figure, the chloroplast-encoded gene is substituted by a nuclear-encoded mitochondrial gene, and vice versa.

mitochondrial genome (Lonsdale, Hodge, Howe, & Stern, 1983) and retrotransposable elements in the *O. sativa* mitochondrial genome (Notsu *et al.*, 2002). However, DNA flow from organelles to the nucleus is most common, and the other translocations seem to have only minor impacts in terms of organization of the nuclear and organellar genomes.

In recent studies, large numbers of mitochondrial and chloroplast DNA fragments that have been translocated into the nuclear genome have been successfully identified using whole-genome information. These are referred to as NUMT (nuclear mitochondrial DNA) and NUPT (nuclear plastid DNA), respectively (Richly & Leister, 2004). In particular, 620-kb and 190-kb mitochondrial DNA sequences and a 131-kb chloroplast DNA sequence were identified in *A. thaliana* chromosome 2, *O. sativa* chromosome 12, and *O. sativa* chromosome 10, respectively (Shahid Masood *et al.*, 2004; Stupar *et al.*, 2001; Ueda, Tsutsumi, & Kadowaki, 2005). For example, the *O. sativa* 190-kb mitochondrial DNA sequence, which corresponds to 38.8% of the whole mitochondrial genome, is interrupted by seven foreign DNA segments. The foreign DNA sequences are composed of 45 kb comprising four segments of retrotransposon origin, and 13 kb comprising three segments of unknown origin. The 190-kb sequence shows more than 99.7% similarity to the current mitochondrial sequence, suggesting that its integration into the nucleus was quite recent. However, several sequences in the 190-kb segment have been rearranged relative to the current mitochondrial sequence. Similar rearrangement of NUMT was observed in the 620-kb *A. thaliana* mitochondrial DNA sequence (Stupar *et al.*, 2001) and in *Z. mays* species (Lough *et al.*, 2008). The four retrotransposons show no mutual sequence similarity and are integrated into different locations, suggesting their integration events were independent, frequent, and quite recent (Ueda *et al.*, 2005). NUMT seems to be easily and rapidly invaded by transposable elements. NUPT has a tendency to be excluded from the nuclear genome (Matsuo, Ito, Yamauchi, & Obokata, 2005). Analysis of gene artificially transferred from the chloroplast to the nucleus (an *in vivo* gene transfer experiment) revealed that NUPT is unstable in the nuclear genome (Sheppard & Timmis, 2009). These data suggest that NUMT might also be easily rearranged. Alternatively, the rearrangement might reflect rapid changes in the present mitochondrial genome.

A large number of NUMTs and NUPTs have been observed in eukaryotic nuclear genomes. However, the mechanism through which they have been integrated into the nuclear genome remains unclear. Recent studies involving *in vivo* gene transfer experiments have shed light on the process (Bock & Timmis, 2008). *In vivo* experiments have succeeded in calculating the frequency of organellar DNA translocations to the nucleus and have clarified their frequencies in yeast *Saccharomyces cerevisiae* mitochondria (Thorsness & Fox, 1990) and *Nicotiana tabacum* chloroplasts (Huang, Ayliffe, & Timmis, 2003; Sheppard *et al.*, 2008; Stegemann, Hartmann, Ruf, & Bock, 2003). As a result, the frequency of chloroplast DNA translocation is now known to be remarkably higher than previously predicted. In particular, a high frequency in the male germ line was observed in *N. tabacum*, with a translocation rate from the chloroplast to the nuclear genome of one in about 16,000 pollen grains (Huang *et al.*, 2003). The frequency of chloroplast DNA translocation in the male line was higher than that in the female germ line (Sheppard *et al.*, 2008) or in green leaves (Stegemann *et al.*, 2003). In *S. cerevisiae*, vacuolar-dependent turnover of abnormal mitochondrial compartments increased the translocation efficiency of mitochondrial DNA to the nucleus (Campbell & Thorsness, 1998). Because drastic mitochondrial and plastidic turnover occurs during male gametogenesis, it is reasonable for male gametogenesis to play a major role in organellar DNA translocation. Recent studies of uniparental inheritance of the nematode *Caenorhabditis elegans* mitochondrial genome revealed that autophagy (mitophagy) regulates paternal mitochondrial degradation during fertilization, resulting in the degradation of paternal mitochondrial DNA (Al Rawi *et al.*, 2011; Sato and Sato, 2011). These observations support the hypothesis that proteomic turnover of mitochondria, which may be related to uniparental inheritance of organellar genomes during male gametogenesis, contributes to DNA translocation. The mechanism of uniparental inheritance of organellar genomes is complex; not only proteomic organelle degradation but also the physical exclusion of the organelle and direct organellar DNA degradation during fertilization seem to be involved (Berger, Hamamura, Ingouff, & Higashiyama, 2008; Nishimura, 2010). Clearly, uniparental inheritance of organellar genomes is not always involved in organellar DNA translocation.

The number of NUMTs and NUPTs found in nuclear genomes is variable among eukaryotic species. In *C. elegans*, just a single NUMT was identified. On the other hand, the *O. sativa* nuclear genome contains thousands of NUMTs (Hazkani-Covo, Zeller, & Martin, 2010; Richly & Leister, 2004). *In vivo* experiments in *S. cerevisiae* (Ricchetti, Fairhead, & Dujon, 1999) and analysis of NUMT and sequence junctions in the primate nucleus indicated that the insertion of organellar DNA might be mediated by non-homologous end joining (NHEJ) (Hazkani-Covo & Covo, 2008). NHEJ is the major mechanism for double-strand break (DSB) repair in plants (Puchta, 2005). Analyses of the relationship between DSB repair and organellar DNA translocation, and between uniparental inheritance and organellar DNA translocation may uncover a detailed mechanism for organellar DNA uptake and provide an answer to the question of why the numbers of NUMTs and NUPTs are drastically different between species in eukaryotes.

For organellar DNA translocation, in addition to the DNA-mediated pathway described above, an RNA-mediated pathway was reported for the cytochrome *c* oxidase subunit 2 gene (Nugent & Palmer, 1991). Recent studies using information from massive numbers of NUMT and NUPT sequences have not yet found any trace translocated from organelles to the nucleus via an RNA-mediated pathway, therefore the DNA-mediated pathway is the major pathway and the RNA-mediated pathway contributes rarely to DNA translocation from organellar genomes to the nucleus.

4. PROMOTER ACQUISITION

The processes by which a transferred gene acquires regulatory elements such as promoter sequences is still unknown, partly because promoter sequences are highly variable compared with protein-coding sequences. It seems most feasible for transferred genes to acquire promoter sequences from preexisting genes, in addition to acquiring a targeting signal. However, a group of genes transferred from the mitochondria to the nucleus contain introns in the 5′-untranslated region (UTR) (Liu & Adams, 2010), suggesting that the transferred genes obtained proper promoter sequences via recombination or other processes to adapt their expression according to their function within the cell.

Whole-genome sequencing projects have been undertaken in many species, including humans (*Homo sapiens*), *S. cerevisiae*, and *A. thaliana*. These projects have revealed that inter- and intrachromosomal duplications occur in a complex manner (Arabidopsis Genome Initiative, 2000; Bailey *et al.*, 2002; Dujon *et al.*, 2004). During evolution, chromosomal rearrangements including inter- and intrachromosomal duplications have occurred incidentally and have sometimes conferred a new function on a gene product or abolished its function, thus increasing genetic diversity (Eichler & Sankoff, 2003). In angiosperms, whole-genome sequencing projects have been conducted in more than ten species (Van Bel *et al.*, 2012). Among them, *A. thaliana* and *O. sativa* genome sequences were built by contig cloning, together with full-length cDNA sequences (Arabidopsis Genome Initiative, 2000; International Rice Genome Sequencing Project, 2005), providing accurate information about nuclear genome organization and allowing us to analyse the process of promoter sequence acquisition in gene transfer events in detail.

Analysis of the *O. sativa* mitochondrial ribosomal protein large subunit 27 gene (*rpl27*), showed the involvement of inter- and intrachromosomal duplications in the acquisition of regulatory elements for a gene transferred from mitochondria to the nucleus (Ueda *et al.*, 2006a). The mitochondrial genome of the heterotrophic flagellate *Reclinomonas americana* contains *rpl27* (Lang *et al.*, 1997), whereas the *rpl27* gene is absent from all plant mitochondrial genomes analysed to date. Detailed analysis of the mitochondrial *rpl27* gene in the *O. sativa* nuclear genome shows that the *rpl27* gene acquired a promoter sequence and 5′-UTR via inter- and intrachromosomal duplications from the *O. sativa spt16*-related (*Osspt16*) gene, which is a homologue of the *S. cerevisiae spt16* gene (Fig. 2.2).

Intrachromosomal duplications (tandem duplication) occurred around *rpl27*. A repeat of seven nucleotides (AATAGTT) was identified at the junction of the duplicated sequences and the same repeat was also identified at the 5′ and the 3′ ends of the duplicated sequences. It is possible that this 7-bp repeat was involved in the intrachromosomal recombination event or is a footprint of the sequence duplication (Fig. 2.2). Illegitimate recombination is mediated by topoisomerase I, which recognizes small repeat sequences then nicks the DNA and ligates the nicked DNA (Sherratt & Wigley, 1998). Analysis of illegitimate recombination in *S. cerevisiae* has revealed that there is a tendency for topoisomerase I to recognize sequences containing a particular stretch of nucleotides ((G/C)(A/T)T, hot-spot sequences) (Zhu & Schiestl, 1996). In angiosperms, topoisomerase I extracted from wheat (*Triticum aestivum*) germ preferentially recognized the hot-spot sequence AGTT (Been, Burgess,& Champoux, 1984). In *O. sativa rpl27*, micro-homology of seven nucleotides (AATAGTT) was identified at both ends of the repeats and at the junction of the repeats, and this contains the hot-spot sequence (AGTT) observed in wheat germ. The AATAGTT repeat is

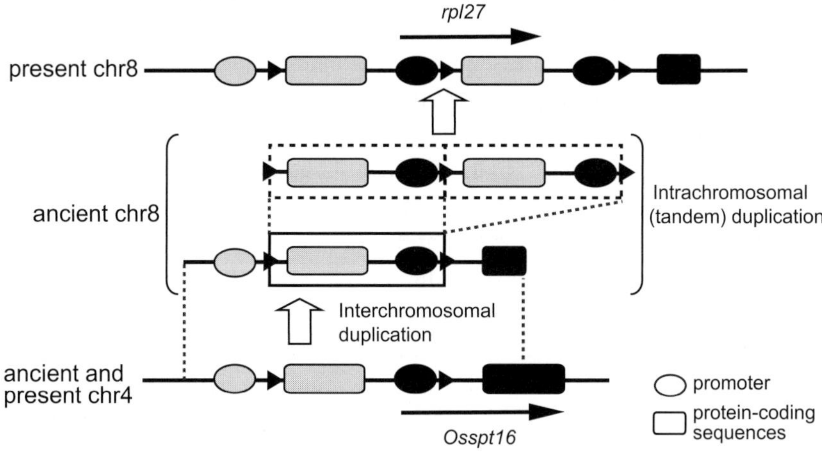

Figure 2.2 Schematic representation of promoter acquisition by the mitochondrial ribosomal protein L27 gene (*rpl27*) mediated by inter- and subsequent intra-chromosomal duplications in *O. sativa*. Horizontal arrows indicate transcribed regions of *rp27* and *O. sativa spt16*-related (*Osspt16*) gene identified from the full-length cDNA sequence. Protein-coding sequences and promoters are indicated by boxes and ovals, respectively. Black and gray colours indicate *Osspt16* and *rpl27*. Black dotted lines and boxes indicate the regions duplicated. Black triangles show tandem repeats (AATAGTT) that are likely to have been involved in the intrachromosomal duplication. Details are shown in Ueda *et al.* (2006a).

implicated in the recombination as described above, resulting in promoter shuffling involving *rpl27*, as well as in *rpl27* sequence duplication (Ueda *et al.*, 2006a).

Although functional analyses have not been conducted, several examples of gene fusion are known to have occurred via genomic rearrangement even after the divergence of *A. thaliana* and *O. sativa* (Nakamura, Itoh, & Martin, 2007). Not only fusions between partial coding sequences but also between a promoter and a protein-coding sequence have occurred repeatedly and may provide proper regulatory elements for a gene during evolution as in the case of *rpl27*.

Transposable elements have also induced genomic rearrangement during the evolution of life (Wessler, 2006). They occasionally provide a transferred gene from mitochondria with regulatory elements. The 5′-UTR region of the *O. sativa* ribosomal protein L6 gene was derived from a transposable element (Kubo, Fujimoto, Arimura, Hirai, & Tsutsumi, 2008).

As mentioned above, promoter shuffling via chromosomal rearrangement including tandem duplications and transposition of transposable elements plays an important role in the acquisition of regulatory elements for proper expression after gene transfer from the mitochondria to the nucleus.

5. TARGETING SIGNAL ACQUISITION

To complete gene transfer, the transferred gene must acquire a targeting signal to enable its protein to be delivered to the mitochondria. Many genes transferred from the mitochondria to the nucleus in various angiosperm species have been isolated, and several mechanisms for the acquisition of a targeting signal have been reported.

One mechanism is the acquisition of an existing presequence via duplication, as in the *O. sativa* ribosomal protein small subunit 11 gene (*rps11*) (Kadowaki, Kubo, Ozawa, & Hirai, 1996), *Z. mays* ribosomal protein large subunit 5 gene (Sandoval *et al.*, 2004), and *A. thaliana* succinate dehydrogenase subunit 3 gene (Adams, Rosenblueth, Qiu, & Palmer, 2001b) (Fig. 2.3A). Similar duplications have been found in genes transferred from the chloroplast to the nucleus; many genes also hijacked transit peptides from preexisting genes encoding chloroplast-targeted proteins (e.g. the ribosomal protein large subunit 32 gene (*rpl32*) (Ueda *et al.*, 2007) and ribosomal protein large subunit 9 gene (Arimura *et al.*, 1999)). In addition, the acquisition of an existing presequence by alternative splicing has been reported for the *O. sativa* and *Z. mays* ribosomal protein large subunit 14 genes (Fig. 2.3A) (Figueroa, Gomez, Holuigue, Araya, & Jordana, 1999; Kubo, Harada, Hirai, & Kadowaki, 1999). Chloroplast *rpl32* has been transferred to the nucleus in *Bruguiera gymnorrhiza*, and its chloroplast targeting signal (called a transit peptide) was also acquired by alternative splicing in the chloroplast Cu–Zn superoxide dismutase gene (*sod-1*) (Cusack and Wolfe, 2007). In contrast, alternative splicing is silenced in *Populus*, and *rpl32* and *sod-1* are encoded at different loci in the nucleus. In nuclear-encoded *rpl32*, protein-coding sequences encoding the SOD-1 domain have become highly divergent but sequences for transit peptide still show high similarity with *sod-1* in *Populus* (Ueda *et al.*, 2007). Comparative analysis of *rpl32* in

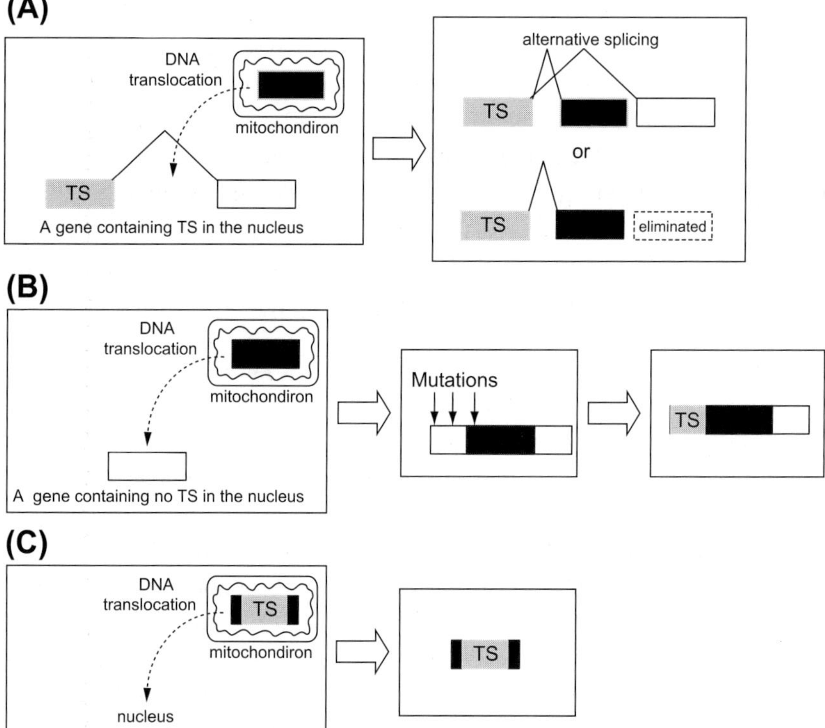

Figure 2.3 Examples of the acquisition of mitochondrial targeting signals. (A) Acquisition of a mitochondrial targeting signal from an existing one. (B) Acquisition of a presequence from an irrelevant fragment via mutations. Point mutations within a transferred gene occasionally generate a targeting signal. (C) No acquisition of a mitochondrial targeting signal because of the presence of a latent mitochondrial targeting signal in the gene in the mitochondrial genome. TS, targeting signal.

Malpighiales revealed that the inactivation of alternative splicing and their independent presence on different loci have occurred lineage specifically (Ueda, unpublished data). These results suggest that alternative splicing might be an initial step in the acquisition of a presequence via duplication. The integration of a transferred gene into a duplicated gene for an organelle protein seems to be the major driving force contributing targeting signals to genes newly transferred from organelles to the nucleus.

Another possible scenario is that a presequence may be derived from DNA not originally containing a targeting signal (Fig. 2.3B). This can occur by accumulation of point mutations in the coding sequence of the transferred gene without the acquisition of an obvious N-terminal extension as a presequence, such as in the *O. sativa* ribosomal protein S10 gene (*rps10*) (Kubo *et al.*, 2000a; Kubo, Arimura, Tsutsumi, & Kadowaki, 2003; Murcha *et al.*, 2005). Similarly, the presequence of the potato cytochrome *c* oxidase subunit 1 gene was generated by duplication and subsequent accumulation of point mutations in a sequence with a protein structure resembling that of a presequence (Long, de Souza, Rosenberg, & Gillbert, 1996). Experimental evidence that sequences having no relationship with targeting signals (so-called irrelevant sequences) could function as targeting signals has also been reported (Baker & Schatz, 1987; Lucattini, Likic, & Lithgow, 2004). In the case of transit peptides, a frameshift mutation generated a transit peptide through duplication and subsequent shifting of the reading frame of a preexisting irrelevant protein gene (Ueda, Fujimoto, Arimura, Tsutsumi, & Kadowaki, 2006b). Thus, mutations in irrelevant DNA fragments have undoubtedly generated targeting signals for organelles during evolution.

Generally, an obvious N-terminal extension for the targeting signal is present in transferred genes in comparison with genes that remain in the mitochondrial genome. By contrast, genes in which a distinguishable targeting signal could not be found have been also reported (e.g. *rps10* in several angiosperms (Adams, Daley, Qiu, Whelan, & Palmer, 2000; Kubo *et al.*, 2000a; Murcha *et al.*, 2005), and ribosomal protein L2 gene (*rpl2*) in *Glycine max* and *Medicago truncatula* (Adams, Ong, & Palmer, 2001a)). A dual targeting signal for mitochondria and chloroplasts has been discovered in the ribosomal protein small subunit S16 gene (RPS16), which lacks an N-terminal extension (Ueda *et al.*, 2008b). Approximately 75% of the nuclear-encoded mitochondrial ribosomal proteins in *A. thaliana* have N-terminal extensions (Bonen & Calixte, 2006). The remaining ribosomal proteins may have an internal targeting signal without an obvious N-terminal extension.

Ribosomal proteins containing an internal targeting signal have also been found in *S. cerevisiae* (Matsushita & Isono, 1993).

The genes for RPL2 and RPS4 in the *A. thaliana* mitochondrial genome contain information for protein targeting to the mitochondria. Similarly, the genes for RPL2 and RPS19 in the *O. sativa* mitochondrial genome contain information for protein targeting to mitochondria. These results suggest that targeting information already existed in some of genes in the plant mitochondrial genome before the transfer event to the nucleus occurred (Fig. 2.3C) (Ueda, Fujimoto, Arimura, Tsutsumi, & Kadowaki, 2008a).

It is therefore possible that some genes encoded in the mitochondrial genome have already acquired a targeting signal before translocation. Angiosperms that are still undergoing the process of gene transfer will enable us to confirm the above hypothesis and to unveil the hidden new scenario.

6. GENE SUBSTITUTION

It is probable that proteins translated from transferred genes occasionally change destinations within the cell and become targeted to a different organelle (Fig. 2.1B). Non-transferred genes from organelles are also candidates for gene substitution; if their functions fit in a different organelle, they can become new genes with the same function but located in a different organelle. There are a few instances of gene substitution among organellar ribosomal proteins such as the large subunit 10 gene (*rpl10*) (Kubo & Arimura, 2010), large subunit 21 gene (Gallois, Achard, Green, & Mache, 2001; Martin, Lagrange, Li, Bisanz-Seyer, & Mache, 1990), large subunit 23 gene (Bubunenko, Schmidt, & Subramanian, 1994), small subunit 8 gene (Adams, Daley, Whelan, & Palmer, 2002; Chang, Szick-Miranda, Pan, & Bailey-Serres, 2005), and *rps16* (Ueda *et al.*, 2008b). In the case of the small subunit 13 gene (*rps13*) (Adams *et al.*, 2002; Mollier, Hoffmann, Debast, & Small, 2002), cytoplasmic ribosomal protein small subunit 15A gene substituted mitochondrial *rps13*. Although gene substitution is rarely reported compared with the large number of cases of gene transfer to date, gene substitution also facilitates gene loss from organellar genomes.

In particular, *rps16* is a fascinating example for the analysis of gene substitution among the above genes. Dual-targeted RPS16 proteins, which are thought to be of mitochondrial origin, are widely found in angiosperms, and putatively dual-targeted RPS16 proteins are found in several gymnosperms. These results suggest that gene substitution of chloroplast-encoded *rps16* began at an early stage of seed plant evolution. The gene substitution of chloroplast-encoded *rps16* is also an interesting situation among angiosperms. For example, chloroplast-encoded *rps16* is functional in *N. tabacum* (Fleischmann *et al.*, 2011) and is lost in the genus *Populus* (Okumura *et al.*, 2006; Steane, 2005). *A. thaliana* retains *rps16*, however, as a pseudogene (Roy, Ueda, Kadowaki, & Tsutsumi, 2010). These results indicate that the gene substitution of chloroplast-encoded *rps16* is in an intermediate state in several angiosperms. Comparative analysis of chloroplast-encoded *rps16* in close relatives of *A. thaliana* revealed when gene substitution ended because the status of their chloroplast-encoded *rps16* genes was different, and their divergence time was analysed in detail. Judging from their divergence estimation and the status of their chloroplast-encoded *rps16* genes, gene substitution could continue for more than 126 million years to complete gene substitution in *A. thaliana* (Roy *et al.*, 2010), in contrast to the frequent and rapid occurrence of gene transfer during angiosperm evolution (Adams *et al.*, 2000; Adams & Palmer, 2003).

Except for *rpl10* and *rps16*, genes involved in gene substitution encode unitargeted proteins. Dual targeting appears to be a benefit to the cell as a way to save the energy it takes to produce different proteins for different organelles. However, dual-targeted organellar ribosomal proteins have been rarely reported, suggesting that the generation of unitargeting is more frequent than that of dual targeting, such as via alternative transcriptional initiation (Obara, Sumi, & Fukuda, 2002) and alternative translational initiation (Christensen *et al.*, 2005; Watanabe *et al.*, 2001). Dual targeting of a protein might create conflict in the cell, which could be the explanation for the long time for completion of *rps16* gene substitution. Mass analysis of proteins extracted from *A. thaliana* green leaves detected a single RPS16 protein (AtRPS16-1, AT4G34620) although two plastid-targeted RPS16 proteins (AtRPS16-1, plastid targeted; AtRPS16-2, dual targeted to plastids and mitochondria) exist in the *A. thaliana* nuclear genome (Olinares, Ponnala, & van Wijk, 2010; Ueda *et al.*, 2008b). The mass analysis implies that dual-targeted RPS16 proteins in angiosperms might ultimately become unitargeted proteins like other ribosomal proteins involved in gene substitution and that present dual-targeted RPS16 proteins might be poised at an intermediate status relative to the end of chloroplast *rps16* gene substitution in angiosperms (Fig. 2.1B).

Comparative analysis of chloroplast-encoded *rps16* genes in close relatives of *A. thaliana* also revealed the strong correlation of gene loss with self-compatibility. Outcrossing plants (*Arabidopsis arenosa* and *Arabidopsis lyrata*) tend to have retained, and inbreeding plants (*Arabis hirsuta*, *A. thaliana*, *Crucihimalaya lasiocarpa* and *Olimarabidopsis pumila*) tend to have lost, *rps16* in their chloroplast genomes. The epistatic model predicts that self-pollinating reproduction maintains cyto–nuclear gene combinations and increases the response to selection of epistatic combinations (Wade & Goodnight, 2006), potentially facilitating gene transfer. Conversely, outcrossing tends to break apart adaptive cyto–nuclear gene combinations, potentially reducing the amount of adaptive transfer in outcrossing lineages (Brandvain & Wade, 2009). In short, it predicts that the level of inbreeding is positively associated with the level of functional transfer (and loss) of organellar genes (Brandvain, Barker, & Wade, 2007), so self-compatibility may correlate positively with the loss of *rps16* from the chloroplast genome.

7. PERSPECTIVES ON GENE TRANSFER ANALYSIS

Comparative analyses of mitochondrial genomes in angiosperms, in particular of genes for ribosomal proteins, have shed light on the mechanisms of gene transfer and substitution from the mitochondrial to the nuclear genome, including gene translocation, acquisition of promoter elements for expression and a targeting signal, and elimination of the original sequence from the organellar genome. However, the mechanisms of gene transfer are still not completely understood. In particular, the acquisition of the promoter elements is rarely revealed because analysis of promoter elements generally requires whole-genome sequences. Next-generation sequencing now enables us to obtain whole-genome sequences in various species more easily. Combined with coexpression analysis, this should allow the mechanisms for acquisition of promoter elements in transferred genes to be revealed in the near future, as well as facilitate our understanding of the acquisition of targeting signals. Gene transfer may correlate with reproduction systems, as shown by the increased DNA translocation during male gametogenesis and gene loss in inbreeding plants. The discrepancies in gene content of angiosperm mitochondrial genomes seem to reflect the divergence of reproduction systems in each species. Therefore, analyses of the relationship between gene transfer and the molecular mechanisms of reproduction are expected and should serve to answer long-standing questions regarding the molecular mechanisms of gene transfer, such as integration of organellar DNA into the nucleus, the expression adaptation of transferred genes, and so on.

Since the discovery of chloroplast DNA in 1962 (Ris & Plaut, 1962) and mitochondrial DNA in 1963 (Nass & Nass, 1963), many researchers have addressed the evolution and organization of organellar genomes, and have been finding mysteries in biology and genetics, such as the process and reason for gene transfer. Recent studies using comparative genomic analysis together with mass DNA sequencing have greatly accelerated discovery of the gene transfer mechanisms after endosymbiosis, raising the possibility that the divergence of reproduction system may be a driving force to facilitate gene transfer. Further efforts are required to understand unknown driving forces to achieve gene transfer and unravel the whole aspect of gene transfer.

REFERENCES

Adams, K. L., Daley, D. O., Qiu, Y. L., Whelan, J., & Palmer, J. D. (2000). Repeated, recent and diverse transfers of a mitochondrial gene to the nucleus in flowering plants. *Nature, 408*, 354–357.

Adams, K. L., Daley, D. O., Whelan, J., & Palmer, J. D. (2002). Genes for two mitochondrial ribosomal proteins in flowering plants are derived from their chloroplast or cytosolic counterparts. *The Plant Cell, 14*, 931–943.

Adams, K. L., Ong, H. C., & Palmer, J. D. (2001a). Mitochondrial gene transfer in pieces: fission of the ribosomal protein gene *rpl2* and partial or complete gene transfer to the nucleus. *Molecular Biology and Evolution, 18*, 2289–2297.

Adams, K. L., & Palmer, J. D. (2003). Evolution of mitochondrial gene content: gene loss and transfer to the nucleus. *Molecular Phylogenetics and Evolution, 29*, 380–395.

Adams, K. L., Rosenblueth, M., Qiu, Y. L., & Palmer, J. D. (2001b). Multiple losses and transfers to the nucleus of two mitochondrial succinate dehydrogenase genes during angiosperm evolution. *Genetics, 158*, 1289–1300.

Al Rawi, S., Louvet-Vallee, S., Djeddi, A., Sachse, M., Culetto, E., Hajjar, C., et al. (2011). Postfertilization autophagy of sperm organelles prevents paternal mitochondrial DNA transmission. *Science, 334*, 1144–1147.

Arabidopsis Genome Initiative. (2000). Analysis of the genome sequence of the flowering plant *Arabidopsis thaliana*. *Nature, 408*, 796–815.

Arimura, S., Takusagawa, S., Hatano, S., Nakazono, M., Hirai, A., & Tsutsumi, N. (1999). A novel plant nuclear gene encoding chloroplast ribosomal protein S9 has a transit peptide related to that of rice chloroplast ribosomal protein L12. *FEBS Letters, 450*, 231–234.

Bailey, J. A., Gu, Z., Clark, R. A., Reinert, K., Samonte, R. V., Schwartz, S., et al. (2002). Recent segmental duplications in the human genome. *Science, 297*, 1003–1007.

Baker, A., & Schatz, G. (1987). Sequences from a prokaryotic genome or the mouse dihydrofolate reductase gene can restore the import of a truncated precursor protein into yeast mitochondria. *Proceedings of the National Academy of Sciences of the United States of America, 84*, 3117–3121.

Been, M. D., Burgess, R. R., & Champoux, J. J. (1984). Nucleotide sequence preference at rat liver and wheat germ type 1 DNA topoisomerase breakage sites in duplex SV40 DNA. *Nucleic Acids Research, 12*, 3097–3114.

Berger, F., Hamamura, Y., Ingouff, M., & Higashiyama, T. (2008). Double fertilization - caught in the act. *Trends in Plant Science, 13*, 437–443.

Bock, R., & Timmis, J. N. (2008). Reconstructing evolution: gene transfer from plastids to the nucleus. *Bioessays, 30*, 556–566.

Bonen, L., & Calixte, S. (2006). Comparative analysis of bacterial-origin genes for plant mitochondrial ribosomal proteins. *Molecular Biology and Evolution, 23*, 701–712.

Boore, J. L. (1999). Animal mitochondrial genomes. *Nucleic Acids Research, 27*, 1767–1780.

Boore, J. L., & Brown, W. M. (1998). Big trees from little genomes: mitochondrial gene order as a phylogenetic tool. *Current Opinion in Genetics & Development, 8*, 668–674.

Brandvain, Y., Barker, M. S., & Wade, M. J. (2007). Gene co-inheritance and gene transfer. *Science, 315*, 1685.

Brandvain, Y., & Wade, M. J. (2009). The functional transfer of genes from the mitochondria to the nucleus: the effects of selection, mutation, population size and rate of self-fertilization. *Genetics, 182*, 1129–1139.

Brennicke, A., Grohmann, L., Hiesel, R., Knoop, V., & Schuster, W. (1993). The mitochondrial genome on its way to the nucleus: different stages of gene transfer in higher plants. *FEBS Letters, 325*, 140–145.

Breton, S., Stewart, D. T., Shepardson, S., Trdan, R. J., Bogan, A. E., Chapman, E. G., et al. (2011). Novel protein genes in animal mtDNA: a new sex determination system in freshwater mussels (Bivalvia: Unionoida)? *Molecular Biology and Evolution, 28*, 1645–1659.

Bubunenko, M. G., Schmidt, J., & Subramanian, A. R. (1994). Protein substitution in chloroplast ribosome evolution. A eukaryotic cytosolic protein has replaced its organelle homologue (L23) in spinach. *Journal of Molecular Biology, 240*, 28–41.

Bullerwell, C. E., & Gray, M. W. (2004). Evolution of the mitochondrial genome: protist connections to animals, fungi and plants. *Current Opinion in Microbiology, 7*, 528–534.

Campbell, C. L., & Thorsness, P. E. (1998). Escape of mitochondrial DNA to the nucleus in *yme1* yeast is mediated by vacuolar-dependent turnover of abnormal mitochondrial compartments. *Journal of Cell Science, 111*, 2455–2464.

Chang, I. F., Szick-Miranda, K., Pan, S., & Bailey-Serres, J. (2005). Proteomic characterization of evolutionarily conserved and variable proteins of *Arabidopsis* cytosolic ribosomes. *Plant Physiology, 137*, 848–862.

Christensen, A. C., Lyznik, A., Mohammed, S., Elowsky, C. G., Elo, A., Yule, R., et al. (2005). Dual-domain, dual-targeting organellar protein presequences in *Arabidopsis* can use non-AUG start codons. *The Plant Cell, 17*, 2805–2816.

Cusack, B. P., & Wolfe, K. H. (2007). When gene marriages don't work out: divorce by subfunctionalization. *Trends in Genetics, 23*, 270–272.

Dujon, B., Sherman, D., Fischer, G., Durrens, P., Casaregola, S., Lafontaine, I., et al. (2004). Genome evolution in yeasts. *Nature, 430*, 35–44.

Eichler, E. E., & Sankoff, D. (2003). Structural dynamics of eukaryotic chromosome evolution. *Science, 301*, 793–797.

Figueroa, P., Gomez, I., Holuigue, L., Araya, A., & Jordana, X. (1999). Transfer of *rps14* from the mitochondrion to the nucleus in maize implied integration within a gene encoding the iron-sulphur subunit of succinate dehydrogenase and expression by alternative splicing. *The Plant Journal, 18*, 601–609.

Fleischmann, T. T., Scharff, L. B., Alkatib, S., Hasdorf, S., Schottler, M. A., & Bock, R. (2011). Nonessential plastid-encoded ribosomal proteins in tobacco: a developmental role for plastid translation and implications for reductive genome evolution. *The Plant Cell, 23*, 3137–3155.

Gallois, J. L., Achard, P., Green, G., & Mache, R. (2001). The *Arabidopsis* chloroplast ribosomal protein L21 is encoded by a nuclear gene of mitochondrial origin. *Gene, 274*, 179–185.

Gray, M. W. (1992). The endosymbiont hypothesis revisited. *International Review of Cytology, 141*, 233–357.

Hazkani-Covo, E., & Covo, S. (2008). Numt-mediated double-strand break repair mitigates deletions during primate genome evolution. *PLoS Genetics, 4*, e1000237.

Hazkani-Covo, E., Zeller, R. M., & Martin, W. (2010). Molecular poltergeists: mitochondrial DNA copies (*numts*) in sequenced nuclear genomes. *PLoS Genetics, 6*, e1000834.

Huang, C. Y., Ayliffe, M. A., & Timmis, J. N. (2003). Direct measurement of the transfer rate of chloroplast DNA into the nucleus. *Nature, 422*, 72–76.

International Rice Genome Sequencing Project. (2005). The map-based sequence of the rice genome. *Nature, 436*, 793–800.

Kadowaki, K., Kubo, N., Ozawa, K., & Hirai, A. (1996). Targeting presequence acquisition after mitochondrial gene transfer to the nucleus occurs by duplication of existing targeting signals. *The EMBO Journal, 15*, 6652–6661.

Kubo, N., & Arimura, S. (2010). Discovery of the *rpl10* gene in diverse plant mitochondrial genomes and its probable replacement by the nuclear gene for chloroplast RPL10 in two lineages of angiosperms. *DNA Research, 17*, 1–9.

Kubo, N., Arimura, S., Tsutsumi, N., & Kadowaki, K. (2003). Involvement of N-terminal region in mitochondrial targeting of rice RPS10 and RPS14 proteins. *Plant Science, 164*, 1047–1055.

Kubo, N., Fujimoto, M., Arimura, S., Hirai, M., & Tsutsumi, N. (2008). Transfer of rice mitochondrial ribosomal protein L6 gene to the nucleus: acquisition of the 5′-untranslated region via a transposable element. *BMC Evolutionary Biology, 8*, 314.

Kubo, N., Harada, K., Hirai, A., & Kadowaki, K. (1999). A single nuclear transcript encoding mitochondrial RPS14 and SDHB of rice is processed by alternative splicing: common use of the same mitochondrial targeting signal for different proteins. *Proceedings of the National Academy of Sciences of the United States of America, 96*, 9207–9211.

Kubo, N., Jordana, X., Ozawa, K., Zanlungo, S., Harada, K., Sasaki, T., et al. (2000a). Transfer of the mitochondrial *rps10* gene to the nucleus in rice: acquisition of the 5′ untranslated region followed by gene duplication. *Molecular and General Genetics, 263*, 733–739.

Kubo, T., Nishizawa, S., Sugawara, A., Itchoda, N., Estiati, A., & Mikami, T. (2000b). The complete nucleotide sequence of the mitochondrial genome of sugar beet (*Beta vulgaris* L.) reveals a novel gene for tRNA(Cys)(GCA). *Nucleic Acids Research, 28*, 2571–2576.

Lang, B. F., Burger, G., O'Kelly, C. J., Cedergren, R., Golding, G. B., Lemieux, C., et al. (1997). An ancestral mitochondrial DNA resembling a eubacterial genome in miniature. *Nature, 387*, 493–497.

Liu, S. L., & Adams, K. L. (2010). Dramatic change in function and expression pattern of a gene duplicated by polyploidy created a paternal effect gene in the Brassicaceae. *Molecular Biology and Evolution, 27*, 2817–2828.

Long, M., de Souza, S. J., Rosenberg, C., & Gilbert, W. (1996). Exon shuffling and the origin of the mitochondrial targeting function in plant cytochrome c1 precursor. *Proceedings of the National Academy of Sciences of the United States of America, 93*, 7727–7731.

Lonsdale, D. M., Hodge, T. P., Howe, C. J., & Stern, D. B. (1983). Maize mitochondrial DNA contains a sequence homologous to the ribulose-1,5-bisphosphate carboxylase large subunit gene of chloroplast DNA. *Cell, 34*, 1007–1014.

Lough, A. N., Roark, L. M., Kato, A., Ream, T. S., Lamb, J. C., Birchler, J. A., et al. (2008). Mitochondrial DNA transfer to the nucleus generates extensive insertion site variation in maize. *Genetics, 178*, 47–55.

Lucattini, R., Likic, V. A., & Lithgow, T. (2004). Bacterial proteins predisposed for targeting to mitochondria. *Molecular Biology and Evolution, 21*, 652–658.

Martin, W. (2003). Gene transfer from organelles to the nucleus: frequent and in big chunks. *Proceedings of the National Academy of Sciences of the United States of America, 100*, 8612–8614.

Martin, W., Lagrange, T., Li, Y. F., Bisanz-Seyer, C., & Mache, R. (1990). Hypothesis for the evolutionary origin of the chloroplast ribosomal protein L21 of spinach. *Current Genetics, 18*, 553–556.

Matsuo, M., Ito, Y., Yamauchi, R., & Obokata, J. (2005). The rice nuclear genome continuously integrates, shuffles, and eliminates the chloroplast genome to cause chloroplast-nuclear DNA flux. *The Plant Cell, 17*, 665–675.

Matsushita, Y., & Isono, K. (1993). Mitochondrial transport of mitoribosomal proteins, YmL8 and YmL20, in *Saccharomyces cerevisiae*. *European Journal of Biochemistry, 214*, 577–585.

Mollier, P., Hoffmann, B., Debast, C., & Small, I. (2002). The gene encoding *Arabidopsis thaliana* mitochondrial ribosomal protein S13 is a recent duplication of the gene encoding plastid S13. *Current Genetics, 40*, 405–409.

Murcha, M. W., Rudhe, C., Elhafez, D., Adams, K. L., Daley, D. O., & Whelan, J. (2005). Adaptations required for mitochondrial import following mitochondrial to nucleus gene transfer of ribosomal protein S10. *Plant Physiology, 138*, 2134–2144.

Nakamura, Y., Itoh, T., & Martin, W. (2007). Rate and polarity of gene fusion and fission in *Oryza sativa* and *Arabidopsis thaliana*. *Molecular Biology and Evolution, 24*, 110–121.

Nass, M. M., & Nass, S. (1963). Intramitochondrial fibers with DNA characteristics. I. Fixation and electron staining reactions. *Journal of Cell Biology, 19*, 593–611.

Nishimura, Y. (2010). Uniparental inheritance of cpDNA and the genetic control of sexual differentiation in *Chlamydomonas reinhardtii*. *Journal of Plant Research, 123*, 149–162.

Notsu, Y., Masood, S., Nishikawa, T., Kubo, N., Akiduki, G., Nakazono, M., et al. (2002). The complete sequence of the rice (*Oryza sativa* L.) mitochondrial genome: frequent DNA sequence acquisition and loss during the evolution of flowering plants. *Molecular Genetics & Genomics, 268*, 434–445.

Nugent, J. M., & Palmer, J. D. (1991). RNA-mediated transfer of the gene *coxII* from the mitochondrion to the nucleus during flowering plant evolution. *Cell, 66*, 473–481.

Obara, K., Sumi, K., & Fukuda, H. (2002). The use of multiple transcription starts causes the dual targeting of *Arabidopsis* putative monodehydroascorbate reductase to both mitochondria and chloroplasts. *Plant and Cell Physiology, 43*, 697–705.

Oda, K., Yamato, K., Ohta, E., Nakamura, Y., Takemura, M., Nozato, N., et al. (1992). Gene organization deduced from the complete sequence of liverwort *Marchantia polymorpha* mitochondrial DNA. A primitive form of plant mitochondrial genome. *Journal of Molecular Biology, 223*, 1–7.

Okumura, S., Sawada, M., Park, Y. W., Hayashi, T., Shimamura, M., Takase, H., et al. (2006). Transformation of poplar (*Populus alba*) plastids and expression of foreign proteins in tree chloroplasts. *Transgenic Research, 15*, 637–646.

Olinares, P. D., Ponnala, L., & van Wijk, K. J. (2010). Megadalton complexes in the chloroplast stroma of *Arabidopsis thaliana* characterized by size exclusion chromatography, mass spectrometry, and hierarchical clustering. *Molecular & Cellular Proteomics, 9*, 1594–1615.

Puchta, H. (2005). The repair of double-strand breaks in plants: mechanisms and consequences for genome evolution. *Journal of Experimental Botany, 56*, 1–14.

Ricchetti, M., Fairhead, C., & Dujon, B. (1999). Mitochondrial DNA repairs double-strand breaks in yeast chromosomes. *Nature, 402*, 96–100.

Richly, E., & Leister, D. (2004). NUMTs in sequenced eukaryotic genomes. *Molecular Biology and Evolution, 21*, 1081–1084.

Ris, H., & Plaut, W. (1962). Ultrastructure of DNA-containing areas in chloroplasts of *Chlamydomonas*. *Journal of Cell Biology, 13*, 383–391.

Roy, S., Ueda, M., Kadowaki, K., & Tsutsumi, N. (2010). Different status of the gene for ribosomal protein S16 in the chloroplast genome during evolution of the genus *Arabidopsis* and closely related species. *Genes & Genetic Systems, 85*, 319–326.

Sandoval, P., Leon, G., Gomez, I., Carmona, R., Figueroa, P., Holuigue, L., et al. (2004). Transfer of RPS14 and RPL5 from the mitochondrion to the nucleus in grasses. *Gene, 324*, 139–147.

Sato, M., & Sato, K. (2011). Degradation of paternal mitochondria by fertilization-triggered autophagy in *C. elegans* embryos. *Science, 334*, 1141–1144.

Shahid Masood, M., Nishikawa, T., Fukuoka, S., Njenga, P. K., Tsudzuki, T., & Kadowaki, K. (2004). The complete nucleotide sequence of wild rice (*Oryza nivara*) chloroplast genome: first genome wide comparative sequence analysis of wild and cultivated rice. *Gene, 340*, 133–139.

Sheppard, A. E., Ayliffe, M. A., Blatch, L., Day, A., Delaney, S. K., Khairul-Fahmy, N., et al. (2008). Transfer of plastid DNA to the nucleus is elevated during male gametogenesis in tobacco. *Plant Physiology, 148*, 328–336.

Sheppard, A. E., & Timmis, J. N. (2009). Instability of plastid DNA in the nuclear genome. *PLoS Genetics, 5*, e1000323.

Sherratt, D. J., & Wigley, D. B. (1998). Conserved themes but novel activities in recombinases and topoisomerases. *Cell, 93*, 149–152.

Sloan, D. B., Alverson, A. J., Storchova, H., Palmer, J. D., & Taylor, D. R. (2010). Extensive loss of translational genes in the structurally dynamic mitochondrial genome of the angiosperm *Silene latifolia*. *BMC Evolutionary Biology, 10*, 274.

Steane, D. A. (2005). Complete nucleotide sequence of the chloroplast genome from the Tasmanian blue gum, *Eucalyptus globulus* (Myrtaceae). *DNA Research, 12*, 215–220.

Steemans, P., Herisse, A. L., Melvin, J., Miller, M. A., Paris, F., Verniers, J., et al. (2009). Origin and radiation of the earliest vascular land plants. *Science, 324*, 353.

Stegemann, S., Hartmann, S., Ruf, S., & Bock, R. (2003). High-frequency gene transfer from the chloroplast genome to the nucleus. *Proceedings of the National Academy of Sciences of the United States of America, 100*, 8828–8833.

Stupar, R. M., Lilly, J. W., Town, C. D., Cheng, Z., Kaul, S., Buell, C. R., et al. (2001). Complex mtDNA constitutes an approximate 620-kb insertion on *Arabidopsis thaliana* chromosome 2: implication of potential sequencing errors caused by large-unit repeats. *Proceedings of the National Academy of Sciences of the United States of America, 98*, 5099–5103.

Thorsness, P. E., & Fox, T. D. (1990). Escape of DNA from mitochondria to the nucleus in *Saccharomyces cerevisiae*. *Nature, 346*, 376–379.

Ueda, M., Arimura, S., Yamamoto, M. P., Takaiwa, F., Tsutsumi, N., & Kadowaki, K. (2006a). Promoter shuffling at a nuclear gene for mitochondrial RPL27. Involvement of interchromosome and subsequent intrachromosome recombinations. *Plant Physiology, 141*, 702–710.

Ueda, M., Fujimoto, M., Arimura, S., Murata, J., Tsutsumi, N., & Kadowaki, K. (2007). Loss of the rpl32 gene from the chloroplast genome and subsequent acquisition of a preexisting transit peptide within the nuclear gene in *Populus*. *Gene, 402*, 51–56.

Ueda, M., Fujimoto, M., Arimura, S., Tsutsumi, N., & Kadowaki, K. (2006b). Evidence for transit peptide acquisition through duplication and subsequent frameshift mutation of a preexisting protein gene in rice. *Molecular Biology and Evolution, 23*, 2405–2412.

Ueda, M., Fujimoto, M., Arimura, S., Tsutsumi, N., & Kadowaki, K. (2008a). Presence of a latent mitochondrial targeting signal in gene on mitochondrial genome. *Molecular Biology and Evolution, 25*, 1791–1793.

Ueda, M., Nishikawa, T., Fujimoto, M., Takanashi, H., Arimura, S., Tsutsumi, N., et al. (2008b). Substitution of the gene for chloroplast RPS16 was assisted by generation of a dual targeting signal. *Molecular Biology and Evolution, 25*, 1566–1575.

Ueda, M., Tsutsumi, N., & Kadowaki, K. (2005). Translocation of a 190-kb mitochondrial fragment into rice chromosome 12 followed by the integration of four retrotransposons. *International Journal of Biological Sciences, 1*, 110–113.

Unseld, M., Marienfeld, J. R., Brandt, P., & Brennicke, A. (1997). The mitochondrial genome of *Arabidopsis thaliana* contains 57 genes in 366,924 nucleotides. *Nature Genetics, 15*, 57–61.

Van Bel, M., Proost, S., Wischnitzki, E., Movahedi, S., Scheerlinck, C., Van de Peer, Y., et al. (2012). Dissecting plant genomes with the PLAZA comparative genomics platform. *Plant Physiology, 158*, 590–600.

Wade, M. J., & Goodnight, C. J. (2006). Cyto-nuclear epistasis: two-locus random genetic drift in hermaphroditic and dioecious species. *Evolution, 60*, 643–659.

Watanabe, N., Che, F. S., Iwano, M., Takayama, S., Yoshida, S., & Isogai, A. (2001). Dual targeting of spinach protoporphyrinogen oxidase II to mitochondria and chloroplasts by alternative use of two in-frame initiation codons. *Journal of Biological Chemistry, 276*, 20474–20481.

Wessler, S. R. (2006). Transposable elements and the evolution of eukaryotic genomes. *Proceedings of the National Academy of Sciences of the United States of America, 103*, 17600–17601.

Wolstenholme, D. R. (1992). Animal mitochondrial DNA: structure and evolution. *International Review of Cytology, 141*, 173–216.

Zhu, J., & Schiestl, R. H. (1996). Topoisomerase I involvement in illegitimate recombination in *Saccharomyces cerevisiae*. *Molecular and Cellular Biology, 16*, 1805–1812.

CHAPTER THREE

The Role of Horizontal Transfer in Shaping the Plant Mitochondrial Genome

Jeffrey P. Mower[*,†,1] **Kanika Jain**[*,‡] **and Nancy J. Hepburn**[*,†]
[*]Center for Plant Science Innovation, University of Nebraska, Lincoln, NE, USA
[†]Department of Agronomy and Horticulture, University of Nebraska, Lincoln, NE, USA
[‡]School of Biological Sciences, University of Nebraska, Lincoln, NE, USA
[1]Corresponding author. Email: jpmower@unl.edu

Contents

Abstract

Plant mitochondrial genomes are highly active in transferring genetic material between species. They have also gained and lost significant portions of their genomes by intracellular transfer with the plastid and nuclear genomes. Examples exist for all of these types and directions of transfer, and it is now clear that these processes have

ISSN 0065-2296,
Doi: http://dx.doi.org/10.1016/B978-0-12-394279-1.00003-X

greatly contributed to the current composition of plant mitochondria today. Despite major advances in our understanding of horizontal transfer over the last 20 years, evidence for the mechanisms of transfer is still limited, although direct contact *via* parasitism, epiphytic interactions, and grafting likely facilitates interspecific transmission. The elucidation of the extent and mechanisms of horizontal transfer involving plant mitochondria may prove important for developing a plant mitochondrial transformation system and may offer clues to the extent of nuclear horizontal gene transfer in plants.

1. INTRODUCTION

Horizontal transfer is the transmission of genetic material between evolutionarily distinct genomes by mechanisms other than organismal propagation *via* sexual or asexual means. By evolutionarily distinct, we mean genomes not only from different species but also from different cellular compartments that trace their origins to distinct species. Thus, our definition encompasses horizontal (or lateral) gene transfer in the strict sense, which specifically deals with interspecific transfer, as well as intracellular transfer, the transmission of genetic material among the mitochondrial, plastid, and nuclear genomes.

By broadly defining horizontal transfer to include interspecific and intracellular transfer events, it then becomes easy to subsume several processes that blur the rather artificial distinction between interspecific and intercompartmental transfer, such as endosymbiotic gene transfer (the functional transfer of organellar genes to the nuclear genome) as well as several examples of transfer events that cross both species and compartmental boundaries. However, our definition excludes other evolutionary processes that can lead to incongruence between molecular and organismal phylogenies through sexual means. This includes introgression, which results from hybridization and recombination between distinct species or lines of a single species, and lineage sorting, which involves the vertical transmission of multiple alleles that undergo differential loss in subsequent generations or descendant species.

In this chapter, the various types of horizontal transfer that have affected plant mitochondrial genomes are reviewed, some of the recent literature and current controversies are discussed, and several unresolved questions and future prospects in the field are highlighted.

2. ENDOSYMBIOTIC GENE TRANSFER

The extensive transfer of organellar genes to the nuclear genome is one of the most influential evolutionary forces that have shaped the organellar and nuclear genomes of all eukaryotes. Mitochondrial-to-nuclear gene transfer is covered in great detail in this book (see Chapter 2) and elsewhere (Kleine, Maier, & Leister, 2009; Timmis, Ayliffe, Huang, & Martin, 2004), so this is not covered in this chapter. Nevertheless, we do want to point out how endosymbiotic gene transfer blurs the distinction often made between interspecific and intracellular transfer. In the initial stages, as the endosymbiont became more dependent on the protoeukaryote host, genes were likely transferred from the endosymbiont genome into the host nuclear genome. At this point, the host and endosymbiont are clearly distinct species and the transfer events are interspecific. Over time, however, as the endosymbiont became an integral component of the host cell, its status shifted from a highly dependent endosymbiont to a fully essential organelle, which at the same time shifted the designation of gene transfer from interspecific to intracellular. The precise delimitation between endosymbiont and organelle is an unresolved debate (Keeling & Archibald, 2008; Theissen & Martin, 2006), raising an equally debatable delimitation between interspecific and intracellular transfers. This is a strong reason to eliminate the artificial distinction between the two processes and instead group them as two variations of a single unifying process, horizontal gene transfer (HGT).

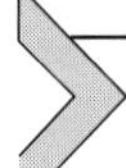

3. INTERSPECIFIC TRANSFER

3.1. Gene Transfers

The mitochondrial genome in plants generally contains 20–40 protein-coding genes, three ribosomal RNAs, and an incomplete set of transfer RNAs. Because of this restricted gene set, there is little opportunity for HGT between plant mitochondria to introduce novel genes. Most interspecific transfer events result in duplication of one or more genes in the recipient species, at least initially. Over time, it is expected that one or the other copy will degrade into a pseudogene or get deleted from the genome. Loss of the horizontal copy would result in a silent HGT event, one that may be difficult or impossible to detect and with no lasting effect on the genome. On the other hand, loss of the native copy would presumably be preceded by activation of the foreign copy, resulting in functional replacement of the native copy by the foreign copy. Prior to loss of either copy, the coexistence of homologous but non-identical sequences within a genome provides an opportunity for genetic interaction between them, through homologous recombination and gene conversion. These processes will generate chimeric mitochondrial genes that are composed of both native and foreign sequences. All of these various outcomes of HGT are depicted in Fig. 3.1.

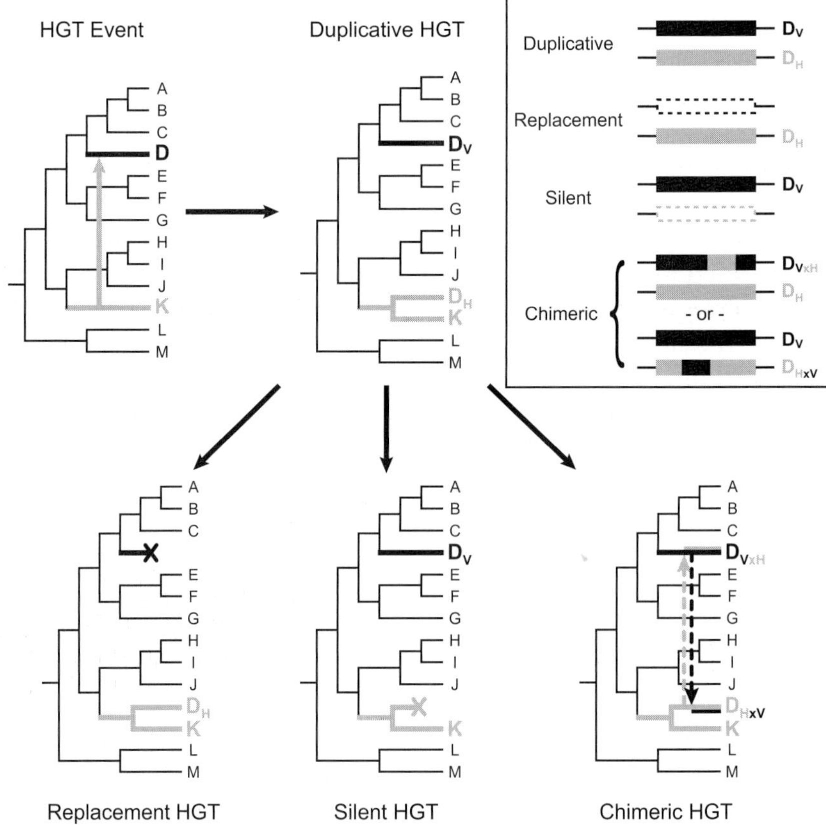

Figure 3.1 Gene trees representing the various evolutionary outcomes of HGT from lineage K to lineage D are shown. The solid grey arrow indicates the HGT event. The X marks indicate gene loss or pseudogenization. The dashed arrows indicate gene conversion events, which could occur in either or both directions. The inset shows the status of the vertical copy (D_V) and the horizontal copy (D_H) in lineage D. Filled boxes indicate active functional genes. Unfilled hatched boxes indicate gene loss or pseudogenization.

Although gene transfer between plant mitochondria usually results in gene duplication, this is not always the case. Sometimes it is possible for a mitochondrial genome to recapture a gene that was previously lost to the nucleus. Over time, a recaptured gene could either functionally replace the nuclear copy or degrade into a pseudogene.

In this section, the literature on plant-to-plant mitochondrial HGT is reviewed and the evolutionary outcomes of these events is discussed.

3.1.1. Multiple transfers, multiple outcomes

Several studies are particularly instructive with regard to the various outcomes of HGT in plant mitochondria. In the initial report of interspecific gene transfer in plants, five different cases of HGT were identified (Bergthorsson, Adams, Thomason, & Palmer, 2003). Three of these cases involved the recapture of two ribosomal protein genes, *rps2* and *rps11*. These two genes were lost early in eudicot evolution, such that all core eudicots should lack these genes in their mitochondrial genomes. Thus, it was surprising when members of three different eudicot genera (*Actinidia*, *Betula* and *Lonicera*) were shown to contain one of these genes based on Southern blot analysis (Adams, Qiu, Stoutemyer, & Palmer, 2002). Phylogenetic analysis verified recapture HGT in all three lineages, including multiple species in the Betulaceae and Caprifoliaceae (Bergthorsson *et al.*, 2003). Some of the recovered foreign genes were clearly pseudogenes indicating a silent HGT event. Other transferred genes appeared intact (without internal stop codons or frameshifting indels), although it remains uncertain whether any of these recaptured genes are functional. Another HGT case reported by Bergthorsson *et al.* (2003) involved the formation of a chimeric *rps11* gene in *Sanguinaria canadensis*, a phenomenon that is discussed more fully in Section 3.1.4. An earlier finding of two phylogenetically distinct copies of the *atp1* gene in *Amborella trichopoda* (Barkman *et al.*, 2000) was reinterpreted by Bergthorsson *et al.* (2003) as a strong signal that one copy was horizontally acquired from an unidentified eudicot.

The discovery of a duplicated *atp1* gene in *Amborella* resulting from HGT turned out to be a harbinger of much more extensive duplicative HGT into this unique plant. An expanded analysis identified 20 additional genes that had also been horizontally acquired from various angiosperm and moss donors (Bergthorsson, Richardson, Young, Goertzen, & Palmer, 2004). In most cases, the horizontally acquired copy was still intact, and at least three are transcribed, RNA edited, and potentially functional. On the other hand, several transferred gene copies are clearly pseudogenes. To date, this genome contains the most extensive amount of HGT found for any plant, although the reasons for this massive integration of foreign DNA are currently unclear.

A third study that has reported multiple resolutions of an HGT event involves the tRNA-Cys gene in angiosperms (Kitazaki *et al.*, 2011). Most plants contain either a native mitochondrial copy (*trnC1*) or a copy derived from intracellular transfer of a plastid *trnC* gene (*pt-trnC*). In *Beta vulgaris*, however, the *trnC1* gene is a pseudogene and *pt-trnC* is not present. Instead, this genome contains a distinct *trnC* gene (*trnC2*) of unknown origin (Kubo *et al.*, 2000). To determine the evolutionary origin of this gene in *B. vulgaris*, Kitazaki *et al.* (2011) looked for the three versions of the *trnC* gene across a diverse collection of land plants. They found that most species contain the native *trnC1* gene. However, the native gene has been replaced by *pt-trnC* in some plants, representing several cases of replacement HGT *via* intracellular gene transfer. Even more surprising was their finding of the unusual *trnC2* gene in three disparate angiosperm lineages (*B. vulgaris*, *Citrullus lanatus* and *Vigna radiata*), all three of which also contain a copy of the *trnC1* gene. The authors suggested that the *trnC2* gene was horizontally acquired from an unknown donor early in angiosperm history and then retained in these few lineages. Functional analysis of the native and foreign *trnC* genes in these three lineages showed alternative retention strategies: in *B. vulgaris*, the *trnC1* gene became a pseudogene and *trnC2* was functionally retained; in *C. lanatus*, *trnC2* became non-functional and *trnC1* remained functional; and in *V. radiata*, both genes are still intact and potentially functional.

3.1.2. Possible replacement transfers

In addition to the *trnC* examples just discussed, several other studies have provided speculative evidence of replacement transfers *via* HGT. In the parasitic family, Rafflesiaceae, at least two genes (*atp1* and *nad1*) were apparently acquired from their host plants in genus *Tetrastigma* (Barkman *et al.*, 2007; Davis & Wurdack, 2004). A vertically transmitted, Rafflesiaceae-like copy was not obtained for either gene. The foreign *atp1* gene appears to be intact, transcribed, and RNA edited (Barkman *et al.*, 2007). This suggests that the gene may produce functional proteins and implies that a native *atp1* gene may be absent from the genome. For the foreign *nad1* gene, however, no functional analysis was performed (Davis & Wurdack, 2004). Thus, we do not know if it is functional or if a native and functional *nad1* gene simply went undetected in their analysis.

Similar to the Rafflesiaceae *nad1* example, there are several additional studies in which a horizontally acquired gene was the only copy recovered, including a foreign *atp1* sequence in the parasitic *Mitrastema yamamotoi* (Barkman *et al.*, 2007) and two putatively foreign copies of *nad2* in *A. trichopoda* (Bergthorsson *et al.*, 2004). These sequences appear intact, but no functional analysis was performed to verify expression. These genes may indeed be additional examples of replacement HGT, but more extensive analyses will need to be performed to verify functionality and the absence of a functional vertical copy in their respective mitochondrial genomes.

3.1.3. Non-functional silent transfers

Although plant mitochondrial HGT has the potential for functional gene replacement or recapture, there are very few verified examples of such events. A survey of the literature shows that HGT among plants often results in the pseudogenization of the transferred copy. Numerous examples exist where a horizontally acquired gene is present as an unexpressed pseudogene, including the *nad1* gene and intron found in several *Gnetum* species (Won & Renner, 2003), the *nad1* and *matR* genes in *Botrychium virginianum* (Davis, Anderson, & Wurdack, 2005), the *atp1*, *atp6* and *matR* genes in several *Plantago* species (Mower, Stefanović, Young, & Palmer, 2004; Mower *et al.*, 2010), and the *cox2* gene in *Magnolia tripetala* (Hepburn, Schmidt, & Mower, in press).

The fact that so many horizontally acquired genes have degraded into pseudogenes indicates that there is little selective pressure to maintain these foreign sequences in plant mitochondrial genomes. This suggests that horizontally transferred genes will also be eliminated from genomes over time with no detrimental effects to the recipient plant. These conclusions imply that HGT is more common than the literature suggests, because there is a limited window of opportunity to detect HGT before the transferred sequence gets lost from the genome or becomes degraded beyond recognition of the detection method.

3.1.4. Chimeric gene formation

Even though plant mitochondrial HGT often appears to be an evolutionary dead end for the transferred gene, an emerging trend in the literature is that the foreign copies can interact with their native homologues to create chimeric genes *via* gene conversion (Barkman *et al.*, 2007; Bergthorsson *et al.*, 2003; Hao *et al.*, 2010; Hepburn *et al.*, in press; Mower *et al.*, 2010). In *Hedychium coronarium*, for example, a native and foreign version of an *nad1* exon, intron, and intron-encoded maturase (*matR*) are both present and clearly chimeric (Hao *et al.*, 2010). The mostly native copy is probably functional because it is transcribed and RNA edited, whereas the mostly foreign copy appears to be a pseudogene. Additional examples of coexisting versions of mitochondrial genes can be found in *Plantago* and *Magnolia*, where one copy is fully native and the other contains fragments of native and foreign DNA (Hepburn *et al.*, in press; Mower *et al.*, 2010). In these two cases, the chimeric genes with foreign DNA are not expressed.

In some species, the chimeric gene was the only copy recovered in the analysis. For example, in *S. canadensis* and *Ternstroemia gymanthera*, a chimeric gene is transcribed and RNA edited, and thus may be the only copy remaining in the genome (Bergthorsson *et al.*, 2003; Hao *et al.*, 2010). In other species with chimeric genes, such as *Pilostyles thurberi* and *Boesenbergia rotunda*, it is not clear if the chimeric genes are functional or whether additional copies exist in their genomes today (Barkman *et al.*, 2007; Hao *et al.*, 2010).

To explain the evolutionary effects of HGT and gene conversion on gene structure and function, the duplicative HGT–differential gene conversion (DH-DC) model was proposed (Hao *et al.*, 2010). In this model, interspecific (or intracellular) transfer creates duplicate but non-identical copies of a gene in a genome. These coexisting duplicates provide opportunities for gene conversion, resulting in gene sequences with a mosaic evolutionary history for one or both copies. Over time, it is unlikely that both copies will be functionally maintained. One of the copies may become lost through pseudogenization, genomic deletion, or complete gene conversion by the other copy, leaving a single gene behind that may exhibit a chimeric structure. Which copy is retained could be completely random. However, if this differential gene conversion process created novel substitutional combinations, these may be selectively retained or eliminated depending on their adaptive or detrimental effects.

3.2. Intron Transfers

3.2.1. The cox1 *group I intron: horizontal gain or stochastic loss?*

The first and most thoroughly studied case of plant mitochondrial horizontal transfer involves the *cox1* group I intron, with Dombrovska–Qiu designation cox1i729 (Dombrovska & Qiu, 2004). This intron is found sporadically in various angiosperm lineages (Cusimano, Zhang, & Renner, 2008; Sanchez-Puerta, Cho, Mower, Alverson, & Palmer, 2008; Sanchez-Puerta *et al.*, 2011).

A highly similar intron (>70% sequence identity) is also present at the same position in several diverse fungi, suggesting that the original intron donor may have belonged to one of these fungal lineages (Cusimano *et al.*, 2008; Seif *et al.*, 2005). Similar introns are also present in the same position in all three sequenced liverworts and in some green algae and fungi (Férandon *et al.*, 2010; Liu, Xue, Wang, Li, & Qiu, 2011; Seif *et al.*, 2005; Turmel, Otis, & Lemieux, 2007), but they are highly divergent relative to the angiosperm intron and may not be orthologous. Aside from angiosperms and liverworts, no other land plant group contains an intron at this position in the *cox1* gene.

Most researchers agree on a horizontal origin of this angiosperm intron from fungi, but there is disagreement regarding its subsequent evolution within angiosperms. The prevailing view for more than 15 years is that the intron has been horizontally transferred numerous times during angiosperm evolution (Adams, Clements, & Vaughn, 1998; Barkman *et al.*, 2007; Cho & Palmer, 1999; Cho, Qiu, Kuhlman, & Palmer, 1998; Sanchez-Puerta *et al.*, 2008; Sanchez-Puerta *et al.*, 2011; Vaughn, Mason, Sper-Whitis, Kuhlman, & Palmer, 1995). Recently, however, an alternative model invoking stochastic loss with mostly or completely vertical transfer was proposed (Cusimano *et al.*, 2008). These two alternative hypotheses are illustrated in Fig. 3.2. Both models seek to explain three main observations: (1) the intron is sporadically distributed among angiosperms, (2) the presence of the intron generally correlates with the presence of diagnostic nucleotide substitutions in the downstream exon sequence (i.e. the co-conversion tract (CCT)), and (3) parts of the intron phylogeny are incongruent with organismal relationships. Where the two models differ is in the interpretation of these observations.

There is no doubt that the intron has a patchy distribution among angiosperms. For instance, in the most species-rich analysis of this intron, Sanchez-Puerta *et al.* (2008) found that 25% (162 of 640) of the sampled angiosperms contained the intron. At a finer scale, the most comprehensive examination within a single family found the intron in only 4% (17 of 429) of Solanaceae species (Sanchez-Puerta *et al.*, 2011). From a parsimony perspective, the sporadic intron distribution is better explained by a model involving numerous horizontal gains rather than an all-loss model. This simple conclusion does not take into account the relative ease with which introns could be lost or horizontally acquired. It could certainly be argued that intron losses might be more evolutionarily frequent than horizontal acquisitions (Cusimano *et al.*, 2008), although mitochondrial intron loss is generally rare, at least in the angiosperm lineage (Mower, Sloan, & Alverson,

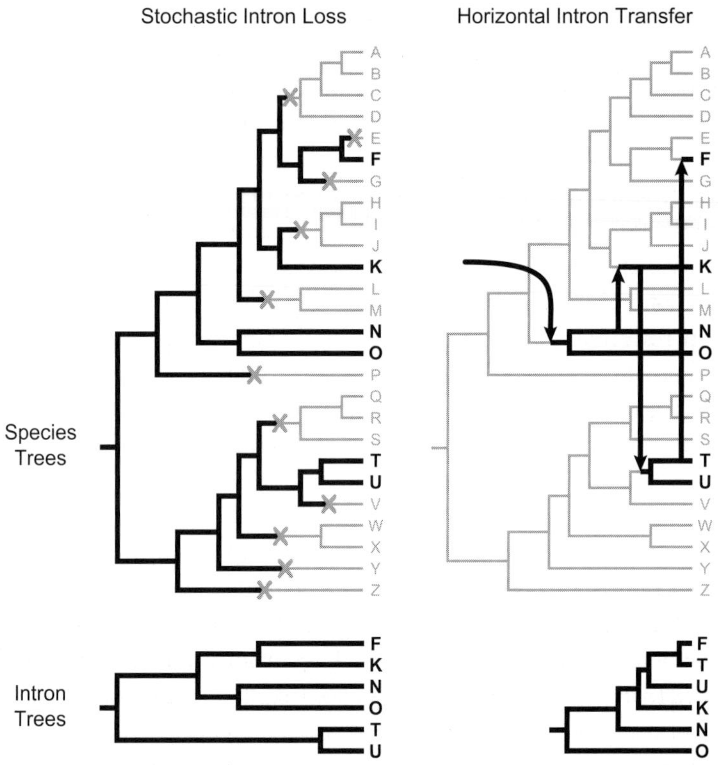

Figure 3.2 Two alternative models are shown to explain the highly sporadic distribution of an intron. The first model involves stochastic loss of the intron from various lineages (marked by an X). The second model involves multiple horizontal gains of an intron (indicated by the arrows). The donor for the initial horizontal intron transfer is unknown. Phylogenetic evaluation of the existing introns should efficiently discriminate between the two models. The stochastic loss model should produce a tree consistent with organismal phylogeny, whereas the horizontal transfer model should be inconsistent.

2012). That is, nearly all other mitochondrial introns are present in nearly all completely sequenced angiosperm mitochondrial genomes. In addition to this *cox1* intron, the only other introns with significant variation among angiosperms are from the *cox2* gene (Joly, Brouillet, & Bruneau, 2001, Kudla, Albertazzi, Blazevic, Hermann, & Bock, 2002, Hepburn *et al.*, in press), although their distribution is not nearly as patchy as the distribution of the *cox1* intron.

It is also clear that there is a strong association between the presence of the *cox1* intron and a CCT, a series of up to eight diagnostic nucleotide substitutions in the downstream exon, which are thought to be generated from the insertion process of the intron (Cusimano *et al.*, 2008, Sanchez-Puerta *et al.*, 2008; Sanchez-Puerta *et al.*, 2011). In the Cusimano *et al.* (2008) data set, 107 of 110 intron-containing species (97%) have at least two of the CCT diagnostic substitutions, and 87% contain at least five such substitutions. Conversely, only two of the 68 intron-lacking species (3%) have at least two of the CCT diagnostic nucleotides (ignoring the *Asimina triloba* sequence, which was misannotated as intronless). Overall, the correlated presence/absence of the intron and a CCT is striking. This is the pattern expected from the horizontal gain model; that is, only those species that gained the intron should have a CCT. For the stochastic loss model, which argues that most or all angiosperm lineages ancestrally had the intron (and therefore also a CCT), it is more difficult to explain the near complete absence of a CCT in the intron-lacking species. Twelve of the 68 intron-lacking species (18%) do contain a single CCT diagnostic substitution, which was suggested to be evidence of a degraded CCT signal (Cusimano *et al.*, 2008). However, they fail to provide a biological basis for the dramatic convergent evolution that must be inferred to eliminate, in nearly identical fashion, the multi-nucleotide CCT signal from the dozens to hundreds of lineages that independently lost their introns (Sanchez-Puerta *et al.*, 2008).

The third major point of contention between the two models comes from the prevalence and underlying cause of phylogenetic incongruence in *cox1* intron phylogenies. All broad-scale studies performed to date have observed unusual relationships in intron phylogenies, whereas exon phylogenies generally agree with organismal relationships (Barkman *et al.*, 2007; Cho *et al.*, 1998; Cusimano *et al.*, 2008; Sanchez-Puerta *et al.*, 2008; Sanchez-Puerta *et al.*, 2011). Most studies have argued that the intron incongruence is substantial, statistically significant, and indicative of widespread horizontal transfer (Barkman *et al.*, 2007; Cho *et al.*, 1998; Sanchez-Puerta *et al.*, 2008; Sanchez-Puerta *et al.*, 2011). In contrast, Cusimano *et al.* (2008) argue that the intron phylogenies are mostly congruent and thus consistent with vertical transmission. They attribute the observed incongruence to limited sequence variation or phylogenetic artefacts such as long-branch attraction. Although these methodological issues may explain some of the incongruent intron relationships, they do not appear sufficient to explain many of the strongly supported conflicts present in their tree or in other analyses. For instance, three of their most abundantly sampled orders (Lamiales, Malpighiales and Zingiberales) are scattered in numerous different clades throughout the tree, often with strong bootstrap support separating the various members. This incongruence is hard to reconcile with their conclusion of largely or completely vertical inheritance of the intron. Furthermore, they provide no statistical evaluation to support their qualitative conclusion of general congruence, and we doubt that any statistical evaluation would support their conclusion.

Overall, we find the stochastic loss model proposed by Cusimano *et al.* (2008) to be incompatible with the observed data. The patchy intron distribution, correlation between the intron and CCTs, and significant phylogenetic incongruence is more consistent with widespread horizontal transfer among angiosperms. Of course, the horizontal transfer model does not preclude the possibility of any vertical transmission. Once a horizontal transfer event is established, it is fully expected that the intron will be vertically transmitted just like any other genetic element. Vertical transmission of the intron is apparent within some families, such as in Araceae, Orchidaceae, Plantaginaceae and Solanaceae (Cho & Palmer, 1999; Cusimano *et al.*, 2008; Inda, Pimentel, & Chase, 2010; Sanchez-Puerta *et al.*, 2008; Sanchez-Puerta *et al.*, 2011).

3.2.2. Other examples of interspecific intron transfer

Another example of horizontal intron transfer involves nad1i77 (Won & Renner, 2003), a group II intron in the *nad1* gene that is specific to seed plants (Mower *et al.*, 2012). Several species of *Gnetum* have two non-identical copies of this intron, one of which shares more similarity to the homologous angiosperm intron than to the native *Gnetum* copy (Won & Renner, 2003). Both intron copies are flanked by *nad1* exonic sequences, so in this case, it does not look like the intron was transferred as a mobile genetic element. No functional maturase gene was detected in this intron, further arguing against mobile transfer. Instead, the intron was most likely transferred as a part of a larger fragment of DNA. In all *Gnetum* species with the angiosperm-like copy, the upstream exon contains a frameshifting indel indicating that it is not functional. The lack of rate acceleration in the foreign copy suggests that it may reside in the mitochondrial genome, although this has yet to be verified by experimental analysis or complete mitochondrial genome sequencing.

More generally, many, if not all, plant mitochondrial (and plastid) introns may have had a horizontal origin. Unless the organelles inherited and maintained one or more group I and group II introns from their alpha-proteobacterial or cyanobacterial ancestors, then the first organelle introns must have been acquired horizontally. Looking across the present-day diversity of introns in land plants and green algae, it is apparent that intron content is highly lineage specific, suggesting frequent gain and loss of introns over time. It is certainly clear that some introns have been spread *via* intragenomic (Laroche & Bousquet, 1999; Ohyama & Takemura, 2008) or intracellular (see Section 4.2) processes. However, there is mounting evidence of interspecific transfer as well, such as the *cox1* intron described above and several other examples in green algae (Brouard, Otis, Lemieux, & Turmel, 2010; Turmel *et al.*, 1999a). The relative contributions of intragenomic, intracellular and interspecific transfers in the evolution of plant organellar introns is not known.

4. INTRACELLULAR TRANSFER

In plants, intracellular transfer (also called intercompartmental transfer) encompasses the exchange of genetic material between the nuclear, mitochondrial and plastid genomes. Currently, there are two well-established acronyms, NUMTs (pronounced noo-mites) for nuclear copies of mitochondrial DNA and NUPTs (noo-peets) for nuclear copies of plastid DNA (Lopez, Yuhki, Masuda, Modi, & O'Brien, 1994; Timmis *et al.*, 2004). To facilitate discussion of the different possible directions of intracellular transfer, we introduce four new acronyms to describe the remaining types of intracellular transfer: MIPTs (mee-peets) for mitochondrial DNA of plastid origin, MINCs (meenks) for mitochondrial DNA of nuclear origin, PLMTs (play-mites) for plastid DNA of mitochondrial origin, and PLNCs (planks) for plastid DNA of nuclear origin (Fig. 3.3).

Here, we intentionally define these acronyms in the broadest sense. Our definitions do not make any presumptions about the size (small fragment *vs* entire genome) or genetic function (gene *vs* intron *vs* intergenic) of the transferred piece, the mechanism of transfer (DNA- *vs* RNA-mediated), or the functional outcomes of the transfer (functional gene *vs* inactive genomic fragment). We consider any segment of a genome that traces its origin to another genomic compartment (rather than to another species directly) as an intracellular transfer event that can be categorized by one of the six acronyms in Fig. 3.3.

4.1. NUMTs and MINCs: Shared Segments of Mitochondrial and Nuclear DNA

Complete sequencing of many plant mitochondrial and nuclear genomes has shown that MINCs and NUMTs are abundant. However, because both

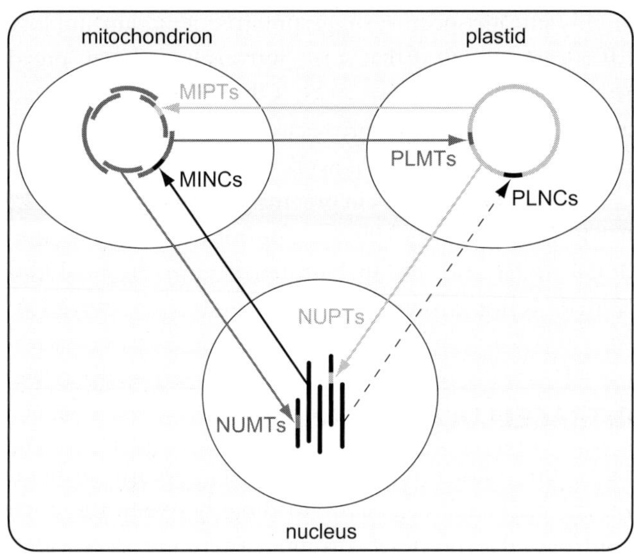

Figure. 3.3 Types of intracellular transfer. NUMTs and PLMTs indicate nuclear and plastid DNA of mitochondrial origin. MIPTs and NUPTs indicate mitochondrial and nuclear DNA of plastid origin. MINCs and PLNCs indicate mitochondrial and plastid DNA of nuclear origin. The dashed arrow indicates that there is limited evidence for PLNCs.

genomes contain large stretches of non-coding and generally featureless DNA, it is often difficult to polarize the direction of transfer (Alverson, Rice, Dickinson, Barry, & Palmer, 2011; Notsu *et al.*, 2002). In *Oryza sativa*, for example, 43 different segments totalling 48 kb (13%) of the mitochondrial genome shared similarity with the nuclear genome (Notsu *et al.*, 2002). Although several of the shared nuclear–mitochondrial segments could be positively classified as MINCs or NUMTs because they contained recognizable mitochondrial genes or nuclear retrotransposons, the direction of transfer for the majority of segments was unclear. Even greater uncertainty exists in the large cucurbit mitochondrial genomes. The 1.7-Mb genome of *Cucumis sativus* contains 535 kb of shared mitochondrial–nuclear DNA, but only 21 kb could be absolutely classified as MINCs (Alverson *et al.*, 2011). In the even larger mitochondrial genome from *Cucumis melo*, over 1.2 Mb of its 2.7-Mb genome shared similarity with the nuclear genome, but no attempt was made to distinguish between NUMTs or MINCs (Rodriguez-Moreno *et al.*, 2011). It is likely MINCs account for a large part of the mitochondrial genome expansion in both cucurbits, and probably for the expanded mitochondrial genome sizes of angiosperms in general.

Despite the overall difficulties in polarizing transfer events, MINCs exhibiting obvious nuclear sequences have been found in many angiosperms, generally making up 1–5% of their genomes. These MINCs are typically identified by homology to nuclear transposable elements, but occasionally they carry remnants of nuclear genes (Alverson *et al.*, 2010, 2011). Surprisingly, a search for MINCs using known nuclear genes came up virtually empty in *Silene conica*, whose 11.3-Mb mitochondrial genome is the largest organellar genome from any organism (Sloan *et al.*, 2012). This suggests either that (1) the huge size expansion in *S. conica* is not due to an accumulation of MINCs, (2) MINCs are not usually derived from genic sequences, or (3) the nuclear and/or mitochondrial copies degrade rapidly and become unrecognizable. Regardless of the uncertainties of the prevalence of nuclear-to-mitochondrial transfer, there are no known examples of a MINC resulting in the functional transfer of a gene into the mitochondrial genome. Thus, from an evolutionary standpoint, MINCs most likely provide little if any selective advantage to the mitochondrial genome.

In stark contrast, NUMTs have played an extremely important role in the evolution of plant mitochondrial genomes, causing wholesale gene loss after functional transfers into the nuclear genome, a process that is still ongoing in plants today (Adams *et al.*, 2002). NUMTs also generate a broad diversity of inactive sequences in the nuclear genome. The extent and evolutionary importance of mitochondrial DNA transfer to the nucleus have been reviewed in detail in this book (see Chapter 2) and elsewhere (Kleine *et al.*, 2009; Timmis *et al.*, 2004).

4.2. MIPTs and PLMTs: Shared Segments of Mitochondrial and Plastid DNA

In contrast to the difficulties in distinguishing NUMTs from MINCs, it is relatively straightforward to distinguish MIPTs from PLMTs because of the high gene density in the plastid genome. Extensive sequencing of plant mitochondrial and plastid genomes has shown that MIPTs are a universal feature of mitochondrial genomes in angiosperms, whereas PLMTs are extraordinarily rare. Among the 200+ sequenced land plant plastomes, the only report of a potential PLMT was from *Daucus carota*, in which a small genomic segment showed greater similarity to plant mitochondrial genomes than to other chloroplast genomes (Goremykin, Salamini, Velasco, & Viola, 2009). In general, the plastid genome appears highly resistant to the acquisition of foreign DNA.

The mitochondrial genome, on the other hand, is very receptive to foreign DNA, at least in angiosperms. The first identified MIPT was a 12-kb fragment from *Zea mays* (Stern & Lonsdale, 1982), and subsequent sequencing of >30 angiosperm mitochondrial genomes has shown that most species contain ~1–10% of identifiable plastid DNA. MIPTs are also present in some non-angiosperm vascular plants, including the gymnosperm *Cycas taitungenis* and the quillwort *Isoetes engelmannii* but not the spikemoss *Selaginella moellendorffii* (Chaw *et al.*, 2008; Grewe, Viehoever, Weisshaar, & Knoop, 2009; Hecht, Grewe, & Knoop, 2011). Expanded sequencing from gymnosperms, lycophytes and ferns is necessary to delimit the prevalence of MIPTs across vascular plants. MIPTs are absent from all seven currently sequenced bryophyte mitochondrial genomes, suggesting a resistance to the mitochondrial import or genomic integration of foreign DNA in bryophytes.

Although most identified MIPTs appear to be non-functional integrations of plastid genomic fragments, it is clear that this process can at least occasionally have evolutionary impact on the mitochondrial genome. Several mitochondrion-encoded tRNAs in angiosperms and gymnosperms trace their origin to the plastid genome (Maréchal-Drouard *et al.*, 1990; Mower *et al.*, 2012). Many large MIPTs also contain one or more plastid genes that are largely or completely intact, although these genes are not likely to be functional, and most degrade rather quickly. For example, the plastid *rbcL* gene has been transferred into the mitochondrial genome numerous times during angiosperm evolution, and all evaluated sequences were clearly pseudogenes (Cummings, Nugent, Olmstead, & Palmer, 2003). However, these plastid pseudogenes can still affect the mitochondrial genome. Similar to the examples of gene conversion after interspecific transfer, there are now several cases in which a functional mitochondrial gene has been gene converted by its homologous plastid gene. Segments of the mitochondrial *atp1* gene have been converted by the plastid *atpA* gene in several angiosperm lineages (Hao & Palmer, 2009). Likewise, in the mitochondrial genome of *Silene latifolia*, the small subunit ribosomal RNA gene contains a fragment derived from the plastid homologue of this gene (Sloan, Alverson, Storchova, Palmer, & Taylor, 2010).

The greatly expanded mitochondrial genome sizes in some angiosperms do not seem to result from an overabundant MIPT content. For example, in *Silene*, a genus with species exhibiting both the largest as well as one of the smallest sequenced angiosperm mitochondrial genomes, plastid integrants are universally low, from 2.5 kb (1.0%) in the 253-kb genome from

S. latifolia to only 35 kb (0.3%) in the 11.3-Mb genome from *S. conica* (Sloan *et al.*, 2012). In the Cucurbitaceae, another group with diverse genome sizes, MIPTs also do not correlate with size. The small 379-kb genome from *C. lanatus* has 23 kb of identifiable MIPTs, which corresponds to 6.1% of the whole genome (Alverson *et al.*, 2010), an intermediate percentage among angiosperms. In *Cucumis sativus*, its intermediately sized genome of 983 kb contains 113 kb (12%) of MIPTs, the largest amount recorded from any plant so far sequenced (Alverson *et al.*, 2010). Conversely, the large 2.74-Mb genome from *Cucumis melo* has only 39 kb of MIPTs, constituting a very small proportion (1.4%) of its total size (Rodriguez-Moreno *et al.*, 2011). Thus, although MIPTs are a universal feature of angiosperms, they are at most a minor component in all species sequenced to date, and the variability among species does not account for the massive size variation among species.

Mitochondrial and plastid genomes in most green algae are generally resistant to foreign DNA integrations. However, accumulating reports are indicating that selfish mobile genetic elements can be transferred between the two organellar genomes. Plastid introns in several diverse species including *Oltmannsillopsis viridis*, *Monomastix* sp., and *Pseudendoclonium akinetum* are most similar to introns in their respective mitochondrial genomes (Pombert, Otis, Lemieux, & Turmel, 2005; Pombert, Beauchamp, Otis, Lemieux, & Turmel, 2006; Turmel, Gagnon, O'Kelly, Otis, & Lemieux, 2009). Also, several types of short dispersed repeats are shared between the plastid and mitochondrial genomes from *P. akinetum* and *Volvox carteri* (Pombert *et al.*, 2005; Smith & Lee, 2009). These shared mitochondrial–plastid sequences indicate that direct intracellular transfer has occurred. However, because the direction of transfer is not clear in any of these algal examples, it is not possible to definitively label them as PLMTs or MIPTs.

4.3. NUPTs and PLNCs: Shared Segments of Nuclear and Plastid DNA

For completeness, Fig. 3.3 depicts NUPTs and PLNCs, but because they have no bearing on mitochondrial genome evolution in plants, they are not discussed in any detail here. Furthermore, there is no convincing evidence for PLNCs in the literature, to the best of our knowledge. However, undiscovered examples may lurk in the unusually large plastid genomes of some green algae such as *Acetabularia mediterranea* and *Floydiella terrestris* (Brouard *et al.*, 2010; Leible, Berger, & Schweiger, 1989). Examples may already exist in several algal plastid genomes with novel genes of unknown origin (Douglas & Penny, 1999; Turmel, Otis, & Lemieux, 1999b; Turmel *et al.*, 2009) or in *V. carteri*, whose plastid and nuclear genomes contain highly similar repeats that were likely spread *via* intracellular transfer (Smith & Lee, 2009). Depending on the direction of transfer, these shared plastid–nuclear sequences represent either NUPTs or PLNCs. For details on NUPTs, we defer to several recent reviews (Kleine *et al.*, 2009; Timmis *et al.*, 2004).

5. INTERSPECIFIC AND INTER COMPARTMENTAL TRANSFER

Some cases of horizontal transfer cannot be classified as either interspecific or intracellular because they cross both the species and the compartmental barriers. In *Phaseolus vulgaris*, for instance, a 190-bp fragment from the chloroplast *trn*A intron is embedded in a mitochondrial gene that causes cytoplasmic male sterility when present at high stoichiometry (Arrieta-Montiel *et al.*, 2001; Woloszynska, Bocer, Mackiewicz, & Janska, 2004). In addition to the plastid *trnA* intron fragment, this chimeric gene, labelled *pvs* for *P. vulgaris* sterility, also contains fragments of nuclear sequences. Surprisingly, sequence and phylogenetic analyses indicated that the *pvs-trnA* sequence did not arise through direct intracellular transfer from the *P. vulgaris* chloroplast (Woloszynska *et al.*, 2004). Instead, *pvs-trnA* showed higher sequence identity and phylogenetic affinity to monocots and magnoliids, suggesting it was acquired from the plastid genome of a non-eudicot. The *Phaseolus* scenario involving interspecific and intercompartmental transfer was compared with three other examples (in *Citrus jambhiri*, *Helianthus annuus* and *Zea mays*) in which *trnA* intron MIPTs were shown to be virtually identical to their respective chloroplast sequences, clearly indicating intracellular transfer events in these three distinct lineages.

Additional examples of interspecific and intercompartmental transfer can be found in algae. The plastid genomes of *Pyraminomonas parkeae* and *F. terrestris* contain introns that look similar to mitochondrial introns from diverse species (Brouard *et al.*, 2010; Turmel *et al.*, 2009), while several *Nephroselmis olivacea* mitochondrial introns exhibit similarity to plastid introns from various sources (Turmel *et al.*, 1999a). In other algae, such as *Oedogonium cardiacum* and *Heterosigma akashiwo*, several unusual plastid open reading frames are present that lack obvious homologues in other plastomes;

instead they share similarity with mitochondrial genes from other species (Brouard, Otis, Lemieux, & Turmel, 2008; Cattolico *et al.*, 2008).

In many cases, it is not known whether a horizontally acquired fragment of plant mitochondrial DNA was integrated into the mitochondrial genome of the recipient. Given the prevalence of NUMTs and NUPTs in plant nuclear genomes, it is certainly plausible that at least some horizontally transferred fragments of mitochondrial DNA could have integrated instead in the nucleus, whereas the plastid genome is an unlikely target due to its general resistance to foreign DNA insertions. To date, relatively few studies have examined the genomic integration point, but all so far have confirmed mitochondrion-to-mitochondrion transfer by showing that the foreign fragments are transcribed and RNA edited like other mitochondrial genes (Barkman *et al.*, 2007; Bergthorsson *et al.*, 2003; Bergthorsson *et al.*, 2004; Hao *et al.*, 2010) or are present in similar copy numbers to known mitochondrial genes (Hepburn *et al.*, in press; Mower *et al.*, 2010). Nevertheless, a more thorough evaluation of known and undiscovered cases of horizontal transfers involving plant mitochondria will be needed to determine the prevalence of interspecific and intercompartmental transfer.

6. MECHANISMS OF TRANSFER

For a horizontal transfer event to be successful, genetic material must be (1) released from the donor lineage, (2) transmitted from donor to recipient, (3) taken up by the recipient, (4) integrated into the recipient's genome (in meristematic tissue for plants), and (5) fixed throughout the recipient species. Most or all of these steps are likely to pose significant challenges that must be overcome for a transfer event to be successful and detectable. Thus, for every successful transfer event in the plant world, there were likely tens, hundreds, or even thousands of additional attempts that failed to fully run this horizontal transfer gauntlet.

However, an emerging pattern in the plant mitochondrial literature is that direct physical contact between plants may facilitate several steps of the horizontal transfer process. There are now numerous studies describing HGT involving parasitic plants (Barkman *et al.*, 2007; Davis & Wurdack, 2004; Davis *et al.*, 2005; Mower *et al.*, 2004; Mower *et al.*, 2010; Nickrent *et al.*, 2004; Park, Manen, & Schneeweiss, 2007; Yoshida, Maruyama, Nozaki, & Shirasu, 2010). This is probably not coincidence but instead suggests that the haustorial connections between host and parasitic plants may assist transmission and cellular uptake. Haustoria enable host–parasite exchange of macromolecules (including RNA), viruses, and phytoplasmas through direct connections between xylem, and in some interactions, between cytoplasm through sieve pores or plasmodesmata (Westwood, Roney, Khatibi, & Stromberg, 2009). It is possible that these same connections between parasite and host also occasionally permit the transmission of mitochondrial DNA. It is also possible that the transferred material was packaged—in a mitochondrion, virus, phytoplasma, or some other agent—and then transferred through the haustorium. The large-scale transfer of several mitochondrial genes from an unknown species of *Cuscuta* into *Plantago*, coupled with the finding that DNA rather than RNA was transferred, suggests that an intact mitochondrion may have been the vectoring agent (Mower *et al.*, 2010).

In addition to plant parasitism, several other types of plant-to-plant interactions are also beginning to find support in the literature as routes of horizontal transfer. Grafting between different species has been linked to interspecific transmission of plastid genes or entire genomes (Stegemann & Bock, 2009; Stegemann, Keuthe, Greiner, & Bock, 2012). The finding of numerous moss genes in *Amborella* suggests that epiphytic interactions may have enabled these transfers (Bergthorsson *et al.*, 2004). The tendency for the *cox1* intron to be spread among more closely related species suggests that illegitimate pollination might also play an important role (Sanchez-Puerta *et al.*, 2008; Sanchez-Puerta et al., 2011).

Once in the host cytoplasm, *via* haustorial connections or some other transfer process, it is likely that the mechanisms promoting the frequent intracellular transfer of genetic material are coopted to integrate horizontally acquired nucleic acids into the host genomes. The prevalence of NUMTs, NUPTs, MINCs and MIPTs in the nuclear and mitochondrial genomes of many different plants clearly indicates that foreign DNA is actively imported by the mitochondrion and nucleus, integrated into their genomes, and stably maintained over evolutionary time. The details of these mechanisms promoting intracellular transfer are currently limited, although it has been shown that intact plant mitochondria are capable of DNA uptake, at least *in vitro* (Koulintchenko, Konstantinov, & Dietrich, 2003). Both the nuclear and mitochondrial genomes could be integration points of transferred material. So far, however, only a mitochondrial location has been verified, based on active RNA editing or quantitative analysis by polymerase chain reaction (Barkman *et al.*, 2007; Bergthorsson *et al.*, 2003; Bergthorsson *et al.*, 2004; Hao *et al.*, 2010; Hepburn *et al.*, in press; Mower *et al.*, 2010).

A more uncertain question relates to why these horizontally transferred genomic fragments ever get fixed in a species at all. In many cases involving plant mitochondria, the transferred genes show obvious signs of pseudogenization. The lack of maintenance of transferred genes suggests that they were never functionally important to the recipient plant. Presumably, then, the fixation of transfer occurred by genetic drift rather than positive selection, which implies that most such transfer events never reach fixation.

7. EVOLUTIONARY SIGNIFICANCE

The most profound evolutionary impact of horizontal transfer on mitochondrial genomes has resulted from the functional transfer of mitochondrial genes to the nucleus through endosymbiotic gene transfer (see Chapter 2). As a result of this process (as well as gene loss), the original endosymbiont gene repertoire was reduced from >1000 protein-coding genes to the current day total of 20–40 in most land plants. The variability in gene content stems from the fact that intracellular gene transfer is still ongoing in many plants, usually involving ribosomal protein or succinate dehydrogenase genes (Adams *et al.*, 2002). Occasionally, however, genes from other functional categories, such as subunits from the ATP synthase, cytochrome *c* oxidase, or NADH dehydrogenase complexes, have also been functionally transferred to the nucleus (Hecht *et al.*, 2011; Kobayashi, Knoop, Fukuzawa, Brennicke, & Ohyama, 1997; Li, Wang, Liu, & Qiu, 2009; Nugent & Palmer, 1991). It is likely that the ongoing creation of non-functional NUMTs (and NUPTs) in many eukaryotes is a vestige of the evolutionary pressure to place mitochondrial proteins under nuclear regulatory control.

Although gene loss has predominated throughout mitochondrial evolution, intracellular transfer events have also led to the acquisition of numerous plastid and nuclear genes, most of which are clearly non-functional. Mitochondrial acquisitions of plastid tRNA genes currently represent the only known examples of functional gene replacement or recapture in plant mitochondrial genomes (Kitazaki *et al.*, 2011; Maréchal-Drouard *et al.*, 1990). These plastid-derived mitochondrial tRNAs are widespread in angiosperms and gymnosperms but have not been observed in other land plant groups (Mower *et al.*, 2012). A comprehensive evaluation of mitochondrial transcription in *Oryza sativa* revealed several genes of nuclear origin that are conserved among some monocots and appear to be expressed

(Fujii *et al.*, 2011). A close examination of their results suggests that some integrated plastid genes may also be expressed in the mitochondrial genome (J.P.M., unpublished results), although this apparent transcriptional signal could be due to plastid RNA contamination of their mitochondrial RNA preparations. Whether these expressed nuclear and/or plastid sequences in the *Oryza* mitochondrial genome are translated into proteins and have a biological function warrants further investigation.

The evolutionary impact of interspecific transfer is less obvious. The recapture of lost mitochondrial *rps2* and *rps11* genes may be the best possible examples of functional gene gains *via* HGT (Bergthorsson *et al.*, 2003). Although some of them appear intact, their functionality has yet to be assayed. In general, transferred genes appear to be degraded pseudogene fragments of no obvious importance. Nevertheless, these horizontally acquired sequences can effect lasting changes on their host genomes through gene conversion interactions between native and foreign genes, which will increase genetic diversity in either or both sequences (Barkman *et al.*, 2007; Bergthorsson *et al.*, 2003; Hao & Palmer, 2009; Hao *et al.*, 2010; Hepburn *et al.*, in press; Mower *et al.*, 2010; Sloan *et al.*, 2010). In addition, HGT and gene conversion also seems capable of driving intron loss in plant mitochondrial genes (Hepburn *et al.*, in press). Whether any of these gene conversion changes have had evolutionary impacts is currently unknown.

8. FUTURE PROSPECTS

Since the first reports of horizontal transfer of mitochondrial introns in angiosperms and green algae (Turmel *et al.*, 1995; Vaughn *et al.*, 1995), it is now evident that horizontal transfer can be a common feature of these green plant lineages, whereas other groups (such as bryophytes) seem resistant to the integration of foreign DNA. Angiosperms and green algae also have the densest sequencing of mitochondrial genes, introns, and complete genomes. Therefore, it is currently difficult to determine whether the apparent disparity in horizontal transfer frequency among lineages is due to a mechanistic difference in the uptake or genomic integration of foreign sequences, to a difference in the prevalence of parasitic or epiphytic interactions, or simply to the difference in sequencing density.

Given this uncertainty, it will be important for future studies to continue to sample genes and genomes from as many diverse species as possible, but more importantly, there is a need to study particularly active donor–

recipient pairs to begin to determine the processes that enable horizontal transfer among species as well as the conditions that promote the acceptance and maintenance of transferred material. The multigene transfers reported for *A. trichopoda* (Bergthorsson *et al.*, 2004), Rafflesiaceae (Barkman *et al.*, 2007; Davis & Wurdack, 2004), and several species of *Plantago* (Mower *et al.*, 2004, Mower *et al.*, 2010) make them especially promising candidates in this regard. Complete mitochondrial genome sequencing is currently underway for *Amborella* and *Plantago* (Richardson & Palmer, 2007; Mower *et al.*, 2010), which will enable a genome-wide evaluation of the extent and evolutionary significance of horizontal transfer and may also provide evidence of the mechanism of transfer.

An increased understanding of the horizontal transfer process in plant mitochondria could have broader impacts as well. One of the major limitations in the plant mitochondrial field is the lack of a transformation system to genetically manipulate the genome. Yet all of the evidence for intracellular and interspecific transfer indicates that many plants have natural mechanisms in place to integrate foreign DNA into the mitochondrial genome. By learning how plant mitochondria naturally acquire foreign sequences, it should be possible to manipulate these processes for mitochondrial transformation, analogous to the successful manipulation of the *Agrobacterium* infection mechanism to create a stable and efficient transformation system for the plant nuclear genome (reviewed in Gelvin, 2003).

The frequency of interspecific transfer of mitochondrial DNA among plants and the proclivity of the nuclear genome to integrate mitochondrial and plastid DNA implies that interspecific transfers of nuclear DNA may also be evolutionarily frequent. There is evidence of extensive horizontal transmission of nuclear transposable elements (reviewed in Schaack, Gilbert, & Feschotte, 2010). Reports are also emerging for functional nuclear gene transfers involving parasitic plants and fungi (Richards *et al.*, 2009; Yoshida *et al.*, 2010), the same two groups often implicated as donors and/or recipients in mitochondrial transfer events. We believe that many of the biggest and most evolutionarily significant findings on plant horizontal transfer will involve the interspecific transmission of genes between nuclear genomes. Unlike mitochondria, whose genomes with limited gene repertoires offer few opportunities to take advantage of novel gene acquisitions, there could be a huge evolutionary payoff for the transfer of novel nuclear genes or even entire pathways among species. The exponentially expanding sequence data from plant nuclear genomes should facilitate this search, although the plant nuclear genome presents additional challenges in discriminating true horizontal transfer from other processes that can also lead to phylogenetic incongruence, such as introgression, lineage sorting of ancient alleles, hybrid speciation, and differential retention of orthologues and paralogues among species after gene and whole-genome duplication events.

In summary, the past two decades have seen amazing progress on our understanding of horizontal transfer in plants, but many discoveries and challenges remain. We fully expect that the next two decades of research will be as surprising, exciting, and informative.

ACKNOWLEDGEMENTS

Research in the Mower Laboratory on plant horizontal transfer is supported by the National Science Foundation and by the University of Nebraska-Lincoln.

REFERENCES

Adams, K. L., Clements, M. J., & Vaughn, J. C. (1998). The *Peperomia* mitochondrial *coxI* group I intron: timing of horizontal transfer and subsequent evolution of the intron. *Journal of Molecular Evolution, 46*, 689–696.

Adams, K. L., Qiu, Y. L., Stoutemyer, M., & Palmer, J. D. (2002). Punctuated evolution of mitochondrial gene content: high and variable rates of mitochondrial gene loss and transfer to the nucleus during angiosperm evolution. *Proceedings of the National Academy of Sciences of the United States of America, 99*, 9905–9912.

Alverson, A. J., Rice, D. W., Dickinson, S., Barry, K., & Palmer, J. D. (2011). Origins and recombination of the bacterial-sized multichromosomal mitochondrial genome of cucumber. *The Plant Cell, 23*, 2499–2513.

Alverson, A. J., Wei, X., Rice, D. W., Stern, D. B., Barry, K., & Palmer, J. D. (2010). Insights into the evolution of mitochondrial genome size from complete sequences of *Citrullus lanatus* and *Cucurbita pepo* (Cucurbitaceae). *Molecular Biology and Evolution, 27*, 1436–1448.

Arrieta-Montiel, M., Lyznik, A., Woloszynska, M., Janska, H., Tohme, J., & Mackenzie, S. (2001). Tracing evolutionary and developmental implications of mitochondrial stoichiometric shifting in the common bean. *Genetics, 158*, 851–864.

Barkman, T. J., Chenery, G., McNeal, J. R., Lyons-Weiler, J., Ellisens, W. J., Moore, G., et al. (2000). Independent and combined analyses of sequences from all three genomic compartments converge on the root of flowering plant phylogeny. *Proceedings of the National Academy of Sciences of the United States of America, 97*, 13166–13171.

Barkman, T. J., McNeal, J. R., Lim, S. H., Coat, G., Croom, H. B., Young, N. D., et al. (2007). Mitochondrial DNA suggests at least 11 origins of parasitism in angiosperms and reveals genomic chimerism in parasitic plants. *BMC Evolutionary Biology, 7*, 248.

Bergthorsson, U., Adams, K. L., Thomason, B., & Palmer, J. D. (2003). Widespread horizontal transfer of mitochondrial genes in flowering plants. *Nature, 424*, 197–201.

Bergthorsson, U., Richardson, A. O., Young, G. J., Goertzen, L. R., & Palmer, J. D. (2004). Massive horizontal transfer of mitochondrial genes from diverse land plant donors to the basal angiosperm *Amborella*. *Proceedings of the National Academy of Sciences of the United States of America, 101*, 17747–17752.

Brouard, J. S., Otis, C., Lemieux, C., & Turmel, M. (2008). Chloroplast DNA sequence of the green alga *Oedogonium cardiacum* (Chlorophyceae): unique genome architecture, derived characters shared with the Chaetophorales and novel genes acquired through horizontal transfer. *BMC Genomics, 9*, 290.

Brouard, J. S., Otis, C., Lemieux, C., & Turmel, M. (2010). The exceptionally large chloroplast genome of the green alga *Floydiella terrestris* illuminates the evolutionary history of the Chlorophyceae. *Genome Biology and Evolution, 2*, 240–256.

Cattolico, R. A., Jacobs, M. A., Zhou, Y., Chang, J., Duplessis, M., Lybrand, T., et al. (2008). Chloroplast genome sequencing analysis of *Heterosigma akashiwo* CCMP452 (West Atlantic) and NIES293 (West Pacific) strains. *BMC Genomics, 9*, 211.

Chaw, S. M., Shih, A. C., Wang, D., Wu, Y. W., Liu, S. M., & Chou, T. Y. (2008). The mitochondrial genome of the gymnosperm *Cycas taitungensis* contains a novel family of short interspersed elements, Bpu sequences, and abundant RNA editing sites. *Molecular Biology and Evolution, 25*, 603–615.

Cho, Y., & Palmer, J. D. (1999). Multiple acquisitions *via* horizontal transfer of a group I intron in the mitochondrial *cox1* gene during evolution of the Araceae family. *Molecular Biology and Evolution, 16*, 1155–1165.

Cho, Y., Qiu, Y.-L., Kuhlman, P., & Palmer, J. D. (1998). Explosive invasion of plant mitochondria by a group I intron. *Proceedings of the National Academy of Sciences of the United States of America, 95*, 14244–14249.

Cummings, M. P., Nugent, J. M., Olmstead, R. G., & Palmer, J. D. (2003). Phylogenetic analysis reveals five independent transfers of the chloroplast gene *rbcL* to the mitochondrial genome in angiosperms. *Current Genetics, 43*, 131–138.

Cusimano, N., Zhang, L.-B., & Renner, S. S. (2008). Reevaluation of the *cox1* group I intron in Araceae and angiosperms indicates a history dominated by loss rather than horizontal transfer. *Molecular Biology and Evolution, 25*, 265–276.

Davis, C. C., Anderson, W. R., & Wurdack, K. J. (2005). Gene transfer from a parasitic flowering plant to a fern. *Proceedings of the Royal Society B: Biological Sciences, 272*, 2237–2242.

Davis, C. C., & Wurdack, K. J. (2004). Host-to-parasite gene transfer in flowering plants: phylogenetic evidence from Malpighiales. *Science, 305*, 676–678.

Dombrovska, O., & Qiu, Y.-L. (2004). Distribution of introns in the mitochondrial gene *nad1* in land plants: phylogenetic and molecular evolutionary implications. *Molecular Phylogenetics and Evolution, 32*, 246–263.

Douglas, S. E., & Penny, S. L. (1999). The plastid genome of the cryptophyte alga, *Guillardia theta*: complete sequence and conserved synteny groups confirm its common ancestry with red algae. *Journal of Molecular Evolution, 48*, 236–244.

Férandon, C., Moukha, S., Callac, P., Benedetto, J. P., Castroviejo, M., & Barroso, G. (2010). The *Agaricus bisporus cox1* gene: the longest mitochondrial gene and the largest reservoir of mitochondrial group I introns. *PLoS ONE, 5*, e14048.

Fujii, S., Toda, T., Kikuchi, S., Suzuki, R., Yokoyama, K., Tsuchida, H., et al. (2011). Transcriptome map of plant mitochondria reveals islands of unexpected transcribed regions. *BMC Genomics, 12*, 279.

Gelvin, S. B. (2003). *Agrobacterium*-mediated plant transformation: the biology behind the "gene-jockeying" tool. *Microbiology and Molecular Biology Reviews, 67*, 16–37.

Goremykin, V. V., Salamini, F., Velasco, R., & Viola, R. (2009). Mitochondrial DNA of *Vitis vinifera* and the issue of rampant horizontal gene transfer. *Molecular Biology and Evolution, 26*, 99–110.

Grewe, F., Viehoever, P., Weisshaar, B., & Knoop, V. (2009). A *trans*-splicing group I intron and tRNA-hyperediting in the mitochondrial genome of the lycophyte *Isoetes engelmannii*. *Nucleic Acids Research, 37*, 5093–5104.

Hao, W., & Palmer, J. D. (2009). Fine-scale mergers of chloroplast and mitochondrial genes create functional, transcompartmentally chimeric mitochondrial genes. *Proceedings of the National Academy of Sciences of the United States of America, 106*, 16728–16733.

Hao, W., Richardson, A. O., Zheng, Y., & Palmer, J. D. (2010). Gorgeous mosaic of mitochondrial genes created by horizontal transfer and gene conversion. *Proceedings of the National Academy of Sciences of the United States of America, 107*, 21576–21581.

Hecht, J., Grewe, F., & Knoop, V. (2011). Extreme RNA editing in coding islands and abundant microsatellites in repeat sequences of *Selaginella moellendorffii* mitochondria: the root of frequent plant mtDNA recombination in early tracheophytes. *Genome Biology and Evolution, 3*, 344–358.

Hepburn, N. J., Schmidt, D. W. & Mower, J. P. in press. Loss of two introns from the Magnolia tripetala mitochondrial cox2 gene implicates horizontal gene transfer and gene conversion as a novel mechanism of intron loss. *Molecular Biology and Evolution*.

Inda, L. A., Pimentel, M., & Chase, M. W. (2010). Contribution of mitochondrial *cox1* intron sequences to the phylogenetics of tribe Orchideae (Orchidaceae): do the distribution and sequence of this intron in orchids also tell us something about its evolution? *Taxon, 59*, 1053–1064.

Joly, S., Brouillet, L., & Bruneau, A. (2001). Phylogenetic implications of the multiple losses of the mitochondrial *coxII.i3* intron in the angiosperms. *International Journal of Plant Sciences, 162*, 359–373.

Keeling, P. J., & Archibald, J. M. (2008). Organelle evolution: what's in a name? *Current Biology, 18*, R345–R347.

Kitazaki, K., Kubo, T., Kagami, H., Matsumoto, T., Fujita, A., Matsuhira, H., et al. (2011). A horizontally transferred tRNA(Cys) gene in the sugar beet mitochondrial genome: evidence that the gene is present in diverse angiosperms and its transcript is aminoacylated. *The Plant Journal, 68*, 262–272.

Kleine, T., Maier, U. G., & Leister, D. (2009). DNA transfer from organelles to the nucleus: the idiosyncratic genetics of endosymbiosis. *Annual Review of Plant Biology, 60*, 115–138.

Kobayashi, Y., Knoop, V., Fukuzawa, H., Brennicke, A., & Ohyama, K. (1997). Interorganellar gene transfer in bryophytes: the functional *nad7* gene is nuclear encoded in *Marchantia polymorpha*. *Molecular and General Genetics, 256*, 589–592.

Koulintchenko, M., Konstantinov, Y., & Dietrich, A. (2003). Plant mitochondria actively import DNA *via* the permeability transition pore complex. *EMBO Journal, 22*, 1245–1254.

Kubo, T., Nishizawa, S., Sugawara, A., Itchoda, N., Estiati, A., & Mikami, T. (2000). The complete nucleotide sequence of the mitochondrial genome of sugar beet (*Beta vulgaris* L.) reveals a novel gene for tRNA^{Cys}(GCA). *Nucleic Acids Research, 28*, 2571–2576.

Kudla, J., Albertazzi, F. J., Blazevic, D., Hermann, M., & Bock, R. (2002). Loss of the mitochondrial *cox2* intron 1 in a family of monocotyledonous plants and utilization of mitochondrial intron sequences for the construction of a nuclear intron. *Molecular Genetics and Genomics, 267*, 223–230.

Laroche, J., & Bousquet, J. (1999). Evolution of the mitochondrial *rps3* intron in perennial and annual angiosperms and homology to *nad5* intron 1. *Molecular Biology and Evolution, 16*, 441–452.

Leible, M. B., Berger, S., & Schweiger, H. G. (1989). The plastome of *Acetabularia mediterranea* and *Batophora oerstedii*: inter- and intraspecific variability and physical properties. *Current Genetics, 15*, 355–361.

Li, L., Wang, B., Liu, Y., & Qiu, Y. L. (2009). The complete mitochondrial genome sequence of the hornwort *Megaceros aenigmaticus* shows a mixed mode of conservative yet dynamic evolution in early land plant mitochondrial genomes. *Journal of Molecular Evolution, 68*, 665–678.

Liu, Y., Xue, J. Y., Wang, B., Li, L., & Qiu, Y. L. (2011). The mitochondrial genomes of the early land plants *Treubia lacunosa* and *Anomodon rugelii*: dynamic and conservative evolution. *PLoS ONE, 6*, e25836.

Lopez, J. V., Yuhki, N., Masuda, R., Modi, W., & O'Brien, S. J. (1994). Numt, a recent transfer and tandem amplification of mitochondrial DNA to the nuclear genome of the domestic cat. *Journal of Molecular Evolution, 39*, 174–190.

Maréchal-Drouard, L., Guillemaut, P., Cosset, A., Arbogast, M., Weber, F., Weil, J.-H., et al. (1990). Transfer RNAs of potato (*Solanum tuberosum*) mitochondria have different genetic origins. *Nucleic Acids Research, 18*, 3689–3696.

Mower, J. P., Sloan, D. B., & Alverson, A. J. (2012). Plant mitochondrial genome diversity: the genomics revolution. In J. F. Wendel, J. Greilhuber, J. Dolezel, & I. J. Leitch (Eds.), *Plant genome diversity Vol. 1. Plant genomes, their residents, and their evolutionary dynamics*. Vienna: Springer.

Mower, J. P., Stefanović, S., Hao, W., Gummow, J. S., Jain, K., Ahmed, D., et al. (2010). Horizontal acquisition of multiple mitochondrial genes from a parasitic plant followed by gene conversion with host mitochondrial genes. *BMC Biology, 8*, 150.

Mower, J. P., Stefanović, S., Young, G. J., & Palmer, J. D. (2004). Gene transfer from parasitic to host plants. *Nature, 432*, 165–166.

Nickrent, D. L., Blarer, A., Qiu, Y. L., Vidal-Russell, R., & Anderson, F. E. (2004). Phylogenetic inference in Rafflesiales: the influence of rate heterogeneity and horizontal gene transfer. *Evolutionary Biology, 4*, 40.

Notsu, Y., Masood, S., Nishikawa, T., Kubo, N., Akiduki, G., Nakazono, M., et al. (2002). The complete sequence of the rice (*Oryza sativa* L.) mitochondrial genome: frequent DNA sequence acquisition and loss during the evolution of flowering plants. *Molecular Genetics and Genomics, 268*, 434–445.

Nugent, J. M., & Palmer, J. D. (1991). RNA-mediated transfer of the gene *coxII* from the mitochondrion to the nucleus during flowering plant evolution. *Cell, 66*, 473–481.

Ohyama, K., & Takemura, M. (2008). Molecular evolution of mitochondrial introns in the liverwort *Marchantia polymorpha*. *Proceedings of the Japan Academy, Series B, Physical and Biological Sciences, 84*, 17–23.

Park, J. M., Manen, J. F., & Schneeweiss, G. M. (2007). Horizontal gene transfer of a plastid gene in the non-photosynthetic flowering plants *Orobanche* and *Phelipanche* (Orobanchaceae). *Molecular Phylogenetics and Evolution, 43*, 974–985.

Pombert, J. F., Beauchamp, P., Otis, C., Lemieux, C., & Turmel, M. (2006). The complete mitochondrial DNA sequence of the green alga *Oltmannsiellopsis viridis*: evolutionary trends of the mitochondrial genome in the Ulvophyceae. *Current Genetics, 50*, 137–147.

Pombert, J. F., Otis, C., Lemieux, C., & Turmel, M. (2005). The chloroplast genome sequence of the green alga *Pseudendoclonium akinetum* (Ulvophyceae) reveals unusual structural features and new insights into the branching order of chlorophyte lineages. *Molecular Biology and Evolution, 22*, 1903–1918.

Richards, T. A., Soanes, D. M., Foster, P. G., Leonard, G., Thornton, C. R., & Talbot, N. J. (2009). Phylogenomic analysis demonstrates a pattern of rare and ancient horizontal gene transfer between plants and fungi. *The Plant Cell, 21*, 1897–1911.

Richardson, A. O., & Palmer, J. D. (2007). Horizontal gene transfer in plants. *Journal of Experimental Botany, 58*, 1–9.

Rodriguez-Moreno, L., Gonzalez, V. M., Benjak, A., Marti, M. C., Puigdomenech, P., Aranda, M. A., et al. (2011). Determination of the melon chloroplast and mitochondrial genome sequences reveals that the largest reported mitochondrial genome in plants contains a significant amount of DNA having a nuclear origin. *BMC Genomics, 12*, 424.

Sanchez-Puerta, M. V., Abbona, C. C., Zhuo, S., Tepe, E. J., Bohs, L., Olmstead, R. G., et al. (2011). Multiple recent horizontal transfers of the *cox1* intron in Solanaceae and extended co-conversion of flanking exons. *BMC Evolutionary Biology, 11*, 277.

Sanchez-Puerta, M. V., Cho, Y., Mower, J. P., Alverson, A. J., & Palmer, J. D. (2008). Frequent, phylogenetically local horizontal transfer of the *cox1* group I Intron in flowering plant mitochondria. *Molecular Biology and Evolution, 25*, 1762–1777.

Schaack, S., Gilbert, C., & Feschotte, C. (2010). Promiscuous DNA: horizontal transfer of transposable elements and why it matters for eukaryotic evolution. *Trends in Ecology and Evolution, 25*, 537–546.

Seif, E., Leigh, J., Liu, Y., Roewer, I., Forget, L., & Lang, B. F. (2005). Comparative mitochondrial genomics in zygomycetes: bacteria-like RNase P RNAs, mobile elements and a close source of the group I intron invasion in angiosperms. *Nucleic Acids Research, 33*, 734–744.

Sloan, D. B., Alverson, A. J., Chuckalovcak, J. P., Wu, M., McCauley, D. E., Palmer, J. D., et al. (2012). Rapid evolution of enormous, multichromosomal genomes in flowering plant mitochondria with exceptionally high mutation rates. *PLoS Biology, 10*, e1001241.

Sloan, D. B., Alverson, A. J., Storchova, H., Palmer, J. D., & Taylor, D. R. (2010). Extensive loss of translational genes in the structurally dynamic mitochondrial genome of the angiosperm *Silene latifolia. Evolutionary Biology, 10*, 274.

Smith, D. R., & Lee, R. W. (2009). The mitochondrial and plastid genomes of *Volvox carteri*: bloated molecules rich in repetitive DNA. *BMC Genomics, 10*, 132.

Stegemann, S., & Bock, R. (2009). Exchange of genetic material between cells in plant tissue grafts. *Science, 324*, 649–651.

Stegemann, S., Keuthe, M., Greiner, S., & Bock, R. (2012). Horizontal transfer of chloroplast genomes between plant species. *Proceedings of the National Academy of Sciences of the United States of America, 109*, 2434–2438.

Stern, D. B., & Lonsdale, D. M. (1982). Mitochondrial and chloroplast genomes of maize have a 12-kilobase DNA sequence in common. *Nature, 299*, 698–702.

Theissen, U., & Martin, W. (2006). The difference between organelles and endosymbionts. *Current Biology, 16*, R1016–R1017.

Timmis, J. N., Ayliffe, M. A., Huang, C. Y., & Martin, W. (2004). Endosymbiotic gene transfer: organelle genomes forge eukaryotic chromosomes. *Nature Reviews Genetics, 5*, 123–135.

Turmel, M., Cote, V., Otis, C., Mercier, J. P., Gray, M. W., Lonergan, K. M., et al. (1995). Evolutionary transfer of ORF-containing group I introns between different subcellular compartments (chloroplast and mitochondrion). *Molecular Biology and Evolution, 12*, 533–545.

Turmel, M., Gagnon, M. C., O'Kelly, C. J., Otis, C., & Lemieux, C. (2009). The chloroplast genomes of the green algae *Pyramimonas*, *Monomastix*, and *Pycnococcus* shed new light on the evolutionary history of prasinophytes and the origin of the secondary chloroplasts of euglenids. *Molecular Biology and Evolution, 26*, 631–648.

Turmel, M., Lemieux, C., Burger, G., Lang, B. F., Otis, C., Plante, I., et al. (1999a). The complete mitochondrial DNA sequences of *Nephroselmis olivacea* and *Pedinomonas minor*. Two radically different evolutionary patterns within green algae. *The Plant Cell, 11*, 1717–1730.

Turmel, M., Otis, C., & Lemieux, C. (1999b). The complete chloroplast DNA sequence of the green alga *Nephroselmis olivacea*: insights into the architecture of ancestral chloroplast genomes. *Proceedings of the National Academy of Sciences of the United States of America, 96*, 10248–10253.

Turmel, M., Otis, C., & Lemieux, C. (2007). An unexpectedly large and loosely packed mitochondrial genome in the charophycean green alga *Chlorokybus atmophyticus*. *BMC Genomics, 8*, 137.

Vaughn, J. C., Mason, M. T., Sper-Whitis, G. L., Kuhlman, P., & Palmer, J. D. (1995). Fungal origin by horizontal transfer of a plant mitochondrial group I intron in the chimeric *coxI* gene of *Peperomia. Journal of Molecular Evolution, 41*, 563–572.

Westwood, J. H., Roney, J. K., Khatibi, P. A., & Stromberg, V. K. (2009). RNA translocation between parasitic plants and their hosts. *Pest Management Science, 65*, 533–539.

Woloszynska, M., Bocer, T., Mackiewicz, P., & Janska, H. (2004). A fragment of chloroplast DNA was transferred horizontally, probably from non-eudicots, to mitochondrial genome of *Phaseolus*. *Plant Molecular Biology, 56*, 811–820.

Won, H., & Renner, S. S. (2003). Horizontal gene transfer from flowering plants to *Gnetum*. *Proceedings of the National Academy of Sciences of the United States of America, 100*, 10824–10829.

Yoshida, S., Maruyama, S., Nozaki, H., & Shirasu, K. (2010). Horizontal gene transfer by the parasitic plant *Striga hermonthica*. *Science, 328*, 1128.

CHAPTER FOUR

Mitochondrial Genome Evolution and Gynodioecy

Pascal Touzet[1]
Laboratoire de Génétique et Evolution des Populations Végétales, UMR CNRS 8198, Université des Sciences et Technologies de Lille - Lille1, Villeneuve d'Ascq cedex, France
[1]Corresponding author. E-mail: pascal.touzet@univ-lille1.fr

Contents

Abstract

Gynodioecy is a breeding system frequently encountered in flowering plants. It consists of the co-occurrence of hermaphrodites and females in populations. Gynodioecy is generally under nuclear-cytoplasmic control, which involves mitochondrial sterilizing genes and nuclear genes that restore male fertility. Sterilizing mitochondrial genomes have been described in crops in which cytoplasmic male sterility (CMS) is cryptic, i.e. not maintained in populations of wild relative species. However, the isolation of sterilizing genes has led to the definition of a profile that can help to find candidate genes in CMSs found in gynodioecious species. We discuss the expected effect of two alternative evolutionary dynamics of gynodioecy on mitochondrial diversity and describe the pattern of diversity observed at the gene and genome levels. On the basis of whole sequence analyses of mitochondrial genomes in beet and maize, we suggest that CMS mitochondrial genomes might exhibit a faster evolution rate, and a clue to its cause might be found in male sterility itself.

Advances in Botanical Research, Volume 63
ISSN 0065-2296,
http://dx.doi.org/10.1016/B978-0-12-394279-1.00004-1

1. EVOLUTIONARY DYNAMICS OF GYNODIOECY

1.1. What is Gynodioecy?

Gynodioecy, a breeding system commonly found in flowering plants, consists of the occurrence in the same species of two sexual morphs: hermaphrodite individuals and female plants that have lost the ability to produce viable pollen. In European flora, it can occur in up to 7% of species (Richards, 1997). Seen as a transitional step towards dioecy (two morphs: one male, one female) or a stable breeding system, it has been under investigation by a large line of evolutionary biologists going back to Darwin (1877). The maintenance of females with theoretically lower fitness due to the loss of one way to transmit their genes might seem a mystery at first glance, but this can be explained as soon as females exhibit more or better seedlings than hermaphrodites. This phenomenon has been called female compensation or female advantage. This female advantage can be the direct consequence of the energy saved from pollen production or be the result of self-fertilization prevention when species are self-compatible. Females are obligate outcrossers, and therefore might produce better seedlings than self-compatible hermaphrodites that can partially produce progeny obtained by selfing and thus exhibit a deprived fitness due to homozygous deleterious alleles (Dufay & Billard, 2012; Thompson & Tarayre, 2000). Gynodioecy can be determined by nuclear genes as in the case of the Virginian wild strawberry (Ashman, 1999) but is generally believed to be under nuclear–cytoplasmic control; i.e. cytoplasmic male-sterilizing (CMS) genes and nuclear male fertility restorer loci (see also Chapter 5). Gynodioecy maintenance is possible under less stringent conditions when it is under nucleo-cytoplasmic (NC) control: a minimum female fecundity advantage is sufficient for CMS genomes to spread in populations, whereas female seed production must be at least twice as high as hermaphroditic seed production in the case of nuclear male sterility (Lewis, 1941).

Empirical studies show that, as expected, female advantage is often detected in gynodioecious species and the highest values of female advantage are found in the case of nuclear gynodioecy (Dufay & Billard, 2012). This female advantage can consist of more flowers, higher fruit set, higher total seed production or heavier seeds with a better germination rate (Dufay & Billard 2012; Shykoff, Kolokotronis, Collin, & Lopez-Villavicencio, 2003).

The NC determinism of gynodioecy is, however, more often assumed than tested adequately. It has been clearly demonstrated in such species as

Silene vulgaris (Charlesworth & Laporte, 1998; Taylor, Olson, & McCauley, 2001), *Silene nutans* (Garraud, Brachi, Dufay, Touzet, & Shykoff, 2011), *S. acaulis* (Städler & Delph, 2002), wild beet (Dufay, Cuguen, Arnaud, & Touzet, 2009; Laporte *et al.*, 1998; Touzet, Hueber, Bürkholz, Barnes, & Cuguen, 2004), thyme (Belhassen *et al.*, 1991; Charlesworth & Laporte, 1998) and plantains (de Haan, Mateman, Van Dijk, & Van Damme, 1997; Van Damme, Hundscheid, Ivanovic, & Koelewijn, 2004).

When gynodioecy is under nuclear–cytoplasmic control, it can be interpreted as the result of a genomic conflict (Cosmides & Tooby, 1981). Since cytoplasmic genomes are transmitted only through seeds, any mutation that disrupts pollen production will be selected, as long as it maximizes its transmission through female compensation. On the other hand, this cytoplasmic context will create a selective pressure in favour of any nuclear gene that counteracts the effect of cytoplasmic sterilizing genes and consequently restore male transmission. Therefore, the dynamics are similar to the arms race found in host/pathogen interactions, where the CMS genome can be seen as the virulent pathogen, and the nuclear male fertility restorer allele the specific resistance allele of the host (Touzet & Budar, 2004).

Male sterility has been found and used in numerous crop species for hybrid production on a large scale (Schnable & Wise, 1998). In most studied crop systems, male sterility genes are mitochondrial. Therefore, it is reasonable to assume that NC gynodioecy is nuclear–mitochondrial. Thus, CMS in crops has been proposed as a model to study nuclear–mitochondrial interactions. The molecular characterization of a male fertility restorer locus in petunia has been a cornerstone in the study of a large nuclear gene family involved in the regulation of organelle gene expression, the PPR gene family (see Chapter 10) (Bentolila, Alfonso, & Hanson, 2002; Touzet & Budar, 2004).

1.2. Evolutionary Forces

1.2.1. Drift and selection in mitochondria

A brief view of the different evolutionary forces that are involved in the evolution and diversity of mitochondrial genomes is given before considering the possible effect of gynodioecy. As pointed in Lynch's seminal book (Lynch, 2007), genome evolution, when one considers content and structure, must be understood in the light of population genetics principles. In particular, the role of non-adaptive processes such as mutation and random genetic drift might be essential to understand the evolution of genome complexity, such as the variation of genome size of prokaryotes and eukaryotes, the occurrence or absence of introns, the proliferation or not of mobile elements, or the maintenance and diversification of duplicates in nuclear genomes (Lynch & Conery, 2003). Random genetic drift is the effect of chance on allele frequency over time, similar to a sample bias in the transmission of gametes in the descendent population. Random genetic drift is predominant in small size populations. The intensity of random genetic drift is quantified by what is called the genetic effective size of a population: when the genetic effective size is small, the intensity of random genetic drift is high. Therefore, one expects in this case that the selection will be less efficient. The genetic effective size depends on the size of the population, but also on the breeding system (the way individuals mate) since variation in reproductive success among individuals will reduce the effective population size.

Considering the importance of non-adaptive processes, Lynch, Koskella, and Schaack (2006) proposed an explanation called the mutation pressure hypothesis (also called the mutational hazard hypothesis or the mutational burden hypothesis (MBH)) to understand the contrasting features of mitochondrial genomes found in plants compared with those found in animals. It has been known for a long time that animal mitochondrial genomes are compact (around 16 kb) and structurally strongly conserved, that is, gene order is stable among species, and they exhibit a high mutation rate. In contrast, mitochondrial plant genomes can reach megabases in size in Cucurbitaceae or *Silene* (Alverson *et al.*, 2010; Sloan *et al.*, 2012; Ward, Anderson, & Bendich, 1981), are structurally labile even at the species level but exhibit a very low mutation rate (Muse, 2000; Palmer & Herbon, 1988; Wolfe, Li, Sharp, 1987).

The MBH postulates that non-coding DNA insertion near a gene "increases the susceptibility of a gene to degenerative changes by increasing the size of the mutational target" (Lynch, 2007). The hazard of this insertion depends on its size and the mutation rate. When the mutation rate is low, the probability of deleterious mutation is low, and therefore the insertion is near neutral. In addition, when drift is high, the selection is particularly inefficient to eliminate the insertion. Considering that the effective population size is globally similar between animal and plant mitochondrial genomes, the difference in mutation rate is postulated to be responsible for the difference in invasiveness between plant and animal mitochondrial genomes; a low mutation rate leads to an inevitable increase in genome size. This hypothesis has recently been tested in a study of mitochondrial genomes of Cucurbitaceae (watermelon and zucchini) (Alverson *et al.*, 2010) and *Silene*

(Sloan *et al.*, 2012). In both studies, contrary to the MBH, large genomes were associated with a higher mutation rate, suggesting that additional forces need to be taken into account.

In addition to the effect of non-adaptive processes, a recent study comparing invertebrate and vertebrate mitochondrial diversities has questioned the relationship between the population size and mitochondrial diversity (Bazin, Glemin, & Galtier, 2006). A positive correlation is expected because genetic drift, which causes a reduction in diversity, is stronger in small populations. While this correlation was found in the case of nuclear diversity, with large populations of invertebrate exhibiting the highest levels of diversity, this correlation was not found with mitochondrial gene diversity. This absence of correlation suggests that in small populations, like those found in vertebrates, genetic drift is predominant and leads to a reduction in diversity. Conversely, in large populations, like those found in invertebrates, recurrent positive selection, i.e. fixation of beneficial variants, also leads to a reduction in diversity through hitchhiking. In a non-recombining genome such as the mitochondrial genome, hitchhiking is assumed to be strong. Consequently, any variant linked to the favourable allele of the gene under selection will also reach fixation in the population. This study generated a lively discussion in the biodiversity community, because mitochondrial diversity has been widely used as a proxy in population genetics studies to estimate the conservation status of a population or species. This work placed new emphasis not only on the role of selection in the evolution of mitochondrial genomes in animals but also on its mode of action, as former studies had mostly shown that mitochondrial genomes were mainly under purifying (negative) selection, i.e. the elimination of deleterious variants (Meiklejohn, Montooth, & Rand, 2007). Large datasets, such as found for animals, are still not available for plants. In addition, it is difficult to apprehend how studies conducted on animal species can provide clues on mitochondrial genome diversity in plants, and in particular whether such positive selection dynamics can occur in a low mutation rate context. However, the bridges that are starting to be built between data from animal and plant species will undoubtedly be fruitful to draw a more general picture of mitochondrial genome evolution (Galtier, 2011).

1.2.2. Theoretical effects of gynodioecy on mitochondrial genome evolution

Theoretical models have proposed two scenarios to explain the evolutionary dynamics of gynodioecy maintenance in populations; i.e. the maintenance of females and hermaphrodites in populations that implies a polymorphism both at the cytoplasmic level (two CMSs or one CMS/one fertile one) and at the nuclear level (restorer and non-restorer alleles segregate in the population). Both scenarios rely on female advantage that promotes the invasion of CMS in populations as long as it is mainly borne by females. This situation is found when the CMS is newly arrived in the population, either by migration or mutation, and the corresponding restorer allele is therefore rare or absent. The scenarios differ in whether the genetic factors controlling nuclear–cytoplasmic gynodioecy are maintained in species over long evolutionary time scales *via* balancing selection or are continually arising and being replaced either locally or globally (epidemic dynamics) (Fig. 4.1). Balancing selection arises when the genetic factor, either cytoplasmic or nuclear factors, are favoured when they are rare. This scenario also implies that restorer alleles are costly when useless, that is, when they are borne by hermaphrodites that do not have the corresponding CMS (Bailey, Delph, & Lively, 2003; Delph, Touzet, & Bailey, 2007; Dufay, Touzet, Maurice, &

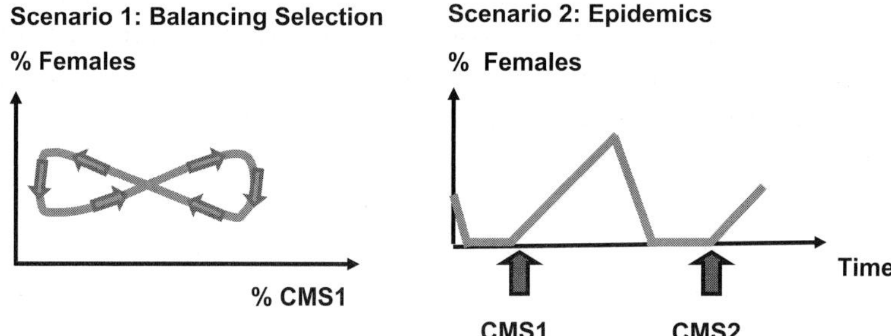

Figure 4.1 Two evolutionary scenarios of gynodioecy. Scenario 1: Balancing selection. Under the hypothesis of two CMSs (CMS1 and CMS2) with their corresponding dominant restorer alleles, R1 and R2, on two independent loci occurring in the same large population. Restorer alleles are assumed to be costly when useless (R1 on CMS2 or R2 on CMS1). Female frequency varies over time (arrows indicate the evolution over time), as the two CMSs and restorer alleles vary in frequency under balancing selection. When CMS1 is rare, R1 is rare since it is costly on CMS2, which is the common cytoplasm. Most individuals on CMS1 are females. Due to female advantage, CMS1 spreads in the population until R1 is selected. With R1 increasing, plants on CMS1 are restored more and more, losing female advantage. Then CMS2 is mostly found on females and thus increases in frequency at the expense of CMS1. These dynamics can induce cycles of female frequency in the population over time (Gouyon *et al.*, 1991). Scenario 2: Epidemics. The variation in female frequency is due to recurrent arrival of new CMSs. In a cycle, CMS1 spreads in the population due to female advantage until R1 is selected. CMS1 and R1 go to fixation. The population is thus hermaphroditic until the arrival of a new CMS (CMS2). *(Frank, 1989).* See the colour plate.

Cuguen, 2007; Gouyon, Vichot, & Van Damme, 1991). This balancing selection leads to cycles that can explain sex ratio variations in a population over time or among populations of gynodioecious species (Dufay *et al.*, 2009). The second scenario imagines the recurrent arrival and loss of CMSs in populations. A new CMS arrives (through migration or mutation) in a hermaphroditic population, invades the population over time due to the higher or better seed production of females, until the restorer allele is selected and fixed. The population becomes hermaphroditic and polymorphism on cytoplasm or restorer loci is no longer maintained. At the arrival of a new CMS, the phenomenon starts all over again. This last scenario postulates that the emergence of new CMSs is frequent in gynodioecious species (Frank, 1989).

The two alternative scenarios will have different effects on any cytoplasmic neutral locus diversity, as these loci will be linked to the male sterility genes, given uniparental inheritance and the absence of recombination (Charlesworth, 2002). In the case of haplotypes being maintained over a long period of time through balancing selection, diversity is expected to be high, because of the accumulation of mutations (Städler & Delph, 2002). Conversely, epidemic dynamics would induce a lower diversity, because, under this scenario, new sterilizing cytoplasms will continuously arise and sweep through populations, replacing the former cytoplasms (Ingvarsson & Taylor, 2002). Theoretical studies on the effect of gynodioecy on the genetic effective size of populations are scarce, but it is expected that the difference in female fitness between females and hermaphrodites might generate a lower genetic effective size of population when compared with the case when all individuals share the same female fitness. Gynodioecy should thus increase the effect of drift (Laporte, Cuguen, & Couvet, 2000). However, in all empirical studies (see below), the effect of selection is expected to overcome the reduction in the effective size of population.

1.3. Cryptic CMS

Numerous biology population studies have been conducted to understand gynodioecy dynamics, mostly by measuring reproductive success between females and hermaphrodites, but most molecular descriptions of mitochondrial male-sterilizing genes and nuclear restorer loci come from CMS systems in crops. In most cases, in such crop models, gynodioecy does not seem to be maintained in wild relative species (e.g. the case of *Helianthus*, Rieseberg, Fossen, Arias, & Carter, 1994). Consequently, crop CMSs that have been described should most probably be categorized as cryptic CMSs. This could be due to many reasons, one of which is that the fitness conditions necessary for gynodioecy to be maintained are not met in terms of value of female advantage or cost of restoration. For example, for a given CMS, one might expect that if the restorer allele is not costly, it will inevitably lead to its fixation and thus to a hermaphroditic population that harbours cryptic CMS. This is why CMS cytoplasms are often revealed after interspecies hybridization (Budar, Touzet, & De Paepe, 2003). In other words, we do not know how the data we get from crops inform of the nature of genetic factors that control a nuclear–cytoplasmic gynodioecy that is stable over time in a given species. However, Fujii, Bond, and Small (2011) studying nucleotide diversity of Rf-like PPR genes in several species and in particular in *Arabidopsis lyrata* and *Arabidopsis thaliana*, species with no known CMS, detected positive selection of Rf-like PPR. The signature of adaptive selection seemed to specifically involve a few codons that could code for amino acid residues in contact with the RNA ligand. This means that interactions between mitochondrial and nuclear genes most probably generate this selective process. It might be indicative of the recurrent emergence of mitochondrial male-sterilizing genes, which would create selective pressure on the large PPR family and lead to its functional diversification, even if gynodioecy has been transitory in the species (Foxe & Wright, 2009; Fujii *et al.*, 2011). It would be intriguing if gynodioecy occurred in a self-fertilizing species like *A. thaliana*, but evidence suggest that its loss of self-incompatibility and thus self-fertilization is recent. Consequently, the traces of selection could be rooted back to when *A. thaliana* was allogamous (Bechsgaard, Castric, Vekemans, & Schierup, 2006). This would also suggest that cryptic CMS might be a common feature of mitochondrial genomes in plants. This could be revealed through interspecies crosses or with genetically distant genotypes.

2. GENOME EVOLUTION AND GYNODIOECY

2.1. At the Gene Content Level: New Genes or New Variants?

2.1.1. Characteristics of CMS genes

a) Cryptic CMS

(i) Most crop CMSs Several reviews have been published on crop CMSs, and therefore the nature of CMS genes that have been identified in crops is not described in detail here but we will insist on the profile that has been drawn. Saumitou-Laprade, Cuguen, and Vernet (1994) pointed out that the CMS genes found in crops were novel genes encoding new open reading frames (ORFs), usually of a chimeric nature, resulting in intragenomic recombination. They are thus generally gain of function variants, in accordance with the theoretical expectations of gynodioecy dynamics.

The CMS genes are usually cotranscribed by essential genes, and as such they can be characterized as opportunistic (Budar *et al.*, 2003). When one considers the nature of the genes in proximity to CMS genes or composed of fragments of essential genes, the over-representation of *atp* genes is striking (Hanson & Bentolila, 2004). How can the selective sieve favour the occurrence of sterilizing genes in proximity to *atp* genes? Do CMS genes benefit from a higher rate of expression or steady state transcript levels of *atp* genes (Giegé, Hoffmann, Binder, & Brennicke, 2000)? Or do they interfere with the colocated gene expression or stability leading to a subsequent decrease in ATP production, directly responsible for the depletion of pollen production?

Generally, the expression of CMS genes is constitutive, with the exception of CMS protein of *Phaseolus vulgaris*, which is specifically accumulated in anther tissues (Sarria, Lyznik, Vallejos, & Mackenzie, 1998). This has raised the question of its specificity of action since its expression does not have any pleiotropic effect besides the loss of male fertility, or at least its effects are not too costly (Dufay *et al.*, 2007).

Two hypotheses that are not mutually exclusive have been proposed. The first hypothesis, built from the study of CMS-T in maize, suggests that "normal anther development is prevented by an interaction between a substance(s) present only in anthers and organelles with altered structures" (Flavell, 1974). The second hypothesis is that CMS factors generate dysfunction of mitochondria. A less efficient mitochondrion will only have visible consequences on pollen production in a developmental step that is particularly energy demanding, as suggested by an increase in the number of mitochondria per cell in tapetum or sporogenous cells in maize (Warmke & Lee, 1978). Up to now, the function of CMS proteins and how they ultimately lead to male sterility, either by energy defect and/or interaction with a specific substance, are unknown (Chase, 2007).

CMS genes usually code for hydrophobic proteins, with predicted *trans*-membrane domains.

In recent years, the research on CMS genes has been conducted through whole sequencing of mitochondrial genomes, due to the reduction in the cost of the methodology. By comparing CMS and non-CMS mitochondrial genomes in a given species, it is possible to list as candidate genes any specific ORFs that fit the profile (usually chimeric, located in the vicinity of essential genes, etc.) (e.g. in rice; Bentolila & Stefanov 2012; Fujii, Kazama, Yamada, & Toriyama, 2010). This strategy has been validated in maize where the CMS genes were detected in CMS-T and CMS-S (Allen *et al.*, 2007). Unfortunately, it did not provide any clues for CMS-C, for which the male-sterilizing gene is still unknown.

(ii) Cryptic CMS in *Mimulus* The *Mimulus guttatus* species complex is composed of morphologically diverse but largely interfertile species. Its centre of diversity is found in western North America. A CMS was revealed in *M. guttatus*, the most common species of the complex, when it was crossed with another species within the species complex: *M. nasutus* (Fishman & Willis, 2006). This CMS can be considered as cryptic, since no male sterility had been described in any populations of *M. guttatus*, all being hermaphroditic. This CMS appears to be fixed (the only one cytoplasm) in the population in Iron Mountain (Oregon) from which the *M. guttatus* maternal parent originates. The CMS was found at a low frequency in only two additional populations in Oregon out of 33 populations (Case & Willis, 2008). We are facing the expected issue of the spread of a CMS cytoplasm that invades the population: it eliminates non-CMS cytoplasms and favours the increase in frequency of restorer alleles until fixation of CMS and the corresponding restorer alleles, resulting in a hermaphroditic population. Remember that to spread, females must benefit from the loss of pollen production by producing more or better seeds. Barr and Fishman (2011) showed that male sterility had negative pleiotropic effects on corolla size, male-sterile plants had smaller flowers, but also positive effects on female traits such as the style length and thus better exposure of style and stigma, and a marginal increase in seed production. The question remains how restorer alleles succeeded in spreading to the whole species until their complete fixation in the species, since only interspecies crosses revealed male sterility.

Mitochondrial gene expression from reproductive tissues between plants on CMS, sterile or fertile (restored), and plants bearing a fertile cytoplasm were compared, assuming that any modification in the transcript pattern of a given essential gene is specific to sterile plants, in particular a larger transcript that would reveal the co-transcription of an additional ORF, following the opportunistic profile of male-sterile genes found in crops. Among the 13 essential genes analysed, only *nad6* met the criteria with a larger additional transcript. The amplification of cDNA with *nad6*-specific reverse primers revealed that length variation of the transcript was generated 5′ to *nad6*. Four unidentified ORFs potentially coding for more than 100 amino acid peptides were detected *in silico* in this 2-kb region. Among them, ORF 141 was the only one that shared partial similarity with an ORF in *Beta vulgaris* and with *cox2*. Transformation experiments are still needed to confirm that the product of one of these ORFs causes male sterility (Case & Willis, 2008).

On the nuclear side, two tight restorer loci, called *Rf1* and *Rf2*, one dominant restorer allele being sufficient for male fertility, were detected by fine mapping (Barr & Fishman, 2010). Each restorer locus spans a physical region containing numerous PPRs, with high homology to each other suggesting recent tandem duplication or transposition, and phylogenetically close to Rf-PPR from petunia or radish. This is clearly a promising path towards the identification of Rf loci in the *Mimulus* cryptic CMS and seems to follow what was drawn from the diverse cryptic CMSs found in crops.

b) Cases of gynodioecious species

(i) Ogura CMS in radish A CMS genome found in cultivated radish by Ogura (1968) was introduced intro *Brassica* germplasm to facilitate hybrid production, first through interspecies hybridization, embryo rescue and back-crossing with *Brassica*, and then using protoplast fusion to keep the mitochondrial radish CMS genome and replace the radish chloroplastic genome with the *Brassica* genome. The male-sterile gene was identified as an ORF called *orf138*, found in proximity to *atp8* (called *orfB* at that time). In rapeseed cybrids, in which *orf138* is physically dissociated from *atp8*, spontaneous fertility reversion was observed through loss of ORF 138, suggesting that the location of a CMS gene in the vicinity of an essential gene is necessary for its maintenance (Bellaoui, Martin-Canadell, Pelletier, & Budar, 1998). *Orf138* codes for a 19-kDa polypeptide found in the mitochondrial membrane (Grelon, Budar, Bonhomme, & Pelletier, 1994). Recently, it has been located in the inner membrane of mitochondria in a larger complex of 750 kDa, most probably only composed of ORF 138. ORF 138 did not seem to interfere with the oxidative phosphorylation complexes (Duroc, Hiard, Vrielynck, Ragu, & Budar, 2009). ORF 138 might form a pore in the inner mitochondrial membrane, similar to the T-URF13 responsible for CMS-T in maize (Rhoads, Levings, & Siedow, 1995).

In vitro analyses on mitochondria isolated from flower buds showed that mitochondria from male-sterile flowers exhibited a higher O_2 consumption, in agreement with the hypothesis of an uncoupling effect of ORF 138, forming a pore in the internal membrane. *In vivo* analyses showed that this uncoupling could be slightly detected on flower buds but not on vegetative tissues. Ogura CMS can be restored by a single locus found in radish, Rfo, which prevents the accumulation of ORF 138 in anthers, through a post-translational mechanism (Bellaoui, Grelon, Pelletier, & Budar, 1999). Cloned by several teams (Brown *et al.*, 2003; Desloire *et al.*, 2003; Koizuka *et al.*, 2003), *Rfo* is a PPR gene lying in a complex locus composed of two additional PPR genes (Uyttewaal *et al.*, 2008). This restorer locus has undergone complex evolution throughout numerous rearrangements (Mora, Rivals, Mireau, & Budar, 2010).

Ogura CMS was found in wild radish populations in Japan. The Ogura CMS was found to be common in this population; its frequency was positively correlated with ratio of the female populations (Murayama, Yahara, & Terachi, 2004). However, measurement of female reproductive traits on females and hermaphrodites did not reveal any clear female advantage (Miyake, Miyake, Terachi, & Yahara, 2009).

Orf138 was also found at low frequency in European populations of wild radish that were all hermaphroditic and known to bear the restorer alleles at high frequency. Further molecular characterization revealed that the European mitochondrial genomes carrying *orf138* did not convey male sterility. Northern blots and circular reverse transcriptase polymerase chain reaction analyses showed that *orf138* gene expression was impaired due to a novel cytoplasm-dependent transcript processing site. It illustrated the possible evolution of a CMS after restorer alleles have been fixed: relaxing selection in favour of the sterilizing cytoplasm and/or or favouring selection against a possible cost associated with a pleiotropic effect of the sterilizing gene (Giancola *et al.*, 2007).

(ii) Gynodioecy in beet Wild beet is a gynodioecious species. Female frequency can reach up to 43% of a population with a median of 13.5% (Dufay *et al.*, 2009). Four CMSs based on the type of restriction fragment length polymorphism have been detected in European populations from a total of 20 different mitotypes: CMS-E (I-12CMS(3)), CMS-G, CMS-H (very rare in populations), and Owen CMS, which is the only CMS used in the breeding of sugar beet (Cuguen *et al.*, 1994).

(a) Owen CMS Owen CMS was introduced early in sugar beet breeding (Owen, 1945). Rare in wild beet populations, it has been used as a marker of gene flow from the field to the wild (Arnaud, Viard, Delescluse, & Cuguen,

2003). Even though it is rare in populations, corresponding restorer alleles seem to occur at high frequency, since most plants bearing Owen CMS in the wild are hermaphrodites (Dufay *et al.*, 2009). In Owen CMS, two loci have been proposed to explain male fertility restoration, *X* and *Z* (Owen, 1945). Genetic mapping revealed a third locus genetically independent of *X* and *Z* (Hjerdin-Panagopoulos, Kraft, Rading, Tuvesson, & Nilsson, 2002). Despite extensive efforts to determine the nature of the sterilizing gene, the molecular determinism is not certain. Yamamoto, Kubo, and Mikami (2005) have shown that among the four specific ORFs detected when the Owen CMS mitochondrial genome was compared with a fertile type (T81-O), only the ORF corresponding to a peculiar 5′-leader sequence of *atp6* (pre-Satp6) was expressed at the protein level. It codes for a 35-kDa polypeptide that is specific to Owen CMS. However, no effect of nuclear restoration was detected on the size or the amount of the preSATP6 polypeptide, and no transformation experiment has validated the sterilizing effect of preSATP6 (Yamamoto *et al.*, 2008).

(b) CMS-E CMS-E is the most frequent CMS found in European populations (Dufay *et al.*, 2009; unpublished data). Its geographic distribution may be much larger as this CMS has also been found by Kubo's team from Hokkaido (Japan) in a population from Pakistan and called I-12CMS(3) (Onodera, Yamamoto, Kubo, & Mikami, 1999; Yamamoto *et al.*, 2008). The occurrence of intermediate hermaphrodites that bear CMS-E suggests that male fertility restoration is controlled by more than one locus. While no significant female advantage could be detected (Boutin, Jean, Valero, & Vernet, 1988; De Cauver, Arnaud, Courseaux, & Dufaÿ, 2011), a possible cost of restoration was suggested (Dufay *et al.*, 2008). Yamamoto *et al.* (2008) demonstrated that the E-specific *orf*129 was transcribed and coded into a specific 12-kDa polypeptide, which accumulated in the mitochondria of flower, root and leaf. Transgenic expression in tobacco of *orf*129 fused with a mitochondrial targeting presequence led to male-sterile plants, demonstrating the sterilizing effect of ORF 129. The question of the effect of restorer loci on this CMS remains, since no effect on the abundance of ORF 129 has been detected when plants were restored.

(c) CMS-G CMS-G is the second most common CMS encountered in wild populations in European coasts (Cuguen *et al.*, 1994; Dufay *et al.*, 2009; unpublished results). Male fertility restoration involves more than one locus (Touzet *et al.*, 2004). Restorer alleles are not frequent in populations (Dufay *et al.*, 2009). Ducos, Touzet, and Boutry (2001) showed that CMS-G exhibited a modified genomic *cox2* sequence that resulted in a truncated

protein at the C terminus and a modified *nad9* that coded for a larger protein. It was shown that the *in vitro* activity of cytochrome *c* oxidase was reduced by 50% in leaves, suggesting a possible effect of the observed mutations on *cox2* on the complex activity. In a recent study, Darracq *et al.* (2011) also found mutations on *cox1* and *cox3,* while no chimeric ORFs specific to CMS-G could be detected. For *cox1*, a mutation at the start codon that is commonly found in other beet genomes could potentially result in the translation of a longer protein with an extended N-terminus. Previous studies did not report a long variant on sodium dodecyl sulphate (SDS)-polyacrylamide gel electrophoresis (PAGE) from *in organello* S-labelled proteins in CMS-G, suggesting that the translated form of CMS-G-*cox1* might be the shortest one, with a size variation that could not be detected under SDS-PAGE conditions (Ducos *et al.*, 2001). However, in the truncated N-terminus, amino acids are expected to be involved in the subunit I/III interface, the D-pathway, or the subunit I/VIIc interface. In addition, two non-synonymous polymorphisms were found on codons that are not associated with any known function. For *cox3*, one non-synonymous polymorphism was detected with no known associated function. The polymorphism of complex IV could be the signature of a co-evolution of subunits that belong to the same complex. These variations on genes that are usually strongly constrained could be the result of a relaxation in selection, enabling the accumulation of non-synonymous mutations, once the sterilizing mutations have been selected through disruption of COX activity. This polymorphism could imply compensatory mutations on COX nuclear genes in male fertility restoration.

2.2. At the Genome Level

2.2.1. Higher diversity of gynodioecious species

The dynamics of gynodioecy are expected to leave a signature in the pattern of neutral diversity of mitochondrial genes. In the case of balancing selection, gynodioecious species have been maintaining sterilizing genomes for long periods of time, and thus have accumulated substitutions. In the case of epidemics, the recurrent selective sweep results in the replacement of old haplotypes (Charlesworth, 2002). Note that the selection acting on the male sterility gene drags the whole genome through hitchhiking since the mitochondrial genome is non-recombining and can thus be considered as a unique linkage group. Theoretically, the chloroplastic genome is expected to be in complete linkage disequilibrium with the mitochondrial genome, as cytoplasmic genomes are co-inherited (Olson & McCauley, 2000). They are thus expected to follow the same dynamics. Conversely, the nuclear genome is not thought to be influenced by gynodioecy.

The genus *Silene* exhibits a diversity of mating systems, including hermaphroditic, gynodioecious and dioecious species (Desfeux, Maurice, Henry, Lejeune, & Gouyon, 1996; Jurgens, Witt, & Gottsberger, 2002). This allows the use of comparative tests to establish whether balancing selection or epidemic dynamics have predominantly affected the evolutionary dynamics of gynodioecy. Previous studies comparing diversity among *Silene* species with different reproductive systems have led to contradictory conclusions. Ingvarsson & Taylor (2002) showed that sequence variation at chloroplast loci within the gynodioecious species *S. vulgaris* is low relative to that in *Silene latifolia*, a closely related dioecious species, whereas the two species did not differ in diversity at a nuclear gene. This supports the hypothesis of epidemic dynamics. Note that even though the target of the selection is the mitochondrial genome, since the chloroplastic genome is co-inherited with the mitochondrial genome, they are expected to follow the same evolutionary dynamics. However, Städler and Delph (2002) found arguments for balancing selection in another gynodioecious *Silene* species, *S. acaulis*, showing high diversity and old haplotypes of the mitochondrial *cob* gene. By comparing polymorphism of two mitochondrial genes (*cox1* and *cob*) in a sample of three gynodioecious *Silene* species, *S. acaulis*, *S. nutans* and *S. vulgaris*, and seven non-gynodioecious *Silene* species, Touzet and Delph (2009) showed that the gynodioecious species harboured divergent haplotypes, while little or no diversity was found in hermaphroditic or dioecious species, supporting also the balancing selection hypothesis. Potential explanations for these conflicting conclusions can be proposed. Ingvarsson and Taylor (2002)'s study relied on a single nuclear locus, the diversity of which could be not representative. They assumed also that the chloroplastic genome was in complete linkage disequilibrium with the mitochondrial genome, and thus reflected faithfully the pattern of diversity found on mitochondrial genes. This may not be completely true if heteroplasmy and recombination occur subsequent to paternal leakage (see below). However, Touzet and Delph (2009)'s study involved comparisons of gynodioecious species with dioecious and hermaphroditic species from distantly related clades within the genus *Silene*. In addition, *S. acaulis* and *S. nutans*, the species that exhibited the highest diversity, are phylogenetically close (Marais *et al.*, 2011). The *Silene* genus is reputed to exhibit a high variation in the mitochondrial mutation rate

(Mower, Touzet, Gummow, Delph, & Palmer, 2007; Sloan, Barr, Olson, Keller, & Taylor, 2008; Sloan, Oxelman, Rautenberg, & Taylor, 2009). Therefore, the difference in diversity could be partly due to different mitochondrial mutation rates among the species studied. However, in studies where the mitochondrial mutation rate is estimated from the divergence between two species from small samples, ancestral polymorphism (which is expected to be the case when there is balancing selection) and recombination (see below) can be missed, generating an apparent mutation rate acceleration (Charlesworth, 2010).

2.2.2. The occurrence of paternal leakage

Although mitochondrial inheritance is largely uniparental, there is evidence of occasional paternal leakage in *Silene*. The occurrence of heteroplasmy and occasional paternal leakage has been documented on the gynodioecious *S. vulgaris* from open pollinated descents harvested on homoplasmic mothers or from controlled crosses (Bentley, Mandel, & McCauley, 2010; McCauley, Bailey, Sherman, & Darnell, 2005; Pearl, Welch, & McCauley, 2009). In addition, using quantitative PCR, heteroplasmic individuals were observed as well as heteroplasmy transmission from heteroplasmic mothers (Bentley *et al.*, 2010; Pearl *et al.*, 2009; Welch, Darnell, & McCauley, 2006). Overall, it appears that paternal leakage is rare, but is variable among fathers in controlled crosses. When it occurred it varied from less than 1% to 100% (Bentley *et al.*, 2010). The variation could be partly attributed to the population they belonged to, suggesting the existence of a genetic component in the phenomenon. In addition, even though heteroplasmy could be transmitted, it was also lost between generations in many cases, in accordance with the theory of vegetative sorting. Paternal leakage and subsequent heteroplasmy can favour interhaplotype recombination. This is observed in mitochondrial genes in *S. vulgaris* (Houliston & Olson, 2006). A recombination signature in mitochondrial genes was also observed in *S. acaulis* (Städler & Delph, 2002) and in *S. nutans* (Touzet & Delph, 2009), two other gynodioecious species. This indicates that paternal leakage might be a general phenomenon in *Silene*. Even if this event is most probably very rare, it could be an important evolutionary force to create recombined genomes and generate genome diversification. In the case of gynodioecy, it could be an important driving mutational force to generate new CMS genes or result in super CMS genomes, accumulating several sterilizing genes (McCauley & Olson, 2008). A theoretical study demonstrated that paternal leakage could facilitate the maintenance of gynodioecy in populations (Wade & McCauley, 2005).

2.2.3. CMS genomes: fast evolving genomes at the sequence and the structural level?

In recent years, analysis of whole sequenced genomes has been the main strategy to find candidate genes for male sterility (Bentolila & Stefanov, 2012; Fujii *et al.*, 2010). Through comparative genomics and phylogenetic approaches, it is also possible to study the origin of CMS genomes in a given species, whether the CMS genomes arose independently or constitute a sterile lineage.

a) The case of beet

A phylogeny based on chloroplastic sequences suggested that the sterile cytoplasms found in wild beet emerged independently from a non-sterile cytoplasm (Fénart, Touzet, Arnaud, & Cuguen, 2006). However, the resolution was very low due to the lack of polymorphism. Darracq *et al.* (2011) sequenced five mitochondrial genomes in addition to two mitochondrial genomes that were published previously (Kubo *et al.*, 2000; Satoh *et al.*, 2004). The seven mitochondrial genomes included three CMSs (Owen CMS, CMS-E and CMS-G) and three non-CMSs from *B. vulgaris*, and one mitochondrial genome from a sister species, *B. macrocarpa*. Through pairwise substitution rates among the mitochondrial genomes, they were able to build phylogenetic trees where two clusters appeared, one composed of the three CMSs and one of the three non-CMSs (Fig. 4.2). In the CMS lineage, two genomes exhibited higher synonymous divergence with *B. macrocarpa*: Owen CMS and CMS-G. This was also true when considering distance based on gene order. A first explanation could be that the CMS genomes are older than expected due to balancing selection that may have favoured their maintenance over larger time scales than non-CMS genomes. This would explain why CMS genomes are found on longer branches in the sequence and rearrangement trees. It would imply that the trees are therefore wrongly rooted with *B. macrocarpa*. However, when Darracq *et al.* (2011) compared chloroplastic sequences that should have followed the same pattern since they are co-transmitted with mitochondrial genomes, they found that chloroplastic sequences related to CMS mitochondrial genomes were no more divergent than non-CMS genomes. Consequently, CMS lineages are probably not older but have a faster rate of evolution at the sequence level (probably through a higher mutation rate in

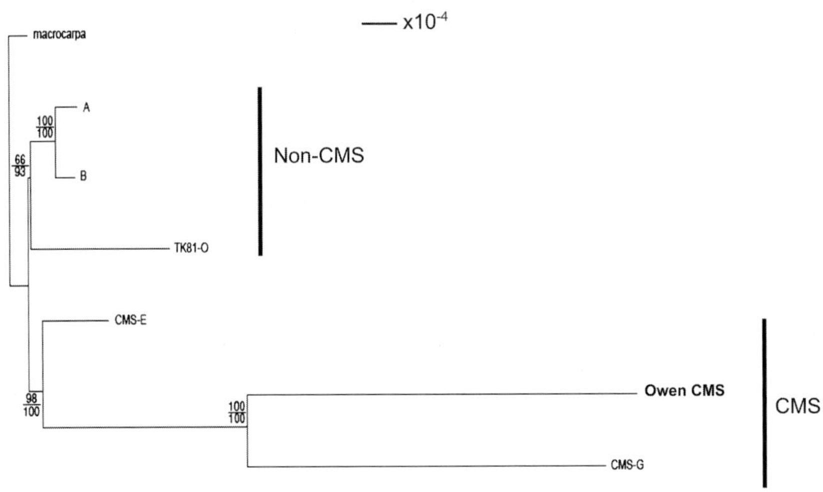

Figure 4.2 Phylogenetic tree of mitochondrial genomes of beet. The tree is based on nucleotide variation on the common sequence among the genomes (a scale of 10^{-4} substitution is given). The CMS and non-CMS clades are indicated. *(modified from Darracq* et al.*, 2011)*

these lineages) and also at the structure level. CMS-E and Owen CMS exhibited additional *atp6* 5′-leader sequences 1 kb in size, which were 88% identical in sequence and with a 1% gap. This could indicate events of recombination among the two genomes, and therefore an ancient episode of heteroplasmy in the species.

b) The case of maize

Allen *et al.* (2007) sequenced five genomes from *Zea mays*, three CMSs (CMS-T, CMS-S and CMC-C) and two fertile genomes, NA and NB. Darracq, Varré, and Touzet (2010), adding three mitochondrial genomes from teosintes, *Z. mays* ssp. *parviglumis*, *Z. perennnis* and *Z.* luxurians (also sequenced by the same research team and deposited in GenBank), analysed the evolution of *Zea* mitochondrial genome structure by concomitantly building a phylogenetic tree based on sequence polymorphism and a phylogenetic tree based on structural rearrangements among genomes. As in beet, both trees were congruent, suggesting that both sources of polymorphism are correlated, i.e. the more divergent a genome is, the more rearranged it is. The phylogenetic relationships among maize mitochondrial genomes suggest that CMS-T and CMS-S are the oldest cytoplasms, while the fertile genomes seem to be derivates (Fig. 4.3). In contrast to the case

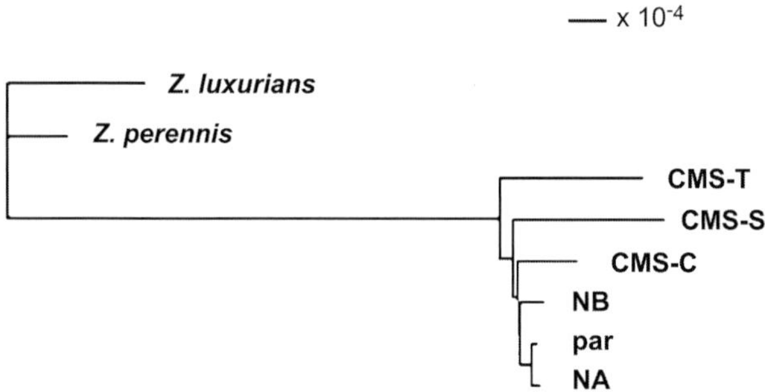

Figure 4.3 Phylogenetic tree of mitochondrial genomes of maize. The tree is based on nucleotide variation on the common sequence among the genomes (a scale of 10^{-4} substitution is given). *(modified from Darracq* et al., *2010)*

encountered in beet, there was not one single CMS lineage. Former studies on mitochondrial and chloroplastic diversity in *Zea* indicated that CMS-S was an old cytoplasm and most likely the result of introgression from teosinte *Z. mays* ssp. *mexicana*. But the phylogenetic location of CMS-T, due to a strong nucleotide divergence and a concomitant rearranged genome, is puzzling because CMS-T shares the same co-inherited chloroplastic genome with CMS-C and NB (Doebley, Renfroe, & Blanton, 1987; Pring & Levings, 1978). Consequently, the high divergence of CMS-T might not have occurred in a molecular clock tempo (as suggested by the rejection of the molecular clock hypothesis in the phylogenetic analysis). Chloroplastic sequence data could shed light on the relative ages of the cytoplasms studied. Similar to what was found in beet, CMS genomes seem to exhibit higher nucleotide divergence, as seen by longer branches on the phylogenetic tree, and the present topology could be due to what is called the phenomenon of long branch attraction (long branches are clustered together by construction). The phylogenetic location of *Zea mays* ssp. *parviglumis* included in the *Zea mays* clade is in agreement with the scenario of recent maize domestication from this teosinte subspecies (Doebley, 2004). Moreover, it strongly suggests that the cytoplasms differentiated before domestication. It would therefore be interesting to assess the frequency of the CMS cytoplasms and the frequency of male sterility in *Zea mays* teosinte populations.

c) The case of rice

Seven different mitochondrial genomes have been sequenced in rice, of which three are CMS: LD-CMS, CW-CMS and WA-CMS (Bentolila & Stefanov, 2012; Fujii *et al.*, 2010; Notsu *et al.*, 2002; Tian, Zheng, Hu, &

Yu, 2006). Fujii *et al.* (2010) sequenced LD-CMS and CW-CMS and suggested that CMS genomes exhibited more rearrangement with the Nipponbare reference that the non-CMS genomes described previously (Tian *et al.*, 2006). However, recently, Bentolila and Stefanov (2012), sequencing two additional mitochondrial genomes, N and WA-CMS, suggested some possible mistakes in the finishing of the fertile genomes by Tian *et al.* (2006) and that the apparent differences observed by Fujii *et al.* (2010) were not conclusive. Therefore, more analyses, similar to the studies conducted on beet, could test whether the fast evolution of CMS genomes in beet (and possibly in maize) is universal and is a general feature of CMS genomes.

2.2.4. Does gynodioecy contribute to the fast evolution of CMS mitochondrial genomes?

We have seen that CMS genomes, when a comparison at the species level was possible, can evolve faster, and that there is correlation between molecular evolution rates and genome rearrangement rates. This correlation has been found at the interspecific level in animal mitochondrial genomes (Xu, Jameson, Tang, & Higgs, 2006). It was proposed that the variation in the accuracy of the replication process in mitochondrial genomes among species, due to deleterious mutations in the nuclear genes involved, could lead to the correlated variation of both rates. Galtier, Jobson, Nabholz, Glemin, and Blier (2009) proposed additional trials to find genetic factors implied in the variation of the mitochondrial mutation rate among animal species, such as genes involved in the regulation of mitochondrial metabolism and redox status, and antioxidant genes regulating the amount of cellular reactive oxygen species (ROS). In plants, the mitochondrial mutation rate is generally low but a high variation in the mitochondrial substitution rate among species has been documented in *Plantago* (Cho, Mower, Qiu, & Palmer, 2004), *Pelargonium* (Parkinson *et al.*, 2005) and in *Silene* (Barr, Keller, Ingvarsson, Sloan, & Taylor, 2007; Mower *et al.*, 2007; Sloan *et al.*, 2008, 2009, 2012). The *Plantago* and *Silene* genera are composed of gynodioecious species (Charlesworth, 2010). This variation in the mitochondrial mutation rate among species might implicate the same genetic factors as hypothesized in animals. However, the variation in evolution rate we are dealing with is taking place at the species level; in other words, in a given species mitochondrial genomes seem to exhibit a higher rate of evolution. This result is reminiscent of a recent study in gynodioecious *S. vulgaris*, in which the variation in the synonymous substitution rate was also described among lineages, suggesting the occurrence of fast evolving mitochondrial genomes (Sloan *et al.*, 2008). The genes involved in DNA repair or any biological process that can prevent mutagenesis are nuclear, whereas the phenomenon that is described must find its cause in the mitochondrial genome. We do not expect any linkage disequilibrium between a nuclear allele promoting mutation and a given mitochondrial genome, because male sterility obligates outcrossing.

If the fast evolving characteristic of CMS genomes is confirmed, this could raise the following chicken and egg question: is the emergence of CMS the result of an episode of increased mutation and rearrangement, leading either to the creation of new ORFs through intragenomic recombination, as generally assumed, or to deleterious mutations on essential mitochondrial genes, in both cases new variants being positively selected as they disrupt pollen production? Or could the fast evolution of CMS genomes be due to male sterility itself, through the dysfunction of mitochondria accumulating ROS, which would generate mutations on the fragile mitochondrial DNA (Richter, Park, & Ames, 1988; Yakes & Van Houten, 1997). In addition, the dysfunction of mitochondria could generate a peculiar context that would relax the selective pressure on essential genes, consequently allowing the accumulation of non-synonymous mutations (Fig. 4.4).

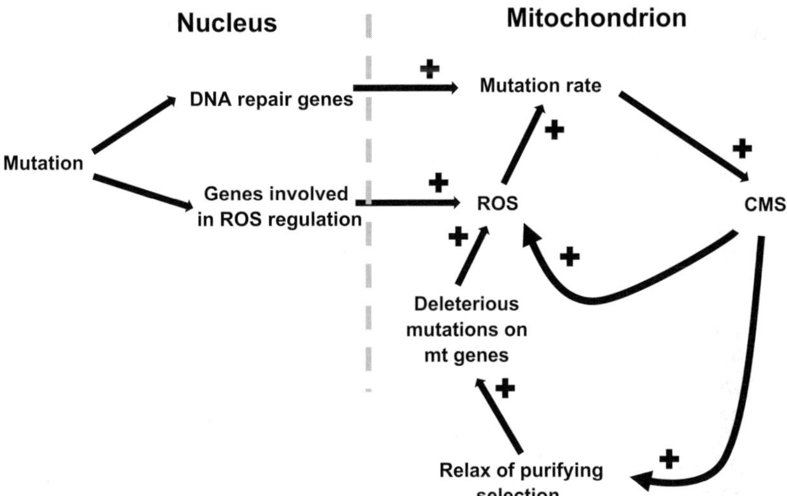

Figure 4.4 Possible links between mitochondrial mutation rate and gynodioecy.

REFERENCES

Allen, J. O., Fauron, C. M., Minx, P., Roark, L., Oddiraju, S., Lin, G. N., et al. (2007). Comparisons among two fertile and three male-sterile mitochondrial genomes of maize. *Genetics, 177*, 1173–1192.

Alverson, A. J., Wei, X. X., Rice, D. W., Stern, D. B., Barry, K., & Palmer, J. D. (2010). Insights into the evolution of mitochondrial genome size from complete sequences of *Citrullus lanatus* and *Cucurbita pepo* (Cucurbitaceae). *Molecular Biology and Evolution, 27*, 1436–1448.

Arnaud, J.-F., Viard, F., Delescluse, M., & Cuguen, J. (2003). Evidence for gene flow *via* seed dispersal from crop to wild relatives in *Beta vulgaris* (Chenopodiaceae): consequences for the release of genetically modified crop species with weedy lineages. *Proceedings of the Royal Society of London B: Biological Sciences, 270*, 1565–1571.

Bailey, M. F., Delph, L. F., & Lively, C. M. (2003). Modeling gynodioecy: novel scenarios for maintaining polymorphism. *American Naturalist, 161*, 762–776.

Barr, C. M., Keller, S. R., Ingvarsson, P. K., Sloan, D. B., & Taylor, D. R. (2007). Variation in mutation rate and polymorphism among mitochondrial genes of *Silene vulgaris*. *Molecular Biology and Evolution, 24*, 1783–1791.

Barr, C. M., & Fishman, L. (2010). The nuclear component of a cytonuclear hybrid incompatibility in *Mimulus* maps to a cluster of pentatricopeptide repeat genes. *Genetics, 184*, 455–465.

Barr, C. M., & Fishman, L. (2011). Cytoplasmic male sterility in *Mimulus* hybrids has pleiotropic effects on corolla and pistil traits. *Heredity, 106*, 886–893.

Bazin, E., Glemin, S., & Galtier, N. (2006). Population size does not influence mitochondrial genetic diversity in animals. *Science, 312*, 570–572.

Bechsgaard, J. S., Castric, V., Vekemans, X., & Schierup, M. H. (2006). The transition to self-compatibility in *Arabidopsis thaliana* and evolution within S-haplotypes over 10 Myr. *Molecular Biology and Evolution, 23*, 1741–1750.

Belhassen, E., Dommée, B., Atlan, A., Gouyon, P. H., Pomente, D., Assouad, M. W., et al. (1991). Complex determination of male sterility in *Thymus vulgaris* L.: genetic and molecular analysis. *Theoretical and Applied Genetics, 82*, 137–143.

Bellaoui, M., Martin-Canadell, A., Pelletier, G., & Budar, F. (1998). Low-copy-number molecules are produced by recombination, actively maintained and can be amplified in the mitochondrial genome of Brassicaceae: relationship to reversion of the male sterile phenotype in some cybrids. *Molecular and General Genetics, 257*, 177–185.

Bellaoui, M., Grelon, M., Pelletier, G., & Budar, F. (1999). The restorer Rfo gene acts post-translationally on the stability of the ORF138 Ogura CMS-associated protein in reproductive tissues of rapeseed cybrids. *Plant Molecular Biology, 40*, 893–902.

Bentley, K. E., Mandel, J. R., & McCauley, D. E. (2010). Paternal leakage and heteroplasmy of mitochondrial genomes in *Silene vulgaris*: evidence from experimental crosses. *Genetics, 185*, 961–968.

Bentolila, S., Alfonso, A. A., & Hanson, M. R. (2002). A pentatricopeptide repeat-containing gene restores fertility to cytoplasmic male-sterile plants. *Proceedings of the National Academy of Sciences of the United States of America, 99*, 10887–10892.

Bentolila, S., & Stefanov, S. (2012). A reevaluation of rice mitochondrial evolution based on the complete sequence of male-fertile and male-sterile mitochondrial genomes. *Plant Physiology, 158*, 996–1017.

Boutin, V., Jean, R., Valero, M., & Vernet, P. (1988). Gynodioecy in *Beta maritima*. *Oecologia Plantarum, 9*, 61–66.

Brown, G. G., Formanova, N., Jin, H., Wargachuk, R., Dendy, C., Patil, P., et al. (2003). The radish Rfo restorer gene of Ogura cytoplasmic male sterility encodes a protein with multiple pentatricopeptide repeats. *The Plant Journal, 35*, 262–272.

Budar, F., Touzet, P., & De Paepe, R. (2003). The nucleo-mitochondrial conflict in cytoplasmic male sterilities revisited. *Genetica, 117*, 3–16.

Case, A. L., & Willis, J. H. (2008). Hybrid male sterility in Mimulus (Phrymaceae) is associated with a geographically restricted mitochondrial rearrangement. *Evolution, 62*, 1026–1039.

Charlesworth, D., & Laporte, V. (1998). The male sterility polymorphism of *Silene vulgaris*. I. Analysis of genetic data from two populations, and comparison with *Thymus vulgaris*. *Genetics, 150*, 1267–1282.

Charlesworth, D. (2002). What maintains male-sterility factors in plant populations. *Heredity, 89*, 408–409.

Charlesworth, D. (2010). Don't forget the ancestral polymorphisms. *Heredity, 105*, 509–510.

Chase, C. D. (2007). Cytoplasmic male sterility: a window to the world of plant mitochondrial-nuclear interactions. *Trends in Genetics, 23*, 81–90.

Cho, Y., Mower, J. P., Qiu, Y. L., & Palmer, J. D. (2004). Mitochondrial substitution rates are extraordinarily elevated and variable in a genus of flowering plants. *Proceedings of the National Academy of Sciences of the United States of America, 101*, 17741–17746.

Cosmides, L. M., & Tooby, J. (1981). Cytoplasmic inheritance and intragenomic conflict. *Journal of Theoretical Biology, 89*, 83–129.

Cuguen, J., Wattier, R., Saumitou-laprade, P., Forcioli, D., Mörchen, M., Van-Dijk, H., et al. (1994). Gynodioecy and mitochondrial DNA polymorphism in natural populations of *Beta vulgaris* ssp *maritima*. *Genetics Selection Evolution, 26*, 87–101.

Darracq, A., Varré, J. S., & Touzet, P. (2010). A scenario of mitochondrial genome evolution in maize based on rearrangement events. *BMC Genomics, 11*, 233.

Darracq, A., Varré, J.-S., Maréchal-Drouard, L., Courseaux, A., Castric, V., Saumitou-Laprade, P., et al. (2011). Structural and content diversity of mitochondrial genome in beet: a comparative genomic analysis. *Genome Biology and Evolution, 3*, 723–736.

Darwin, C. (1877). *The different forms of flowers on plants of the same species*. London: John Murray.

De Cauwer, I., Arnaud, J.-F., Courseaux, A., & Dufaÿ, M. (2011). Sex-specific fitness variation in gynodioecious *Beta vulgaris* ssp. *maritima*: do empirical observations fit theoretical predictions? *Journal of Evolutionary Biology, 24*, 2456–2472.

De Haan, A. A., Mateman, A. C., Van Dijk, P. J., & Van Damme, J. M. M. (1997). New CMS types in *Plantago lanceolata* and their relatedness. *Theoretical and Applied Genetics, 94*, 539–548.

Delph, L. F., Touzet, P., & Bailey, M. F. (2007). Merging theory and mechanism in studies of gynodioecy. *Trends in Ecology and Evolution, 22*, 17–24.

Desfeux, C., Maurice, S., Henry, J.-P., Lejeune, B., & Gouyon, P.-H. (1996). Evolution of reproductive systems in the genus *Silene*. *Proceedings of the Royal Society of London B: Biological Sciences, 263*, 409–414.

Desloire, S., Gherbi, H., Laloui, W., Marhadour, S., Clouet, V., Cattolico, L., et al. (2003). Identification of the fertility restoration locus, Rfo, in radish, as a member of the pentatricopeptide-repeat protein family. *EMBO Reports, 4*, 588–594.

Doebley, J., Renfroe, W., & Blanton, A. (1987). Restriction site variation in the *Zea* chloroplast genome. *Genetics, 117*, 139–147.

Doebley, J. (2004). The genetics of maize evolution. *Annual Review of Genetics, 38*, 37–59.

Ducos, E., Touzet, P., & Boutry, M. (2001). The male sterile G cytoplasm of wild beet displays modified mitochondrial respiratory complexes. *The Plant Journal, 26*, 171–180.

Dufay, M., Touzet, P., Maurice, S., & Cuguen, J. (2007). Modelling the maintenance of male-fertile cytoplasm in a gynodioecious population. *Heredity, 99*, 349–356.

Dufay, M., Vaudey, V., De Cauwer, I., Touzet, P., Cuguen, J., & Arnaud, J.-F. (2008). Variation in pollen production and pollen viability in natural populations of gynodioecious *Beta vulgaris* ssp. *maritima*: evidence for a cost of restoration of male function? *Journal of Evolutionary Biology, 21*, 202–212.

Dufay, M., Cuguen, J., Arnaud, J. F., & Touzet, P. (2009). Sex ratio variation among gynodioecious populations of sea beet: can it be explained by frequency-dependent selection? *Evolution, 63*, 1483–1497.

Dufay, M., & Billard, E. (2012). How much better are females? The occurrence of female advantage, its proximal causes and its variation within and among gynodioecious species. *Annals of Botany, 109*, 505–519.

Duroc, Y., Hiard, S., Vrielynck, N., Ragu, S., & Budar, F. (2009). The Ogura sterility-inducing protein forms a large complex without interfering with the oxidative phosphorylation components in rapeseed mitochondria. *Plant Molecular Biology, 70*, 123–137.

Fénart, S., Touzet, P., Arnaud, J.-F., & Cuguen, J. (2006). Emergence of gynodioecy in wild beet (*Beta vulgaris* ssp. *maritima* L.): a genealogical approach using chloroplastic nucleotide sequences. *Proceedings of the Royal Society of London B: Biological Sciences, 273*, 1391–1398.

Fishman, L., & Willis, J. H. (2006). A cytonuclear incompatibility causes anther sterility in *Mimulus* hybrids. *Evolution, 60*, 1372–1381.

Flavell, R. B. (1974). A model for the mechanism of cytoplasmic male sterility in plants, with special reference to maize. *Plant Science Letters, 3*, 259–263.

Foxe, J. P., & Wright, S. I. (2009). Contrasting patterns of transposable-element insertion polymorphism and nucleotide diversity in autotetraploid and allotetraploid *Arabidopsis* species. *Genetics, 183*, 663–672.

Frank, S. A. (1989). The evolutionary dynamics of cytoplasmic male sterility. *American Naturalist, 133*, 345–376.

Fujii, S., Kazama, T., Yamada, M., & Toriyama, K. (2010). Discovery of global genomic re-organization based on comparison of two newly sequenced rice mitochondrial genomes with cytoplasmic male sterility-related genes. *BMC Genomics, 11*, 209.

Fujii, S., Bond, C. S., & Small, I. D. (2011). Selection patterns on restorer-like genes reveal a conflict between nuclear and mitochondrial genomes throughout angiosperm evolution. *Proceedings of the National Academy of Sciences of the United States of America, 108*, 1723–1728.

Galtier, N., Jobson, R. W., Nabholz, B., Glemin, S., & Blier, P. U. (2009). Mitochondrial whims: metabolic rate, longevity and the rate of molecular evolution. *Biology Letters, 5*, 413–416.

Galtier, N. (2011). The intriguing evolutionary dynamics of plant mitochondrial DNA. *BMC Biology, 9*, 61.

Garraud, C., Brachi, B., Dufay, M., Touzet, P., & Shykoff, J. A. (2011). Genetic determination of male sterility in gynodioecious *Silene nutans*. *Heredity, 106*, 757–764.

Giancola, S., Rao, Y., Chaillou, S., Hiard, S., Martin-Canadell, A., Pelletier, G., et al. (2007). Cytoplasmic suppression of Ogura cytoplasmic male sterility in European natural populations of *Raphanus raphanistrum*. *Theoretical and Applied Genetics, 114*, 1333–1343.

Giegé, P., Hoffmann, M., Binder, S., & Brennicke, A. (2000). RNA degradation buffers asymmetries of transcription in *Arabidopsis* mitochondria. *EMBO Reports, 1*, 164–170.

Gouyon, P. H., Vichot, F., & Van Damme, J. M. M. (1991). Nuclear-cytoplasmic male sterility: single point equilibria *versus* limit cycles. *American Naturalist, 137*, 498–514.

Grelon, M., Budar, F., Bonhomme, S., & Pelletier, G. (1994). Ogura cytoplasmic male-sterility (CMS)–associated orf138 is translated into a mitochondrial-membrane polypeptide in male-sterile *Brassica* cybrids. *Molecular and General Genetics, 243*, 540–547.

Hanson, M. R., & Bentolila, S. (2004). Interactions of mitochondrial and nuclear genes that affect male gametophyte development. *The Plant Cell, 16*(Suppl. 1), S154–169.

Hjerdin-Panagopoulos, A., Kraft, T., Rading, I. M., Tuvesson, S., & Nilsson, N.-O. (2002). Three QTL regions for restoration of *Owen* CMS in sugar beet. *Crop Science, 42*, 540–544.

Houliston, G. J., & Olson, M. S. (2006). Nonneutral evolution of organelle genes in *Silene vulgaris*. *Genetics, 174*, 1983–1994.

Ingvarsson, P. K., & Taylor, D. R. (2002). Genealogical evidence for epidemics of selfish genes. *Proceedings of the National Academy of Sciences of the United States of America, 99*, 11265–11269.

Jurgens, A., Witt, T., & Gottsberger, G. (2002). Pollen grain numbers, ovule numbers and pollen-ovule ratios in Caryophylloideae: correlation with breeding system, pollination, life form, style number, and sexual system. *Sexual Plant Reproduction, 14*, 279–289.

sb:author>N.Koizuka, Imai, R., Fujimoto, H., Hayakawa, T., Kimura, Y., Kohno-Murase, J., et al. (2003). Genetic characterization of a pentatricopeptide repeat protein gene, orf687, that restores fertility in the cytoplasmic male-sterile Kosena radish. *The Plant Journal, 34*, 407–415.

Kubo, T., Nishizawa, S., Sugawara, A., Itchoda, N., Estiati, A., & Mikami, T. (2000). The complete nucleotide sequence of the mitochondrial genome of sugar beet (*Beta vulgaris* L.) reveals a novel gene for tRNACys(GCA). *Nucleic Acids Research, 28*, 2571–2576.

Laporte, V., Merdinoglu, D., Saumitou-Laprade, P., Butterlin, G., Vernet, P., & Cuguen, J. (1998). Identification and mapping of RAPD and RFLP markers linked to a fertility restorer gene for a new source of cytoplasmic male sterility in *Beta vulgaris* ssp *maritima*. *Theoretical and Applied Genetics, 96*, 989–996.

Laporte, V., Cuguen, J., & Couvet, D. (2000). Effective population sizes for cytoplasmic and nuclear genes in a gynodioecious species: the role of the sex determination system. *Genetics, 154*, 447–458.

Lewis, D. (1941). Male sterility in natural populations of hermaphrodite plants: the equilibrium between females and hermaphrodites to be expected with different types of inheritance. *New Phytologist, 40*, 56–63.

Lynch, M., & Conery, J. S. (2003). The origins of genome complexity. *Science, 302*, 1401–1404.

Lynch, M., Koskella, B., & Schaack, S. (2006). Mutation pressure and the evolution of organelle genomic architecture. *Science, 311*, 1727–1730.

Lynch, M. (2007). *The origins of genome architecture*. Sunderland, MA: Sinauer Associates.

Marais, G. A. B., Forrest, A., Kamau, E., Kafer, J., Daubin, V., & Charlesworth, D. (2011). Multiple nuclear gene phylogenetic analysis of the evolution of dioecy and sex chromosomes in the genus *Silene*. *PLoS ONE, 6*. e21915.

McCauley, D. E., Bailey, M. F., Sherman, N. A., & Darnell, M. Z. (2005). Evidence for paternal transmission and heteroplasmy in the mitochondrial genome of *Silene vulgaris*, a gynodioecious plant. *Heredity, 95*, 50–58.

McCauley, D. E., & Olson, M. S. (2008). Do recent findings in plant mitochondrial molecular and population genetics have implications for the study of gynodioecy and cytonuclear conflict? *Evolution, 62*, 1013–1025.

Meiklejohn, C. D., Montooth, K. L., & Rand, D. M. (2007). Positive and negative selection on the mitochondrial genome. *Trends in Genetics, 23*, 259–263.

Miyake, K., Miyake, T., Terachi, T., & Yahara, T. (2009). Relative fitness of females and hermaphrodites in a natural gynodioecious population of wild radish, *Raphanus sativus* L. (Brassicaceae): comparison based on molecular genotyping. *Journal of Evolutionary Biology, 22*, 2012–2019.

Mora, J. R. H., Rivals, E., Mireau, H., & Budar, F. (2010). Sequence analysis of two alleles reveals that intra-and intergenic recombination played a role in the evolution of the radish fertility restorer (Rfo). *BMC Plant Biology, 10*, 35.

Mower, J. P., Touzet, P., Gummow, J. S., Delph, L. F., & Palmer, J. D. (2007). Extensive variation in synonymous substitution rates in mitochondrial genes of seed plants. *BMC Evolutionary Biology, 7*, 135.

Murayama, K., Yahara, T., & Terachi, T. (2004). Variation of female frequency and cytoplasmic male-sterility gene frequency among natural gynodioecious populations of wild radish (*Raphanus sativus* L.). *Molecular Ecology, 13*, 2459–2464.

Muse, S. (2000). Examining rates and patterns of nucleotide substitution in plants. *Plant Molecular Biology, 42*, 25–43.

Notsu, Y., Masood, S., Nishikawa, T., Kubo, N., Akiduki, G., Nakazono, M., et al. (2002). The complete sequence of the rice (*Oryza sativa* L.) mitochondrial genome: frequent DNA sequence acquisition and loss during the evolution of flowering plants. *Molecular Genetics and Genomics, 268*, 434–445.

Ogura, H. (1968). Studies on the new male sterility in Japanese radish, with special reference to utilization of this sterility towards the practical raising of hybrid seeds. *Memoirs of the Faculty of Agriculture Kagoshima University, 6*, 39–78.

Olson, M. S., & McCauley, D. E. (2000). Linkage disequilibrium and phylogenetic congruence between chloroplast and mitochondrial haplotypes in *Silene vulgaris*. *Proceedings of the Royal Society of London B: Biological Sciences, 267*, 1801–1808.

Onodera, Y., Yamamoto, M. P., Kubo, T., & Mikami, T. (1999). Heterogeneity of the atp6 presequences in normal and different sources of male-sterile cytoplasms of sugar beet. *Journal of Plant Physiology, 155*, 656–660.

Owen, F. V. (1945). Cytoplasmically inherited male-sterility in sugar beet. *Journal of Agricultural Research, 71*.

Palmer, J. D., & Herbon, L. A. (1988). Plant mitochondrial DNA evolves rapidly in structure, but slowly in sequence. *Journal of Molecular Evolution, 28*, 87–97.

Parkinson, C. L., Mower, J. P., Qiu, Y. L., Shirk, A. J., Song, K., Young, N. D., et al. (2005). Multiple major increases and decreases in mitochondrial substitution rates in the plant family Geraniaceae. *BMC Evolutionary Biology, 5*, 73.

Pearl, S. A., Welch, M. E., & McCauley, D. E. (2009). Mitochondrial heteroplasmy and paternal leakage in natural populations of *Silene vulgaris*, a gynodioecious plant. *Molecular Biology and Evolution, 26*, 537–545.

Pring, D., & Levings, C., III (1978). Heterogeneity of maize cytoplasmic genomes among male-sterile cytoplasms. *Genetics, 89*, 121–136.

Richards, A. (1997). *Plant breeding systems.* London: Chapman & Hall.

Richter, C., Park, J.-W., & Ames, B. N. (1988). Normal oxidative damage to mitochondrial and nuclear DNA is extensive. *Proceedings of the National Academy of Sciences of the United States of America, 85*, 6465–6467.

Rieseberg, L. H., Fossen, C. V., Arias, D., & Carter, R. L. (1994). Cytoplasmic male sterility in sunflower: origin, inheritance, and frequency in natural populations. *Journal of Heredity, 85*, 233–238.

Rhoads, D., Levings, C., III, & Siedow, J. (1995). URF13, a ligand-gated, poreforming receptor for T-toxin in the inner membrane of cms-T mitochondria. *Journal of Bioenergetics and Biomembranes, 27*, 437–445.

Sarria, R., Lyznik, A., Vallejos, C. E., & Mackenzie, S. A. (1998). A cytoplasmic male sterility-associated mitochondrial peptide in common bean is post-translationally regulated. *The Plant Cell, 10*, 1217–1228.

Satoh, M., Kubo, T., Nishizawa, S., Estiati, A., Itchoda, N., & Mikami, T. (2004). The cytoplasmic male-sterile type and normal type mitochondrial genomes of sugar beet share the same complement of genes of known function but differ in the content of expressed ORFs. *Molecular Genetics and Genomics, 272*, 247–256.

Saumitou-Laprade, P., Cuguen, J., & Vernet, P. (1994). Cytoplasmic male sterility in plants: molecular evidence and the nucleocytoplasmic conflict. *Trends in Ecology & Evolution, 9*, 431–435.

Schnable, P. S., & Wise, R. P. (1998). The molecular basis of cytoplasmic male sterility and fertility restoration. *Trends in Plant Sciences, 3*, 175–180.

Shykoff, J. A., Kolokotronis, S. O., Collin, C. L., & Lopez-Villavicencio, M. (2003). Effects of male sterility on reproductive traits in gynodioecious plants: a meta-analysis. *Oecologia, 135*, 1–9.

Sloan, D. B., Barr, C. M., Olson, M. S., Keller, S. R., & Taylor, D. R. (2008). Evolutionary rate variation at multiple levels of biological organization in plant mitochondrial DNA. *Molecular Biology and Evolution, 25*, 243–246.

Sloan, D. B., Oxelman, B., Rautenberg, A., & Taylor, D. R. (2009). Phylogenetic analysis of mitochondrial substitution rate variation in the angiosperm tribe Sileneae. *BMC Evolutionary Biology, 9*.

Sloan, D. B., Alverson, A. J., Chuckalovcak, J. P., Wu, M., McCauley, D. E., Palmer, J. D., et al. (2012). Rapid evolution of enormous, multichromosomal genomes in flowering plant mitochondria with exceptionally high mutation rates. *PLoS Biology, 10*. e1001241.

Städler, T., & Delph, L. F. (2002). Ancient mitochondrial haplotypes and evidence for intragenic recombination in a gynodioecious plant. *Proceedings of the National Academy of Sciences of the United States of America, 99*, 11730–11735.

Taylor, D. R., Olson, M. S., & McCauley, D. E. (2001). A quantitative genetic analysis of nuclear-cytoplasmic male sterility in structured populations of *Silene vulgaris*. *Genetics, 158*, 833–841.

Thompson, J. D., & Tarayre, M. (2000). Exploring the genetic basis and proximate causes of female fertility advantage in gynodioecious *Thymus vulgaris*. *Evolution, 54*, 1510–1520.

Tian, X., Zheng, J., Hu, S., & Yu, J. (2006). The rice mitochondrial genomes and their variations. *Plant Physiology, 140*, 401–410.

Touzet, P., & Budar, F. (2004). Unveiling the molecular arms race between two conflicting genomes in cytoplasmic male sterility? *Trends in Plant Science, 9*, 568–570.

Touzet, P., Hueber, N., Bürkholz, A., Barnes, S., & Cuguen, J. (2004). Genetic analysis of male fertility restoration in wild cytoplasmic male sterility G of beet. *Theoretical and Applied Genetics, 109*, 240–247.

Touzet, P., & Delph, L. F. (2009). The effect of breeding system on polymorphism in mitochondrial genes of *Silene*. *Genetics, 181*, 631–644.

Uyttewaal, M., Arnal, N., Quadrado, M., Martin-Canadell, A., Vrielynck, N., Hiard, S., et al. (2008). Characterization of *Raphanus sativus* pentatricopeptide repeat proteins encoded by the fertility restorer locus for Ogura cytoplasmic male sterility. *The Plant Cell, 20*, 3331–3345.

Van Damme, J. M. M., Hundscheid, M. P. J., Ivanovic, S., & Koelewijn, H. P. (2004). Multiple CMS–restorer gene polymorphism in gynodioecious *Plantago coronopus*. *Heredity, 93*, 175–181.

Wade, M. J., & McCauley, D. E. (2005). Paternal leakage sustains the cytoplasmic polymorphism underlying gynodioecy but remains invisible by nuclear restorers. *American Naturalist, 166*, 592–602.

Ward, B. L., Anderson, R. S., & Bendich, A. J. (1981). The mitochondrial genome is large and variable in a family of plants (Cucurbitaceae). *Cell, 25*, 793–803.

Warmke, H. E., & Lee, S.-L. L. (1978). Pollen abortion in T cytoplasmic male-sterile corn (*Zea mays*): a suggested mechanism. *Science, 200*.

Welch, M. E., Darnell, M. Z., & McCauley, D. E. (2006). Variable populations within variable populations: quantifying mitochondrial heteroplasmy in natural populations of the gynodioecious plant *Silene vulgaris*. *Genetics, 174*, 829–837.

Wolfe, K. H., Li, W. H., & Sharp, P. M. (1987). Rates of nucleotide substitution vary greatly among plant mitochondrial, chloroplast, and nuclear DNAs. *Proceedings of the National Academy of Sciences of the United States of America, 84*, 9054–9058.

Xu, W., Jameson, D., Tang, B., & Higgs, P. (2006). The relationship between the rate of molecular evolution and the rate of genome rearrangement in animal mitochondrial genomes. *Journal of Molecular Evolution, 63*, 375–392.

Yakes, M. F., & Van Houten, B. (1997). Mitochondrial DNA damage is more extensive and persists longer than nuclear DNA damage in human cells following oxidative stress. *Proceedings of the National Academy of Sciences of the United States of America, 94*, 514–519.

Yamamoto, M. P., Kubo, T., & Mikami, T. (2005). The 5′-leader sequence of sugar beet mitochondrial atp6 encodes a novel polypeptide that is characteristic of Owen cytoplasmic male sterility. *Molecular Genetics and Genomics, 273*, 342–349.

Yamamoto, M. P., Shinada, H., Onodera, Y., Komaki, C., Mikami, T., & Kubo, T. (2008). A male sterility-associated mitochondrial protein in wild beets causes pollen disruption in transgenic plants. *The Plant Journal, 54*, 1027–1036.

CHAPTER FIVE

Cytonuclear Adaptation in Plants

Françoise Budar[1,*,†] **and Sota Fujii**[2,‡]
[*]INRA, UMR1318, Institut Jean-Pierre Bourgin, Versailles, France
[†]AgroParisTech, Institut Jean-Pierre Bourgin, Versailles, France
[‡]University of Kyoto, Oiwakecho, Kitashirakawa, Sakyo-ku, Kyoto, Japan
[1]Corresponding author. E-mail: Francoise.Budar@versailles.inra.fr
[2]Corresponding author. E-mail: sfujii@pmg.bot.kyoto-u.ac.jp

Contents

Abstract

Plants possess compartmentalized genomes that are distributed in the nucleus and in two organelles: the mitochondria (mt) and plastids (pt). The crucial functions of these organelles require interaction between products encoded by the organelle genome and the nucleus. Hence, coadaptation contributes to the evolution of plant genomes, leading to a cooperative coevolution between interacting gene products that are encoded in different compartments. In addition, the different modes of inheritance between the Mendelian nuclear genes and uniparental organelles create a genomic conflict that also contributes to shape the evolution of some mt and nuclear genes: those involved in cytonuclear male sterilities. Pentatricopeptide repeat proteins have been involved in the evolution of the mt-nuclear conflict, but are also suspected to participate in the cooperative coadaptation between the nuclear compartment and organelle genomes. Several lines of evidence indicate that organelle genetic variations contribute to plant adaptation to their environment. So far, very few studies have identified potentially adaptive organelle variants. Nevertheless, it is very likely that cytonuclear coadaptation interferes with cytoplasmic adaptation to the environment.

Advances in Botanical Research, Volume 63
ISSN 0065-2296,
http://dx.doi.org/10.1016/B978-0-12-394279-1.00005-3

Both phenomena must therefore be considered in future work aiming at a better understanding of the evolution of organelle genes and adaptive component.

1. INTRODUCTION

Eukaryotes possess compartmentalized genomes and, among them, the green lineage harbours three genetic compartments, namely nucleus, plastids (pt) and mitochondria (mt). This genetic compartmentalization of plant cells is a consequence of two independent endosymbiosis events: first between an organelle-free organism and an α-proteobacterium, giving rise to the eukaryote lineage; then between a eukaryotic cell and a photosynthetic cyanobacterium, producing the green eukaryotic lineage (Margulis, 2004). During evolution, integration between these formerly distinct organisms occurred both at the genetic and metabolic levels, and transformed the bacterial endosymbionts into organelles (López-García & Moreira, 1999; McFadden, 2001). Metabolic integration resulted in a specialization of the physiological and metabolic roles of the organelles in the plant cell (Gould, Waller, & McFadden, 2008; Searcy, 2003). Genetic integration led to the relocation of a large proportion of the endosymbiont genomes to the host nucleus (Bogorad, 2008; Gray, Burger, & Lang, 1999; Reyes-Prieto, Weber, & Bhattacharya, 2007; see also Chapter 2, this volume). A tiny proportion of the endosymbiont genes was nevertheless retained in the organelle genome (Gray, 1989; Gray, Burger, & Lang, 2001). These are crucial for the functions of the organelles, but are far from being sufficient since mt and pt import more than 90% of their proteomes as products of nuclear genes (Huang *et al.*, 2008; Leister & Kleine, 2008; Millar, Heazlewood, Kristensen, Braun, & Møller, 2005; Yu *et al.*, 2008). Furthermore, physical interactions between nuclear- and organelle-encoded products are necessary for proper genome maintenance and expression, and for the assembly and function of electron transport chain complexes. Thus, the interactions between nuclear- and organelle-encoded products helped shape their evolution. The coevolution of organelle and nuclear genomes led to the coadaptation of genetic compartments in a given species/organism.

In the first part of this chapter, the mechanisms underlying coadaptation between the compartmentalized genomes of plants are presented and examples of experimental evidence for this phenomenon are given. The present knowledge on the genes acting in cytonuclear coadaptation, as revealed by recent studies, is reviewed.

The second part of this chapter considers the contribution of organelle genomes in the adaptation of plants to their environment, a role that has been largely overlooked in organelle studies and, in our opinion, deserves more attention from the scientific community. Using different examples from the literature, we show that this issue has to be understood in the frame of cytonuclear coadaptation.

Addressing questions about coadaptation between genetic compartments, or adaptation to the environment, often involves genetic approaches. In such approaches, mt and pt genomes behave as a single genetic unit because of their joint maternal inheritance (Birky, 2001). Although situations such as biparental transmission of pt (and more rarely mt) or paternal transmission of either pt or mt allow distinction between organelles in their influence on the phenotype being considered, these cases are restricted to a few species or exceptional situations (Bogdanova & Kosterin, 2006; Mogensen, 1996). Therefore in this chapter, some pt examples are discussed and some cases in which the respective roles of pt and mt are not yet fully understood are considered.

2. CYTONUCLEAR COADAPTATION

2.1. The Mechanisms of Coadaptation between Nuclear and Cytoplasmic Genetic Compartments

The first and most obvious consequence of the integration of multiple genomes into eukaryotic cells is the necessity for cooperation between those gene products originating from the nucleus and those originating from the organelle, with respect to gene expression and to the assembly of multimeric structures of the organelle containing components specified by two compartments. These processes involve physical interactions between gene products (proteins or RNAs) originating from the nucleus and the organelle (or between a nuclear-encoded protein and an organelle gene in the case of interaction between RNA polymerases, transcription factors and promoters). If we consider a given interaction of this type, we expect that variation in one partner may modify the interaction properties, creating selective pressure for a countermanding mutation in the other partner, so that proper interaction is restored.

These processes result in coadaptation between interacting gene products encoded in different compartments (Rand, Haney, & Fry, 2004). The driving force for this coadaptation is the necessity for both genetic compartments to ensure their transmission to the next generation through the reproductive success of the plant (Fig. 5.1).

It is generally assumed that the organelle variant is the first to get fixed in a population or species (Rand *et al.*, 2004). In the case of cytoplasmic variants

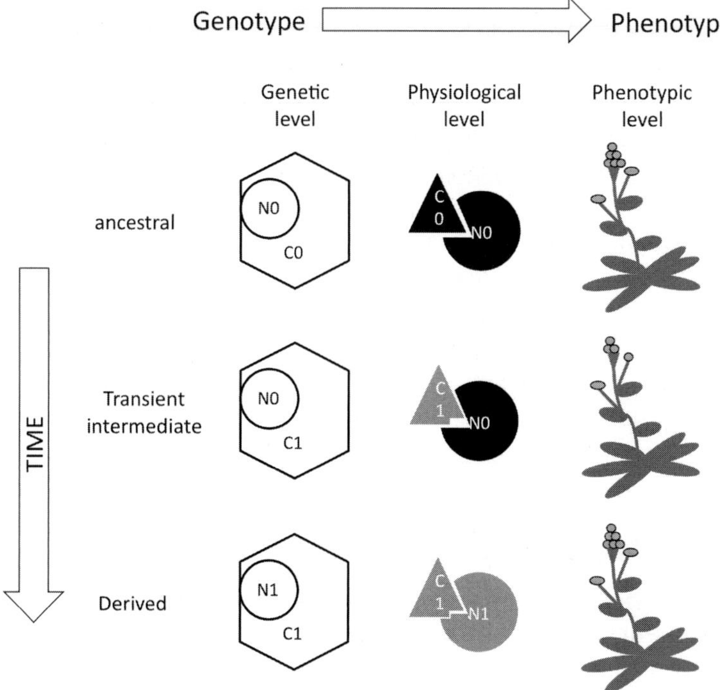

Figure 5.1 Cooperative cytonuclear coadaptation. Each row of the scheme represents one stage of a coevolutionary history, at different integration levels, from the genotype to the plant phenotype. In this figure, we consider the interaction between partners (N and C) that are encoded in the nucleus and an organelle (mt or pt). The nuclear genotype is represented in the circle and is designated by N0 (ancestral allele) or N1 (derived allele). For simplicity, only homozygous individuals are depicted. The cytotype is designated by C0 (ancestral form) and C1 (derived form). The products of these genes are designated with the allele names. At the physiological level, the correct interaction between N and C partners participate in the global phenotype and the fitness of the plant. Transiently, a cytoplasmic variant (C1) may be fixed in a population. The variation may affect the interaction with the N0 partner at the physiological level, and this may affect the global fitness of the plant. It is also possible that, although the N0/C1 interaction is less efficient than the N0/C0 one, the global fitness of the plant in the intermediate stage is not lower than that of the ancestral step, due to new environmental conditions, for example (see text). The N1 new allele for the nuclear-encoded partner restores an optimal interaction with C1 at the physiological level and therefore can be selected because the fitness of plants with the N1/C1 combination will be higher than that of plants with the N0/C1 combination. See the colour plate.

that may have a slightly deleterious effect on fitness, the clonal heredity of organelle genes make their fixation by genetic drift more likely than that of a nuclear variant with a similar effect on the phenotype. However, this has been theorized for animal mt genomes, which have very high mutation rates and small effective population sizes, which favours the fixation of slightly deleterious mutations by genetic drift (Rand *et al.*, 2004). It is likely that such theoretical models need to be adjusted to take into account the peculiarities of plant organelle genomes, and particularly those of plant mt. The size of the effective population for plant organelle genomes depends on the reproductive regime of the plant (Laporte, Cuguen, & Couvet, 2000). Plant mt genomes evolve chiefly through genomic rearrangements and are known to accumulate nucleotide substitutions at a very low rate (Drouin, Daoud, & Xia, 2008; Davila *et al.*, 2011; Palmer & Herbon, 1988). However, fast substitution rates have been reported in some plant mt genomes or genomic regions (Moison *et al.*, 2010; Parkinson *et al.*, 2005; Sloan *et al.*, 2012). As pt genomes have a more classic mode of evolution, although their mutation rate is lower than that reported for animal mt genomes, it is conceivable that they fit better with theoretical models designed for animal mt genomes than for plant mt genomes.

Nevertheless, it is also possible that a new cytoplasm could allow a better response to a new environment (see also Section 3). In this case, although the physiological output of the interaction between the partners is not optimal, the new cytotype (the pt and mt genomes collectively) may be fixed by selection. Subsequently, variants in the nuclear-encoded partner that improve the efficiency of the interaction between partners will be positively selected.

Thus, cooperative coevolution between nuclear and organelle genomes is expected when variations in a complex with components encoded by both genomes have an impact on a fitness-related trait, with the exception of pollen production. Descent through the male lineage does not affect the fitness of organelle genomes because they are maternally inherited. This does not apply to those species for which organelle genomes are inherited paternally or biparentally. The loss of pollen production, which has severe consequences on the fitness of nuclear genes, is sometimes linked with a higher fitness of cytoplasmic genomes, as in some studied cases of nucleo-cytoplasmic gynodioecy (see Chapter 4) (Dufaÿ, Touzet, Maurice, & Cuguen, 2007). When pollen production is considered, the selective forces applied to the genetic compartments of the same organism tend to favour different phenotypes: the mitochondrial genome inducing pollen sterility is favoured, whereas the nuclear genome restoring pollen fertility is favoured. In this genomic conflict, the male sterility cytotype is expected to appear first, and the nuclear partner (*Rf* gene) is selected in response to the new cytotype (Touzet & Budar, 2004). Fertility-restoring nuclear variants can therefore be considered as mutations that enhance the adaptation of nuclear genes to their new cytoplasmic environment. A study on mt polymorphism in *Silene* species with different reproduction regimes showed a correlation between nucleotide diversity and reproduction mode (Touzet & Delph, 2009). It remains to be established if the reproduction mode influences the rate of genetic variation in mt or if gynodioecy is more likely to occur in species with fast-varying mt genomes. In the latter case, the cause(s) of the observed fluctuations in the variation rates of plant mt genomes would remain an issue.

In any case, both cooperative coadaptation and genomic conflict are acting in the coevolution between plant genetic compartments. Hence, both must be considered when addressing questions regarding cytonuclear coadaptation.

2.2. Experimental Evidence for Cytonuclear Coadaptation

The best way to reveal coadaptation is to disrupt it. In the Bateson–Dobzhansky–Muller (BDM) model, hybrid incompatibilities are caused by interactions between genes that have diverged in the two lineages under consideration (Bomblies & Weigel, 2007). When applied to cytonuclear coadaptation, it can lead to asymmetrical results in reciprocal hybrids (Fig. 5.2) (Turelli & Moyle, 2007). Cytonuclear epistasis therefore raises hybridization barriers between divergent genotypes, hence is thought to contribute to speciation (Alcázar, Pecinka, Aarts, Fransz & Koornneef, 2012; Chou & Leu, 2010; Greiner, Rauwolf, Meurer, & Herrmann, 2011; Levin, 2003; Rand *et al.*, 2004; Turelli & Moyle, 2007).

Cytonuclear genetic epistasis at the species level has been well documented by the abnormal phenotypes that arise in interspecific crosses or alloplasmic lines (plants with a nuclear genome from a species carrying the organelle genomes from another species). Pollen sterility is very common among alloplasmic lines, but they also frequently exhibit photosynthetic deficiencies and abnormal development of flowers (Greiner *et al.*, 2011; Linke & Börner, 2005).

As previously mentioned, in most cases the joint heredity of mt and pt genomes precludes the identification of which of these two genomes contributes to the interaction with the nucleus leading to the phenotypes

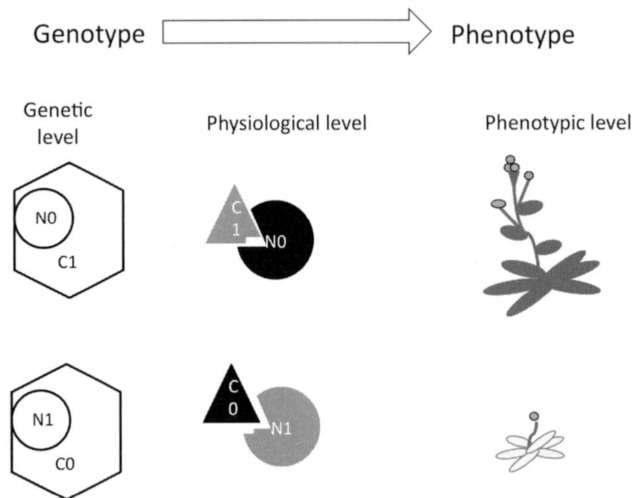

Figure 5.2 Bateson–Dobzhansky–Muller genetic epistasis due to disruption of cytonuclear coadaptation. The symbolism is the same as in Fig. 5.1. Lines that combine either the ancestral nuclear genotype with the derived cytotype (upper row) or the derived nuclear genotype with the ancestral cytotype (lower row) are represented. The former case is similar to the transient step described in Fig. 5.1. In the second case, the interaction between gene products is deeply compromised. This can lead to serious physiological defects that will severely affect the plant phenotype. These combinations can occur in reciprocal F2 hybrids homozygous for parental alleles at the nuclear locus under consideration or in reciprocal alloplasmic lines. For colour version of this figure, the reader is referred to the online version of this book.

observed in alloplasmic lines. The somatic hybridization technique, which allows the production of cybrids (for hybrid cytoplasm), was extensively applied to species from the Brassicaceae and Solanaceae families. In some cases, this made it possible to attribute phenotype alterations to either the mt or pt of alien origin; male sterility and flower morphology alterations were associated with the mt genome, whereas deficiencies in photosynthesis were associated with the pt genome (Belliard, Pelletier, Vedel, & Quetier, 1978; Belliard, Vedel, & Pelletier, 1979; Bonnett, Kofer, Hakansson, & Glimelius, 1991; Galun, Arzee-Gonen, Fluhr, Edelman, & Aviv, 1982; Zubko, Zubko, & Gleba, 2002).

In addition, the analysis of mt and pt profiles of calli derived from somatic hybridization between *Nicotiana plumbaginifolia* and *Solanum tuberosum* showed that pt and mt parental types were correlated with the nuclear composition of the cells (Wolters, Koornneef, & Gilissen, 1993), providing evidence that cytonuclear incompatibilities interfered with the recovery of viable cells.

It was also proposed that some cytoplasmic male sterilities appearing in alloplasmic lines can result from ancient cytoplasmic male sterilities (CMS), in which restorer genes were fixed in the cytoplasm-donor species but absent from the nuclear-donor parent of the alloplasmic line (Budar, 1998). This seems to be the case in male-sterile interspecific hybrids of *Mimulus* (Fishman & Willis, 2006). It is also probably the case when expression of a typical CMS gene is correlated with an alloplasmic CMS, as reported in several alloplasmic lines of *Brassica* species (Ashutosh, Dinesh, Sharma, Prakash, & Bhat, 2008; Landgren, Zetterstrand, Sundberg, & Glimelius, 1996; Shinada, Kikuchi, Fujimoto, & Kishitani, 2006). It is therefore conceivable that the phenotypes of some alloplasmic lines result from a genomic conflict arising from reactivation of an mt CMS gene in which the restorer(s) was fixed in its species of origin, in addition to cytonuclear incompatibility of the BDM type.

In contrast to alloplasmic situations, cases of cytonuclear epistasis within species have been rarely reported. One reason could be that the selection for proper interaction between gene products originating from different compartments prevails over the fixation of variants in either partner. The occurrence of BDM cytonuclear incompatibilities relies on the combination of organelle and nuclear genomes that experienced divergent evolution for a sufficiently long period. It can therefore be expected that cytonuclear epistasis between intraspecific variants will be observed only when quite distantly related genotypes are used. Some examples have been reported in plants.

Recently, F2 plants from crosses between a European *Arabidopsis lyrata* subspecies (*petraea*) and a North American subspecies (*lyrata*), and carrying the cytoplasm of the former, were reported to be partially male sterile (Leppälä & Savolainen, 2011). In this case, it is conceivable that the *petraea* cytoplasm retains a CMS gene in which restorer genes are fixed in this subspecies. The CMS is expressed when the *petraea* cytoplasm is confronted with alleles from the *lyrata* subspecies, which did not fix restorers. Similarly, CMS can be discovered after intraspecific crosses in crops, such as the *nap* CMS in rapeseed (Thompson, 1972), and most CMS observed in rice (Fujii, Kazama, & Toriyama, 2008).

A case of negative epistasis was observed between pt of *Pisum sativum* ssp. *elatius* and the nuclear genome of cultivated pea. It resulted in sterility, chlorophyll deficiency, leaf colour variegation, and modified leaf morphology (Bogdanova & Berdnikov, 2001; Bogdanova, Galieva, & Kosterin, 2009). In this case, the cytoplasmic compartment responsible for the negative interaction could be identified due to exceptional paternal transmission of pt (Bogdanova, 2007). In *Arabidopsis thaliana*, coadaptation between nuclear loci and organelle genomes was suggested from the distribution of parental alleles in reciprocal recombinant inbred line (RIL) populations: some regions of the nuclear genome appeared to preferentially fix different parental alleles according to the accession from which they inherited their cytoplasm (Törjék *et al.*, 2006). A study on reciprocal F2 families derived from *A. thaliana* natural accessions showed that germination in challenging conditions is under the influence of interactions between cytoplasm- and nuclear-encoded factors and indicated that the coadaptation between the genetic compartments was disrupted in some F2s (Moison *et al.*, 2010).

These findings raise the question of the nature of genetic variations underlying the coadaptations revealed by the observed phenotypes, and of the mechanism(s) that led to their fixation in specific lineages.

2.3. Molecular Actors in Cytonuclear Coadaptation

As described in the above sections, when the simplest BDM model is applied to the coevolution of nucleus and organelle, there must be a pair of genes responsible for the incompatibility: one encoded in the nucleus and the other in the organelle. Although cytonuclear incompatibilities have long been known via the application of hybrid and cybrid technologies, only a few genes involved in such cytonuclear BDM incompatibilities have been identified. Some known examples of molecular factors underlying cytonuclear coadaptation are listed in this section.

Regarding cytonuclear incompatibility resulting from genomic conflict, a few molecular biological studies elucidated components involved in CMS (Budar, Touzet, & De Paepe, 2003; Fujii & Toriyama, 2008; Hanson & Bentolila, 2004; Kubo, Kitazaki, Matsunaga, Kagami, & Mikami, 2011; Schnable & Wise, 1998). Several mt sterility-inducing genes have been identified in different CMS systems (Hanson & Bentolila, 2004; Kubo *et al.*, 2011). Formally identified nuclear *Rf* genes, that is, genes allowing male fertility despite the presence of the mt CMS gene, are rare. Nevertheless, the pentatricopeptide repeat (PPR) protein family is largely overrepresented among the *Rf* gene products identified so far (Bentolila, Alfonso, & Hanson, 2002; Brown *et al.*, 2003; Desloire *et al.*, 2003; Kazama & Toriyama, 2003; Koizuka *et al.*, 2003; Komori *et al.*, 2004; Wang *et al.*, 2006). Details on the pattern of evolutionary variability in the PPR superfamily and its possible contribution to cytonuclear coadaptation are discussed later in this chapter.

Several studies indicate that PPR proteins bind RNA (Okuda, Nakamura, Sugita, Shimizu, & Shikanai, 2006; Okuda, Myouga, Motohashi, Shinozaki, & Shikanai, 2007; Prikryl, Rojas, Schuster, & Barkan, 2011; Zhelyazkova *et al.*, 2011), and most are known to be involved in the post-transcriptional RNA modification of organelle genes (for reviews, see Delannoy, Stanley, Bond, & Small, 2007; Fujii & Small, 2011; Saha, Prasad, & Srinivasan, 2007; Schmitz-Linneweber & Small, 2008; Small & Peeters, 2000). Although the fine details of their molecular functions remain to be elucidated, PPRs encoded by *Rf* genes undoubtedly suppress the accumulation of the CMS gene product in mitochondria (Fujii & Small, 2011; Gillman, Bentolila, & Hanson, 2007; Kazama, Nakamura, Watanabe, Sugita, & Toriyama, 2008; Savir, Noor, Milo, & Tlusty, 2010; Uyttewaal *et al.*, 2008) via interaction with the CMS gene transcript. In a number of CMS systems in which only the mt sterility-inducing gene is identified, male fertility restoration by nuclear *Rf* very often coincides with an alteration of the expression of the mt sterility gene by a post-transcriptional event (Bergelson & Roux, 2010; Kubo *et al.*, 2011). This latter observation suggests that the involvement of PPR proteins in restoration of fertility might be shared by a number of CMS systems.

PPR encoding *Rf* (*Rf-PPR*) genes have been isolated from a wide range of angiosperm species such as *Petunia* (Bentolila *et al.*, 2002; Fujii, Bond, & Small, 2011), *Raphanus/Brassica* (Brown *et al.*, 2003; Desloire *et al.*, 2003; Koizuka *et al.*, 2003) and *Oryza* (Kazama & Toriyama, 2003; Komori *et al.*, 2004; Wang *et al.*, 2006). In species such as *Mimulus* (Barr & Fishman, 2010), *Sorghum* (Jordan *et al.*, 2011) and *A. lyrata* (Leppälä & Savolainen, 2011), *Rf* genes or QTLs are mapped within a region containing clusters of tandemly duplicated PPR genes, a characteristic genomic structure also shared among *Petunia, Raphanus* and *Oryza Rf* loci (Bentolila *et al.*, 2002; Desloire *et al.*, 2003; Komori *et al.*, 2004). In addition, these loci seem to evolve through duplication and unequal crossover events (Hernandez Mora, Rivals, Mireau, & Budar, 2010; Kato *et al.*, 2007). These peculiar signatures are a reminder of the birth-and-death evolution of disease resistance genes (Touzet & Budar, 2004). The conflict between the mt-encoded CMS gene and the nucleus-encoded *Rf* locus (cf. Section 2.1) can be compared with the coevolutionary arms race between the host disease resistance (*R*) gene and the parasite avirulence (*Avr*) gene. The interaction of *R* and *Avr* genes is a well-known example of a gene-to-gene relationship, and extremely rapid evolution is observed in these loci (Bergelson, Kreitman, Stahl, & Tian, 2001). Several lines of evidence support the resemblance of *Rf*-CMS to the *R-Avr* relationship. First, as already mentioned, *Rf-PPR* genes often form a genomic cluster, meaning highly identical *Rf-PPR*-like genes are found in their proximity (Geddy & Brown, 2007; Kato *et al.*, 2007). Gene clustering in chromosomes results from frequent gene duplications or gene conversions, and this genomic structure is also the hallmark of *R* genes (Ellis, Dodds, & Pryor, 2000). Second, high rates of non-synonymous codon substitutions were observed in *Rf-PPR* genes (Geddy & Brown, 2007; Fujii *et al.*, 2011), in a manner comparable with that of *R* genes. Third, similar to *Rf*-PPR, *R* genes encode a short tandem repeat motif protein known as the leucine-rich repeat (LRR). LRR proteins can recognize specific protein targets via a combination of LRR modules, which resembles the capacity of PPR proteins to target specific RNA sequences (see below). Positive selection on *Rf-PPR* genes is a possible signal of conflicting evolution between nucleus and mt.

Therefore, our knowledge on mt CMS and nuclear *Rf* genes is in agreement with the view that *Rf-PPR* genes provide an adaptive response of the nucleus to mt male sterility-inducing variants. Nevertheless, the question arises on the role of *Rf-PPR*-like genes in species in which no CMS has been described. Most of the *Arabidopsis* mutants of *Rf-PPR*-like genes lose mt gene 5′ end formation, without exhibiting any obvious phenotypic effects (Hölzle *et al.*, 2011; Jonietz, Forner, Hölzle, Thuss, & Binder, 2010; Jonietz, Forner, Hilderbrandt, & Binder, 2011). It is conceivable that these genes in the *Arabidopsis* genome provided a source of novel *Rf-PPR* genes by gene conversion or unequal chromosome crossovers, generated in response to mt genome alterations. Maintenance of genomic clusters has been proposed to be important for fast adaptive evolution of *R* genes against rapidly evolving pathogens (Bergelson *et al.*, 2001).

Rf-PPR genes are only a subset of the PPR genes found in plant genomes (see Chapter 10). In land plants, 100–1000 PPR genes are present in a single genome, and are recognized as one of the largest family of protein types in photosynthetic organisms (Fujii & Small, 2011; O'Toole *et al.*, 2008; Schmitz-Linneweber & Small, 2008; Small & Peeters, 2000). Enrichment of whole-genome information has now allowed us to expect that each angiosperm species carries approximately 400–600 PPR genes. Protein subcellular localization programs predict that approximately 80% of the PPR proteins are targeted to mitochondria or chloroplasts (Lurin *et al.*, 2004). From numerous *Arabidopsis* mutant studies, the molecular functions of *Rf-PPR* genes usually do not overlap with other members of this family, as disruption of most other PPR genes is sufficient to confer a defective phenotype (often lethality) to the mutant (Delannoy *et al.*, 2007; Lurin *et al.*, 2004; Schmitz-Linneweber & Small, 2008). It is considered that each of the tandem-aligned 35 amino acid PPR motifs within a PPR protein functions as a single RNA-nucleotide-binding adaptor module, providing the flexibility for this protein family to target variable RNA sequences by shuffling motif alignment (Filipovska & Rackham, 2012; Kobayashi *et al.*, 2011; Prikryl *et al.*, 2011; Schmitz-Linneweber & Small, 2008; Small & Peeters, 2000).

As a result of this mode of action through specific interaction with organelle RNA, PPR proteins seem to possess the required characteristics to operate in cytonuclear coadaptation. The presence or absence of a PPR is strongly correlated with its target RNA sequences in the organellar genome (Hayes & Mulligan, 2011). The loss of requirement for organelle RNA modification results in the pseudogenization of the PPR gene, and in some cases neofunctionalization may occur due to the relaxed selective pressure (Hayes & Mulligan, 2011).

The loss of organelle RNA editing events would probably be one of the most common cases of BDM incompatibility between angiosperm species, given that this post-transcriptional RNA modification event has been experiencing ongoing loss processes (Fujii & Small, 2011; Tillich, Lehwark, Morton, & Maier, 2006; Tillich *et al.*, 2009). Cytidine to uridine (C-to-U) RNA editing (or less frequent U-to-C RNA editing) has evolved in land plant species independently from similar types of editing events in metazoan nuclear RNA (Gray, 1996; Gray, 2009; Shikanai, 2006; Takenaka, Verbitskiy, Van Der Merwe, Zehrmann, & Brennicke, 2008). Many RNA editing events play important roles in gene expression by generating start codons at ACG sites, correcting codons to encode conserved amino acids and generating required stop codons. It is not unusual for the mutants defective in RNA editing to exhibit severe growth defective phenotypes as a consequence of the loss of their ability to produce proteins of the proper amino acid sequences (Shikanai, 2006; Takenaka *et al.*, 2008). Usually around 30–40 RNA sites in the angiosperm chloroplast genomes and more than 400 sites in the mitochondrial genomes are edited C-to-U (Fujii & Small, 2011; Gray, 2009). Half of the PPR family members (known as the PLS subclass) (Lurin *et al.*, 2004; Rivals, Bruyère, Toffano-Nioche, & Lecharny, 2006) are predicted *a priori* to be involved in C-to-U RNA editing at specific RNA nucleotides (Fujii & Small, 2011; Schmitz-Linneweber & Small, 2008; Shikanai, 2006; Takenaka *et al.*, 2008). A significant proportion of PPR mutants (mainly in *Arabidopsis* and *Oryza*) are defective in post-transcriptional C-to-U RNA editing (Fujii & Small, 2011; Shikanai, 2006; Takenaka *et al.*, 2008).

Several studies have indicated that a significant net loss of C-to-U RNA editing has occurred throughout the evolution of angiosperms especially in pt genomes (Freyer, Kiefer-Meyer, & Kössel, 1997; Jobson & Qiu, 2008; Tillich *et al.*, 2006). The poor conservation of RNA editing sites among extant species has suggested that these losses have occurred independently in each lineage (Fujii & Small, 2011; Tillich *et al.*, 2009). Thus, even among closely related species, sites of RNA editing frequently differ.

One of the best examples of this has been documented in a study of cybrids between *Atropa belladonna* and *Nicotiana tabacum* (Schmitz-Linneweber *et al.*, 2005). Cybrids carrying the *Atropa* nucleus and *Nicotiana* plastid had albinism, and a suppressor mutation in the 264th codon of *atpA* was able to recover chloroplast differentiation (Schmitz-Linneweber *et al.*, 2005).

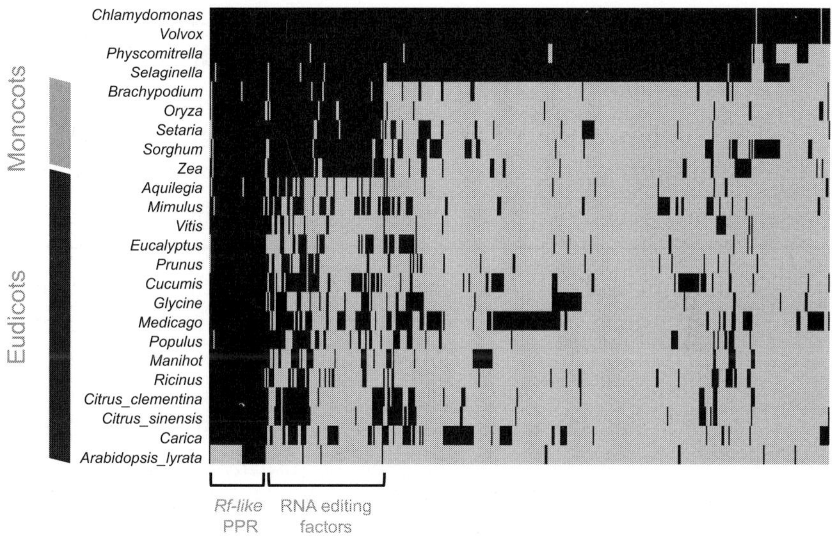

Figure 5.3 *Arabidopsis thaliana* PPR homologues in other species. Homologues of 450 PPR genes in *A. thaliana* were searched against 24 other plant species. Successful reciprocal BLAST hits were considered as indicating the presence of a homologue. The horizontal axis indicates the 450 PPR genes and gray boxes indicate the presence of the gene in a species; the black colour indicates the absence of a corresponding homologue. *Rf-PPR* genes are seemingly missing from most other species; this is because a reciprocal hit between two species was unable to establish the occurrence of a homologue due to frequent gene duplication of this subgroup. *Rf-PPR*-like clusters are ubiquitously present in angiosperm species (Fujii *et al.*, 2011; Willett & Burton, 2003).

The *Atropa* nucleus lacks the ability to perform C-to-U RNA editing at this codon, which is edited in *Nicotiana*, suggesting that a loss (or functional diversification) of the RNA editing factor occurred in *Atropa* corresponding to the loss of the RNA editing site as a result of coadaptation. Hence, it is highly likely that the lack of RNA editing activity at the 264th codon of *atpA* in *Atropa* (Schmitz-Linneweber *et al.*, 2005) is due to loss of a corresponding PPR gene. Database searches show that PLS subclass PPR genes are frequently lost independently in different lineages (Fig. 5.3), in a similar manner to C-to-U RNA editing (Freyer *et al.*, 1997; Fujii & Small, 2011; Tillich *et al.*, 2009).

Therefore, coevolution between PPR genes (encoded in the nucleus) involved in RNA editing, and their target RNA editing site (in organelle transcript) are serious candidates for playing a role in some cytonuclear BDM incompatibilities observed in plants.

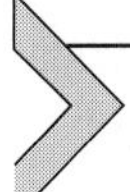

3. THE INVOLVEMENT OF ORGANELLES IN PLANT ADAPTATION TO THEIR ENVIRONMENT

3.1. Evidence for a Cytoplasmic Component in Plant Adaptation

As already mentioned, it can be argued that variants in organelle genes are not necessarily mild deleterious mutations fixed by genetic drift and that selection can also drive the fixation of variants with a positive effect on fitness, as adaptation to new environmental conditions for example. Adaptive variations of animal and human mitochondrial genes have been reported (Mishmar *et al.*, 2003). Their occurrence raises some issues on the use of mt genes in population genetics and evolution studies. In population genetic studies, organelle genes are usually used under neutral models, the fluctuation in variant frequencies depending only on population history events. If organelle gene variations are adaptive, inferences on population genetics might be biased. This issue is discussed in depth in the field of animal and human mitochondrial evolution (Balloux, Handley, Jombart, Liu, & Manica, 2009; Mardulyn, Othmezouri, Mikhailov, & Pasteels, 2011; Meiklejohn, Montooth, & Rand, 2007; Wallace, 2010), but the similar use of pt diversity in plants is rarely addressed.

However, the contribution of variations in mt and/or pt genes to plant adaptation is documented by different types of experimental results.

Cytoplasm capture has been reported between interfertile sympatric species. In the case of the capture of cultivated *Brassica napus* cytoplasm by its wild parental species *B. rapa*, it was shown that the captured cytoplasm was linked to an adaptive advantage of the wild species in the specific habitat of river banks (Allainguillaume *et al.*, 2009). Other cases of cytoplasm capture are conceivably linked to an adaptive advantage of the captured cytoplasm. For example, the distribution of chlorotypes in interfertile species of *Nothofagus* correlates with latitude, but not with nuclear phylogeny, which strongly suggests an adaptation to the geographic environment among the chlorotypes (or cytotypes) of the genus (Acosta & Premoli, 2010). In these two studies, cytoplasm capture was demonstrated using pt markers, but inferring which organelle instigated the capture would necessitate further, and probably difficult, investigations.

Cytoplasmic local adaptation is revealed by observing equivalent nuclear genotypes (or populations) possessing different cytoplasms in the habitat of origin of each cytoplasm. This is usually realized by growing reciprocal crossed progenies in the habitats under consideration. Using this strategy, Sambatti *et al.* (2008) found that sunflower hybrids and backcross progenies produced from *Helianthus annuus* adapted to moist conditions, and *H. petiolaris*, found in dry habitats, have different survival probabilities in each parental habitat, depending on their cytoplasm. Cytoplasmic local adaptation was also detected at the within-species level in *Chamaecrista fasciculata* and *A. lyrata* (Galloway & Fenster, 1999; Galloway & Fenster, 2001; Leinonen, Remington, & Savolainen, 2011).

In the documented cases of cytoplasmic local adaptation, a significant effect of cytonuclear interactions was also observed. In the case of sunflowers, this led to cytonuclear incompatibilities (Sambatti, Ortiz-Barrientos, Baack, & Rieseberg, 2008). This is likely to reflect coadaptation between the locally adapted cytoplasm and its associated nuclear genome.

A cytoplasm effect on drought tolerance was found in a study using reciprocal RIL populations between two natural accessions of *A. thaliana* (Mckay *et al.*, 2008). However, in this case, the effect of the cytoplasm on drought tolerance was puzzling: the cytoplasm of the drought-adapted accession had a negative effect on water-use efficiency compared with that of the accession from the wet habitat (Mckay *et al.*, 2008). Although such an observation deserves further investigation, including measurements of the cytoplasm effect on more global phenotypes, one can speculate that it may reflect constraints on cytoplasmic variation, among which coadaptation with nuclear partners appears likely.

3.2. Molecularly Characterized Examples

Examples of adaptive variations in organelle genes identified at the molecular level are scarce. However, recent studies provide strong evidence of the contribution of the pt-encoded large subunit of ribulose-1,5-biphosphate carboxylase (*rbcL*) amino acid sequence evolution to plant adaptation. Studies have investigated significant chances of positive selection acting on *rbcL* genes in diverse taxa (Iida *et al.*, 2009; Kapralov & Filatov, 2006; Kapralov & Filatov, 2007; Kapralov, Kubien, Andersson, & Filatov, 2011; Miwa *et al.*, 2009; Young, Rickaby, Kapralov, & Filatov, 2012). For example, models allowing positive selection were significantly better at explaining rbcL codon substitution patterns in the Hawaiian island genus *Schiedea* than those allowing only neutral evolution or purifying selection (Kapralov & Filatov, 2006). Furthermore, some of the significant substitutions were correlated with a change in habitat conditions, suggesting adaptation of the photosynthetic ability to a new environment (Kapralov & Filatov, 2006). When applied to the crystal structure of spinach RbcL, residues under high positive selection coincided with those involved in physical interaction with the nuclear-encoded Rubisco activase, possibly suggesting alteration in the regulation of the enzyme complex.

Evolution of C_4 photosynthesis also had a great impact on the molecular adaptation of core photosynthetic modules shared with C_3 photosynthesis. The tenfold increase in CO_2 concentration in C_4 plants compared with C_3 plants would likely relieve the purifying selection pressure on RbcL to maintain high CO_2 affinity (von Caemmerer & Furbank, 2003; Christin *et al.*, 2008). Some C_4 natural variations in *rbcL* are adapted to possess higher catalytic efficiencies and lower substrate specificities. Phylogenetic analysis of *Flaveria*, a genus that includes both C_4 and C_3 species, revealed amino acid sites under adaptive evolution (Kapralov *et al.*, 2011). These residues are on the interface of subunits, and are likely important for hexadecamer formation of the Rubisco complex, eventually affecting the enzyme kinetics.

On the other hand, although Rubisco is the fundamental enzyme complex ubiquitously required for photosynthesis, retention of *rbcL* positive selection was restricted to 1.5–0.41 billion years ago and has been essentially absent in extant algae species (Kapralov & Filatov, 2007; Young *et al.*, 2012). This could possibly be due to the apparent difference between thermal and water regimes between algae and land plant species. Meanwhile, there is a large discrepancy in the GC content and codon usage of pt and mt genomes, between algae and terrestrial plants (Smith, 2009). As the GC content of the genome and codon usage can affect synonymous substitution rates (Duret & Mouchiroud, 1999; Sueoka, 1988), there is a possibility that evidence for algal *rbcL* adaptive evolution may be detected after the correction for such parameters.

At the time of life migration to land, protection against ultraviolet (UV) light must have been one of the most urgently required adaptations (Häder & Sinha, 2005; Takahashi & Ohnishi, 2004). One possible explanation for sequence drift to a higher GC content in pt and mt genomes of basal land plant species is the beneficial effect of avoiding TT dinucleotides in their sequences, because the di-thymidine sequence forms mutagenic pyrimidine dimers on exposure to UV light (Maier *et al.*, 1996; Rozema *et al.*, 2002; Singer & Ames, 1970; Yura, Sulaiman, Hatta, Shionyu, & Go, 2009). Alternatively, selection pressure for thermal stability can also be the cause of GC drive. Several studies indicate the positive correlation between pt and mt genomic GC content and the number of C-to-U RNA editing sites (Fujii & Small, 2011; Kubo *et al.*, 2011; Malek, Lättig, Hiesel, Brennicke, & Knoop, 1996; Smith, 2009; Yura *et al.*, 2009). This link is evident because RNA editing reverses the DNA mutation at the RNA level, making it possible for pt or mt to increase genomic GC content and maintain protein function at the same time. Most RNA editing events occur at the first or second position of the codon and thus are often directly involved in protein primary sequence coding (Bentolila *et al.*, 2002; Desloire *et al.*, 2003; Jobson & Qiu, 2008; Komori *et al.*, 2004).

Overall, whole pt and mt genome-scale adaptive reprogramming occurred on land migration of the plant species, increasing their GC content. Correlatively, PPR proteins involved in organelle RNA editing corrected the mRNA in order to ensure the conservation of proteins at the amino acid sequence level (Fujii & Small, 2011; Hernandez Mora *et al.*, 2010; Kato *et al.*, 2007). Again, adaptive variation in organelle genomes was coupled with cytonuclear coadaptation. An interesting question arises whether PPR recognizing newly edited target sequences existed prior to the appearance of new editing sites, or if the appearance of new C residues in coding sequences drove the selection of specific PPR editing factors. The three-step model of RNA editing evolution hypothesizes the fixation of the RNA editing site by genetic drift (Covello & Gray, 1993; Touzet & Budar, 2004). It is known that some PPR proteins participate in RNA editing at multiple sites

(Schmitz-Linneweber & Small, 2008). Conceivably then, a potentially detrimental T to C mutation in an organelle gene could be tolerated if recognized by a preexisting PPR editing factor capable of recognizing diverse RNA sequences. This might then allow fixation of the mutation by genetic drift.

4. CONCLUSION AND PERSPECTIVES FOR FURTHER RESEARCH

Coadaptation between organelle and nuclear genomes at the species level is widely accepted. The contribution of cytonuclear epistasis to genetic isolation, hence its possible involvement in speciation, has been recognized (Alcázar *et al.*, 2012; Chou & Leu, 2010; Greiner *et al.*, 2011; Levin, 2003). The occurrence of cytonuclear epistasis within species has been documented by recent reports. However, the genetic diversification within a species of coadapted molecular partners encoded in different genetic compartments is probably underestimated so far, except for CMS appearing after intraspecific crosses. They reveal the contribution of the genomic conflict between the nuclear and maternally inherited organelle genomes in the raising of genetic barriers (Alcázar *et al.*, 2012; Geddy & Brown, 2007; Kato *et al.*, 2007). However, it might be difficult to discriminate between the disruption of cooperative coadaptation and reactivation of a genomic conflict when observing maternally inherited male sterility after a cross between distantly related genotypes. It is likely that both mechanisms can act together in the phenotype of a hybrid plant.

The issue remains whether the model proposed for animal genomes, under which coadaptation is driven by variations in the mt genome first, and the subsequent selection of nuclear coadapted variants, is also valid for plant cytonuclear cooperative coadaptation. Better knowledge of the genetic variation occurring in plant organelle genomes both at the species level and within species, and probably a reappraisal of theoretical models, are needed before clarifying this issue.

Our knowledge on the genetic diversity of plant organelles is, in most species, restricted to pt intergenic polymorphisms thought to be neutral and used to infer maternal phylogenies. Obviously the programs based on the use of new generation sequencing (NGS) technologies will provide precious data on the substitutions occurring in plant organelle genomes (e.g. the 1001 genomes project for *A. thaliana*; http://www.1001genomes.org). However, the peculiar mode of evolution of plant mt will probably also necessitate the *de novo* assembly of variant mt genomes, as a large amount of polymorphism in these plant organelles results from rearrangements (see Chapter 9) (Davila *et al.*, 2011). In addition, evidence is accumulating that both mt substitution rates and constraints on mt genome size fluctuate among plant lineages (Sloan *et al.*, 2012). The impact of these fluctuations on cytonuclear coevolution remains to be investigated.

Nevertheless, PPR proteins have been identified as major molecular actors in cytonuclear coadaptation, both in the frame of cooperative coadaptation and genomic conflict. Probably, this also reflects the peculiarities of cytonuclear coadaptation in plants, since the expansion of this protein family is characteristic of the green eukaryotic lineage.

Most of the known examples of molecular interactions underlying coadaptation between plant mt genomes and their nucleus involve PPR proteins and their target mt RNAs, as shown above. In yeast, two recently deciphered cytonuclear BDM incompatibilities involved nuclear-encoded factors that are required for the proper expression of specific mt genes (Chou, Hung, Lin, Lee, & Leu, 2010; Lee *et al.*, 2008).

In animals, several reports suggested that cytonuclear incompatibilities resulted from impaired electron transport chain function, due to poorly matched mt- and nuclear-encoded subunits (Barrientos, Müller, Dey, Eienberg, & Moraes, 2000; Blier, Dufresne, & Burton, 2001; Sackton, Haney, & Rand, 2003; Wu, Schmidt, Goodman, & Grossman, 2000). In addition, in the case of the marine copepod, *Tigriopus californicus*, the decrease in complex IV (cytochrome oxidase) efficiency of unfit hybrids could be traced to single amino acid polymorphisms in the nuclear-encoded cytochrome *c* apoprotein and corresponding sequence variants of the mt-encoded subunit II of cytochrome oxidase (Harrison & Burton, 2006). In laboratory evolved populations of *T. californicus*, mt-nuclear negative epistasis was found to depend on environmental conditions, namely the temperature regime (Galloway & Fenster, 1999; Galloway & Fenster, 2001; Leinonen *et al.*, 2011; Willett & Burton, 2003). However, in this species, in which populations evolve in almost strict isolation, hybrid breakdown in fitness appears to involve more complex BDM incompatibilities than simple two-factor cytonuclear epistasis (Willett, 2011).

Although evidence for the contribution of cytoplasmic variation in plant adaptation to the environment is accumulating, mainly from ecological studies, this contribution has been neglected in most studies reported so far on plant adaptation. Regarding this aspect, research on animal mt evolution is several steps ahead. Nevertheless, for plants also, a key issue for environmental adaptation is bioenergetics (Wallace, 2010). This should motivate us to pay more attention to organelle variation with respect to plant adaptation to new environments, particularly in the context of global climate change. In this context, a recent study on adaptation to climate in *A. thaliana* indicated the involvement of genes whose functions were related to photosynthesis and energy metabolism, among others (Hancock *et al.*, 2011). Both cases of organelle GC content and PPR editing factors and of the *rbcL* gene mentioned above (Section 3.2) are demonstrative examples that the issues of cytonuclear adaptation and adaptive variation in organelle genes are entangled (Barrientos *et al.*, 2000; Blier *et al.*, 2001; Fujii & Small, 2011; Sackton *et al.*, 2003; Savir *et al.*, 2010; Wu *et al.*, 2000). Therefore, coadaptation with nuclear genes will have to be carefully considered when addressing the contribution of organelle variants in plant adaptation.

Exploration of the adaptive features of mt–nuclear coevolution in plants will require a combination of approaches and collaborative efforts between scientific disciplines. In addition to exploration of the diversity in organelle and nuclear genes and thorough genetic analysis of their epistatic interactions, a comprehensive analysis of the physiological impact of poorly matched genetic combinations is highly desirable. The adaptive nature of the traced polymorphisms will also necessitate the evaluation of their impact on fitness in realistic ecological environments (Bergelson & Roux, 2010). In this respect, deciphering the contributions of mt– (or pt–) nuclear epistatic interactions to fitness-related traits in varying environments represents an exciting challenge.

In addition, such studies are likely to provide precious knowledge for breeders. The impact of nuclear–cytoplasm interactions has been reported to be significant in a wide range of traits of interest in several crops. For instance, cytonuclear interactions and cytoplasmic variation were found to influence yield and low-temperature tolerance in rice (Harrison & Burton, 2006; Tao *et al.*, 2004). At the moment, the potential of organelle genetic variations and cytonuclear combination in breeding is mostly restricted to the use of CMS in hybrid seed production. It is most likely that exploitation of genetic resources for crop improvement and breeding strategies will benefit from increased knowledge of the cytonuclear component in the adaptive response of plants to their environment.

ACKNOWLEDGEMENTS

The kind help of Professor Gregory G. Brown in improving the manuscript is gratefully acknowledged.

REFERENCES

Acosta, M. C., & Premoli, A. C. (2010). Evidence of chloroplast capture in South American *Nothofagus* (subgenus *Nothofagus*, Nothofagaceae). *Molecular Phylogenetics and Evolution, 54*, 235–242.

Alcázar, R., Pecinka, A., Aarts, M. G. M., Fransz, P. F., & Koornneef, M. (2012). Signals of speciation within *Arabidopsis thaliana* in comparison with its relatives. *Current Opinion in Plant Biology, 15*, 205–211.

Allainguillaume, J., Harwood, T., Ford, C. S., Cuccato, G., Norris, C., Allender, C. J., et al. (2009). Rapeseed cytoplasm gives advantage in wild relatives and complicates genetically modified crop biocontainment. *New Phytologist, 183*, 1201–1211.

Ashutosh Kumar, P., Dinesh Kumar, V., Sharma, P. C., Prakash, S., & Bhat, S. R. (2008). A novel orf108 co-transcribed with the atpA gene is associated with cytoplasmic male sterility in *Brassica juncea* carrying *Moricandia arvensis* cytoplasm. *Plant and Cell Physiology, 49*, 284–289.

Balloux, F., Handley, L.-J. L., Jombart, T., Liu, H., & Manica, A. (2009). Climate shaped the worldwide distribution of human mitochondrial DNA sequence variation. *Proceedings of the Royal Society B: Biological Sciences, 276*, 3447–3455.

Barr, C. M., & Fishman, L. (2010). The nuclear component of a cytonuclear hybrid incompatibility in *Mimulus* maps to a cluster of pentatricopeptide repeat genes. *Genetics, 184*, 455–465.

Barrientos, A., Müller, S., Dey, R., Wienberg, J., & Moraes, C. T. (2000). Cytochrome c oxidase assembly in primates is sensitive to small evolutionary variations in amino acid sequence. *Molecular Biology and Evolution, 17*, 1508–1519.

Belliard, G., Pelletier, G., Vedel, F., & Quétier, F. (1978). Morphological characteristics and chloroplast DNA distribution in different cytoplasmic parasexual hybrids of *Nicotiana tabacum*. *Molecular and General Genetics, 165*, 231–237.

Belliard, G., Vedel, F., & Pelletier, G. (1979). Mitochondrial recombination in cytoplasmic hybrids of *Nicotiana tabacum* by protoplast fusion. *Nature, 281*, 401–403.

Bentolila, S., Alfonso, A. A., & Hanson, M. R. (2002). A pentatricopeptide repeat-containing gene restores fertility to cytoplasmic male-sterile plants. *Proceedings of the National Academy of Sciences of the United States of America, 99*, 10887–10892.

Bergelson, J., & Roux, F. (2010). Towards identifying genes underlying ecologically relevant traits in *Arabidopsis thaliana*. *Nature Reviews Genetics, 11*, 867–879.

Bergelson, J., Kreitman, M., Stahl, E. A., & Tian, D. (2001). Evolutionary dynamics of plant R-genes. *Science, 292*, 2281–2285.

Birky, C. W. (2001). The inheritance of genes in mitochondria and chloroplasts: laws, mechanisms, and models. *Annual Review of Genetics, 35*, 125–148.

Blier, P. U., Dufresne, F., & Burton, R. S. (2001). Natural selection and the evolution of mtDNA-encoded peptides: evidence for intergenomic co-adaptation. *Trends in Genetics, 17*, 400–406.

Bogdanova, V. S. (2007). Inheritance of organelle DNA markers in a pea cross associated with nuclear-cytoplasmic incompatibility. *Theoretical and Applied Genetics, 114*, 333–339.

Bogdanova, V. S., & Berdnikov, V. (2001). Observation of a phenomenon resembling hybrid dysgenesis, in a wild pea subspecies *Pisum sativum* ssp. *elatius*. *Pisum Genetics, 33*, 5–8.

Bogdanova, V. S., & Kosterin, O. E. (2006). A case of anomalous chloroplast inheritance in crosses of garden pea involving an accession of wild subspecies. *Doklady Biological Sciences, 406*, 44–46.

Bogdanova, V. S., Galieva, E. R., & Kosterin, O. E. (2009). Genetic analysis of nuclear-cytoplasmic incompatibility in pea associated with cytoplasm of an accession of wild subspecies *Pisum sativum* subsp. *elatius* (Bieb.) Schmahl. *Theoretical Applied Genetics, 118*, 801–809.

Bogorad, L. (2008). Evolution of early eukaryotic cells: genomes, proteomes, and compartments. *Photosynthesis Research, 95*, 11–21.

Bomblies, K., & Weigel, D. (2007). Hybrid necrosis: autoimmunity as a potential gene-flow barrier in plant species. *Nature Reviews Genetics, 8*, 382–393.

Bonnett, H. T., Kofer, W., Håkansson, G., & Glimelius, K. (1991). Mitochondrial involvement in petal and stamen development studied by sexual and somatic hybridization of *Nicotiana* species. *Plant Science, 80*, 119–130.

Brown, G. G., Formanová, N., Jin, H., Wargachuk, R., Dendy, C., Patil, P., et al. (2003). The radish Rfo restorer gene of Ogura cytoplasmic male sterility encodes a protein with multiple pentatricopeptide repeats. *The Plant Journal, 35*, 262–272.

Budar, F. (1998). What can we learn about the interactions between the nuclear and mitochondrial genomes by studying cytoplasmic male sterilities? In I. M. Moller, P. Gardestrom, K. Glimelius, & E. Glaser (Eds.), *Plant Mitochondria: From Gene to Function* (pp. 49–55) Leiden: Backhyus Publishers.

Budar, F., Touzet, P., & De Paepe, R. (2003). The nucleo-mitochondrial conflict in cytoplasmic male sterilities revisited. *Genetica, 117*, 3–16.

Caemmerer, von, S., & Furbank, R. T. (2003). The C(4) pathway: an efficient CO(2) pump. *Photosynthesis Research, 77*, 191–207.

Chou, J.-Y., & Leu, J.-Y. (2010). Speciation through cytonuclear incompatibility: insights from yeast and implications for higher eukaryotes. *Bioessays, 32*, 401–411.

Chou, J.-Y., Hung, Y.-S., Lin, K.-H., Lee, H.-Y., & Leu, J.-Y. (2010). Multiple molecular mechanisms cause reproductive isolation between three yeast species. *PLoS Biology, 8*. e1000432.

Christin, P.-A., Salamin, N., Muasya, A. M., Roalson, E. H., Russier, F., & Besnard, G. (2008). Evolutionary switch and genetic convergence on rbcL following the evolution of C4 photosynthesis. *Molecular Biology and Evolution, 25*, 2361–2368.

Covello, P. S., & Gray, M. W. (1993). On the evolution of RNA editing. *Trends in Genetics, 9*, 265–268.

Davila, J. I., Arrieta-Montiel, M. P., Wamboldt, Y., Cao, J., Hagmann, J., Shedge, V., et al. (2011). Double-strand break repair processes drive evolution of the mitochondrial genome in *Arabidopsis*. *BMC Biology, 9*, 64.

Delannoy, E., Stanley, W., Bond, C., & Small, I. (2007). Pentatricopeptide repeat (PPR) proteins as sequence-specificity factors in post-transcriptional processes in organelles. *Biochemical Society Transactions, 35*, 1643–1647.

Desloire, S., Gherbi, H., Laloui, W., Marhadour, S., Clouet, V., Cattolico, L., et al. (2003). Identification of the fertility restoration locus, Rfo, in radish, as a member of the pentatricopeptide-repeat protein family. *EMBO Reports, 4*, 588–594.

Drouin, G., Daoud, H., & Xia, J. (2008). Relative rates of synonymous substitutions in the mitochondrial, chloroplast and nuclear genomes of seed plants. *Molecular Phylogenetics and Evolution, 49*, 827–831.

Dufaÿ, M., Touzet, P., Maurice, S., & Cuguen, J. (2007). Modelling the maintenance of male-fertile cytoplasm in a gynodioecious population. *Heredity, 99*, 349–356.

Duret, L., & Mouchiroud, D. (1999). Expression pattern and, surprisingly, gene length shape codon usage in *Caenorhabditis*, *Drosophila*, and *Arabidopsis*. *Proceedings of the National Academy of Sciences of the United States of America, 96*, 4482–4487.

Ellis, J., Dodds, P., & Pryor, T. (2000). Structure, function and evolution of plant disease resistance genes. *Current Opinion in Plant Biology, 3*, 278–284.

Filipovska, A., & Rackham, O. (2012). Modular recognition of nucleic acids by PUF, TALE and PPR proteins. *Molecular BioSystems, 8*, 699–708.

Fishman, L., & Willis, J. H. (2006). A cytonuclear incompatibility causes anther sterility in *Mimulus* hybrids. *Evolution, 60*, 1372–1381.

Freyer, R., Kiefer-Meyer, M. C., & Kössel, H. (1997). Occurrence of plastid RNA editing in all major lineages of land plants. *Proceedings of the National Academy of Sciences of the United States of America, 94*, 6285–6290.

Fujii, S., & Small, I. (2011). The evolution of RNA editing and pentatricopeptide repeat genes. *New Phytologist, 191*, 37–47.

Fujii, S., & Toriyama, K. (2008). Genome barriers between nuclei and mitochondria exemplified by cytoplasmic male sterility. *Plant and Cell Physiology, 49*, 1484–1494.

Fujii, S., Kazama, T., & Toriyama, K. (2008). Molecular studies on cytoplasmic male sterility-associated genes and restorer genes in rice. In H. Hirano, Y. Sano, A. Hirai, & T. Sasaki (Eds.), *Rice Biology in the Genomics Era* (pp. 205–215). Berlin, Heidelberg: Springer.

Fujii, S., Bond, C. S., & Small, I. D. (2011). Selection patterns on restorer-like genes reveal a conflict between nuclear and mitochondrial genomes throughout angiosperm evolution. *Proceedings of the National Academy of Sciences of the United States of America, 108*, 1723–1728.

Galloway, L. F., & Fenster, C. B. (1999). The effect of nuclear and cytoplasmic genes on fitness and local adaptation in an annual legume, *Chamaechrista fasciculata*. *Evolution, 53*, 1734–1743.

Galloway, L. F., & Fenster, C. B. (2001). Nuclear and cytoplasmic contributions to intraspecific divergence in an annual legume. *Evolution, 55*, 488–497.

Galun, E., Arzee-Gonen, P., Fluhr, R., Edelman, M., & Aviv, D. (1982). Cytoplasmic hybridization in *Nicotiana*: mitochondrial DNA analysis in progenies resulting from fusion between protoplasts having different organelle constitutions. *Molecular and General Genetics, 186*, 50–56.

Geddy, R., & Brown, G. G. (2007). Genes encoding pentatricopeptide repeat (PPR) proteins are not conserved in location in plant genomes and may be subject to diversifying selection. *BMC Genomics, 8*, 130.

Gillman, J. D., Bentolila, S., & Hanson, M. R. (2007). The petunia restorer of fertility protein is part of a large mitochondrial complex that interacts with transcripts of the CMS-associated locus. *The Plant Journal, 49*, 217–227.

Gould, S. B., Waller, R. F., & McFadden, G. I. (2008). Plastid evolution. *Annual Review of Plant Biology, 59*, 491–517.

Gray, M. W. (1989). Origin and evolution of mitochondrial DNA. *Annual Review of Cell Biology, 5*, 25–50.

Gray, M. W. (1996). RNA editing in plant organelles: a fertile field. *Proceedings of the National Academy of Sciences of the United States of America, 93*, 8157–8159.

Gray, M. W. (2009). RNA editing in plant mitochondria: 20 years later. *IUBMB Life, 61*, 1101–1104.

Gray, M. W., Burger, G., & Lang, B. F. (1999). Mitochondrial evolution. *Science, 283*, 1476–1481.

Gray, M. W., Burger, G., & Lang, B. F. (2001). The origin and early evolution of mitochondria. *Genome Biology, 2*. reviews1018.1–reviews1018.5.

Greiner, S., Rauwolf, U., Meurer, J., & Herrmann, R. G. (2011). The role of plastids in plant speciation. *Molecular Ecology, 20*, 671–691.

Hancock, A. M., Brachi, B., Faure, N., Horton, M. W., Jarymowycz, L. B., Sperone, F. G., et al. (2011). Adaptation to climate across the *Arabidopsis thaliana* genome. *Science, 334*, 83–86.

Hanson, M. R., & Bentolila, S. (2004). Interactions of mitochondrial and nuclear genes that affect male gametophyte development. *The Plant Cell, 16*(Suppl), S154–S169.

Harrison, J. S., & Burton, R. S. (2006). Tracing hybrid incompatibilities to single amino acid substitutions. *Molecular Biology and Evolution, 23*, 559–564.

Hayes, M. L., & Mulligan, R. M. (2011). Pentatricopeptide repeat proteins constrain genome evolution in chloroplasts. *Molecular Biology and Evolution, 28*, 2029–2039.

Häder, D.-P., & Sinha, R. P. (2005). Solar ultraviolet radiation-induced DNA damage in aquatic organisms: potential environmental impact. *Mutation Research, 571*, 221–233.

Hernandez Mora, J. R., Rivals, E., Mireau, H., & Budar, F. (2010). Sequence analysis of two alleles reveals that intra-and intergenic recombination played a role in the evolution of the radish fertility restorer (Rfo). *BMC Plant Biology, 10*, 35.

Hölzle, A., Jonietz, C., Törjek, O., Altmann, T., Binder, S., & Forner, J. (2011). A restorer of fertility-like PPR gene is required for 5′-end processing of the nad4 mRNA in mitochondria of *Arabidopsis thaliana. The Plant Journal, 65*, 737–744.

Huang, S., Taylor, N. L., Narsai, R., Eubel, H., Whelan, J., & Millar, A. H. (2008). Experimental analysis of the rice mitochondrial proteome, its biogenesis, and heterogeneity. *Plant Physiology, 149*, 719–734.

Iida, S., Miyagi, A., Aoki, S., Ito, M., Kadono, Y., & Kosuge, K. (2009). Molecular adaptation of rbcL in the heterophyllous aquatic plant Potamogeton. *PLoS ONE, 4*. e4633.

Jobson, R., & Qiu, Y. (2008). Did RNA editing in plant organellar genomes originate under natural selection or through genetic drift? *Biology Direct, 3*, 43.

Jonietz, C., Forner, J., Hölzle, A., Thuss, S., & Binder, S. (2010). RNA processing factor2 is required for 5′ end processing of nad9 and cox3 mRNAs in mitochondria of *Arabidopsis thaliana. The Plant Cell, 22*, 443–453.

Jonietz, C., Forner, J., Hildebrandt, T., & Binder, S. (2011). RNA processing factor 3 is crucial for the accumulation of mature ccmC transcripts in mitochondria of *Arabidopsis thaliana* accession Columbia. *Plant Physiology, 157*, 1430–1439.

Jordan, D. R., Klein, R. R., Sakrewski, K. G., Henzell, R. G., Klein, P. E., & Mace, E. S. (2011). Mapping and characterization of Rf (5): a new gene conditioning pollen fertility restoration in A (1) and A (2) cytoplasm in sorghum (*Sorghum bicolor* (L.) Moench). *Theoretical and Applied Genetics, 123*, 383–396.

Kapralov, M. V., & Filatov, D. A. (2006). Molecular adaptation during adaptive radiation in the Hawaiian endemic genus *Schiedea. PLoS ONE, 1*, e8.

Kapralov, M. V., & Filatov, D. A. (2007). Widespread positive selection in the photosynthetic Rubisco enzyme. *BMC Evolutionary Biology,* 7, 73.

Kapralov, M. V., Kubien, D. S., Andersson, I., & Filatov, D. A. (2011). Changes in Rubisco kinetics during the evolution of C4 photosynthesis in *Flaveria* (Asteraceae) are associated with positive selection on genes encoding the enzyme. *Molecular Biology and Evolution, 28*, 1491–1503.

Kato, H., Tezuka, K., Feng, Y. Y., Kawamoto, T., Takahashi, H., Mori, K., et al. (2007). Structural diversity and evolution of the Rf-1 locus in the genus *Oryza. Heredity, 99*, 516–524.

Kazama, T., & Toriyama, K. (2003). A pentatricopeptide repeat-containing gene that promotes the processing of aberrant atp6 RNA of cytoplasmic male-sterile rice. *FEBS Letters, 544*, 99–102.

Kazama, T., Nakamura, T., Watanabe, M., Sugita, M., & Toriyama, K. (2008). Suppression mechanism of mitochondrial ORF79 accumulation by Rf1 protein in BT-type cytoplasmic male sterile rice. *The Plant Journal, 55*, 619–628.

Kobayashi, K., Kawabata, M., Hisano, K., Kazama, T., Matsuoka, K., Sugita, M., et al. (2011). Identification and characterization of the RNA binding surface of the pentatricopeptide repeat protein. *Nucleic Acids Research*. http://dx.doi.org/10.1093/nar/gkr1084.

Koizuka, N., Imai, R., Fujimoto, H., Hayakawa, T., Kimura, Y., Kohno-Murase, J., et al. (2003). Genetic characterization of a pentatricopeptide repeat protein gene, orf687, that restores fertility in the cytoplasmic male-sterile Kosena radish. *The Plant Journal, 34*, 407–415.

Komori, T., Ohta, S., Murai, N., Takakura, Y., Kuraya, Y., Suzuki, S., et al. (2004). Map-based cloning of a fertility restorer gene, Rf-1, in rice (*Oryza sativa* L.). *The Plant Journal, 37*, 315–325.

Kubo, T., Kitazaki, K., Matsunaga, M., Kagami, H., & Mikami, T. (2011). Male sterility-inducing mitochondrial genomes: how do they differ? *Critical Reviews in Plant Sciences, 30*, 378–400.

Landgren, M., Zetterstrand, M., Sundberg, E., & Glimelius, K. (1996). Alloplasmic male-sterile *Brassica* lines containing *B. tournefortii* mitochondria express an ORF 3′ of the atp6 gene and a 32 kDa protein. *Plant Molecular Biology, 32*, 879–890.

Laporte, V., Cuguen, J., & Couvet, D. (2000). Effective population sizes for cytoplasmic and nuclear genes in a gynodioecious species. The role of the sex determination system. *Genetics, 154*, 447–458.

Lee, H.-Y., Chou, J.-Y., Cheong, L., Chang, N.-H., Yang, S.-Y., & Leu, J.-Y. (2008). Incompatibility of nuclear and mitochondrial genomes causes hybrid sterility between two yeast species. *Cell, 135*, 1065–1073.

Leinonen, P. H., Remington, D. L., & Savolainen, O. (2011). Local adaptation, phenotypic differentiation, and hybrid fitness in diverged natural populations of *Arabidopsis lyrata. Evolution, 65*, 90–107.

Leister, D., & Kleine, T. (2008). Towards a comprehensive catalog of chloroplast proteins and their interactions. *Cell Research, 18*, 1081–1083.

Leppälä, J., & Savolainen, O. (2011). Nuclear-cytoplasmic interactions reduce male fertility in hybrids of *Arabidopsis lyrata* subspecies. *Evolution, 65*, 2959–2972.

Levin, D. (2003). The cytoplasmic factor in plant speciation. *Systematic Botany, 28*, 5–11.

Linke, B., & Börner, T. (2005). Mitochondrial effects on flower and pollen development. *Mitochondrion, 5*, 389–402.

López-Garćia, P., & Moreira, D. (1999). Metabolic symbiosis at the origin of eukaryotes. *Trends in Biochemical Sciences, 24*, 88–93.

Lurin, C., Andrés, C., Aubourg, S., Bellaoui, M., Bitton, F., Bruyère, C., et al. (2004). Genome-wide analysis of *Arabidopsis* pentatricopeptide repeat proteins reveals their essential role in organelle biogenesis. *The Plant Cell, 16*, 2089–2103.

Maier, R. M., Zeltz, P., Kössel, H., Bonnard, G., Gualberto, J. M., & Grienenberger, J. M. (1996). RNA editing in plant mitochondria and chloroplasts. *Plant Molecular Biology, 32*, 343–365.

Malek, O., Lättig, K., Hiesel, R., Brennicke, A., & Knoop, V. (1996). RNA editing in bryophytes and a molecular phylogeny of land plants. *EMBO Journal, 15*, 1403–1411.

Mardulyn, P., Othmezouri, N., Mikhailov, Y. E., & Pasteels, J. M. (2011). Conflicting mitochondrial and nuclear phylogeographic signals and evolution of host-plant shifts in the boreo-montane leaf beetle *Chrysomela lapponica. Molecular Phylogenetics and Evolution, 61*, 686–696.

Margulis, L. (2004). Serial endosymbiotic theory (SET) and composite individuality. Transition from bacterial to eukaryotic genomes. *Microbiology Today, 31*, 172–174.

McFadden, G. I. (2001). Chloroplast origin and integration. *Plant Physiology, 125*, 50–53.

Mckay, J. K., Richards, J. H., Nemali, K. S., Sen, S., Mitchell-Olds, T., Boles, S., et al. (2008). Genetics of drought adaptation in *Arabidopsis thaliana* II. QTL analysis of a new mapping population, KAS-1 x TSU-1. *Evolution, 62*, 3014–3026.

Meiklejohn, C. D., Montooth, K. L., & Rand, D. M. (2007). Positive and negative selection on the mitochondrial genome. *Trends in Genetics, 23*, 259–263.

Millar, A. H., Heazlewood, J. L., Kristensen, B. K., Braun, H.-P., & Møller, I. M. (2005). The plant mitochondrial proteome. *Trends in Plant Science, 10*, 36–43.

Mishmar, D., Ruiz-Pesini, E., Golik, P., Macaulay, V., Clark, A. G., Hosseini, S., et al. (2003). Natural selection shaped regional mtDNA variation in humans. *Proceedings of the National Academy of Sciences of the United States of America, 100*, 171–176.

Miwa, H., Odrzykoski, I. J., Matsui, A., Hasegawa, M., Akiyama, H., Jia, Y., et al. (2009). Adaptive evolution of rbcL in *Conocephalum* (Hepaticae, bryophytes). *Gene, 441*, 169–175.

Mogensen, H. L. (1996). The hows and whys of cytoplasmic inheritance in seed plants. *American Journal of Botany, 83*, 383–404.

Moison, M., Roux, F., Quadrado, M., Duval, R., Ekovich, M., Lê, D.-H., et al. (2010). Cytoplasmic phylogeny and evidence of cyto-nuclear co-adaptation in *Arabidopsis thaliana*. *The Plant Journal, 63*, 728–738.

O'Toole, N., Hattori, M., Andres, C., Iida, K., Lurin, C., Schmitz-Linneweber, C., et al. (2008). On the expansion of the pentatricopeptide repeat gene family in plants. *Molecular Biology and Evolution, 25*, 1120–1128.

Okuda, K., Nakamura, T., Sugita, M., Shimizu, T., & Shikanai, T. (2006). A pentatricopeptide repeat protein is a site recognition factor in chloroplast RNA editing. *Journal of Biological Chemistry, 281*, 37661–37667.

Okuda, K., Myouga, F., Motohashi, R., Shinozaki, K., & Shikanai, T. (2007). Conserved domain structure of pentatricopeptide repeat proteins involved in chloroplast RNA editing. *Proceedings of the National Academy of Sciences of the United States of America, 104*, 8178–8183.

Palmer, J. D., & Herbon, L. A. (1988). Plant mitochondrial DNA evolves rapidly in structure, but slowly in sequence. *Journal of Molecular Evolution, 28*, 87–97.

Parkinson, C. L., Mower, J. P., Qiu, Y.-L., Shirk, A. J., Song, K., Young, N. D., et al. (2005). Multiple major increases and decreases in mitochondrial substitution rates in the plant family Geraniaceae. *BMC Evolutionary Biology, 5*, 73.

Prikryl, J., Rojas, M., Schuster, G., & Barkan, A. (2011). Mechanism of RNA stabilization and translational activation by a pentatricopeptide repeat protein. *Proceedings of the National Academy of Sciences of the United States of America, 108*, 415–420.

Rand, D. M., Haney, R. A., & Fry, A. J. (2004). Cytonuclear coevolution: the genomics of cooperation. *Trends in Ecology & Evolution, 19*, 645–653.

Reyes-Prieto, A., Weber, A. P. M., & Bhattacharya, D. (2007). The origin and establishment of the plastid in algae and plants. *Annual Review of Genetics, 41*, 147–168.

Rivals, E., Bruyère, C., Toffano-Nioche, C., & Lecharny, A. (2006). Formation of the *Arabidopsis* pentatricopeptide repeat family. *Plant Physiology, 141*, 825–839.

Rozema, J., Björn, L. O., Bornman, J. F., Gaberscik, A., Häder, D.-P., Trost, T., et al. (2002). The role of UV-B radiation in aquatic and terrestrial ecosystems–an experimental and functional analysis of the evolution of UV-absorbing compounds. *Journal of Photochemistry and Photobiology B: Biology, 66*, 2–12.

Sackton, T. B., Haney, R. A., & Rand, D. M. (2003). Cytonuclear coadaptation in *Drosophila*: disruption of cytochrome c oxidase activity in backcross genotypes. *Evolution, 57*, 2315–2325.

Saha, D., Prasad, A. M., & Srinivasan, R. (2007). Pentatricopeptide repeat proteins and their emerging roles in plants. *Plant Physiology and Biochemistry, 45*, 521–534.

Sambatti, J. B. M., Ortiz-Barrientos, D., Baack, E. J., & Rieseberg, L. H. (2008). Ecological selection maintains cytonuclear incompatibilities in hybridizing sunflowers. *Ecology Letters, 11*, 1082–1091.

Savir, Y., Noor, E., Milo, R., & Tlusty, T. (2010). Cross-species analysis traces adaptation of Rubisco toward optimality in a low-dimensional landscape. *Proceedings of the National Academy of Sciences of the United States of America, 107*, 3475–3480.

Schmitz-Linneweber Small, C. I., & Small, I. (2008). Pentatricopeptide repeat proteins: a socket set for organelle gene expression. *Trends in Plant Science, 13*, 663–670.

Schmitz-Linneweber, C., Kushnir, S., Babiychuk, E., Poltnigg, P., Herrmann, R. G., & Maier, R. M. (2005). Pigment deficiency in nightshade/tobacco cybrids is caused by the failure to edit the plastid ATPase alpha-subunit mRNA. *The Plant Cell, 17*, 1815–1828.

Schnable, P., & Wise, R. (1998). The molecular basis of cytoplasmic male sterility and fertility restoration. *Trends in Plant Science, 3*, 175–180.

Searcy, D. G. (2003). Metabolic integration during the evolutionary origin of mitochondria. *Cell Research, 13*, 229–238.

Shikanai, T. (2006). RNA editing in plant organelles: machinery, physiological function and evolution. *Cellular and Molecular Life Sciences, 63*, 698–708.

Shinada, T., Kikuchi, Y., Fujimoto, R., & Kishitani, S. (2006). An alloplasmic male-sterile line of *Brassica oleracea* harboring the mitochondria from *Diplotaxis muralis* expresses a novel chimeric open reading frame, orf72. *Plant and Cell Physiology, 47*, 549–553.

Singer, C. E., & Ames, B. N. (1970). Sunlight ultraviolet and bacterial DNA base ratios. *Science, 170*, 822–825.

Sloan, D. B., Alverson, A. J., Chuckalovcak, J. P., Wu, M., McCauley, D. E., Palmer, J. D., et al. (2012). Rapid evolution of enormous, multichromosomal genomes in flowering plant mitochondria with exceptionally high mutation rates. *PLoS Biology, 10*. e1001241.

Small, I. D., & Peeters, N. (2000). The PPR motif - a TPR-related motif prevalent in plant organellar proteins. *Trends in Biochemical Sciences, 25*, 46–47.

Smith, D. R. (2009). Unparalleled GC content in the plastid DNA of *Selaginella*. *Plant Molecular Biology, 71*, 627–639.

Sueoka, N. (1988). Directional mutation pressure and neutral molecular evolution. *Proceedings of the National Academy of Sciences of the United States of America, 85*, 2653–2657.

Takahashi, A., & Ohnishi, T. (2004). The significance of the study about the biological effects of solar ultraviolet radiation using the exposed facility on the international space station. *Biological Sciences in Space, 18*, 255–260.

Takenaka, M., Verbitskiy, D., Van Der Merwe, J. A., Zehrmann, A., & Brennicke, A. (2008). The process of RNA editing in plant mitochondria. *Mitochondrion, 8*, 35–46.

Tao, D., Hu, F., Yang, J., Yang, G., Yang, Y., Xu, P., et al. (2004). Cytoplasm and cytoplasm-nucleus interactions affect agronomic traits in japonica rice. *Euphytica, 135*, 129–134.

Thompson, K. (1972). Cytoplasmic male-sterility in oil-seed rape. *Heredity, 29*, 253–257.

Tillich, M., Lehwark, P., Morton, B. R., & Maier, U. G. (2006). The evolution of chloroplast RNA editing. *Molecular Biology and Evolution, 23*, 1912–1921.

Tillich, M., Sy, V. L., Schulerowitz, K., Haeseler, von, A., Maier, U. G., & Schmitz-Linneweber, C. (2009). Loss of matK RNA editing in seed plant chloroplasts. *BMC Evolutionary Biology, 9*, 201.

Touzet, P., & Budar, F. (2004). Unveiling the molecular arms race between two conflicting genomes in cytoplasmic male sterility? *Trends in Plant Science, 9*, 568–570.

Touzet, P., & Delph, L. F. (2009). The effect of breeding system on polymorphism in mitochondrial genes of Silene. *Genetics, 181*, 631–644.

Törjék, O., Witucka-Wall, H., Meyer, R. C., Korff, von, M., Kusterer, B., Rautengarten, C., et al. (2006). Segregation distortion in *Arabidopsis* C24/Col-0 and Col-0/C24 recombinant inbred line populations is due to reduced fertility caused by epistatic interaction of two loci. *Theoretical and Applied Genetics, 113*, 1551–1561.

Turelli, M., & Moyle, L. C. (2007). Asymmetric postmating isolation: Darwin's corollary to Haldane's rule. *Genetics, 176*, 1059–1088.

Uyttewaal, M., Arnal, N., Quadrado, M., Martin-Canadell, A., Vrielynck, N., Hiard, S., et al. (2008). Characterization of *Raphanus sativus* pentatricopeptide repeat proteins encoded by the fertility restorer locus for Ogura cytoplasmic male sterility. *The Plant Cell, 20*, 3331–3345.

Wallace, D. C. (2010). The epigenome and the mitochondrion: bioenergetics and the environment. *Genes & Development, 24*, 1571–1573.

Wang, Z., Zou, Y., Li, X., Zhang, Q., Chen, L., Wu, H., et al. (2006). Cytoplasmic male sterility of rice with boro II cytoplasm is caused by a cytotoxic peptide and is restored by two related PPR motif genes via distinct modes of mRNA silencing. *The Plant Cell, 18*, 676–687.

Willett, C. S. (2011). The nature of interactions that contribute to postzygotic reproductive isolation in hybrid copepods. *Genetica, 139*, 575–588.

Willett, C. S., & Burton, R. S. (2003). Environmental influences on epistatic interactions: viabilities of cytochrome c genotypes in interpopulation crosses. *Evolution, 57*, 2286–2292.

Wolters, A. M., Koornneef, M., & Gilissen, L. J. (1993). The chloroplast and mitochondrial DNA type are correlated with the nuclear composition of somatic hybrid calli of *Solanum tuberosum* and *Nicotiana plumbaginifolia*. *Current Genetics, 24*, 260–267.

Wu, W., Schmidt, T. R., Goodman, M., & Grossman, L. I. (2000). Molecular evolution of cytochrome c oxidase subunit I in primates: is there coevolution between mitochondrial and nuclear genomes? *Molecular Phylogenetics and Evolution, 17*, 294–304.

Young, J. N., Rickaby, R. E. M., Kapralov, M. V., & Filatov, D. A. (2012). Adaptive signals in algal Rubisco reveal a history of ancient atmospheric carbon dioxide. *Philosophical Transactions of the Royal Society B: Biological Sciences, 367*, 483–492.

Yu, Q.-B., Li, G., Wang, G., Sun, J.-C., Wang, P.-C., Wang, C., et al. (2008). Construction of a chloroplast protein interaction network and functional mining of photosynthetic proteins in *Arabidopsis thaliana*. *Cell Research, 18*, 1007–1019.

Yura, K., Sulaiman, S., Hatta, Y., Shionyu, M., & Go, M. (2009). RESOPS: a database for analyzing the correspondence of RNA editing sites to protein three-dimensional structures. *Plant and Cell Physiology, 50*, 1865–1873.

Zhelyazkova, P., Hammani, K., Rojas, M., Voelker, R., Vargas-Suárez, M., Börner, T., et al. (2011). Protein-mediated protection as the predominant mechanism for defining processed mRNA termini in land plant chloroplasts. *Nucleic Acids Research*. http://dx.doi.org/10.1093/nar/gkr1137.

Zubko, K., Zubko, I., & Gleba, Y. (2002). Self-fertile cybrids *Nicotiana tabacum* (+ *Hyoscyamus aureus*) with a nucleo-plastome incompatibility. *Theoretical and Applied Genetics, 105*, 822–828.

CHAPTER SIX

Mitochondrial Genomes of Photosynthetic Euglenids and Alveolates

Pavel Flegontov and Julius Lukeš[1]
Biology Centre, Institute of Parasitology, Czech Academy of Sciences and Faculty of Science, University of South Bohemia, České Budějovice (Budweis), Czech Republic
[1]Corresponding author. E-mail: jula@paru.cas.cz

Contents

Abstract

Euglenida belong to the eukaryotic supergroup Excavata, the members of which possess the most varied mitochondrial genomes in terms of their structure and gene content. Heterotrophic protists represent the majority of Excavata, as only the Euglenida contain a green plastid, apparently acquired by secondary endosymbiosis. The sister group of Euglenida, the mostly parasitic Kinetoplastida, have an extremely complex mitochondrial DNA (kinetoplast DNA), which is usually composed of thousands of mutually interlocked DNA circles. Most mRNAs encoded by this genome are rendered translatable only after they undergo intricate editing via insertions and/or deletions of uridines. The mitochondrial DNA of the other sister group, Diplonemida, is unique as its transcripts must be massively *trans*-spliced before translation. None of these complex mechanisms has so far been found in the mitochondrial genome and transcriptome of *Euglena gracilis*, the best studied member of Euglenida. Its mitochondrial DNA exists in the form of numerous differently sized linear fragments. Their

Advances in Botanical Research, Volume 63
ISSN 0065-2296,
http://dx.doi.org/10.1016/B978-0-12-394279-1.00006-5

significant fraction is non-coding and full of various repeats, which intersperse fragments of a handful of protein-coding genes. Mostly photosynthetic dinoflagellates and parasitic apicomplexans with a relic plastid constitute a large and diverse group within alveolates. All species of this group share the most reduced mitochondrial genome found, containing just three, and in some cases probably two, protein-coding genes along with highly fragmented rRNA genes, and no tRNA genes. Mitochondrial genomes of dinoflagellates and those of smaller groups within the apicomplexa–dinoflagellata assemblage, perkinsids and chromerids, in all cases have a recombining, highly scrambled sequence, and frequently demonstrate other non-canonical features in structure and expression: fused genes, extensive RNA editing, *trans*-splicing, 5′ oligoU caps, loss of start and stop codons, extensive translational frameshifting. Some of these oddities apparently appeared in several groups independently, probably due to relaxed selective constraints in tiny organellar genomes.

1. MITOCHONDRIAL GENOMES OF EUGLENIDS

1.1. Phylogeny of Euglenida

The Euglenida, a group of protists, have been intensely studied throughout most of the twentieth century. This interest was stimulated by their apparent ecological significance, as well as the ease with which they can be cultivated in a simple and cheap medium. In the era predating molecular biology, the species *Euglena gracilis* was the subject of numerous physiological studies. After the advent of molecular biology, it served as a model protist, because the genome of its green plastid was the second plastid genome to be completely sequenced (Hallick *et al.*, 1993). However, surprisingly little has been known about its mitochondrial and nuclear genomes.

It has been well established that the Euglenida belong to the superkingdom Excavata (Fig. 6.1), which arguably represents the earliest branch of the eukaryotic tree (Cavalier-Smith, 2010). Within this morphologically and genetically extremely diverse group of single-celled eukaryotes, Euglenida, along with their two sister groups Diplonemida and Kinetoplastida, constitute the phylum Euglenozoa. The following morphological features unite these free-living, commensalic and parasitic flagellates: (1) with very few exceptions of aflagellar stages, all cells carry at least a single flagellum equipped with a prominent structure called a paraflagellar rod; (2) a morphologically pronounced flagellar pocket; and (3) a single, usually reticulated, mitochondrion with tubular cristae (Adl *et al.*, 2005). Moreover, the most prominent common molecular features include polycistronic transcription, massive *trans*-splicing and, with very few exceptions, the absence of introns (Lukeš, Hashimi, & Zíková, 2005; Lukeš, Leander, &

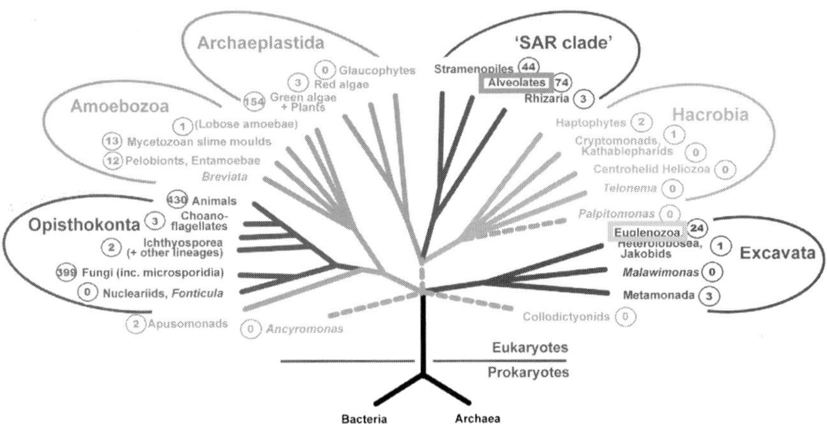

Figure 6.1 Current view of eukaryotic diversity; both protist groups dealt within this chapter are highlighted in colour. The tree is based on Roger and Simpson (2009). The numbers of whole-genome sequencing projects for a given group are shown in circles. See the colour plate.

Keeling, 2009). Although kinetoplastids and diplonemids have a single, often reticulated, mitochondrion in their cells, the situation is slightly more complex in euglenids. Although most species also carry one, usually large, mitochondrion, *Peranema* and likely some other species contain multiple small organelles (Hall, 2005; Roy, Faktorová, Lukeš, & Burger, 2007).

A unique and important feature of euglenids is their acquisition of a green plastid via secondary endosymbiosis (Archibald & Keeling, 2002). Despite earlier claims (Hannaert *et al.*, 2003), the kinetoplastids did not seem to have harboured a plastid in the course of their evolutionary history (Leander, 2004; Simpson, Stevens, & Lukeš J, 2006). The numerous fully sequenced genomes available for the kinetoplastid flagellates belonging to the genera *Trypanosoma* and *Leishmania* (El-Sayed *et al.*, 2005) strongly support this scenario, although these mostly parasitic protists seem to have acquired a handful of plastid-derived genes by horizontal gene transfer (Týč, Long, Jirků,& Lukeš, 2010). No complete genome is so far available for Diplonemida to convincingly address this important question for this least known group, although the most studied species, *Diplonema papillatum*, is currently the subject of a whole-genome initiative (Č. Vlček and G. Burger, personal communication). Current phylogenetic analyses based on numerous nuclear-encoded genes identify diplonemids as the most closely related group to euglenids (Simpson, Gill, Callahan, Litaker, & Roget, 2004), a relationship that is certainly not supported by the structure of their mitochondrial genomes (see below).

1.2. Mitochondrial Genomes of Euglenids

Although the mitochondrial DNA of euglenids is poorly known and does not seem to be very complex (Spencer & Gray, 2011), this certainly does not apply to its sister groups. The Kinetoplastida belong to organisms with the best studied mitochondrial DNA, also termed kinetoplast (k) DNA. The extremely complex kDNA of *Trypanosoma brucei* is composed of well-described maxicircles and minicircles interconnected into a single network, as well as of hundreds of proteins responsible for the maintenance and replication of this network (Lukeš *et al.*, 2005; Stuart, Schnaufer, Ernst, & Panigrahi, 2005). A similar situation holds for the mitochondrial proteome of this causative agent of African sleeping sickness, as up to 1000 proteins have been identified by numerous methods in its single mitochondrion (Panigrahi *et al.*, 2009).

The initial studies on *E. gracilis* on what was likely mitochondrial DNA were performed by Ray and Hanawalt (1965), and further characterization of this DNA occurred in the 1970s. Several authors have shown that the mitochondrial DNA of *E. gracilis* is represented by heterogeneously sized linear molecules (Nass, Schori, Ben-Shaul, & Edelman, 1974), which, however, may have high genomic complexity. Based on hybridization experiments, it was estimated that this complexity may reach up to 70 kb (Crouse, Vandrey, & Stutz, 1974; Talen, Sanders, & Flavell, 1974), at that time a very high level for an organellar genome. Although it was of general interest to learn more about this genome, especially for comparison with the intensely studied mitochondrial genomes of kinetoplastid parasites, a hiatus in any progress in this field lasted for more than two decades.

It was only in 1997 that a more detailed analysis of this long-ignored genome was initiated. First, electron microscopy and hybridization experiments confirmed both the presence of heterogeneous linear molecules, with the peak reaching size around 5 kb, and the overall complexity of this organellar genome of ~70 kb (Yasuhira & Simpson, 1997). The same authors were unable to resolve mitochondrial DNA molecules under the conditions of a pulse-field gel, even after γ-irradiation. This applied only to the mitochondrial DNA, as RNA from the mitochondrion was easily recovered, with four prominent bands likely representing ribosomal (r) RNA (Yasuhira & Simpson, 1997). Another approach to obtaining mitochondrial DNA and RNA from *E. gracilis* was recently adopted by Spencer and Gray (2011), who purified intact mitochondrial vesicles from lyzed cells.

The AT rich mitochondrial DNA from this preparation was comprised of differently sized linear molecules with peaks around 4.0 and 7.5 kb. Several laboratories thus confirmed the composition of *E. gracilis* mitochondrial DNA from variously sized linear molecules, yet these results remain to be reconciled with the fact that this mitochondrial DNA cannot enter the gel under pulse-field conditions. Unusual conformations have been implied, such as two- or three-dimensional networks of DNA molecules or unusually tight associations with proteins (Yasuhira & Simpson, 1997) or even branched molecules (Spencer & Gray, 2011), yet no data are so far available to support any of these claims.

The laboratories of Larry Simpson and Jean-Michel Grienenberger both established the first sequence information from the mitochondrial genome of *E. gracilis*. In both cases, part of the open reading frame (ORF) of cytochrome *c* oxidase subunit 1 (*cox1*) was obtained, revealing several unexpected features (Yasuhira & Simpson, 1997; Tessier, van der Speck, Gualberto, & Grienenberger, 1997). Since it was difficult to clone mitochondrial DNA fragments, *cox1* sequences were obtained using a 5′ and 3′ RACE (rapid amplification of cDNA ends) protocol. Invariably, only fragments of the *cox1* ORF were retrieved, mostly representing the 5′ or 3′ part of this highly conserved gene. A full-size gene was never obtained (Yasuhira & Simpson, 1997).

This lack of a full-size ORF initially implied the existence of a process similar to RNA editing of kinetoplastid flagellates, due to which parts of the mitochondrial genetic information are encrypted at the DNA level. Since the extensive insertions and/or deletions of uridines in the kDNA transcripts are specified by small abundant and heterogeneous molecules called guide RNAs (Blum, Bakalara, & Simpson, 1990), a search for homologous RNA species was also performed in the *E. gracilis* organelle. Guide RNAs can be tracked down relatively easily due to their ability to be capped *in vitro* by the activity of guanylyltransferase and radioactively labelled GTP (Blum *et al.*, 1990). However, a thorough search proved that these specialized uridine-tailed RNA molecules are not present either in *E. gracilis* (Yasuhira & Simpson, 1997) or in *D. papillatum* (D.A. Maslov, personal communication). Therefore, sufficient evidence is now available to conclude that the uridine insertion/deletion type of RNA editing, pervasive in the sister group of euglenids, the Kinetoplastida, is apparently absent from the *E. gracilis* mitochondrion. This lack of editing in euglenids was used to support a scenario whereby this enigmatic and complex process emerged relatively late in evolution. Furthermore, trypanosomes and leishmanias are known to use the non-canonical TGA triplet to encode tryptophan in their mitochondria, which is a substantial departure from the universal genetic code, where this triplet is used as one of three stop codons. Throughout the *E. gracilis cox1* gene, only the TGG triplet is invariably used to specify tryptophan (Yasuhira & Simpson, 1997; Tessier *et al.*, 1997). More recent deposition into GenBank of two protein-coding gene sequences from the *E. gracilis* mitochondrial genome, namely cytochrome *c* oxidase subunit 2 (cox2) and NADH dehydrogenase subunit 6 (nad6) (unpublished data), further supports the notion that RNA editing mechanistically similar to the process known from kinetoplastids is most likely lacking in the euglenids.

In combination with the apparent absence of guide RNAs, these data indicate that the structure and transcription of the mitochondrial DNA of euglenids may significantly differ from that of the kinetoplastids. However, in two euglenid species, *Petalomonas cantuscigni* and *P. mediocanellata*, an electron-dense mitochondrial inclusion body was observed that was reminiscent of the kDNA disk of kinetoplastids (Leander, Triemer, & Farmer MA, 2001), although a similar structure was not observed in the mitochondria of other euglenids, such as *Peranema trichophorum* and *Entosiphon sulcatum* (Roy *et al.*, 2007). Moreover, using 4′,6-diamidino-2-phenylindole (DAPI) stain, Roy *et al.* (2007) showed that the inclusion body in *P. cantuscygni* does not contain DNA, so its proposed homology with the kDNA disk (Leander *et al.*, 2001) is unlikely. The abundant mitochondrial DNA of *P. cantuscygni* is distributed throughout the organelle in a network-like pattern, reminiscent of the related diplonemids (see below). The other two euglenids investigated, *Peranema* and *Entosiphon*, contain a significantly smaller amount of DNA in their mitochondria, which seems to be scattered in the form of multiple fluorescent spheres or agglomerates (Roy *et al.*, 2007), similar to structures reported earlier in *E. gracilis* (Hayashi and Ueda, 1989; Hayashi-Isimaru, Ueda, & Nonaka, 1993).

Using buoyant density CsCl–bisbenzimide gradient centrifugation, attempts were made to separate the mitochondrial and nuclear DNA of these three euglenids. Inspection of the AT rich fraction of *P. cantuscygni* DNA by electron microscopy revealed the presence of linear molecules, as well as small and large circles. It was proposed that the various molecules observed could be produced by replication via the rolling circle mechanism. Sequencing of the mitochondrial DNA-enriched fraction had so far detected only a fragment of subunit 6 of the ATPase gene (Roy *et al.*, 2007).

Preliminary small-scale sequence analyses of mitochondrial fractions of *Peranema* and *Entosiphon* failed to identify any protein-coding regions. However, this comparative analysis showed that multi-chromosome mitochondrial genomes are likely widespread in the euglenid flagellates (Roy *et al.*, 2007).

Sequencing of several linear mitochondrial chromosomes of *E. gracilis* revealed a common interesting feature. Regardless of whether the sequenced molecule contained fragments of the small subunit mitoribosomal RNAs (SSU rRNA), *cox1*, *cox2*, or *cox3* coding regions, they were flanked on both the 5′ and 3′ ends by highly conserved repeats, which have been proposed to play a role in replication, recombination and/or transcription (Spencer & Gray, 2011). Small subunit (SSU) and large subunit (LSU) rRNA genes are bipartide, and all four fragments can be capped by the action of guanylyltransferase. Therefore, they are very likely independently transcribed. This situation has been used as an argument to support a scenario of rRNA evolution, according to which ancestral ribosomes rRNA fragments were held together by inter- and intramolecular interactions (Boer & Gray, 1988).

Gene fragments constituting the mitochondrial genome of *E. gracilis* along with respective full-size copies led Spencer and Gray (2011) to postulate a model to explain the emergence of guide RNAs, which provide information for editing of the mitochondrial transcripts in related kinetoplastid flagellates. Initially, by illegitimate recombination via the flanking conserved repeats, a gene fragment becomes located on a small circle. If this circular DNA happens to contain an origin of replication and a promoter, which is opposite to the gene fragment, a small antisense RNA complementary to the fragment of the parental gene will be produced. It is such small RNAs that, in collaboration with several intricate protein complexes, execute the exact insertions/deletions of uridine residues into mitochondrial mRNAs of all kinetoplastids studied so far (Lukeš *et al.*, 2005). Despite its speculative nature, this model explains for the first time the emergence of the unique guide RNA-directed editing machinery and also the complex kDNA network, earlier dubbed an evolutionary improbable structure (Lukeš *et al.*, 2002).

1.3. Mitochondrial Genomes of Diplonemids

The uniquely complex nature of the mitochondrial genomes of euglenids and kinetoplastids, supporting the anything goes postulation for mitochondrial genomes (Burger, Gray, & Lang, 2003; Gray, Lang, & Burger G, 2004), is surpassed by what has been uncovered so far in the mitochondrion of the third group: the non-photosynthetic diplonemids. In the mitochondrial genome of *D. papillatum*, a model species representing this least common, usually commensalic group, genes are invariably fragmented. Each gene fragment, called a module, is individually located on a circular chromosome belonging to one of two types labelled A and B (Marande, Lukeš, & Burger G, 2005). Thousands of these minicircles of conserved structure are freely dispersed throughout the mitochondrial lumen (Marande *et al.*, 2005). Separate non-overlapping precursor RNAs are assembled into a mature transcript via extensive *trans*-splicing (Marande & Burger, 2007), the mechanism of which remains unknown. However, it is obvious that the mechanism must be highly complex, because, for example, in the case of *cox1*, it is able to *trans*-splice together, in an orderly manner, nine separately transcribed fragments. To complicate matters even further, at least in one case, six uridines are inserted between two modules, implying that splicing and editing machineries may exist next to each other (Marande & Burger, 2007). This en-block insertion of uridines is highly conserved among diplonemids; it was recently encountered in three other members of this group (*D. ambulator*, *Diplonema* sp. and *Rhynchopus euleeides*) (Kiethega, Turcotte, & Burger, 2011). So far, ten protein-coding genes were predicted to be assembled from three to 12 modules ranging in size from 60 to 350 bp. However, with the exception of a 3′ module of the LSU rRNA, the remaining LSU and all SSU rRNA fragments remain elusive. These combined features qualify the mitochondrial genome of *D. papillatum* as truly the most bizarre (Vlček *et al.*, 2011).

1.4. Conclusion

Comparative studies of the mitochondrial genomes of various euglenozoans led to the prediction that in the common evolutionary ancestor of kinetoplastids, diplonemids and euglenids, rampant genome fragmentation produced via the neutral evolutionary pathway dramatically different, unique and highly complex organellar genomes and transcriptomes (Flegontov, Gray, Burger, & Lukeš, 2011). Furthermore, the euglenozoan mitochondrial genomes share several unique features with homologous genomes of dinoflagellates, a totally unrelated group of mostly photosynthetic protists (Keeling *et al.*, 2005). These shared characteristics further reinforce the theory of cascades of convergent evolution between both ecologically important lineages (Lukeš *et al.*, 2009).

2. MITOCHONDRIAL GENOMES OF PHOTOSYNTHETIC ALVEOLATES

2.1. Phylogeny of Alveolata

The Alveolata includes three well-established and well-known groups: predatory Ciliata, parasitic Apicomplexa with a relic non-photosynthetic plastid in most species, and their sister group Dinoflagellata (Leander & Keeling, 2004), which are either photosynthetic, predatory or parasitic (Hackett, Anderson, Erdner, Bhattacharya, 2004). It is hypothesized that all alveolates (Fig. 6.1), heterokonts, and related groups united in the taxon. Chromalveolata (Adl *et al.*, 2005) got their secondary plastids in a single endosymbiotic event from a red alga (Archibald, 2009; Janouškovec, Horák, Oborník, Lukeš, & Keeling, 2010; Keeling, 2009), with subsequent partial or complete plastid losses in some lineages.

Recently, several new groups have been recognized within the Apicomplexa–Dinoflagellata assemblage. A non-parasitic predatory group, Colpodellida (Berney, Fahrni, & Pawlowski, 2004; Brugerolle, 2002; Kuvardina *et al.*, 2002; Leander, Kuvardina, Aleshin, Mylnikov, & Keeling, 2003; Mylnikov, 2009) was placed within Apicomplexa, previously an exclusively parasitic group. Colpodellids have an apical complex, an eponymous diagnostic feature of Apicomplexa, a complex of organelles used for cell invasion or predation. A typical apical complex is composed of rhoptries and micronemes (extrusive organelles) enveloped by a microtubule-formed conoid (Leander and Keeling, 2003).

Another new twig on the apicomplexan stem is Chromerida, lacking the complete apical complex (Oborník *et al.*, 2011) and containing a fully functional secondary plastid (Janouškovec, Horák, Oborník, Lukeš, Keeling *et al.*, 2010). Thus, this group is the closest photosynthetic relative of Apicomplexa (Keeling, 2008; Moore *et al.*, 2008; Okamoto & McFadden, 2008). The first species of Chromerida described was an alga, *Chromera velia*, which lives in association with corals, but very likely also has a free-living stage (Moore *et al.*, 2008; Oborník *et al.*, 2011; Weatherby, Murray, Carter, & Šlapeta, 2011). The second species isolated as CCMP3155 and described recently as *Vitrella brassicaformis* (Oborník *et al.*, 2012) forms a distinct lineage unexpectedly distant from *C. velia* (Janouškovec *et al.*, 2010; Oborník *et al.*,

2012). Chromerida and Colpodellida may be more closely related to each other than they are to the crown apicomplexans (Moore *et al.*, 2008).

The dinoflagellate branch has come under study more recently for several reasons, the main one being the paramount ecological significance of these protists. The tentative branching order within this species-rich group is as follows (Bachvaroff, Handy, Place, & Delwiche, 2011; Gómez, López-García, Nowaczyk, & Moreira, 2009, Gómez, Moreira, & López-García, 2010; Hoppenrath & Leander, 2010; Saldarriaga, Taylor, Cavalier-Smith, Menden-Deuer, Keeling, 2004; Skovgaard, Meneses, & Angélico, 2009): (1) Perkinsozoa (Perkinsidae, Perkinsea), intracellular parasites of bivalve molluscs (*Perkinsus*) and protists (*Cryptophagus, Parvilucifera, Rastrimonas*), probably have a relic plastid (Fernandes Robledo *et al.*, 2011), share spliced leader RNAs with crown dinoflagellates (Joseph *et al.*, 2010; Zhang, Campbell, Sturm, Dungan, & Lin *et al.*, 2011) and an apical complex with crown apicomplexans and colpodellids (Leander and Keeling, 2003); (2) Ellobiopsida (*Ellobiopsis, Thalassomyces, Ellobiocystis, Parallobiopsis*), a group with uncertain position, mostly ectoparasites of crustaceans (Gómez et al., 2009); (3) Marine Alveolate Group I, intracellular parasites of fish eggs (*Ichthyodinium*) and marine protists (*Duboscquella*) (Groisillier, Massana, Valentin, Vaulot, & Guillou *et al.*, 2006; Harada, Ohtsuka, & Horiguchi, 2007; Skovgaard *et al.*, 2009); (4) Syndiniales or Marine Alveolate Group II (*Amoebophrya, Hematodinium, Syndinium*), endoparasitic dinoflagellates without plastids, found mainly in crustaceans and protists, such as other dinoflagellates (Groisillier *et al.*, 2006; Guillou *et al.*, 2008; Skovgaard, Massana, Balagué, & Saiz, 2005; Stentiford & Shields, 2005); (5) *Oxyrrhis*, a group with uncertain position relative to Syndiniales, a predatory dinoflagellate with some evidence of a relic plastid (Bachvaroff *et al.*, 2011; Jackson, Gornik, & Waller, 2011; Saldarriaga *et al.*, 2004; Slamovits & Keeling, 2011); (6) Noctilucales, an early-branching group of photosynthetic dinoflagellates (Gómez *et al.*, 2010); (7) Dinophyceae (Dinokaryota), core dinoflagellates (Hoppenrath & Leander, 2010; Saldarriaga *et al.*, 2004) with secondary or tertiary plastids (Oborník, Janouškovec, Chrudimský, & Lukeš, 2009). Groups (3)–(7) have been united in the taxon Dinoflagellata (Saldarriaga *et al.*, 2004; Skovgaard *et al.*, 2005).

With the exception of a medically important apicomplexan parasite, *Cryptosporidium*, which has mitosomes (Keithly, Langreth, Buttle, & Mannella, 2005), all groups of the Apicomplexa–Dinoflagellata branch investigated so far contain a conventional mitochondrion. Mitochondrial genomes were studied mainly in core dinoflagellates and core apicomplexans. Recent studies focused also on *Hematodinium* (Jackson *et al.*, 2011), *Perkinsus* (Masuda, Matsuzaki, & Kita, 2010; Zhang *et al.*, 2011), *Chromera* (Flegontov *et al.*, unpublished data) and *Vitrella* (Janouškovec *et al.*, unpublished data). The published results on dinoflagellates and *Perkinsus* and our unpublished findings on chromerids are discussed in this section.

2.2. Mitochondrial Genomes in Ciliates and Parasitic Apicomplexans

Ciliata, the basal group of Alveolata, have linear-mapping mitochondrial genomes with a normal gene number: two rRNAs, seven tRNAs, 21 protein-coding genes of known function, and 22 ciliate-specific ORFs (Brunk, Lee, Tran, Li, 2003; Burger *et al.*, 2000). Both large subunit (LSU) and small subunit (SSU) rRNAs are split into two separately encoded fragments; many tRNA genes are lost from the mitochondrial genome, with the corresponding tRNAs imported from the cytosol (Rusconi & Cech, 1996). Alternative start codons AU(U/A), (G/U)UG are used in at least 7 of 43 protein-coding genes (Burger *et al.*, 2000; Edqvist, Burger, Gray, 2000). UGA encodes tryptophan, UAG is unassigned, and so the only stop codon remaining in use is UAA (Burger *et al.*, 2000).

Apicomplexa diverged much further from the canonical mitochondrial genome structure. The genome is reduced to just three protein-coding genes, cytochrome oxidase subunits 1 (*cox1*) and 3 (*cox3*), and apocytochrome *b* (*cob*), arranged on circularly permutated linear molecules (Feagin, 1992). *Cox2*, universally present in other mitochondrial genomes, is transferred to the nucleus (Waller & Keeling, 2006). Apicomplexans apparently lack genes coding for subunits of complex I (NADH dehydrogenase) of the respiratory chain (Gardner, Hall, Fung, White, & Berriman, 2002). Ribosomal RNAs are highly fragmented: 23 LSU and SSU fragments were found in *Plasmodium*, but some functionally important parts of the rRNAs are missing in this set of fragments (Feagin, Mericle, Werner, & Morris, 1997; Kairo, Fairlamb, Gobright, & Nene, 1994). Such absences might be accounted for by targeting of small rRNA fragments from the cytoplasm.

A precedent for such an rRNA import exists, as mitochondrial import of cytosolic 5S rRNA has been demonstrated in mammals (Entelis, Kolesnikova, Dogan, Martin, & Tarassov, 2001). Moreover, all tRNAs are imported into the apicomplexan mitochondria from the cytosol (Esseiva, Naguleswaran, Hemphill, & Schneider, 2004). It has been suggested that tRNA-fMet is imported even from the plastid (Barbrook, Howe, & Purton, 2006a; Howe & Purton, 2007), but evidence for this type of import is still lacking. In *cox1* and *cox3* genes, AUA or AUU are used as start codons, AUG is used in *cob* (Feagin, 1992; Kairo *et al.*, 1994; Rehkopf, Gillespie, Harrell, & Feagin, 2000). Stop codon usage is identical to ciliates; the only codon used is UAA (Rehkopf *et al.*, 2000). All transcripts, including rRNA fragments, are oligoadenylated (Gillespie, Salazar, Rehkopf, & Feagin, 1999; Rehkopf *et al.*, 2000).

2.3. Mitochondrial Genomes in Dinokaryota

Mitochondrial genomes of core dinoflagellates, Dinokaryota, have the same extremely reduced coding capacity of three protein-coding genes and fragmented rRNAs, but have accumulated numerous oddities in genome architecture and expression (Nash, Nisbet, Barbrook, & Howe, 2008; Waller & Jackson, 2009). Therefore, it is safe to assume that dinoflagellate mitochondrial DNA evolved from an apicomplexan-like state in the last common ancestor (LCA) of Apicomplexa and Dinokaryota. Expansion of actively recombining non-coding DNA increased genome size greatly, created truncated pseudogenes and tiny gene fragments, and put remaining full-length genes in dozens of sequence contexts. Recombination-driven rearrangements are common in the mtDNA of land plants as well (Knoop, 2004), however, never to the extent observed in dinoflagellates.

Mitochondrial DNA of dinoflagellates is revealed by Southern blot analysis as a pool of heterogeneous molecules, 6–10 kb and longer (Chaput, Wang, & Morse, 2002; Jackson *et al.*, 2007; Nash *et al.*, 2007; Norman & Gray, 2001). Preliminary pulse-field gel electrophoresis experiments suggest that an upper chromosome size limit for *Amphidinium carterae* is ~30 kb (Nash *et al.*, 2008). In the same species, non-coding DNA content is estimated at 85% (Nash *et al.*, 2007), and it also seems high in other species. Non-coding regions are rich in inverted repeats capable of forming stem-loop structures: ~50–150 bp repeats ~10–30 bp apart in *A. carterae* (Nash *et al.*, 2007), >6 bp repeats no more than 5 bp apart in *Karlodinium micrum* (Jackson *et al.*, 2007) or >9 bp repeats no more than 50 bp apart in *Crypthecodinium cohnii* and *K. micrum* (Jackson *et al.*, 2007; Norman & Gray, 2001). No sequence conservation in the inverted repeats is apparent between species apart from a higher than average GC content. In other organisms, stem-loop structures are thought to play a role in the control of mitochondrial replication (Arunkumar & Nagaraju, 2006), transcript stability (Kuhn, Tengler, & Binder S, 2001) and genome recombination (Bartoszewski, Katzir, & Havey, 2004), but their role in dinoflagellates remains unknown.

Limited shotgun cloning, polymerase chain reaction (PCR) surveys, and Southern blot analyses in several species (Chaput *et al.*, 2002; Imanian & Keeling, 2007; Jackson *et al.*, 2007; Kamikawa, Nishimura, & Sako, 2009; Nash *et al.*, 2007; Norman & Gray, 2001) all point in one direction: the genome structure in Dinokaryota is chaotic, with full-length genes, truncated genes, small gene fragments, and non-coding DNA mixed in numerous arrangements. Sequence divergence in gene fragments is always negligible, suggesting active ongoing recombination. It appears that gene arrangements may not be completely random; deeper genome sequencing is needed to address this issue. In *A. carterae*, *cox3* and *cob* are usually arranged head-to-head with variable spacers. *Cob* and *cox1* were amplified tail-to-tail with only one spacer. No single DNA molecule containing all three genes was shown by either restriction digestion followed by Southern blot analysis or by PCR (Nash *et al.*, 2007). In *C. cohnii* and *K. micrum*, a pool of small gene fragments contained *cox1* and *cox3* sequences, but *cob* sequences were lacking (Jackson *et al.*, 2007). In *K. micrum*, the following arrangements have been encountered in PCR amplicons: *cox1-cob*, *cox1-cox3*, *cob-cob*, and *cob-cox3* (Jackson *et al.*, 2007). In *Alexandrium catenella cob-cox1*, *cox1-cox1* intergenic spacers of random structure were sequenced. Moreover, some *cox1-cob* spacers contained *cox3* copies considered pseudogenes due to the lack of a conserved region at the 3′ end (Kamikawa *et al.*, 2009).

Despite the abundance of truncated gene copies, in some studies only full-length transcripts were revealed in expressed sequence tag (EST) datasets (Nash *et al.*, 2007), with RACE (Kamikawa *et al.*, 2009) and Northern blot analysis (Norman & Gray, 2001). On the other hand, apparently non-functional transcripts of pseudogenes or gene fragments were found in other studies: transcripts of *cox1* with insertions, some of them containing *cob* fragments (Imanian & Keeling, 2007); polycistronic transcripts with protein-coding and rRNA gene fragments (Jackson *et al.*, 2007); a long transcript matching apparently non-coding DNA (Jackson *et al.*, 2007); *cox3* transcript truncated at the 3′ end, with fragments of *cox1* and *cob* (Chaput *et al.*, 2002). In the latter species, *Lingulodinium polyedrum* (previous name *Gonyaulax polyedra*), *cob* and *cox3* probes hybridized to a smear of transcripts (Chaput *et al.*, 2002). Orderly transcription with defined promoters upstream of genes is difficult to imagine in such a disordered genomic system where a gene can be flanked by dozens of totally different sequences. Promoters located within genes would lead to transcription of many truncated gene copies. In our view, the most reasonable assumption based on the data available is that (almost) all mitochondrial DNA in dinoflagellates is transcribed, but quickly degraded and therefore not visible in some experimental setups. Mature transcripts are most likely generated by cleavage at both ends and reach detectable concentrations.

Similar to apicomplexans, all transcripts in the dinoflagellate mitochondria, including rRNA fragments, are oligoadenylated (Chaput *et al.*, 2002; Jackson *et al.*, 2007; Kamikawa, Inagaki, & Sako, 2007, 2009; Nash *et al.*, 2007). Transcripts of *cox3* require *trans*-splicing in all Dinokaryota that have been investigated (Jackson *et al.*, 2007; Nash *et al.*, 2007; Waller & Jackson, 2009). In *K. micrum*, mature *cox3* transcripts are 839 nt in length, but some cDNAs are oligoadenylated at 712 nt. The genome contains a single 712-nt-long ORF immediately followed by a stop codon and *cox3* fragments including positions 718–839. Transcripts of the long and the short fragments are apparently *trans*-spliced taking six As from the oligoA tail of the long fragment (Jackson *et al.*, 2007). The *cox3* transcript in a basally branching species *A. carterae* lacks these oligoA-derived nucleotides (Nash *et al.*, 2007). The splicing mechanism remains totally unknown, and no intron-like sequences have been identified so far (Jackson *et al.*, 2007; Waller & Jackson, 2009).

Another striking feature of the dinoflagellate mitochondrial genetic system is extensive RNA editing. All three protein-coding genes are edited (and some rRNA fragments; see below), but editing was investigated mostly in *cob* and *cox1*. Predominant changes are A-G (~50%); U-C and C-U changes are also common. However, almost all possible changes were detected: G-A, U-G, G-U, G-C, C-G, A-U, U-A, A-C (Gray, 2003; Jackson *et al.*, 2007; Lin, Zhang, Spencer, Norman, & Gray, 2002; Lin, Zhang, & Gray, 2008; Zhang & Lin, 2005; Zhang & Lin, 2008; Zhang, Bhattacharya, Maranda, & Lin, 2008). Moreover, G-C substitutions seem to be unique to dinoflagellates. Such versatility of an editing system is totally unprecedented (Gray, 2003). Editing occurs mostly at the first and second codon positions usually affecting 2–3% of the nucleotide sequence and up to 6% in *cox3* of *K. micrum* (Jackson *et al.*, 2007). Ile-Val and Phe-Leu amino acid changes are most common as a result of editing (Waller & Jackson, 2009). It is particularly intriguing that editing is not progressing in the 5′ to 3′ direction or vice versa (Nash *et al.*, 2007) and that the editing sites are distributed in clusters (Waller & Jackson, 2009). Only full-length mature transcripts are edited in *K. micrum* (Jackson *et al.*, 2007). Sometimes, editing eliminates in-frame UAG stop codons, unassigned in alveolates, such as two stop codons in *cox1* of *A. carterae* or one stop codon in *cox3* of *K. micrum* (Jackson *et al.*, 2007; Lin *et al.*, 2002; Nash *et al.*, 2007; Zhang & Lin, 2005). New editing sites are constantly evolving, but some sites are apparently highly conserved (Jackson *et al.*, 2007). The mechanism of RNA editing in dinoflagellates and its possible significance remain completely unknown. Apparently, (de)amination should be involved in relatively frequent transitions and base excision replacement in more rare transversions. Editing increases the GC content, and this may be important for the use of nucleus-encoded tRNAs imported into the mitochondrion (Waller & Jackson, 2009). Few sites in *A. carterae* gene fragments match post-edited transcripts suggesting that a guide RNA-like mechanism might be involved (Nash *et al.*, 2007), yet no such sites were found in *K. micrum* (Jackson *et al.*, 2007).

Extensive editing has been reported in ~25 species of Dinokaryota (Lin *et al.*, 2008; Zhang *et al.*, 2008): *Adenoides eludens* (Zhang *et al.*, 2007), *Akashiwo sanguinea* (Zhang *et al.*, 2007; Zhang et al., 2008), *Alexandrium tamarense, A. affine. A. catenella* (Kamikawa *et al.*, 2009; Zhang and Lin, 2005; Zhang *et al.*, 2005, 2007, 2008), *Ceratium longipes, Ceratocorys horrida* (Zhang *et al.*, 2007), *Dinophysis acuminata* (Zhang *et al.*, 2008), *Durinskia baltika* (Imanian & Keeling, 2007), *Gambierdiscus toxicus, Gonyaulax cochlea, Gymnodinium simplex* (Zhang *et al.*, 2007), *Karenia brevis* (Zhang *et al.*, 2005, 2007, 2008), *Karlodinium micrum* (Jackson *et al.*, 2007; Zhang & Lin, 2005; Zhang *et al.*, 2005, 2007, 2008), *Pfiesteria shumwayae, P. piscidida* (Zhang & Lin, 2002; Zhang & Lin, 2005; Zhang *et al.*, 2005, 2007, 2008), *Prorocentrum minimum, P. micans* and other *Prorocentrum* spp. (Lin, Zhang, & Jiao, 2006; Zhang & Lin, 2005; Zhang *et al.*, 2005, 2007, 2008), *Protoceratium reticulatum* (Zhang *et al.*, 2007, 2008), *Pyrocystis lunula, P. noctiluca* (Zhang *et al.*, 2007), *Pyrodinium bahamense* (Zhang *et al.*, 2005), *Scrippsiella* sp. and *S. sweeneyae* (Zhang *et al.*, 2005, 2007, 2008). Editing seems to be missing in *Noctiluca scintillans* (Zhang & Lin, 2008; Zhang *et al.*, 2007), a member of Noctilucales, and in basally branching species of Dinokaryota, *Heterocapsa triquetra* and *H. rotundata* (Zhang & Lin, 2008; Zhang *et al.*, 2005, 2007, 2008). Moreover, editing is not very extensive in some other basally branching species: *A. carterae, A. operculatum* (Nash *et al.*, 2007; Zhang & Lin, 2008; Zhang et al. 2007, 2008), *C. cohnii, Symbiodinium* sp. and *S. microadriaticum* (Zhang & Lin, 2005; Zhang *et al.*, 2005, 2007, 2008). These observations suggest that RNA editing appeared and spread within Dinokaryota.

The general consensus is that, in Dinokaryota AUG, start codons are missing in the conserved 5′ regions of transcripts, and it is difficult to determine which codons act as alternative starts (Nash *et al.*, 2008; Waller & Jackson, 2009). Potential AUG occurs in *cox1* of *C. cohnii*, but these positions are not conserved (Jackson *et al.*, 2007). In *A. catenella*, isoleucine AU(A/U/C) or leucine (U/C)UG codons were found near the start of *cob* and *cox1* transcripts (Kamikawa *et al.*, 2009), and AUG was found near the start of the *cob* transcript. This AUG codon is also conserved in five other *Alexandrium* species and *Gonyaulax* sp. (Kamikawa *et al.*, 2008), and another potential AUG occurs in the *cob* of *K. micrum*, but sequence conservation starts upstream of this codon (Jackson *et al.*, 2007). In summary, it is still not clear whether AUG is used in some cases, and which alternative start codons are utilized.

Reliance on stop codons is also relaxed in Dinokaryota: oligoadenylation in *cox1* and *cob* occurs before UAG and UAA codons in *A. carterae* (Nash *et al.*, 2007), *A. catenella* (Kamikawa *et al.*, 2009), *Pfiesteria piscicida*, *Prorocentrum minimum*, *L. polyedrum* and *Karenia brevis* (Jackson *et al.*, 2007). In *cox3*, a UAA codon is generated by the oligoadenylation process itself in *A. carterae* (Nash *et al.*, 2007) and *K. micrum* (Jackson *et al.*, 2007). Creation of termination codons in such a manner is not unprecedented, as similar oligoadenylation is also known to occur in human mitochondria (Chrzanowska-Lightowlers, Temperley, Smith, Seneca, & Lightowlers *et al.*, 2004), and stop codons are occasionally, but not systematically, absent in plant mitochondria (Raczynska *et al.*, 2006). The mechanism of stop-codon-free termination in dinoflagellates remains to be established. To explain the available data, reliance on tmRNA-like molecules or special ribosome release factors has been implied (Nash *et al.*, 2008; Waller & Jackson, 2009). Furthermore, it was also proposed that oligoA translation producing lysine stretches can be tolerated in *cox1* and *cob*, because positively charged amino acids frequently occur at the C-termini of these proteins in other eukaryotes; however, *cox3* lacks a positively charged C-terminus and requires a defined terminator (Waller & Jackson, 2009).

The ribosomal RNA fragmentation pattern in Dinokaryota remains very similar to that of apicomplexans (Nash *et al.*, 2008; Waller & Jackson, 2009). However, few rRNA fragments have been identified in dinoflagellates probably due to the limited depth of genome/transcriptome sequencing: LSUA, LSUD, LSUE, LSUF, LSUG, and LSU RNA2 were found along with their fragmentary copies in *A. catenella* (Kamikawa *et al.*, 2007; Kamikawa et al., 2009); LSUE, LSUG in *C. cohnii* (Jackson *et al.*, 2007); LSUE, LSUF, LSUG in *Heterocapsa triquetra* (Jackson *et al.*, 2007); LSUA, LSUE, LSUG, LSU RNA2, LSU RNA10, SSU RNA8, and unassigned

RNA7 were found along with some unprocessed precursors or alternative variants in *K. micrum* (Jackson *et al.*, 2007). RNA editing of the same type as in protein-coding transcripts was demonstrated for LSUE in *A. catenella* (Kamikawa *et al.*, 2007) and for LSUA, LSUG in *K. micrum* (Jackson *et al.*, 2007). Surprisingly, rRNA genes were not found in 33 kb of shotgun clones and in PCR products of *A. carterae* mitochondrial DNA (Nash *et al.*, 2007). Import of some, but not all, rRNA fragments from the cytosol was proposed for apicomplexans, but remains purely hypothetical (see above). Import of tRNAs from the plastid into the mitochondrion was suggested for *Plasmodium* (Barbrook *et al.*, 2006a), yet again, experimental data to support this claim is lacking. In principle, the same might be true in dinoflagellates because tRNA-fMet is one of only a handful of tRNAs encoded in the dinoflagellate chloroplast genome (Barbrook, Santucci, Plenderleith, Hiller, & Howe, 2006b; Nelson *et al.*, 2007).

The predicted amino acid sequences of dinoflagellate *cox1*, *cox3* and *cob* show substitutions at several functionally important sites that are conserved in most other organisms (Nash *et al.*, 2008). Despite this sequence divergence and all the molecular oddities described for the dinoflagellate mitochondria, respiratory complexes III and IV activity was detected in *A. catenella* (Kamikawa *et al.*, 2009) suggesting that even such a messy genetic system can produce a properly functioning conventional respiratory chain.

2.4. Mitochondrial Genomes in Other Dinoflagellate Groups, Perkinsids, and Chromerids

Some results obtained from the dinoflagellates branching off prior to Dinokaryota, and also from perkinsids and chromerids, are presented. Mitochondrial DNA of *Hematodinium* sp. shares many features with core dinoflagellates (Jackson *et al.*, 2011), although it contains much less inverted repeats, since tightly packed fragments of different genes were found on genomic amplicons. A transcriptome assembly of 454 reads suggests that gene fragments and non-coding DNA are transcribed, but only full-length mature transcripts were detected with Northern blot analysis. This finding is in line with less extensive results available for Dinokaryota (see above), suggesting that everything is transcribed in these genomes (which can be detected by RNAseq) but probably quickly degraded, and hence not detectable in Northern blots. Unlike all Dinokaryota, the *cox3* transcript is not *trans*-spliced.

Typical dinoflagellate-type RNA editing was demonstrated for all three *Hematodinium* genes: A-G, U-C and C-U changes predominate, but A-U,

G-A and C-G also occur. Apparently non-functional fragmentary transcripts are also edited, unlike in *K. micrum* (Jackson *et al.*, 2007). Editing sites are not conserved between *Hematodinium* and Dinokaryota which, together with the absence of editing in *Heterocapsa* and *Noctiluca*, suggests that RNA editing arose independently in Syndiniales and Dinokaryota. The AUG start codon is not used in *Hematodinium* and the AUU triplet apparently does not take over its function. In *cob* and *cox1*, oligoadenylation occurs prior to the UAA stops. Cases of premature oligoadenylation have also been detected. Unlike in Dinokaryota, in this protest, the UAA stop in *cox3* is encoded. The set of rRNA fragments in *Hematodinium* is the largest found to date, especially regarding the SSU fragments: LSUA, LSUD, LSUE, LSUF, LSUG, LSU RNA2, LSU RNA10; SSUA, SSUB, SSUD, SSUF, SSU RNA8; unassigned RNA6 and RNA7. However, this exceptionally high number of fragments may be the consequence of deep transcriptome sequencing (Jackson *et al.*, 2011). Although variable transcripts were observed in 454 reads for SSUA and SSUB, only single bands were detected by Northern blot analysis. The apicomplexan pattern of fragmentation is conserved in *Hematodinium* as well. Although ongoing opportunity for fragmentation exists due to active recombination, further rRNA disassembly would presumably affect its ability to self-associate (Jackson *et al.*, 2011).

The predatory dinoflagellate *Oxyrrhis marina* with an uncertain phylogenetic position has been extensively studied by sequencing its mitochondrial DNA and EST library (Slamovits & Keeling, 2011; Slamovits, Saldarriaga, Larocque, Keeling, 2007). Its mitochondrial DNA encodes the *cox1*, *cox3* and *cob* genes but, remarkably, *cob* and *cox3* were always detected fused in one ORF. The authors showed that the *cob–cox3* fusion transcript is not processed, but fusion at the protein level was not investigated. The situation may be similar to that described in the distantly related *Acanthamoeba castellanii*, namely that a *cox1–cox2* fusion transcript does not necessarily result in a fusion protein (Lonergan & Gray, 1996). Moreover, *cox3* is very divergent in *O. marina*. *cox3* is absent from the mitochondrial genome of ciliates (Brunk *et al.*, 2003; Burger *et al.*, 2000) and there is no evidence of a mitochondrion-targeted homologue in the *Tetrahymena thermophila* nuclear genome (Eisen *et al.*, 2006).

In *Oxyrrhis*, two different genes were never found on the same EST or mitochondrial genomic fragment; many gene copies probably occur in closely-spaced tandem repeats of the same gene. Fragments of *cob* and *cox3*, but not of *cox1*, were found. Inverted repeats typical for core dinoflagellates are few in *Oxyrrhis*, which is reminiscent of the situation in *Hematodinium*.

Genes with different flanking sequences were found, but only mature polyA transcripts of uniform size can be identified like in many Dinokaryota and in *Hematodinium*. No RNA editing was found in *O. marina* and it was noted that the gene sequences correspond more to the post-edited sequence in Dinokaryota (Slamovits *et al.*, 2007; Zhang & Lin, 2008; Zhang *et al.*, 2007, 2008). Another feature unique to this species is 5′ oligoU caps of 8–9 uridines on mitochondrial mRNAs added by an unknown machinery. Not surprisingly, AUG is not used, with AUU being the more likely start codon. Oligoadenylation occurs prior to UAA stop in *cox1*, UAA stop in *cob–cox3* is created by oligoadenylation like in most Dinokaryota. Finally, oligoadenylated LSUE, LSUG, and LSU RNA10 were identified (Slamovits *et al.*, 2007; Zhang & Lin, 2008; Zhang *et al.*, 2007, 2008).

Mitochondrial DNA of *Perkinsus marinus* is incompletely characterized and its genomic architecture remains unknown (Masuda *et al.*, 2010; Zhang *et al.*, 2011). As in dinoflagellates, *cox1* hybridized to a smear of genomic molecules <10 kb in length. The picture is further complicated by short fragments of mitochondrial DNA being found in the nuclear genome. There are no signs of RNA editing, at least in *cob* and *cox1* (data for *cox3* are unavailable) and no canonical start and stop codons were identified (Jackson *et al.*, 2011; Masuda *et al.*, 2010). What is really remarkable about the *Perkinsus* mitochondrion is very extensive translational frameshifting. Frameshifts occur at all AGG (Gly) and CCC (Pro) codons. AGG and CCC usually occur in the motives UAGGY and CCCCUA, respectively, and the codons are read as AGGY or CCCCU, generating +1 or +2 frameshifts. The mechanism responsible for such frequent frameshifting remains unknown: ribosome stalling or special quadruplet- or quintuplet-decoding tRNAs might be involved. There are ten frameshifts in *cox1* (eight AGGY, two CCCCU) and seven in *cob* (six AGGY, one UAGGC). Frameshifts are not unprecedented in the mitochondrial genomes, but only one or a maximum of two occur per gene (Beckenbach, Robson, & Crozier, 2005; Milbury & Gaffney, 2008; Russell & Beckenbach, 2008). High frameshift density in *Perkinsus* must involve a unique and very efficient frameshifting mechanism. Conservation of the *cox1* sequence in *P. chesapeaki*, *P. olseni* and *P. honshuensis* implies the presence of this mechanism in all these species (Zhang *et al.*, 2011).

We are now completing a deep sequencing study of *Chromera velia* mitochondrial DNA with Illumina and 454 technologies, and of transcriptome with Illumina (Flegontov *et al.*, unpublished data). The organellar genome is essentially dinoflagellate-like (i.e. rich in non-coding

DNA, short GC-rich interspersed repeats apparently acting as recombination hot spots), gene fragments and truncated genes are abundant, canonical start and stop codons are probably missing, rRNAs are fragmented, and all transcripts are oligoadenylated. According to the RNAseq data, the entire genome seems to be transcribed at a high rate but, as in *Hematodinium*, only full-length transcripts were visualized by Northern blot analysis. Conserved *cox1* is found in the genome along with many truncated copies and fragments. Non-fragmented *cox3* is hardly recognizable at the sequence level and is invariably fused into one ORF with a 5′ fragment of *cox1*. Preliminary data from the mitochondrial genome of a distantly related chromerid, *Vitrella brassicaformis*, show a more orderly genome (Flegontov *et al.*, unpublished data) with *cob* and *cox1* fused in one ORF.

2.5. Conclusions

The evolutionary picture that emerges from this review is that of convergent evolution of radically reduced mitochondrial genomes. Tendencies for the loss of start and stop codons, rRNA fragmentation and tRNA loss appear in the ciliates, the most basal alveolate group. Ribosomal RNA fragmentation apparently reaches the maximum permissible level in the last common ancestor of the Apicomplexa–Dinoflagellata group (ADLCA) and remains constant in different lineages. Suspected cases of outright mitochondrial rRNA relocation to the nucleus are an appealing possibility but remain unverified. Oligoadenylation of all transcripts including rRNA fragments became established in the ADLCA. The most significant event that very likely happened in the ADLCA was a drastic reduction of the coding capacity: complete loss of tRNAs and most protein-coding genes except for *cox1*, *cox3* and *cob*. We suggest that this genome reduction, relaxing of selective constraints, was a prerequisite for the development of other striking molecular features independently in different lineages of the Apicomplexa–Dinoflagellata group. Loss of canonical start and stop codons occurred apparently separately in the perkinsid–dinoflagellate branch and in *Chromera* (core apicomplexans retain AUG in one gene and UAA stops in all genes). Mitochondrial genome scrambling follows the same phylogenetic pattern since unscrambling of a messy genome in the ADLCA leading to an orderly genome in core apicomplexans is difficult to imagine. RNA editing of the same type appeared probably separately in *Hematodinium* (Syndiniales) and within Dinokaryota (see the earlier discussion). Gene fusions appeared several times: *cox1–cox3* in *Chromera*, *cob–cox1* in *Vitrella*, *cob–cox3* in

Oxyrrhis. Molecular features unique to some groups include extensive frameshifting in *Perkinsus* and oligoU 5′ caps in *Oxyrrhis*.

These evolutionary pathways, in our view, are best explained by the constructive neutral evolution framework (Flegontov *et al.*, 2011; Lukeš, Archibald, Keeling, Doolittle, & Gray, 2011). Pre-existing enzymatic activities can be recruited for RNA editing, DNA recombination activities for genome scrambling, special tRNAs for frameshifting as long as their effects are tolerated, without any immediate selective benefits. Highly reduced genomes should be most tolerant of any kind of evolutionary tinkering. Hence, the abundance of molecular oddities in mitochondrial genomes of the Apicomplexa–Dinoflagellata group is not that surprising after all.

ACKNOWLEDGEMENTS

This work was supported by the Grant Agency of the Czech Republic (204/09/1667 and P305/11/2179) and the Praemium Academiae award to J.L., who is a Fellow of the Canadian Institute for Advanced Research.

REFERENCES

Adl, S. M., Simpson, A. G. B., Farmer, M. A., Andersen, R. A., Anderson, O. R., Barta, J. R., et al. (2005). The new higher level classification of eukaryotes with emphasis on the taxonomy of protists. *Journal of Eukaryotic Microbiology, 52*, 399–451.

Archibald, J. M. (2009). The puzzle of plastid evolution. *Current Biology, 19*, R81–R88.

Arunkumar, K. P., & Nagaraju, J. (2006). Unusually long palindromes are abundant in mitochondrial control regions of insects and nematodes. *PLoS ONE, 1*. e110.

Archibald, J. M., & Keeling, P. J. (2002). Recycled plastids: a "green movement" in eukaryotic evolution. *Trends in Genetics, 18*, 577–584.

Bachvaroff, T. R., Handy, S. M., Place, A. R., & Delwiche, C. F. (2011). Alveolate phylogeny inferred using concatenated ribosomal proteins. *Journal of Eukaryotic Microbiology, 58*, 223–233.

Barbrook, A. C., Howe, C. J., & Purton, S. (2006a). Why are plastid genomes retained in non-photosynthetic organisms? *Trends in Plant Science, 11*, 101–108.

Barbrook, A. C., Santucci, N., Plenderleith, L. J., Hiller, R. G., & Howe, C. J. (2006b). Comparative analysis of dinoflagellate chloroplast genomes reveals rRNA and tRNA genes. *BMC Genomics, 7*, 297.

Bartoszewski, G., Katzir, N., & Havey, M. J. (2004). Organization of repetitive DNAs and the genomic regions carrying ribosomal RNA, cob, and atp9 genes in the cucurbit mitochondrial genomes. *Theoretical and Applied Genetics, 108*, 982–992.

Beckenbach, A. T., Robson, S. K., & Crozier, R. H. (2005). Single nucleotide +1 frameshifts in an apparently functional mitochondrial cytochrome b gene in ants of the genus *Polyrhachis*. *Journal of Molecular Evolution, 60*, 141–152.

Berney, C., Fahrni, J., & Pawlowski, J. (2004). How many novel eukaryotic 'kingdoms'? Pitfalls and limitations of environmental DNA surveys. *BMC Biology, 2*, 13.

Blum, B., Bakalara, N., & Simpson, L. (1990). A model for RNA editing in kinetoplastid mitochondria: guide RNA molecules transcribed from maxicircle DNA provide the edited information. *Cell, 60*, 189–198.

Boer, P. H., & Gray, M. W. (1988). Scrambled ribosomal RNA gene pieces in *Chlamydomonas reinhardtii* mitochondrial DNA. *Cell, 55*, 399–411.

Brugerolle, G. (2002). *Colpodella vorax*: ultrastructure, predation, life-cycle, mitosis, and phylogenetic relationships. *European Journal of Protistology, 38*, 113–125.

Brunk, C. F., Lee, L. C., Tran, A. B., & Li, J. (2003). Complete sequence of the mitochondrial genome of *Tetrahymena thermophila* and comparative methods for identifying highly divergent genes. *Nucleic Acids Research, 31*, 1673–1682.

Burger, G., Zhu, Y., Littlejohn, T. G., Greenwood, S. J., Schnare, M. N., Lang, B. F., et al. (2000). Complete sequence of the mitochondrial genome of *Tetrahymena pyriformis* and comparison with *Paramecium aurelia* mitochondrial DNA. *Journal of Molecular Biology, 297*, 365–380.

Burger, G., Gray, M. W., & Lang, B. F. (2003). Mitochondrial genomes – anything goes. *Trends in Genetics, 19*, 709–716.

Cavalier-Smith, T. (2010). Kingdoms protozoa and chromista and the eozoan root of the eukaryotic tree. *Biology Letters, 6*, 342–345.

Chaput, H., Wang, Y., & Morse, D. (2002). Polyadenylated transcripts containing random gene fragments are expressed in dinoflagellate mitochondria. *Protist, 153*, 111–122.

Chrzanowska-Lightowlers, Z. M., Temperley, R. J., Smith, P. M., Seneca, S. H., & Lightowlers, R. N. (2004). Functional polypeptides can be synthesized from human mitochondrial transcripts lacking termination codons. *Biochemical Journal, 377*, 725–731.

Crouse, E. J., Vandrey, J. P., & Stutz, E. (1974). Hybridization studies with RNA and DNA isolated from *Euglena gracilis* chloroplasts and mitochondria. *FEBS Letters, 42*, 262–266.

Edqvist, J., Burger, G., & Gray, M. W. (2000). Expression of mitochondrial protein-coding genes in *Tetrahymena pyriformis*. *Journal of Molecular Biology, 297*, 381–393.

Eisen, J. A., Coyne, R. S., Wu, M., Wu, D., Thiagarajan, M., Wortman, J. R., et al. (2006). Macronuclear genome sequence of the ciliate *Tetrahymena thermophila*, a model eukaryote. *PLoS Biology, 4*, e286.

El-Sayed, N. M., et al. (2005). Comparative genomics of trypanosomatid parasitic protozoa (and 44 co-authors). *Science, 309*, 404–409.

Entelis, N. S., Kolesnikova, O. A., Dogan, S., Martin, R. P., & Tarassov, I. A. (2001). 5S rRNA and tRNA import into human mitochondria. Comparison of *in vitro* requirements. *Journal of Biological Chemistry, 276*, 45642–45653.

Esseiva, A. C., Naguleswaran, A., Hemphill, A., & Schneider, A. (2004). Mitochondrial tRNA import in *Toxoplasma gondii*. *Journal of Biological Chemistry, 279*, 42363–42368.

Feagin, J. E. (1992). The 6-kb element of *Plasmodium falciparum* encodes mitochondrial cytochrome genes. *Molecular and Biochemical Parasitology, 52*, 145–148.

Feagin, J. E., Mericle, B. L., Werner, E., & Morris, M. (1997). Identification of additional rRNA fragments encoded by the *Plasmodium falciparum* 6 kb element. *Nucleic Acids Research, 25*, 438–446.

Fernández Robledo, J. A., Caler, E., Matsuzaki, M., Keeling, P. J., Shanmugam, D., Roos, D. S., et al. (2011). The search for the missing link: a relic plastid in *Perkinsus*? *International Journal of Parasitology, 41*, 1217–1229.

Flegontov, P., Gray, M. W., Burger, G., & Lukeš, J. (2011). Gene fragmentation: a key to mitochondrial genome evolution in Euglenozoa? *Current Genetics, 57*, 225–232.

Gardner, M. J., Hall, N., Fung, E., White, O., Berriman, M., et al. (2002). Genome sequence of the human malaria parasite *Plasmodium falciparum*. *Nature, 419*, 498–511.

Gillespie, D. E., Salazar, N. A., Rehkopf, D. H., & Feagin, J. E. (1999). The fragmented mitochondrial ribosomal RNAs of *Plasmodium falciparum* have short A tails. *Nucleic Acids Research, 27*, 2416–2422.

Gómez, F., López-García, P., Nowaczyk, A., & Moreira, D. (2009). The crustacean parasites *Ellobiopsis* Caullery, 1910 and *Thalassomyces* Niezabitowski, 1913 form a monophyletic divergent clade within the Alveolata. *Systematic Parasitology, 74*, 65–74.

Gómez, F., Moreira, D., & López-García, P. (2010). Molecular phylogeny of noctilucoid dinoflagellates (Noctilucales, Dinophyceae). *Protist, 161*, 466–478.

Gray, M. W. (2003). Diversity and evolution of mitochondrial RNA editing systems. *IUBMB Life, 55*, 227–233.

Gray, M. W., Lang, B. F., & Burger, G. (2004). Mitochondria of protists. *Annual Review of Genetics, 38*, 477–524.

Hall, R. P. (2005). Reaction of certain cytoplasmic inclusions to vital dyes and their relation to mitochondria and golgi apparatus in the flagellate *Peranema trichophorum*. *Journal of Morphology, 48*, 105–121.

Groisillier, A., Massana, R., Valentin, K., Vaulot, D., & Guillou, L. (2006). Genetic diversity and habitats of two enigmatic marine alveolate lineages. *Aquatic Microbial Ecology, 42*, 277–291.

Guillou, L., Viprey, M., Chambouvet, A., Welsh, R. M., Kirkham, A. R., Massana, R., et al. (2008). Widespread occurrence and genetic diversity of marine parasitoids belonging to Syndiniales (Alveolata). *Environmental Microbiology, 10*, 3349–3365.

Hackett, J., Anderson, D., Erdner, D., & Bhattacharya, D. (2004). Dinoflagellates: a remarkable evolutionary experiment. *American Journal of Botany, 91*, 1523–1534.

Hallick, R. B., Hong, L., Drager, R. G., Favreau, M. R., Monfort, A., Orsat, B., et al. (1993). Complete sequence of *Euglena gracilis* chloroplast DNA. *Nucleic Acids Research, 21*, 3537–3544.

Hannaert, V., Saavedra, E., Duffieux, F., Szikora, J.-P., Rigden, D. J., Michels, P. A. M., et al. (2003). Plant-like traits associated with metabolism of *Trypanosoma* parasites. *Proceedings of the National Academy of Sciences of the United States of America, 100*, 1067–1071.

Harada, A., Ohtsuka, S., & Horiguchi, T. (2007). Species of the parasitic genus *Duboscquella* are members of the enigmatic marine alveolate group I. *Protist, 158*, 337–347.

Hayashi, Y., & Ueda, K. (1989). The shape of the mitochondria and the number of mitochondrial nucleoids during the cell cycle of *Euglena gracilis*. *Journal of Cell Science, 93*, 565–570.

Hayashi-Isimaru, Y., Ueda, K., & Nonaka, M. (1993). Detection of DNA in the nucleoids of chloroplasts and mitochondria in *Euglena gracilis* by immunoelectron microscopy. *Journal of Cell Science, 105*, 1159–1164.

Hoppenrath, M., & Leander, B. S. (2010). Dinoflagellate phylogeny as inferred from heat shock protein 90 and ribosomal gene sequences. *PLoS ONE, 5*. e13220.

Howe, C. J., & Purton, S. (2007). The little genome of apicomplexan plastids: its raison d'etre and a possible explanation for the 'delayed death' phenomenon. *Protist, 158*, 121–133.

Imanian, B., & Keeling, P. J. (2007). The dinoflagellates *Durinskia baltica* and *Kryptoperidinium foliaceum* retain functionally overlapping mitochondria from two evolutionarily distinct lineages. *BMC Evolutionary Biology, 7*, 172.

Jackson, C. J., Norman, J. E., Schnare, M. N., Gray, M. W., Keeling, P. J., & Waller, R. F. (2007). Broad genomic and transcriptional analysis reveals a highly derived genome in dinoflagellate mitochondria. *BMC Biology, 5*, 41.

Jackson, C. J., Gornik, S. G., & Waller, R. F. (2011). The mitochondrial genome and transcriptome of the basal dinoflagellate *Hematodinium* sp.: character evolution within the highly derived mitochondrial genomes of dinoflagellates. *Genome Biology and Evolution* 59–72. http://dx.doi.org/10.1093/gbe/evr122.

Janouškovec, J., Horák, A., Oborník, M., Lukeš, J., & Keeling, P. J. (2010). A common red algal origin of the apicomplexan, dinoflagellate, and heterokont plastids. *Proceedings of the National Academy of Sciences of the United States of America, 107*, 10949–10954.

Joseph, S. J., Fernández-Robledo, J. A., Gardner, M. J., El-Sayed, N. M., Kuo, C.-H., Schott, E. J., et al. (2010). The alveolate *Perkinsus marinus*: biological insights from EST gene discovery. *BMC Genomics, 11*, 228.

Kairo, A., Fairlamb, A. H., Gobright, E., & Nene, V. (1994). A 7.1 kb linear DNA molecule of *Theileria parva* has scrambled rDNA sequences and open reading frames for mitochondrially encoded proteins. *EMBO Journal, 13*, 898–905.

Kamikawa, R., Inagaki, Y., & Sako, Y. (2007). Fragmentation of mitochondrial large subunit rRNA in the dinoflagellate *Alexandrium catenella* and the evolution of rRNA structure in alveolate mitochondria. *Protist, 158*, 239–245.

Kamikawa, R., Hosoi-Tanabe, S., Yoshimatsu, S., Oyama, K., Masuda, I., & Sako, Y. (2008). Development of a novel molecular marker on the mitochondrial genome of a toxic dinoflagellates, *Alexandrium* spp., and its application in single-cell PCR. *Journal of Applied Phycology, 20*, 153–159.

Kamikawa, R., Nishimura, H., & Sako, Y. (2009). Analysis of the mitochondrial genome, transcripts, and electron transport activity in the dinoflagellate *Alexandrium catenella* (Gonyaulacales, Dinophyceae). *Phycology Research, 57*, 1–11.

Keeling, P. J., Burger, G., Durnford, D. G., Lang, B. F., Lee, R. W., Pearlman, R. E., et al. (2005). The tree of eukaryotes. *Trends in Ecology & Evolution, 20*, 670–676.

Keeling, P. J. (2008). Bridge over troublesome plastids. *Nature, 451*, 896–897.

Keeling, P. J. (2009). Chromalveolates and the evolution of plastids by secondary endosymbiosis. *Journal of Eukaryotic Microbiology, 56*, 1–8.

Keithly, J. S., Langreth, S. G., Buttle, K. F., & Mannella, C. A. (2005). Electron tomographic and ultrastructural analysis of the *Cryptosporidium parvum* relict mitochondrion, its associated membranes, and organelles. *Journal of Eukaryotic Microbiology, 52*, 132–140.

Kiethega, G. N., Turcotte, M., & Burger, G. (2011). Evolutionarily conserved cox1 trans-splicing without cis-motifs. *Molecular Biology and Evolution, 28*, 2425–2428.

Knoop, V. (2004). The mitochondrial DNA of land plants: peculiarities in phylogenetic perspective. *Current Genetics, 46*, 123–139.

Kuhn, J., Tengler, U., & Binder, S. (2001). Transcript lifetime is balanced between stabilizing stem-loop structures and degradation-promoting polyadenylation in plant mitochondria. *Molecular and Cellular Biology, 21*, 731–742.

Kuvardina, O. N., Leander, B. S., Aleshin, V. V., Mylnikov, A. P., Keeling, P. J., & Simdyanov, T. G. (2002). The phylogeny of colpodellids (Alveolata) using small subunit rRNA gene sequences suggests they are the free-living sister group to apicomplexans. *Journal of Eukaryotic Microbiology, 49*, 498–504.

Leander, B. S., Triemer, R. E., & Farmer, M. A. (2001). Character evolution in heterotrophic euglenids. *European Journal of Protistology, 37*, 337–356.

Leander, B. S., & Keeling, P. J. (2003). Morphostasis in alveolate evolution. *Trends in Ecology & Evolution, 18*, 395–402.

Leander, B. S. (2004). Did trypanosomatid parasites have photosynthetic ancestors? *Trends in Microbiology, 12*, 251–258.

Leander, B. S., & Keeling, P. J. (2004). Early evolutionary history of dinoflagellates and apicomplexans (Alveolata) as inferred from hsp90 and actin phylogenies. *Journal of Phycology, 40*, 341–350.

Leander, B. S., Kuvardina, O. N., Aleshin, V. V., Mylnikov, A. P., & Keeling, P. J. (2003). Molecular phylogeny and surface morphology of *Colpodella edax* (Alveolata): insights into the phagotrophic ancestry of apicomplexans. *Journal of Eukaryotic Microbiology, 50*, 334–340.

Lin, S., Zhang, H., Spencer, D. F., Norman, J. E., & Gray, M. W. (2002). Widespread and extensive editing of mitochondrial mRNAS in dinoflagellates. *Journal of Molecular Biology, 320*, 727–739.

Lin, S., Zhang, H., & Jiao, N. (2006). Potential utility of mitochondrial cytochrome b and its mRNA editing in resolving closely related dinoflagellates: a case study of *Prorocentrum* (Dinophyceae). *Journal of Phycology, 42*, 646–654.

Lin, S., Zhang, H., & Gray, M. W. (2008). RNA editing in dinoflagellates and its implications for the evolutionary history of the editing machinery. In H. C. Smith (Ed.), *RNA and DNA editing: Molecular mechanisms and their integration into biological systems* (pp. 280–309). Hoboken, NJ: John Wiley and Sons, Inc.

Lonergan, K. M., & Gray, M. W. (1996). Expression of a continuous open reading frame encoding subunits 1 and 2 of cytochrome c oxidase in the mitochondrial DNA of *Acanthamoeba castellanii*. *Journal of Molecular Biology, 257*, 1019–1030.

Lowe, C. D., Keeling, P. J., Martin, L. E., Slamovits, C. H., Watts, P. C., & Montagnes, D. J. S. (2011). Who is *Oxyrrhis marina*? Morphological and phylogenetic studies on an unusual dinoflagellate. *Journal of Plankton Research, 33*, 555–567.

Lukeš, J., Guilbride, D. L., Votýpka, J., Zíková, A., Benne, R., & Englund, P. T. (2002). Kinetoplast DNA network: evolution of an improbable structure. *Eukaryotic Cell, 1*, 495–502.

Lukeš, J., Hashimi, H., & Zíková, A. (2005). Unexplained complexity of the mitochondrial genome and transcriptome in kinetoplastid flagellates. *Current Genetics, 48*, 277–299.

Lukeš, J., Leander, B. S., & Keeling, P. J. (2009). Cascades of convergent evolution: the corresponding evolutionary histories of euglenozoans and dinoflagellates. *Proceedings of the National Academy of Sciences of the United States of America, 106*, 9963–9970.

Lukeš, J., Archibald, J. M., Keeling, P. J., Doolittle, W. F., & Gray, M. W. (2011). How a neutral evolutionary ratchet can build cellular complexity. *IUBMB Life, 63*, 528–537.

Marande, W., Lukeš, J., & Burger, G. (2005). Unique mitochondrial genome structure in diplonemids, the sister group of kinetoplastids. *Eukaryotic Cell, 4*, 1137–1146.

Marande, W., & Burger, G. (2007). Mitochondrial DNA as a genomic jigsaw puzzle. *Science, 318*, 415.

Masuda, I., Matsuzaki, M., & Kita, K. (2010). Extensive frameshift at all AGG and CCC codons in the mitochondrial cytochrome c oxidase subunit 1 gene of *Perkinsus marinus* (Alveolata; Dinoflagellata). *Nucleic Acids Research, 38*, 6186–6194.

Milbury, C. A., & Gaffney, P. M. (2008). Complete mitochondrial sequence of the eastern oyster *Crassostrea virginica*. *Marine Biotechnology, 7*, 697–712.

Moore, R. B., Oborník, M., Janouškovec, J., Chrudimský, T., Vancová, M., Green, D. H., et al. (2008). A photosynthetic alveolate closely related to apicomplexan parasites. *Nature, 451*, 959–963.

Mylnikov, A. P. (2009). Ultrastructure and phylogeny of colpodellids (Colpodellida, Alveolata). *The Biological Bulletin, 36*, 582–590.

Nash, E. A., Barbrook, A. C., Edwards-Stuart, R. K., Bernhardt, K., Howe, C. J., & Nisbet, E. R. (2007). Organization of the mitochondrial genome in the dinoflagellate *Amphidinium carterae*. *Molecular Biology and Evolution, 24*, 1528–1536.

Nash, E. A., Nisbet, R. E., Barbrook, A. C., & Howe, C. J. (2008). Dinoflagellates: a mitochondrial genome all at sea. *Trends in Genetics, 24*, 328–335.

Nass, M. M. K., Schori, L., Ben-Shaul, Y., & Edelman, M. (1974). Size and configuration of mitochondrial DNA in *Euglena gracilis*. *Biochimica et Biophysica Acta, 374*, 283–291.

Nelson, M. J., Dang, Y., Filek, E., Zhang, Z., Yu, V. W., Ishida, K., et al. (2007). Identification and transcription of transfer RNA genes in dinoflagellate plastid minicircles. *Gene, 392*, 291–298.

Norman, J. E., & Gray, M. W. (1997). The cytochrome oxidase subunit 1 gene (cox1) from the dinoflagellate, *Crypthecodinium cohnii*. *FEBS Letters, 413*, 333–338.

Norman, J. E., & Gray, M. W. (2001). A complex organization of the gene encoding cytochrome oxidase subunit 1 in the mitochondrial genome of the dinoflagellate,

Crypthecodinium cohnii: homologous recombination generates two different cox1 open reading frames. *Journal of Molecular Evolution, 53*, 351–363.

Oborník, M., Janouškovec, J., Chrudimský, T., & Lukeš, J. (2009). Evolution of the apicoplast and its hosts: from heterotrophy to autotrophy and back again. *International Journal of Parasitology, 39*, 1–12.

Oborník, M., Vancová, M., Lai, D. H., Janouškovec, J., Keeling, P. J., & Lukeš, J. (2011). Morphology and ultrastructure of multiple life cycle stages of the photosynthetic relative of apicomplexa, *Chromera velia*. *Protist, 162*, 115–130.

Oborník, M., Modrý, D., Lukeš, M., Černotíková-Stříbrná, E., Cihlář, J., Tesařová, M., et al. (2012). Morphology, ultrastructure and life cycle of *Vitrella brassicaformis* n. sp., n. gen., a novel chromerid from the Great Barrier Reef. *Protist., 163*, 306–323.

Okamoto, N., & McFadden, G. I. (2008). The mother of all parasites. *Future Microbiology, 3*, 391–395.

Panigrahi, A. K., Ogata, Y., Zikova, A., Anupama, A., Dalley, R. A., Acestor, N., et al. (2009). A comprehensive analysis of *Trypanosoma brucei* mitochondrial proteome. *Proteomics, 9*, 434–450.

Raczynska, K. D., Le Ret, M., Rurek, M., Bonnard, G., Augustyniak, H., & Gualberto, J. M. (2006). Plant mitochondrial genes can be expressed from mRNAs lacking stop codons. *FEBS Letters, 580*, 5641–5646.

Ray, D. S., & Hanawalt, P. C. (1965). Satellite DNA components in *Euglena gracilis* cells lacking chloroplasts. *Journal of Molecular Biology, 11*, 760–768.

Rehkopf, D. H., Gillespie, D. E., Harrell, M. I., & Feagin, J. E. (2000). Transcriptional mapping and RNA processing of the *Plasmodium falciparum* mitochondrial mRNAs. *Molecular and Biochemical Parasitology, 105*, 91–103.

Roger, A. J., & Simpson, A. G. B. (2009). Revisiting the root of the eukaryotic tree. *Current Biology, 19*, R165–R167.

Roy, J., Faktorová, D., Lukeš, J., & Burger, G. (2007). Unusual mitochondrial genome structures throughout the Euglenozoa. *Protist, 158*, 385–396.

Rusconi, C. P., & Cech, T. R. (1996). Mitochondrial import of only one of three nuclear-encoded glutamine tRNAs in *Tetrahymena thermophila*. *EMBO Journal, 15*, 3286–3295.

Russell, R. D., & Beckenbach, A. T. (2008). Recoding of translation in turtle mitochondrial genomes: programmed frameshift mutations and evidence of a modified genetic code. *Journal of Molecular Evolution, 67*, 682–695.

Saldarriaga, J. F., Taylor, F. J. R. M., Cavalier-Smith, T., Menden-Deuer, S., & Keeling, P. J. (2004). Molecular data and the evolutionary history of dinoflagellates. *European Journal of Protistology, 40*, 85–111.

Simspon, A. G. B., Gill, E. E., Callahan, H. A., Litaker, R. W., & Roget, A. G. (2004). Early evolution within kinetoplastids (Euglenozoa), and the late emergence of trypanosomatids. *Protist, 155*, 407–422.

Simpson, A. G. B., Stevens, J. R., & Lukeš, J. (2006). The evolution and diversity of kinetoplastid flagellates. *Trends in Parasitology, 22*, 168–174.

Skovgaard, A., Massana, R., Balagué, V., & Saiz, E. (2005). Phylogenetic position of the copepod-infesting parasite *Syndinium turbo* (Dinoflagellata, Syndinea). *Protist, 156*, 413–423.

Skovgaard, A., Meneses, I., & Angélico, M. M. (2009). Identifying the lethal fish egg parasite *Ichthyodinium chabelardi* as a member of Marine Alveolate Group I. *Environmental Microbiology, 11*, 2030–2041.

Slamovits, C. H., Saldarriaga, J. F., Larocque, A., & Keeling, P. J. (2007). The highly reduced and fragmented mitochondrial genome of the early-branching dinoflagellate *Oxyrrhis marina* shares characteristics with both apicomplexan and dinoflagellate mitochondrial genomes. *Journal of Molecular Biology, 372*, 356–368.

Slamovits, C. H., & Keeling, P. J. (2011). Contributions of *Oxyrrhis marina* to molecular biology, genomics and organelle evolution of dinoflagellates. *Journal of Plankton Research, 33*, 591–602.

Spencer, D. F., & Gray, M. W. (2011). Ribosomal RNA genes in *Euglena gracilis* mitochondrial DNA: fragmented genes in a seemingly fragmented genome. *Molecular Genetics & Genomics, 285*, 19–31.

Stentiford, G. D., & Shields, J. D. (2005). A review of the parasitic dinoflagellates *Hematodinium* species and *Hematodinium*-like infections in marine crustaceans. *Diseases of Aquatic Organisms, 66*, 47–70.

Stuart, K. D., Schnaufer, A., Ernst, N. L., & Panigrahi, A. K. (2005). Complex management: RNA editing in trypanosomes. *Trends in Biochemical Sciences, 30*, 97–105.

Talen, J. L., Sanders, J. P. M., & Flavell, R. A. (1974). Genetic complexity of mitochondrial DNA from *Euglena gracilis*. *Biochimica et Biophysica Acta, 374*, 129–135.

Tessier, L. H., van der Speck, H., Gualberto, J. M., & Grienenberger, J. M. (1997). The cox1 gene from *Euglena gracilis*: a protist mitochondrial gene without introns and genetic code modifications. *Current Genetics, 31*, 208–213.

Týč, J., Long, S., Jirků, M., & Lukeš, J. (2010). YCF45 protein, usually associated with plastids, is targeted into the mitochondrion of *Trypanosoma brucei*. *Molecular and Biochemical Parasitology, 173*, 43–47.

Vlček, C., Marande, W., Teijeiro, S., Lukeš, J., & Burger, G. (2011). Systematically fragmented genes in a multipartite mitochondrial genome. *Nucleic Acids Research, 39*, 979–988.

Waller, R. F., & Jackson, C. J. (2009). Dinoflagellate mitochondrial genomes: stretching the rules of molecular biology. *Bioessays, 31*, 237–245.

Waller, R. F., & Keeling, P. J. (2006). Alveolate and chlorophycean mitochondrial cox2 genes split twice independently. *Gene, 383*, 33–37.

Weatherby, K., Murray, S., Carter, D., & Šlapeta, J. (2011). Surface and flagella morphology of the motile form of *Chromera velia* revealed by field-emission scanning electron microscopy. *Protist, 162*, 142–153.

Yasuhira, S., & Simpson, L. (1996). Guide RNAs and guide RNA genes in the cryptobiid kinetoplastid protozoan, *Trypanoplasma borreli*. *RNA, 2*, 1153–1160.

Yasuhira, S., & Simpson, L. (1997). Phylogenetic affinity of mitochondria of *Euglena gracilis* and kinetoplastids using cytochrome oxidase I and hsp60. *Journal of Molecular Evolution, 44*, 341–347.

Zhang, H., Bhattacharya, D., Maranda, L., & Lin, S. (2008). Mitochondrial *cob* and *cox1* genes and editing of the corresponding mRNAs in *Dinophysis acuminata* from Narragansett Bay, with special reference to the phylogenetic position of the genus *Dinophysis*. *Applied and Environmental Microbiology, 74*, 1546–1554.

Zhang, H., & Lin, S. (2002). Detection and quantification of *Pfiesteria piscicida* by using the mitochondrial cytochrome *b* gene. *Applied and Environmental Microbiology, 68*, 989–994.

Zhang, H., & Lin, S. (2005). Mitochondrial cytochrome b mRNA editing in dinoflagellates: possible ecological and evolutionary associations? *Journal of Eukaryotic Microbiology, 52*, 538–545.

Zhang, H., & Lin, S. (2008). mRNA editing and spliced-leader RNA trans-splicing groups *Oxyrrhis, Noctiluca, Heterocapsa*, and *Amphidinium* as basal lineages of dinoflagellates. *Journal of Phycology, 44*, 703–711.

Zhang, H., Campbell, D. A., Sturm, N. R., Dungan, C. F., & Lin, S. (2011). Spliced leader RNAs, mitochondrial gene frameshifts and multi-protein phylogeny expand support for the genus *Perkinsus* as a unique group of alveolates. *PLoS ONE, 6*. e19933.

CHAPTER SEVEN

Evolution of Mitochondrial Introns in Plants and Photosynthetic Microbes

Linda Bonen[1]
Biology Department, University of Ottawa, Ottawa, Canada
[1]Corresponding author. E-mail: lbonen@uottawa.ca

Contents

Advances in Botanical Research, Volume 63
ISSN 0065-2296,
http://dx.doi.org/10.1016/B978-0-12-394279-1.00007-7

Abstract

Introns within the mitochondrial genes of plants and algae are categorized as group I or group II based on distinctive structural properties that are important for splicing. These families of intron ribozymes are mobile genetic elements that can invade genomes through horizontal transfer and spread copies of themselves during evolution. Fully competent group I and group II introns encode their own machinery for splicing and mobility, but those found in present-day plant/algal mitochondria lack certain features and virtually none have been shown to be self-splicing *in vitro*. Accessory proteins are required for splicing *in vivo* and such machinery can include nuclear-encoded homologues of intronic RNA maturases as well as other RNA-binding proteins, as determined primarily from studies of *Arabidopsis* mutants. Several mitochondrial introns in plants/algae have been disrupted by DNA rearrangements during evolution and the independently-transcribed pieces are expressed through *trans*-splicing. The interplay between RNA editing and splicing can also confer an additional layer of complexity and coevolutionary constraint. This chapter highlights aspects of the evolutionary history of mitochondrial introns in plants and algae, the recruitment of host machinery for splicing, and the relationships between splicing and other expression events in the mitochondrion.

1. INTRODUCTION

The mitochondrial genomes of plants and algae contain genes required for respiratory function as well as for expression, and a subset of them typically possess introns. Such intervening sequences must be precisely removed from precursor RNAs by splicing to generate mature transcripts to perform their functions as protein-coding messenger RNAs, ribosomal RNAs or transfer RNAs. The mitochondrial introns are highly structured and can be classified as either group I or group II based on distinctive features. They belong to families of self-splicing RNAs (ribozymes), mobile genetic elements that have invaded mitochondrial genomes at various times during evolution.

This chapter focuses on the mitochondrial introns found in plants and green algae. Photosynthetic function in these lineages was acquired from a cyanobacterial endosymbiont in a mitochondrion-containing unicellular eukaryotic host (reviewed in Gray, 1992), with green algae (charophytes and chlorophytes) belonging to the progenitor lineage leading to land plants.

Green algae display great diversity in mitochondrial genome size and complexity, but in general their intron content is lower than in land plants. It appears that shortly after divergence of the land plant lineage from the charophytes, their mitochondrial genomes expanded considerably in physical complexity, with a concomitant increase in intron content. More is known about these mitochondrial genetic systems than a second category of eukaryotic microbes that acquired their photosynthetic function through secondary endosymbiotic events (reviewed in Keeling, 2010). Examples of this group are rhodophytes (red algae), chrysophytes (golden algae) and chromalveolates (including brown algae, diatoms and dinoflagellates). Some of these microbes have complex histories and it is not yet known whether their respiratory function is provided solely by preexisting mitochondria in the host eukaryote, whether mitochondria from the invading eukaryotic symbiont are maintained, or whether genes from the two separate origins have undergone lateral transfer events. This chapter focuses on the evolutionary history and behaviour of mitochondrial introns in plants and algae, and discusses the implications of their presence in these mitochondrial genomes.

2. CATEGORIES OF MITOCHONDRIAL INTRONS IN PLANTS AND ALGAE

2.1. Group I Introns

Group I introns are widespread in nature, and have been found in the mitochondria of plants, algae, protists, fungi and several early-diverging lineages of animals, as well as within the genomes of chloroplasts, bacteria, bacteriophages, and the nuclear ribosomal RNA genes from a variety of unicellular eukaryotes (reviewed in Nielsen & Johansen, 2009). Based on conserved structural elements, they have been further divided into 11 subgroups (Michel & Westhof, 1990). Some group I introns contain an open reading frame (ORF), which encodes information for mobility and splicing. Conventional group I ORFs are called homing endonuclease genes (HEGs) and the ones present in plant/algal mitochondrial group I introns typically have distinctive LAGLIDADG amino acid motifs. Group I introns can gain or lose such ORFs quite easily during evolution, and they can be located in different non-core positions within the intron. In a few cases, the homing endonuclease/maturase ORF is fused in frame with the upstream exon and may reflect a long shared history. Such configuration also underscores the complexity of evolutionary constraint on both RNA structure as well as at the protein-coding level. Thus, fully competent group I introns are a composite of a highly structured RNA ribozyme and an ORF encoding mobility and splicing machinery.

2.1.1. Group I intron structure and splicing biochemistry

Group I introns have a distinctive secondary structure composed of ten helical domains (called P1–P10) as shown in Fig. 7.1A. This core structure has been subdivided into three components, designated as the substrate domain (P1-P2), the catalytic domain (P3-P7-P8-P9) and the scaffolding domain (P4-P5-P6) (reviewed in Nielsen & Johansen, 2009). The crystal structure of several group I introns has been determined (although none

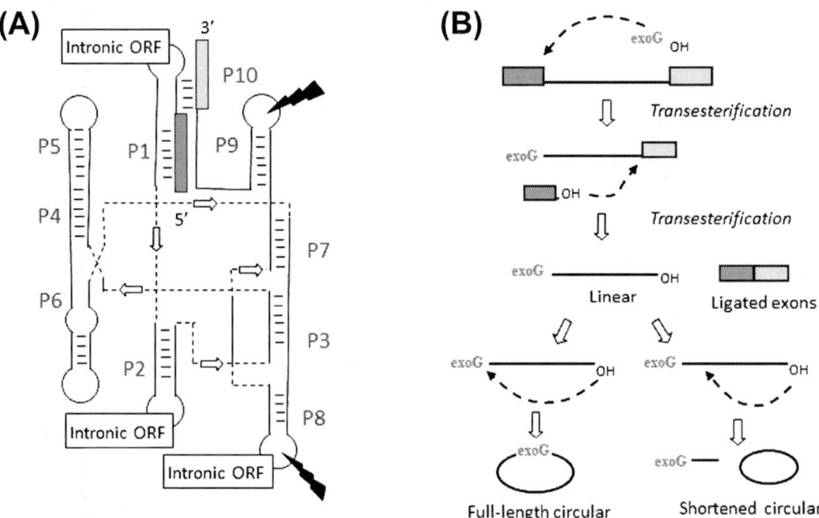

Figure 7.1 (A) Secondary structure model of a classic group I intron, showing exons as green blocks and base pairing within helical domains P1–P10. Intronic ORFs (typically homing endonuclease HEG genes) can be located at various non-core locations, three of which are shown. Lightning rod symbols indicate positions of disruption observed in plant/algal mitochondria that convert the intron from a *cis*-splicing form to a *trans*-splicing form, namely within the P9 loop for *cox1* in the lycophytes, *I. engelmannii* and *S. moellendorffii* (Grewe *et al.*, 2009; Hecht *et al.*, 2011) and within the P4 loop in the green algal *cox1* in *Helicosporidium* (Pombert & Keeling, 2010). These introns are located at different locations within *cox1*. (B) Biochemistry of group I splicing with two trans-esterification steps, the first of which is initiated by an external nucleophile. The intron is liberated as a linear RNA species but can subsequently be circularized as either a full-length or shortened circle. For interpretation of the references to colour in this figure legend, the reader is referred to the online version of this book.

from plant/algal mitochondria), and the positions of two catalytically active Mg^{2+} identified at the active site. The helical regions of group I introns are highly conserved to maintain ribozyme activity, whereas the unstructured loop regions can tolerate large insertions and this is where intronic ORFs are located. Fig. 7.1A shows several such locations, and although group I introns typically do not contain more than one ORF, a *cox1* intron in the entomoparasitic green alga *Helicosporidium* has one in the P8 loop and another in the P4 unstructured region (Pombert & Keeling, 2010).

Group I introns are excised from flanking exons by two trans-esterification steps, with an external guanosine acting as the first nucleophile (Fig. 7.1B). The liberated linear intron can undergo subsequent nucleophilic attack so that the final circularized form is either the full-length intron or one lacking a stretch at the 5' end of the intron. Certain group I introns are effective ribozymes in that they can self-splice in the test tube in the absence of any protein machinery, although accessory splicing factors are believed to be required *in vivo*.

2.2. Group II Introns

Group II introns have been identified in the mitochondria of plants, algae, fungi, protists and several early-diverging animal lineages, as well as in chloroplasts, bacteria and to a lesser extent in archaebacteria (reviewed in Lambowitz & Zimmerly, 2011). Group II intron mobile elements contain an ORF that confers mobility and splicing function. These ORFs have several discrete domains: the reverse transcriptase (RT) domain, a zinc-finger endonuclease (EN) domain, a DNA-binding (D) domain, and an RNA-binding/maturase (X) domain for splicing. Full-length group II ORFs (RT-matR) are typically at least 2 kb in length, which is about three times longer than a typical group I ORF. As retroelements, group II introns move *via* an RNA intermediate that is mediated by the self-encoded reverse transcriptase. Group II introns can move by two reverse splicing mechanisms, namely site-specific allelic retrohoming and ectopic retrotransposition into new non-allelic sites. None of the land plant mitochondrial group II introns have exhibited ribozyme activity *in vitro*, but a *cox1* group II intron from the brown alga *Pylaiella littoralis* can self-splice in the test tube and has been used as a model system for studying ribozyme properties (cf. Costa, Fontaine, Loiseaux-de Goer, & Michel, 1997). Similar to the case for group I introns, splicing *in vivo* requires accessory factors, and mutant studies are shedding light on the complexity of the splicing machinery (see below).

2.2.1. Group II intron structure and splicing biochemistry

Group II introns are distinguished by six helical domains (dI to dVI) extending from a central hub as shown in Fig. 7.2A (reviewed in Bonen & Vogel, 2001; Lambowitz & Zimmerly, 2011; Michel & Ferat, 1995). The highly conserved 34-nt domain V is particularly diagnostic of group II introns and it forms part of the ribozyme core along with sequences in domain I. The binding of two Mg^{2+} ions within domain V is crucial for catalytic function and domain VI is important for providing the attacking nucleophile for the first step of splicing. Sequences within domain I (exon binding sites (EBS)) are complementary to ones at the extreme 3' end of the upstream exon (intron binding sites (IBS)) and this interaction is also important for correct folding. Valuable higher-order structural information has come from X-ray crystallography studies (reviewed in Pyle, 2010). Mitochondrial group II introns have been subdivided into groups IIA and IIB based on particular structural features (Michel & Ferat, 1995) and both types are found in plant/algal mitochondria. The unstructured regions of group II introns are quite amenable to accepting extra sequences, particularly within domain IV. In the lycophyte (an early-diverging lineage of vascular plants), *Selaginella*

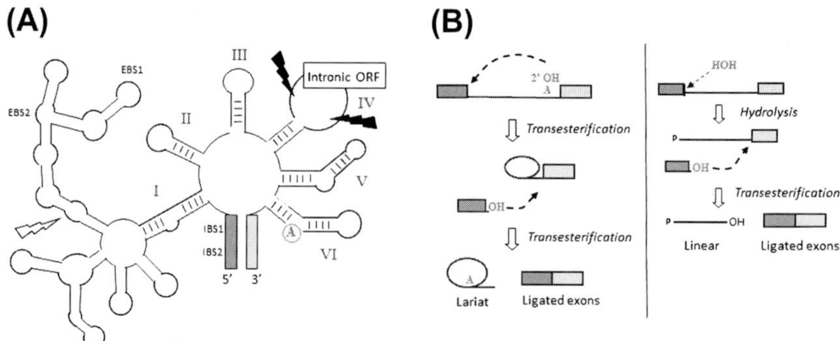

Figure 7.2 (A) Secondary structure model of a classic group II intron, showing exons as green blocks and base pairing within helical domains I–VI. Intronic ORFs (with reverse transcriptase-maturase domains) are located within domain IV, but the reading frame can also extend through domains I–III. Black lightning rod symbols indicate positions of disruption within dIV observed in plant/algal mitochondria that convert the intron from a *cis*-splicing to *trans*-splicing form, and the grey symbol is for an additional break seen in the tripartite *Oenothera berteriana nad2* intron 3 (Knoop *et al.*, 1997). (B) Biochemistry of group II splicing showing the two transesterification steps that generate spliced exons and either a classic lariat excised intron or a linear form if water serves as the initiating nucleophile. Other unusual excised intronic forms have also been observed for certain flowering plant mitochondrial introns that lack conventional group II features (Li-Pook-Than & Bonen, 2006). See the colour plate.

moellendorffii, one of the mitochondrial group II introns within *nad1*, contains another respiratory chain gene, *nad4L* (Hecht, Grewe, & Knoop, 2011).

The biochemistry of group II intron removal from precursor RNAs proceeds through two transesterification reactions, the first being initiated by the 2'OH of the branch point adenosine in domain VI (Fig. 7.2B). In this major pathway, the intron is liberated as a lariat. An alternative hydrolytic pathway has also been observed, in which water acts as the first attacking nucleophile. The biochemistry of the lariat-forming pathway is the same as for nuclear spliceosomal introns. This feature and structural similarities between the group II intron domains and the small nuclear RNAs (snRNAs) of eukaryotic nuclear spliceosomes has prompted the proposal that they share a common evolutionary origin (reviewed in Lambowitz & Zimmerly, 2011). It has been speculated that group II introns in the endosymbiotic respiring α-proteobacterium (which gave rise to the mitochondrion) could have spread to the host genome and been instrumental in triggering the formation of the nuclear envelope, in order to segregate splicing from protein synthesis and prevent premature translation of intron-containing RNAs (Martin & Koonin, 2006).

2.2.2. Degenerate features of plant/algal mitochondrial group II introns

Although most of the group I introns found in plant/algal mitochondria exhibit conventional features (aside from *trans*-splicing, see below), many of the group II introns have degenerate structures or lack certain features, and this is especially evident in flowering plant mitochondria (reviewed in Bonen, 2008, 2010). Some examples include mispairs (e.g. U-U or C-U), which weaken the helical structure of core domains, the absence of a bulged adenosine branch point in domain VI, an exceptionally long domain V loop, and extended linkers between domains V and VI (cf. Carrillo, Chapdelaine, & Bonen, 2001; Li-Pook-Than & Bonen, 2006). Notably certain A-C mispairs within helical regions are converted to canonical A-U pairs by RNA editing (see below). RNA editing in plant mitochondria is the phenomenon whereby certain specific cytidines are post-transcriptionally converted to uridines, for the most part at non-synonymous positions within coding regions but also occasionally within non-coding sequences such as introns (reviewed in Takenaka, Verbitskiy, van der Merwe, Zehrmann, & Brennicke, 2008). Sequence variation among plants at the branch point position of domain VI has also been observed, and this is especially unexpected considering both the importance of the bulged adenosine for splicing and the very low rate of nucleotide substitution typically seen in plant mitochondria (Wolfe, Li, & Sharp, 1987). These observations suggest that such introns are not subject to the evolutionary constraints typical of group II ribozymes and they raise questions about the mechanisms of splicing and compensatory accessory machinery. Several of these atypical introns have been shown to be excised as unusual physical forms (Li-Pook-Than & Bonen, 2006). Deviation from the classic structure appears to have occurred early in seed plant evolution, as there are examples where orthologous introns in earlier-diverging lineages have conventional folding, namely *nad1* intron 1 in the lycophyte, *Isoetes engelmannii* (Carrillo *et al.*, 2001) and *nad4* intron 2 in the multicellular green alga, *Chara vulgaris* (Bonen, 2008). In addition, deviant features have occasionally been seen for a few algal mitochondrial introns. For example, two *trans*-splicing introns in the green alga, *Mesostigma virilis*, have a truncated domain I and/or are missing domain II (Turmel, Otis, & Lemieux, 2002).

2.3. *Trans*-splicing Introns

During the evolutionary history of plants and algae, there have been instances when mitochondrial DNA rearrangements have occurred within introns so that gene segments have been translocated to very distant genomic sites (reviewed in Bonen, 2008; Glanz & Kuck, 2009). Such mutations, however, have not resulted in pseudogenes because independent RNAs are synthesized and these precursor RNAs can reconstitute the correct higher-order structure to enable splicing *in trans*. *Trans*-splicing appears to be at least partially mediated by base pairing between domains, and is presumably facilitated by accessory proteins. There are examples of both group I and group II *trans*-splicing introns in plant/algal mitochondrial genes and comparative analysis has enabled an estimation of the approximate timing of some such gene-disruption events (cf. Malek & Knoop, 1998; Qiu & Palmer, 2004). Insights into the positions at which breaks can be tolerated has also come from studies in which group II intron constructs with artificially introduced breaks and compromised base-pairing ability have been introduced into host bacteria and monitored for faithful splicing (cf. Quiroga, Kronstad, Ritlop, Filion, & Cousineau, 2011).

2.3.1. *Group I intron* trans-*splicing*

Although group I introns are less prevalent in the mitochondrial genomes of plants/algae than are group II introns, several cases of introns requiring *trans*-

splicing have recently been discovered. In the sister lycophytes, *I. engelmannii* (Grewe, Viehoever, Weisshaar, & Knoop, 2009) and *Selaginella moellendorfii* (Hecht *et al.*, 2011), the *cox1* gene contains the same group I *trans*-splicing intron, with the disruption within the P9 loop. The *cox1* gene in the entomoparasitic non-photosynthetic green alga, *Helicosporidium*, also requires *trans*-splicing for expression, but this intron is located at a different position and the break is located within the P8 loop of the intron (Pombert & Keeling, 2010).

2.3.2. *Group II intron* trans*-splicing*

The number of group II introns that have been disrupted by DNA rearrangements in the mitochondria of plants and algae is considerably higher than for group I introns, and the largest sets are in vascular plants. A total of seven introns have been identified to undergo *trans*-splicing (although not all in the same species). They are within NADH dehydrogenase subunit genes, namely *nad1*, *nad2*, *nad5*, except for one within *cox2*. Again all of these genes retain functionality because of *trans*-splicing correction at the RNA level. Conversion from a *cis* to a *trans* state has occurred at different times during evolution, and that timing has been deduced by comparative analysis with counterpart introns from early-diverging plant lineages such as ferns, mosses, and hornworts (reviewed in Knoop, 2004). In *S. moellendorfii*, disruptions within group II introns in the *cob* and *atp9* genes give rise to two additional instances of *trans*-splicing, as well as one within the *cox2* gene (Hecht *et al.*, 2011) which is at the same position as an independently derived counterpart in the eudicot, *Allium cepa* (Kim & Yoon, 2010).

In the case of the fourth (and terminal) *nad1* intron, which incidentally contains the only RNA maturase ORF (*matR*) in seed plant mitochondria, conversions from the *cis* to *trans* state have happened very recently and independently in different lineages of flowering plants (Qiu & Palmer, 2004). The break in some cases is upstream of *matR* and in other cases downstream of it. From a large survey of land plants, it was estimated that the former have occurred about twice as often as the latter. It is unknown as yet why this intron is particularly susceptible to (or tolerant of) breaks. The presence of short direct repeats at the break points of the *trans*-type intron compared with the *cis*-type counterpart (Chapdelaine & Bonen, 1991) suggests that either DNA rearrangement occurred by homologous recombination across such sequences, or that the repeats were created during repair of double-stranded DNA breaks.

Although almost all the identified cases of *trans*-splicing in plant/algal mitochondria appear due to a simple break in the ancestral *cis*-splicing intron, necessitating a bimolecular splicing reaction, in *Oenothera berteriana* mitochondria, the third intron in the *nad5* gene is broken into three pieces, which together reconstitute a structure competent for splicing (Knoop, Altwasser, & Brennicke, 1997). Evolution of this tripartite structure appears to prevent the type of mis-splicing that has been observed for the counterpart bipartite *nad5* introns (Elina & Brown, 2010).

In green algal mitochondria, there are also several examples of *trans*-splicing group II introns. For example, in *M. virilis*, two group II introns in the *nad3* gene are disrupted (Turmel *et al.*, 2002). The locations of these introns indicate evolutionary events specific to this lineage. All three of the group II introns in this mitochondrial genome appear somewhat degenerate.

2.3.3. Other forms of trans-*splicing in photosynthetic microbes*

In the dinoflagellate, *Karlodinium micrum*, *trans*-splicing is required to generate the full-length mitochondrial *cox3* coding sequence, however the exons are not flanked by group I or group II intron features, so there appears to be a novel form of splicing where a stretch of five non-encoded adenosines is inserted at the junction (reviewed in Waller & Jackson, 2009). This is rather reminiscent of the unusual mitochondrial *cox1 trans*-splicing observed in the non-photosynthetic protist, *Diplonema papillatum* (Kiethega, Turcotte, & Burger, 2011).

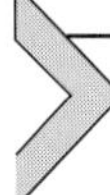

3. DISTRIBUTION OF MITOCHONDRIAL INTRONS IN PLANTS AND PHOTOSYNTHETIC MICROBES

Because group I/II introns are mobile genetic elements, it is perhaps not surprising that their numbers and location vary among the mitochondrial genomes of plants and algae (Fig. 7.3). In general, algae have fewer mitochondrial introns than land plants, but a proportionately larger number of group I introns. Only one group I intron has been identified in seed plant mitochondria and it is present in only certain disparate lineages (see below). Its location is at a position within the *cox1* gene that is known to be a strong homing site and vulnerable to frequent invasion in algae, protists and fungi. In contrast to the variability of mitochondrial introns in early-diverging plant/algal lineages, most of those in seed plants have shared a stable history for more than 200 million years.

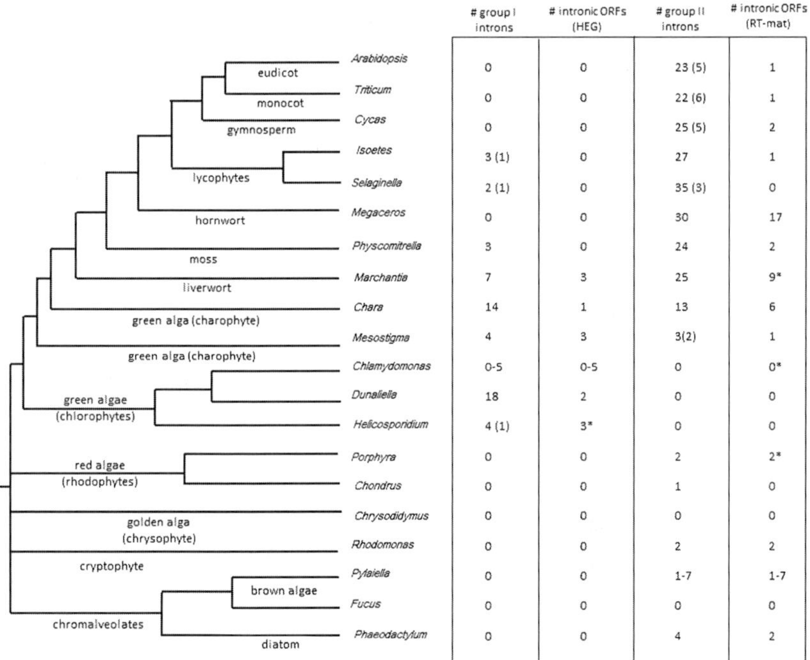

	# group I introns	# intronic ORFs (HEG)	# group II introns	# intronic ORFs (RT-mat)
Arabidopsis	0	0	23 (5)	1
Triticum	0	0	22 (6)	1
Cycas	0	0	25 (5)	2
Isoetes	3 (1)	0	27	1
Selaginella	2 (1)	0	35 (3)	0
Megaceros	0	0	30	17
Physcomitrella	3	0	24	2
Marchantia	7	3	25	9*
Chara	14	1	13	6
Mesostigma	4	3	3(2)	1
Chlamydomonas	0-5	0-5	0	0*
Dunaliella	18	2	0	0
Helicosporidium	4 (1)	3*	0	0
Porphyra	0	0	2	2*
Chondrus	0	0	1	0
Chrysodidymus	0	0	0	0
Rhodomonas	0	0	2	2
Pylaiella	0	0	1-7	1-7
Fucus	0	0	0	0
Phaeodactylum	0	0	4	2

Figure 7.3 Distribution of mitochondrial group I/II introns and their intronic ORFs in selected plants and algae. Complete mitochondrial genome sequences and references are available under accessions numbers: *Arabidopsis thaliana* (NC_001284), *Triticum aestivum* (NC_007579), *C. taitungensis* (NC_010303), *I. engelmannii* (FJ010859, FJ176330, FJ390841), *S. moellendorfii* (JF338143–JF338147), *Megaceros aeignmaticus* (NC_012651), *P. patens* (NC_007945), *M. polymorpha* (NC_001660), *C. vulgaris* (NC_005255), *Mesostigma viride* (NC_008240), *C. reinhardtii* (different strains EU306617–EU306623), *Dunaliella salina* (GQ250045), *Helicosporidium* (GQ339576), *Porphyra purpurea* (NC_002007), *C. crispus* (NC_001677), *C. synuroideus* (NC_002174), *R. salina* (NC_002572), *P. littoralis* (NC_003055), *Fucus vesiculosus* (NC_007683), *P. tricornutum* (HQ840789). Numbers in parentheses indicate *trans*-splicing introns. Asterisks denote the presence of additional free-standing maturase ORF homologues located in mitochondrial intergenic spacers in *C. reinhardtii* (Boer & Gray, 1988), *M. polymorpha* (Oda *et al.*, 1992), *Porphyra purpurea* (Burger *et al.*, 1999) and *Helicosporidium* (Pombert & Keeling, 2010),. Note that the presence of the *I. engelmannii* ORF was reported in Hecht *et al.* (2011). Numbers for *C. reinhardtii* strains (Smith & Lee, 2008) and *P. littoralis* specimens (Ikuta *et al.*, 2008) represent the total number of sites at which introns have been found, but only a subset are present in a given mitochondrial genome. Branch lengths are not to scale and tree topology is according to Rodriguez-Ezpeleta, Philippe, Brinkmann, Becker, and Melkonian, (2007). *Sources include data compiled by Hecht* et al. *(2011) and Turmel* et al. *(2003)*

3.1. Land Plants and Green Algae

There is a marked bias as to which mitochondrial genes are populated by introns when vascular plants are compared with earlier-diverging lineages (Fig. 7.4). In seed plant mitochondria, the vast majority are located within *nad* (NADH dehydrogenase) genes, whereas in non-vascular plants and algae, there are higher numbers in respiratory chain genes such as *cox1* (cytochrome oxidase subunit I) and in ribosomal RNA genes, rather reminiscent of what is seen for fungal mitochondria. Moreover, although vascular and non-vascular land plants typically contain approximately the same number of introns, that is, 20–25, very few are at orthologous sites. For example, the liverwort *Marchantia polymorpha* and seed plants have only one at a common site (namely, *nad2* intron 3), consistent with it having been present in their common ancestor. The moss *Physcomitrella patens*, has 24

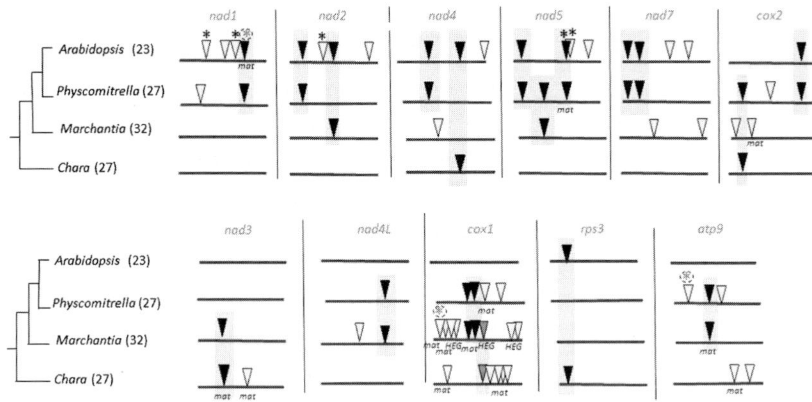

Figure 7.4 Mitochondrial introns that are located at homologous sites in the mitochondrial genomes of selected plants/algae. Representatives of flowering plants (*A. thaliana*, Unseld, Marienfeld, Brandt, & Brennicke, 1997), mosses (*P. patens*, Terasawa *et al.*, 2007), liverworts (*M. polymorpha,* Oda *et al.*, 1992) and green algae (*C. vulgaris*, Turmel *et al.*, 2003). The total number of mitochondrial introns present is shown in parentheses. Only those genes that have at least one shared intron among two or more of these four organisms are shown. Black arrowheads (and shading) indicate shared intron positions, yellow arrows represent non-shared sites and red arrows depict the strong homing site in *cox1*, which is also occupied by an intron in certain flowering plants. Black asterisks indicated *trans*-splicing introns in the plants shown and circled green asterisks indicate that this intron is *trans*-splicing in certain species not shown in the figure, namely, *nad1* in many flowering plants; *atp9* and *cox1* in *S. moellendorfii* (see text). HEG and *matR* depict intronic ORFs of the group I or group II intron type, respectively.. *Data are compiled from various sources including Hecht* et al. *(2011) and Turmel* et al. *(2003), and are depicted as in Terasawa* et al. *(2007).* See the colour plate.

group II introns and three group I introns, of which a subset are at the same location as in flowering plants and another subset at the same position as in *M. polymorpha* mitochondria (Fig. 7.4). The lycophyte *S. moellendorffii* has the largest set of mitochondrial introns (with a total of 37) identified in any plant/alga to date, even though it has one of the smallest gene sets (Hecht *et al.*, 2011).

Systematic surveys of the status of mitochondrial introns in land plants have provided important information about the timing of their acquisition/loss. For example, in a large-scale study of the status of the *nad1* gene, which typically contains four introns in vascular plants, but lacks introns in liverworts and algae (Fig. 7.4), several were traced back to the common ancestor of mosses–hornwort–vascular plants (Dombrovska & Qiu, 2004). This provides strong support for their vertical inheritance over approximately 350–400 million years and arguments have been made that such long retention may reflect the evolution of a functional role in the organelle.

In the evolutionarily divergent groupings of unicellular green algae, there is a broad range in mitochondrial intron number among organisms and in the ratio between group I and group II types (Fig. 7.3), but typically they have few introns and most of them are group I. In *C. vulgaris*, a charophyte from the lineage that is believed to represent the closest green algal relative of land plants, there are about the same total number of introns as in land plants, but about half of them are group I introns (Turmel, Otis, & Lemieux, 2003) concentrated in the *cox1*, *rns* and *rnl* genes (as in many other green algae) and only one group II intron (*rps3*) is at the same location as in seed plants (Fig. 7.4). Only one of the *C. vulgaris* mitochondrial group I introns encodes a LAGLIDADG-type homing endonuclease (within *rns*) although about half of the group II introns contain an RT-maturase ORF. Notably one of the *nad3* intronic ORFs is fused in frame with the upstream exon, which drives its expression, as are a subset of those in *M. polymorpha* mitochondria (Oda *et al.*, 1992).

The chlorophyte green alga *Chlamydomonas reinhardtii* provides an excellent illustration of the dynamic nature of mitochondrial intron gain and loss. Some strains of *C. reinhardtii* have no introns, although they do possess a free-standing RT-mat (*rt1*) gene (Boer & Gray, 1988), whereas others have 1–3 introns at five different sites in the *cob*, *cox1* and ribosomal RNA genes, with the latter introns being quite degenerate (Smith & Lee, 2008). Their mitoribosomes contain discrete short ribosomal RNA segments that are not spliced together into a long classic ribosomal RNA. However, several of these rRNA segments do undergo splicing in certain *C. reinhardtii* strains to generate somewhat longer rRNA segments. Another green alga (chlorophyte) *Dunaliella salina* has an even larger set of group I introns with a total of 18 (Smith *et al.*, 2010), whereas certain non-photosynthetic relatives of these green algae, such as *Prototheca wickerhmanni* and *Acanthamoeba castellium* have only a few introns.

3.2. Other Photosynthetic Microbes

In addition to the charophyte and chlorophyte green algae, which are affiliated with the land plant lineage, there are other algae that have gained photosynthetic function through the secondary acquisition of plastids (reviewed in Keeling, 2010). In general, their mitochondrial genomes contain fewer introns than green algae or land plants (Fig. 7.3). For example, in the red alga, *Chondrus crispus*, there is just a single group II intron and it is located within the tRNA-Ile(GAU) gene. This tRNA gene also contains an intron in some plastids, although not in the *C. crispus* chloroplast, nor is the tRNA sequence very similar to chloroplast counterparts, so a recent intracellular lateral transfer event is considered very unlikely (Leblanc, Kloareg, Loiseaux-de Goer, Boyen, 1995).

In the case of brown algae, of all those examined to date, the only one to contain any mitochondrial introns is *P. littoralis* (cf. Oudot-Le Secq & Green, 2001; Oudot-Le Secq, Loiseaux-de Goer, Stam, & Olsen, 2006). Geographically distinct specimens of this species were found to have a variable overall number of group II introns located at seven sites within the *cox1* and large subunit ribosomal RNA genes (Ikuta, Kawai, Müller, & Ohama, 2008). Some of these introns still possess fully intact intronic ORFs with reverse transcriptase, endonuclease and maturase domains, as well as autocatalytic properties in the test tube (see above), features that are consistent with a very recent group II intron infection and spread in *P. littoralis*.

The mitochondrial genomes of the diatoms examined to date, such as *Phaeodactylum tricornutum*, *Thalassiosira pseudonana* or *Synedra acus*, tend to have only a few introns or none at all (Oudot-Le Secq & Green, 2011) and the ones containing intact intronic ORFs are candidates for particularly recent acquisition by horizontal transfer. The cryptophyte alga, *Rhodomonas salina*, has two group II introns within the *cox1* gene and they have RT-matR ORFs most closely related to those in the brown alga, *P. littoralis* (Hauth, Maier, Lang, & Burger, 2005), whereas another cryptophyte, *Hemiselmis andersenii*, completely lacks mitochondrial introns (Kim *et al.*,

2008). Similarly, the chrysophyte alga, *Chrysodidymus synuroideus*, has a densely packed mitochondrial genome without any introns (Chesnick *et al.*, 2000).

Although little is known as yet about the detailed structure of mitochondrial genes in dinoflagellates, they have a number of bizarre features (reviewed in Waller & Jackson, 2009) including a novel form of *cox3 trans*-splicing (see above). An examination of mitochondrial genes in the dinoflagellates, *Durinskia baltica* and *Kryptoperidinium foliaceum*, revealed the intriguing observation that homologous gene copies with distinctly different heritage have been maintained, that is, the one from the endosymbiont and the one from the host (Imanian & Keeling, 2007). The host-type *cox1* gene contains no introns and, although the gene structure of the endosymbiont one is not yet known, it was suggested that the inability to experimentally isolate a full-length *cox1* gene might relate to the presence of long group II introns, analogous to the situation in certain diatoms. The input from multiple mitochondrial genetic systems raises interesting evolutionary issues regarding opportunity for lateral transfer of introns and chimeric forms of genes.

3.3. Gain/Loss of Mitochondrial Introns during Plant/Algal Evolution

The mobile nature of group I/II introns is mediated by their structural features as well as the protein-coding information they possess. In the case of group I introns, the homing endonuclease facilitates a DNA form of movement, whereas for group II introns, reverse transcriptase enables dispersal *via* an RNA intermediate. In both cases, movement is typically into a homing recognition site in intronless alleles (often crossing species barriers) and group II introns can also integrate into new non-allelic sites by retrotransposition.

3.3.1. Fungus-to-plant horizontal transfer of introns

Although introns found in the mitochondria of seed plants are virtually all group II, there is one group I intron that has repeatedly and independently invaded a homing site in the *cox1* gene (Cho, Qiu, Kuhlman, & Palmer, 1998; Sanchez-Puerta, Cho, Mower, Alverson, & Palmer, 2008). Although this intron is not present in the seed plants shown in Fig. 7.4, its position is depicted by the red triangle for the orthologues in *M. polymorpha* and *C. vulgaris*. Comparative sequence analysis suggests that it was acquired by lateral gene transfer from a fungus (Seif *et al.*, 2005). Since the vast majority of land plants have close association with arbuscular mycorrhizal fungi, they are good candidates for being the donor of this mitochondrial intron and it will be of interest to determine the mode of transfer and triggering events.

3.3.2. Plant-to-plant horizontal transfer of introns

There is an interesting case in the gymnosperm, *Gnetum gnemonoides*, where an extra copy of *nad1* intron 2 and flanking exons is present in its mitochondrial genome and it appears to be of a flowering plant (asterid) origin (Won & Renner, 2003). Similarly, in *Hedychium coronarium*, a ginger family monocot, the fourth intron of *nad1* (containing *matr*) is present in two copies, one *trans*-splicing (native) and the other *cis*-splicing (Hao, Richardson, Zheng, & Palmer, 2010). The latter appears to be of eudicot origin, thus reflecting of eudicot-to-moncot gene transfer. Moreover, comparisons of their *matR* sequences led to the conclusion that the foreign copy is chimeric, having undergone subsequent gene conversion events. This illustrates the complex evolutionary history that mobile introns may have, especially in highly recombinogenic genomes.

3.3.3. Interorganellar (chloroplast–mitochondrion) transfer of introns

Although there are many examples of the integration of chloroplast DNA into mitochondrial genomes in seed plants (including functional tRNA genes, reviewed in Leister & Kleine, 2011), there are few (if any) substantiated cases of any mitochondrial intron being of direct chloroplast origin. Incomplete data sets can make it difficult to ascertain the source of lateral transfer events, and the case of potential chloroplast-to-mitochondrion transfer in the history of the colourless green alga, *Acanthamoeba* (Turmel *et al.*, 1995), was subsequently revised to a probable mitochondrion-to-mitochondrion transfer (Turmel *et al.*, 2002). In the case of a group I intron found at homologous sites in the mitochondrial *atpA* and chloroplast *atp1* genes in the green alga, *Pseudoendoclonium akinetum*, the direction of transfer was inferred to be from the mitochondrion-to-chloroplast (Pombert, Otis, Lemieux, & Turmel, 2005).

3.3.4. Non-functional mitochondrion-to-nucleus transfer of introns

It has long been appreciated that fragments of mitochondrial DNA (including intron sequences) become integrated into the nuclear genome as an ongoing random process (cf. Knoop & Brennicke, 1994). Such translocated sequences, which have been termed numt DNA, appear more common in plants than in animals and these molecular fossils can be useful as markers in timing evolutionary events (Bensasson, Zhang, Hartl, & Hewitt, 2001). Since they are pseudogenes, they typically degenerate over time and are lost; however, it is possible that such sequences can be recruited to perform a function in their new location. For example, in the chlorarachniophyte unicellular alga, *Bigelowiella natans*, a translocated segment of the mitochondrial *cox2* gene has evolved into a spliceosomal intron in a GTPase gene (Curtis & Archibald, 2010). Rather similarly, the mitochondrial *cox2* group II intron contributes to the building of a spliceosomal intron in the nuclear alcohol dehydrogenase gene (*adhB*) in the monocot *Washingtonia robusta* (Kudla, Albertazzi, Blacević, Hermann, & Bock, 2002).

3.3.5. Intron loss through retroprocessing

The phylogenetic distribution of mitochondrial introns in plants and algae provides compelling evidence that while certain introns have been gained during evolution, others have been lost. In seed plants, classic cases are the optional group II introns in *nad4*, *cox2* and *rps3*. For example, *nad4* contains between one and three introns in flowering plants, with all three predicted to have been present in their common ancestor (Gass, Makaroff, & Palmer, 1992). Loss must be precise to avoid disruption of the protein-coding information in the host gene. The favoured model is through RNA-mediated retroprocessing where a spliced (and edited) mRNA is used as a template for cDNA synthesis and subsequent gene conversion of the genomic copy. The extent of co-conversion of sequences flanking the intron position can vary and support for this model is provided by the parallel loss of editing sites close to the intron position, with corroborating data from flowering plants (*nad4*, Geiss, Abbas, & Makaroff, 1994), gymnosperms (*rps3*, Ran, Gao, & Wang, 2010) and lycophytes (*nad4*, Grewe *et al.*, 2011). As an aside, the acquisition of new introns, such as the rampant *cox1* group I intron, is also be accompanied by co-conversion of flanking exon sequences during the double strand break repair process following homing site-specific DNA cleavage (Sanchez-Puerta *et al.*, 2011).

4. LIFE CYCLE OF GROUP I/II INTRON MOBILE GENETIC ELEMENTS

Based on the characteristics of group I/II introns in nature, the progression of these mobile genetic elements through their life cycle can be modelled.

Fig. 7.5 presents a scenario based on observed features of group I/II introns in plant/algal mitochondria to illustrate the hypothetical steps that a ribozyme mobile element (be it group I or group II) might follow starting with invasion of a fully competent form into a mitochondrial genome until it degenerates into a completely host-dependent element or is cleanly lost.

When a conventional group I/II element enters a mitochondrial gene, it is tolerated because it encodes its own machinery (RNA maturase) for precise excision at the RNA level so it is not detrimental to the host gene. Its capacity for mobility (DNA endonuclease and/or reverse transcriptase

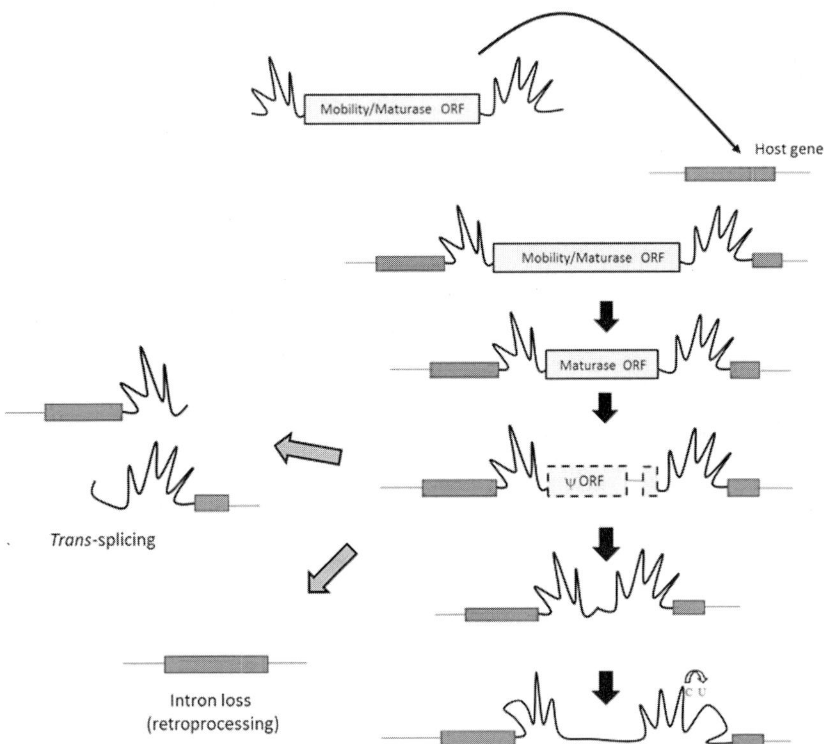

Figure 7.5 Scenario for the life cycle of a group I/II intron. Steps include intron invasion into a new site in the host genome, domestication of the intron, which can lead to degeneration of the intronic ORF and its potential loss (concomitant with dependence on host splicing factors), and loss of core secondary structural features (only some of which can be corrected by RNA editing, as shown by C-to-U arrow). At any point in the life cycle, the intron might be converted into a *trans*-splicing form by genomic rearrangements or it might be cleanly lost *via* retroprocessing. The intron might also acquire extra non-coding or coding sequences unrelated to the original mobile element. For colour version of this figure, the reader is referred to the online version of this book.

activity) enables it to increase its copy number in the genome and sequence similarity among such paralogous introns may allow inferences about their movements (cf. Ohyama & Takemura, 2008). If elements move to intergenic positions, the maturase ORF may persist if it provides a function in the splicing of other introns, and free-standing maturase ORFs have been observed in liverworts (Liu, Xue, Wang, Li, & Qiu, 2011) and certain algal mitochondrial genomes (Boer & Gray, 1988; Burger, Saint-Louis, Gray, & Lang, 1999). It is expected that, over time, the mobility function will be lost, and many mitochondrial intronic ORFs in fact lack these domains. In addition, the intronic maturase ORF may lose its strict self-specificity and assist in the excision of other introns. This has been documented in yeast mitochondria and plant chloroplasts. Within plant/algal mitochondrial genomes, there is almost always at least one intronic ORF (or free-standing non-intronic homologue) present (Fig. 7.3). Over evolutionary time, as the intron becomes further domesticated, the host is expected to make an increasing contribution to splicing competency, so that the RNA maturase ORF is free to degenerate into a pseudogene and be lost. Defective remnant ORFs have been identified within certain mitochondrial introns, for example, within the *cox2* intron in *Ginkgo biloba* (Ahlert, Piepenburg, Kudla, & Bock, 2006) and several in *M. polymorpha* (Ohyama & Takemura, 2008). The successful transfer of intronic RNA maturase genes to the nucleus can also relieve the mitochondrial copies from functional constraint. In flowering plants, there are four copies of the group II *matR*-type gene in the nucleus (Mohr & Lambowitz, 2003), and several have been implicated in the splicing of multiple introns (see below). Host machinery may also include RNA-binding proteins with other cellular functions that have been recruited for secondary moonlighting roles in splicing. Potential conflict due to such gene sharing could be alleviated by subsequent gene duplication and exclusive dedication of one copy to splicing.

During their evolutionary history, plant/algal mitochondrial introns have been seen to undergo insertions or deletions within regions that do not affect the catalytic core structure, and as mentioned above, introns may even acquire other genes, such as the *nad4L* gene within a *nad1* intron in *S. moellendorfii* (Hecht *et al.*, 2011). In the case of the *Cycas taitungensis* (gymnosperm) mitochondrial introns, this enlargement sometimes includes Bpu repetitive elements (Chaw *et al.*, 2008) and in the eudicot cucurbits, which have huge mitochondrial genomes, many introns are inflated in size (Alverson, Rice, Dickinson, Barry, & Palmer, 2011a), although a few appear recalcitrant to expansion (e.g. within *nad5*). Also as discussed above, genomic rearrangements within group I/II introns can disperse flanking exons, but because of secondary structure and higher-order intron folding, splicing is able to occur *in trans*. This could theoretically occur at any stage in the life cycle. It would seem that the generation of a *trans*-splicing form would be an irreversible event, however if subsequent DNA rearrangements fortuitously positioned the segments in close proximity, they might be co-transcribed and spliced *in cis*. In the *Vigna radiata* (mung bean) mitochondrial genome, the *trans*-splicing *nad5* exon 3 and exon 4 are less than 3 kb apart and in the correct order and orientation, so that a co-transcript of this region could theoretically result in a *cis*-splicing intron originating from an ancestral *trans*-splicing organization (Alverson, Zhuo, Rice, Sloan, & Palmer, 2011b).

Over time, the core structure of the intron may accumulate mutations and become more dependent on accessory proteins for folding into a splicing-competent structure. In land plant mitochondria, some such changes may be corrected by RNA editing (e.g. conversion of A-C mispairs to A-U base pairs within helices), although other positions are not editing candidates (e.g. U-U). It is predicted that the longer the residency of an intron within a host genome, the more nondescript it will become, and this degeneracy will be accompanied by a concomitant increase in contribution from accessory host proteins. As mentioned above, certain flowering plant group II introns in particular have quite non-conventional structures and show sequence variation among plant species.

At any point during the group I/II intron life cycle, there might be clean loss of the intron by retroprocessing, with a mature mRNA (lacking the intron) serving as template for generating a cDNA copy by reverse transcription and subsequent integration at the homologous site. On the other hand, it may be that certain introns get locked in due to an intimate relationship with other RNA processing events, for example, if their presence provides *cis*-elements important for exon editing near intron/exon junctions or if they acquire a regulatory function within the mitochondrion.

5. PROTEIN MACHINERY FOR GROUP I/II INTRON SPLICING IN PLANTS AND ALGAE

The splicing machinery involved in processing plant/algal mitochondrial introns can be divided into two categories: RNA maturase functions that are encoded within the intron itself and ones provided by the host. Related to the first type are maturase-type intronic ORFs, which have moved to a free-standing location within the mitochondrial genome or which have been translocated to the nucleus. Experimental evidence from plant/algal introns (as well as those in other systems) indicates that protein machinery is crucial for splicing *in vivo*. For the second category of splicing factors, the simplest model is that host proteins with other functions are recruited to perform additional functions in splicing. Potential roles of host proteins are to stabilize the intron RNA structure, act as chaperones to facilitate RNA folding (or prevent/remove inappropriate folding), or act as transporters to ensure that the milieu is appropriate, for example, regulate Mg^{2+} levels for efficient splicing.

5.1. Group I Intron Splicing Machinery

5.1.1. Group I intron-encoded RNA maturases

A subset of plant/algal group I introns contain an ORF (often identified as such by LAGLIDADG motifs) that provides information for RNA splicing (Fig. 7.3) and there is a direct correlation between the presence of a fully intact intronic ORF and the recent timing of the invasion. As might be expected, unicellular microbes appear more susceptible to waves of infection than multicellular organisms. In flowering plants, the single mitochondrial group I intron, which has been very successful in sporadically invading *cox1* genes, possesses an endonuclease ORF that is predicted to be very active (Sanchez-Puerta *et al.*, 2011).

5.1.2. Group I intron nuclear-encoded splicing machinery

Very little is known about the splicing factors that are imported into the mitochondrion to assist in the splicing of group I introns in plants and algae, but it is anticipated that they will include intron-specific factors as well as factors that act on a number of introns. Very recently acquired group I introns (with their own maturase ORFs) are expected to be relatively self-sufficient. Some splicing factors may be shared between the mitochondrion and chloroplast, or be encoded by paralogous gene copies. In land plant chloroplasts, the single group I intron (located in tRNA-Leu-UAA) is assisted in splicing by the CRM2 protein, which also acts on several chloroplast group II introns. The paralogous CRM3 protein has been shown to be dual-targeted to the mitochondrion (Asakura, Bayraktar, & Barkan, 2008).

5.2. Group II Splicing Machinery

5.2.1. Group II intron-encoded splicing machinery

The number of mitochondrial group II introns that contain RNA maturase ORFs varies among plants and algae, with a higher proportion seen in algae and non-vascular plants (Fig. 7.3). Flowering plants have only a single mitochondrial group II intronic maturase gene (*matR*) and it is located in the final intron of *nad1*. Its biochemical activity has not yet been demonstrated, however it gives rise to stable (and edited) RNA and shows evolutionary constraint on its amino acid sequence (Thomson, Macfarlane, Beagley, & Wolstenholme, 1994), so is believed to be a functional gene involved in splicing. It lacks a canonical initiation codon and the signals for its translation are unknown, but may relate to the importance of keeping its expression in check, given the potential detrimental effect of inappropriate reverse transcriptase (or endonuclease) activities in the organelle. Also the presence of an additional promoter within *matR* raises the possibility that the domain X (splicing function) region may be independently expressed (Farré & Araya, 1999).

5.2.2. Group II intron nuclear-encoded splicing machinery

Splicing factors for group II introns in flowering plant mitochondria have now begun to be identified (reviewed in de Longevialle, Small, & Lurin, 2010). The analysis of *Arabidopsis* mutants defective in complex I of the respiratory chain (since most of the introns are located in *nad* genes) has been particularly powerful in this regard. Even prior to the identification of specific gene products, the importance of nuclear control was known from genetic analysis; for example, *Nicotiana sylvestris* mutants deficient in complex I were identified to be blocked in the splicing of *nad4* intron 1 (Brangeon *et al.*, 2000). Perhaps not surprisingly, nucleus-located homologues of *matR* play an important role in mitochondrial splicing, with At-nMat2 acting on *nad1* intron 2, *nad7* intron 2 and the *cox2* intron (Keren *et al.*, 2009), and At-nMat1a mutants being defective in *nad4* splicing, as well as having altered sensitivity to cellulose biosynthesis inhibitors (Nakagawa & Sakurai, 2006). Also perhaps not surprisingly, the very first splicing factor to be characterized in plant mitochondria was a member of the pentatricopeptide (PPR) multigene family of RNA-binding proteins, which were already appreciated to have important roles in organelle RNA editing (reviewed in Schmitz-Linneweber & Small, 2008; see also this volume, Chapter 10). This OTP43 PPR protein acts on *nad1 trans*-splicing intron 1 (de Longevialle *et al.*, 2007). Several other PPR protein-type mitochondrial splicing factors have been identified from genetic screens; for example, BIR6 mutants resistant to root growth inhibitors are defective in *nad7* intron 1 splicing (Koprivova *et al.*, 2010), and ABO5 mutants with altered root growth response to abscisic acid are defective in *nad1* intron 3 removal (Liu *et al.*, 2010). Non-vascular plants and algae also possess PPR protein genes, although at much lower copy number (reviewed in Schmitz-Linneweber & Small, 2008) and in the moss, *P. patens*, a PPR-DYW type protein has recently been identified to be required for splicing of the group II *cox1* intron 3, but none of the other three *cox1* introns (Ichinose, Tasaki, Sugita, & Sugita, in press). As an aside, one of the PPR proteins (OTP51) involved in splicing of a group II intron in the chloroplast *ycf3* gene, contains two LAGLIDADG motifs characteristic of group I (but not group II) introns near its carboxy-terminus (de Longevialle *et al.*, 2008).

Studies of factors important in chloroplast or yeast mitochondrial splicing are proving helpful in pinpointing candidates involved in plant/algal mitochondrial splicing. The importance of DEAD-box RNA helicases in splicing prompted an examination of *Arabidopsis* mitochondrial counterparts and led to identification of PMH2, which is part of a high molecular weight complex (Matthes *et al.*, 2007) and which improves the efficiency of splicing of introns in *nad1*, *nad5* and *nad7* (Köhler *et al.*, 2010). An *Arabidopsis* Mg^{2+} ion transporter was found to complement a yeast mitochondrial MRS2 splicing mutant (Schock *et al.*, 2000). A non-PPR type of RNA-binding protein, with a PORR domain, and homologous to chloroplast WFT (what's this factor) multigene family has recently been shown to perform a role in splicing of the mitochondrial *rpl2* and *ccmFc* introns in *Arabidopsis* (Colas des Francs-Small *et al.*, 2012) and RUG3, which has a RCC1 domain and is involved in the splicing of *nad2 trans*-intron 2 and *cis*-splicing *nad2* intron 3, may act through protein–protein interactions rather than direct RNA binding (Kühn *et al.*, 2011).

The list of mitochondrial splicing factors will undoubtedly continue to grow, and the emerging picture is that some splicing factors are intron-specific, whereas others have more generalized functions acting on multiple introns. Little is known as yet about the overall complexity (and dynamic nature) of the splicing machinery, but it is anticipated that new specificity factors may be acquired quite easily during evolution, as well as subsequently being replaced by others. It is unknown whether *trans*-splicing introns require a larger number of splicing factors than *cis*-splicing introns, but it is worth noting that for the *C. reinhardtii* chloroplast *psaA* gene, which has two *trans*-splicing group II introns, at least 14 factors (including a small RNA encoded in the plastid genome) are needed (reviewed in Lambowitz & Zimmerly, 2011).

6. RELATIONSHIP BETWEEN SPLICING AND OTHER EVENTS IN THE MITOCHONDRION

6.1. RNA Editing and Splicing

When a mobile genetic element enters a plant/algal mitochondrial gene as an intron, in addition to the necessity of it being precisely excised from the precursor RNA, it must not interfere with any other RNA maturation events. One such important RNA processing step in land plant mitochondria is RNA editing. It is typically an early event since intron-containing transcripts show editing; however, certain sites located very close to the intron/exon junctions show no editing in precursors, although they are fully edited in the spliced RNA (cf. Li-Pook-Than *et al.*, 2007). This may reflect steric hindrance in accessing editing sites or the creation of editing recognition *cis*-elements through splicing. Coevolution of these processes may lock them into each other and confer selective pressure against intron loss (Kagami, Nagano, Takahashi, Mikami, & Kubo, 2012).

Editing within introns (albeit at fewer sites than in exons) illustrates their adaptation to the host environment, and recent deep sequencing analysis suggests that the extent of intron editing may be greater than initially expected. A study of grape (*Vitis vinifera*) mitochondrial transcripts identified 44 edits in group II introns and tRNAs (Picardi *et al.*, 2010). Many edits in plant mitochondrial group II introns are located at sites within helical structures, which can improve base pairing by converting A-C mispairs to A-U base pairs (reviewed in Bonen, 2008). Lycophytes, which have very extensive mitochondrial editing, provide particularly striking examples for group II introns (Grewe *et al.*, 2011) and structural improvement of a *cox1* group I intron through editing has also been observed in the fern, *Osmunda regalis* (Bégu, Castandet, & Araya, 2011). Notably in this case, both C-to-U and U-to-C editing restore conventional base pairing in helical regions. *In organello* electroporation systems have been informative in assessing the importance of intron editing for splicing by monitoring the processing of engineered constructs that are introduced into isolated mitochondria (cf. Farré & Araya, 2002). This experimental approach has also revealed that intronless constructs show reduced editing (Castandet, Choury, Bégu,

Jordana, & Araya, 2010). As more is learnt about precursor RNA structural features, new understanding about splicing and its relationship to other RNA processing events will undoubtedly follow. Insights on this subject are also coming from studies on cold-grown wheat and rice seedlings (Kurihara-Yonemoto & Handa, 2001; Kurihara-Yonemoto & Kubo, 2010). Relatively higher amounts of mitochondrial intron-containing precursor RNAs compared with mature mRNAs were seen, suggesting that the efficiency of splicing is reduced in the cold. This may be due to reduced enzymatic activity and/or inappropriate intron RNA structure. RNA editing also was affected by low temperature, with certain sites close to intron/exon junctions being under-edited in precursor RNA molecules.

7. FUTURE PERSPECTIVES

An impressive amount of mitochondrial genome sequence data from a broad range of multicellular and unicellular photosynthetic organisms has now been collected and it is providing new insights into the intriguing behaviour of group I and group II introns. Surprises include the presence of *trans*-splicing mechanisms for both group I and group II introns, the highly invasive nature of certain group I homing introns in contrast to the relative stability of many group II introns, and the degeneration of ribozyme features of mitochondrial introns in plants. Landmark events in the evolution of particular introns (such as the transition from the *cis*- to *trans*-splicing form) are also proving useful in elucidating phylogenetic relationships among distantly related groups of plants or algae (e.g. Groth-Malonek, Pruchner, Grewe, & Knoop, 2005; Qiu, Cho, Cox, & Palmer, 1998; Regina & Quagliariello, 2010).

The most clear-cut impact of the presence of introns in mitochondrial genomes is their contribution to increased genome size. Although seemingly costly in energetic terms, plant/algal mitochondrial genomes are remarkably flexible in tolerating huge variations in genome length. The presence of mobile genetic elements also brings with it the opportunity for host genome rearrangements (mediated by reverse transcriptase and/or DNA endonuclease activities). Although this could have deleterious consequences if not under appropriate control, it can also generate genetic diversity and it can mediate streamlining events such as retroprocessing of host genes. It has been proposed that the absence of mitochondrial RNA editing activity in the *Marchantia* lineage (unlike other liverworts or land plants) reflects such a secondary loss (cf. Groth-Malonek, Wahrmund, Polsakiewicz, & Knoop, 2007).

It will be of interest to learn what triggers the spread of group I and group II introns. Although this topic has not yet been addressed in plants or algae, studies with bacteria have shown that environmental stressors such as nutrient starvation (e.g. glucose deprivation in *Escherichia coli*) can promote group II intron mobility (Coros, Piazza, Chalamcharla, Smith, & Belfort, 2009). Under such stress conditions, the favoured mechanism appears to be by ectopic retrotransposition rather than site-specific retrohoming. Thus, the chance of survival of the element might be improved through remodelling of the genome. Since group II intron mobility is RNA mediated, conditions favouring elevated levels of excised intron may assist in its successful spread. Mitochondrial DNA rearrangements involving introns have also been implicated in certain types of mitochondrial dysfunction, for example, *nad7* in *N. sylvestris* cytoplasmic male sterility (CMS) mutants (Pla, Mathieu, de Paepe, Chetrit, & Vedel, 1995) and *nad4-nad7* fusion in maize non-chromosomal stripe mutants (Marienfeld & Newton, 1994).

One possible attribute of plant/algal introns is that, over time, they evolve a regulatory role in the mitochondrion. As yet, no such function has been demonstrated nor are there documented cases of alternative splicing in the mitochondria of plants or algae. However the extreme bias of introns in NADH dehydrogenase genes in flowering plants has fuelled speculation of an underlying regulatory connection. This is particularly attractive since complex I of the respiratory chain can be bypassed in favour of alternative respiratory pathways in plants under certain environmental conditions or developmental states. Much has now been learnt about the distribution and properties of plant/algal mitochondrial group I and II introns, and no doubt there is much more to be learnt about their biological impact and behaviour.

ACKNOWLEDGEMENTS

I greatly appreciate the many helpful discussions about introns that I have had with members of my research group, and I also gratefully acknowledge financial support from NSERC Canada (Natural Sciences and Engineering Council of Canada).

REFERENCES

Ahlert, D., Piepenburg, K., Kudla, J., & Bock, R. (2006). Evolutionary origin of a plant mitochondrial group II intron from a reverse transcriptase/maturase-encoding ancestor. *Journal of Plant Research, 119*, 363–371.

Alverson, A. J., Rice, D. W., Dickinson, S., Barry, K., & Palmer, J. D. (2011a). Origins and recombination of the bacterial-sized multichromosomal mitochondrial genome of cucumber. *The Plant Cell, 23*, 2499–2513.

Alverson, A. J., Zhuo, S., Rice, D. W., Sloan, D. B., & Palmer, J. D. (2011b). The mitochondrial genome of the legume *Vigna radiata* and the analysis of recombination across short mitochondrial repeats. *PLoS One, 6*, e16404.

Asakura, Y., Bayraktar, O. A., & Barkan, A. (2008). Two CRM protein subfamilies cooperate in the splicing of group IIB introns in chloroplasts. *RNA, 14*, 2319–2332.

Bégu, D., Castandet, B., & Araya, A. (2011). RNA editing restores critical domains of a group I intron in fern mitochondria. *Current Genetics, 57*, 317–325.

Bensasson, D., Zhang, D., Hartl, D. L., & Hewitt, G. M. (2001). Mitochondrial pseudogenes: evolution's misplaced witnesses. *Trends in Ecology and Evolution, 16*, 314–321.

Boer, P. H., & Gray, M. W. (1988). Genes encoding a subunit of respiratory NADH dehydrogenase (ND1) and a reverse transcriptase-like protein (RTL) are linked to ribosomal RNA gene pieces in *Chlamydomonas reinhardtii* mitochondrial DNA. *EMBO Journal, 7*, 3501–3508.

Bonen, L. (2008). *Cis-* and *trans-*splicing of group II introns in plant mitochondria. *Mitochondrion, 8*, 26–34.

Bonen, L. (2010). RNA splicing in plant mitochondria. In F. Kempken (Ed.), *Plant mitochondria. Advances in plant biology*, 1 (pp. 131–155).

Bonen, L., & Vogel, J. (2001). The ins and outs of group II introns. *Trends in Genetics, 17*, 322–331.

Brangeon, J., Sabar, M., Gutierres, S., Combettes, B., Bove, J., Gendy, C., et al. (2000). Defective splicing of the first *nad4* intron associated with lack of several complex I subunits in the *Nicotiana sylvestris* NMS1 nuclear mutant. *The Plant Journal, 21*, 269–280.

Burger, G., Saint-Louis, D., Gray, M. W., & Lang, B. F. (1999). Complete sequence of the mitochondrial DNA of the red alga *Porphyra purpurea.* Cyanobacterial introns and shared ancestry of red and green algae. *Plant Cell, 11*, 1675–1694.

Carrillo, C., Chapdelaine, Y., & Bonen, L. (2001). Variation in sequence and RNA editing within core domains of mitochondrial group II introns among plants. *Molecular and General Genetics, 264*, 595–603.

Castandet, B., Choury, D., Bégu, D., Jordana, X., & Araya, A. (2010). Intron RNA editing is essential for splicing in plant mitochondria. *Nucleic Acids Research, 38*, 7112–7121.

Chapdelaine, Y., & Bonen, L. (1991). The wheat mitochondrial gene for subunit I of the NADH dehydrogenase complex: a *trans-*splicing model for this gene-in-pieces. *Cell, 65*, 465–472.

Chaw, S. M., Shih, A. C., Wang, D., Wu, Y. W., Liu, S. M., & Chou, T. Y. (2008). The mitochondrial genome of the gymnosperm *Cycas taitungensis* contains a novel family of short interspersed elements, Bpu sequences, and abundant RNA editing sites. *Molecular Biology and Evolution, 25*, 603–615.

Chesnick, J. M., Goff, M., Graham, J., Ocampo, C., Lang, B. F., Seif, E., et al. (2000). The mitochondrial genome of the stramenopile alga *Chrysodidymus synuroideus.* Complete sequence, gene content and genome organization. *Nucleic Acids Research, 28*, 2512–2518.

Cho, Y., Qiu, Y. L., Kuhlman, P., & Palmer, J. D. (1998). Explosive invasion of plant mitochondria by a group I intron. *Proceedings of the National Academy of Sciences of the United States of America, 95*, 14244–14249.

Colas des Francs-Small, C., Kroeger, C. C., Zmudjak, T., Ostersetzer-Biran, M., Rahimi, O., Small, N., et al. (2012). A PORR domain protein required for *rpl2* and *ccmFc* intron splicing and for the biogenesis of c-type cytochromes in *Arabidopsis* mitochondria. *The Plant Journal, 69*, 996–1005.

Coros, C. J., Piazza, C. L., Chalamcharla, V. R., Smith, D., & Belfort, M. (2009). Global regulators orchestrate group II intron retromobility. *Molecular Cell, 34*, 250–256.

Costa, M., Fontaine, J. M., Loiseaux-de Goer, S., & Michel, F. (1997). A group II self-splicing intron from the brown alga *Pylaiella littoralis* is active at unusually low magnesium concentrations and forms populations of molecules with a uniform conformation. *Journal of Molecular Biology, 274*, 353–364.

Curtis, B. A., & Archibald, J. M. (2010). A spliceosomal intron of mitochondrial DNA origin. *Current Biology, 20*, R919–R920.

de Longevialle, A. F., Meyer, E. H., Andres, C., Taylor, N. L., Lurin, C., Millar, A. H., et al. (2007). The pentatricopeptide repeat gene OTP43 is required for *trans*-splicing of the mitochondrial *nad1* intron 1 in *Arabidopsis thaliana. The Plant Cell, 19*, 3256–3265.

de Longevialle, A. F., Hendrickson, L., Taylor, N. L., Delannoy, E., Lurin, C., Badger, M., et al. (2008). The pentatricopeptide repeat gene OTP51 with two LAGLIDADG motifs is required for the cis-splicing of plastid ycf3 intron 2 in *Arabidopsis thaliana. The Plant Journal, 56*, 157–168.

de Longevialle, A. F., Small, I. D., & Lurin, C. (2010). Nuclearly encoded splicing factors implicated in RNA splicing in higher plant organelles. *Molecular Plant, 3*, 691–705.

Dombrovska, O., & Qiu, Y. L. (2004). Distribution of introns in the mitochondrial gene *nad1* in land plants: phylogenetic and molecular evolutionary implications. *Molecular Phylogenetics and Evolution, 32*, 246–263.

Elina, H., & Brown, G. G. (2010). Extensive mis-splicing of a bi-partite plant mitochondrial group II intron. *Nucleic Acids Research, 38*, 996–1008.

Farré, J. C., & Araya, A. (1999). The *mat-r* open reading frame is transcribed from a non-canonical promoter and contains an internal promoter to co-transcribe exons nad1e and nad5III in wheat mitochondria. *Plant Molecular Biology, 40*, 959–967.

Farré, J. C., & Araya, A. (2002). RNA splicing in higher plant mitochondria: determination of functional elements in group II intron from a chimeric *cox II* gene in electroporated wheat mitochondria. *The Plant Journal, 29*, 203–213.

Gass, D. A., Makaroff, C. A., & Palmer, J. D. (1992). Variable intron content of the NADH dehydrogenase subunit 4 gene of plant mitochondria. *Current Genetics, 21*, 423–430.

Geiss, K. T., Abbas, G. M., & Makaroff, C. A. (1994). Intron loss from the NADH dehydrogenase subunit 4 gene of lettuce mitochondrial DNA: evidence for homologous recombination of a cDNA intermediate. *Molecular and General Genetics, 243*, 97–105.

Glanz, S., & Kuck, U. (2009). *Trans*-splicing of organelle introns - a detour to continuous RNAs. *BioEssays, 31*, 921–934.

Gray, M. W. (1992). The endosymbiont hypothesis revisited. *International Reviews in Cytology, 141*, 233–357.

Grewe, F., Viehoever, P., Weisshaar, B., & Knoop, V. (2009). A *trans*-splicing group I intron and tRNA-hyperediting in the mitochondrial genome of the lycophyte *Isoetes engelmannii. Nucleic Acids Research, 37*, 5093–5104.

Grewe, F., Herres, S., Viehöver, P., Polsakiewicz, M., Weisshaar, B., & Knoop, V. (2011). A unique transcriptome: 1782 positions of RNA editing alter 1406 codon identities in mitochondrial mRNAs of the lycophyte *Isoetes engelmannii. Nucleic Acids Research, 39*, 2890–2902.

Groth-Malonek, M., Pruchner, D., Grewe, F., & Knoop, V. (2005). Ancestors of trans-splicing mitochondrial introns support serial sister group relationships of hornworts and mosses with vascular plants. *Molecular Biology and Evolution, 22*, 117–1125.

Groth-Malonek, M., Wahrmund, U., Polsakiewicz, M., & Knoop, V. (2007). Evolution of a pseudogene: exclusive survival of a functional mitochondrial *nad7* gene supports *Haplomitrium* as the earliest liverwort lineage and proposes a secondary loss of RNA editing in Marchantiidae. *Molecular Biology and Evolution, 24*, 1068–1074.

Hao, W., Richardson, A. O., Zheng, Y., & Palmer, J. D. (2010). Gorgeous mosaic of mitochondrial genes created by horizontal transfer and gene conversion. *Proceedings of the National Academy of Sciences of the United States of America, 107*, 21576–21581.

Hauth, A. M., Maier, U. G., Lang, B. F., & Burger, G. (2005). The *Rhodomonas salina* mitochondrial genome: bacteria-like operons, compact gene arrangement and complex repeat region. *Nucleic Acids Research, 33*, 4433–4442.

Hecht, J., Grewe, F., & Knoop, V. (2011). Extreme RNA editing in coding islands and abundant microsatellites in repeat sequences of *Selaginella moellendorffii* mitochondria: the root of frequent plant mtDNA recombination in early tracheophytes. *Genome Biology and Evolution, 3*, 344–358.

Ichinose, M, Tasaki E, Sugita C, Sugita M. A PPR-DYW protein is required for splicing of a group II intron of cox1 pre-mRNA in *Physcomitrella patens. The Plant Journal*. In press.

Ikuta, K., Kawai, H., Müller, D. G., & Ohama, T. (2008). Recurrent invasion of mitochondrial group II introns in specimens of *Pylaiella littoralis* (brown alga), collected worldwide. *Current Genetics, 53*, 207–216.

Imanian, B., & Keeling, P. J. (2007). The dinoflagellates *Durinskia baltica* and *Kryptoperidinium foliaceum* retain functionally overlapping mitochondria from two evolutionarily distinct lineages. *BMC Evolutionary Biology,* 7, 172.

Kagami, H., Nagano, H., Takahashi, Y., Mikami, T., & Kubo, T. (2012). Is RNA editing implicated in group II intron survival in the angiosperm mitochondrial genome? *Genome, 55*, 75–79.

Keeling, P. J. (2010). The endosymbiotic origin, diversification and fate of plastids. *Philosophical Transactions of the Royal Society of London B: Biological Sciences, 365*, 729–748.

Keren, I., Bezawork-Geleta, A., Kolton, M., Maayan, I., Belausov, E., Levy, M., et al. (2009). AtnMat2, a nuclear-encoded maturase required for splicing of group-II introns in *Arabidopsis* mitochondria. *RNA, 15*, 2299–2311.

Kiethega, G. N., Turcotte, M., & Burger, G. (2011). Evolutionarily conserved *cox1* trans-splicing without cis-motifs. *Molecular Biology and Evolution, 28*, 2425–2428.

Kim, E., Lane, C. E., Curtis, B. A., Kozera, C., Bowman, S., & Archibald, J. M. (2008). Complete sequence and analysis of the mitochondrial genome of *Hemiselmis andersenii* CCMP644 (Cryptophyceae). *BMC Genomics, 9*, 215.

Kim, S., & Yoon, M. K. (2010). Comparison of mitochondrial and chloroplast genome segments from three onion (*Allium cepa* L.) cytoplasm types and identification of a *trans*-splicing intron of cox2. *Current Genetics, 56*, 177–188.

Knoop, V. (2004). The mitochondrial DNA of land plants: peculiarities in phylogenetic perspective. *Current Genetics, 46*, 123–139.

Knoop, V., & Brennicke, A. (1994). Promiscuous mitochondrial group II intron sequences in plant nuclear genomes. *Journal of Molecular Evolution, 39*, 144–150.

Knoop, V., Altwasser, M., & Brennicke, A. (1997). A tripartite group II intron in mitochondria of an angiosperm plant. *Molecular and General Genetics, 255*, 269–276.

Köhler, D., Schmidt-Gattung, S., & Binder, S. (2010). The DEAD-box protein PMH2 is required for efficient group II intron splicing in mitochondria of *Arabidopsis thaliana. Plant Molecular Biology, 72*, 459–467.

Koprivova, A., des Francs-Small, C. C., Calder, G., Mugford, S. T., Tanz, S., Lee, B. R., et al. (2010). Identification of a pentatricopeptide repeat protein implicated in splicing of intron 1 of mitochondrial *nad7* transcripts. *Journal of Biological Chemistry, 285*, 32192–32199.

Kudla, J., Albertazzi, F. J., Blazević, D., Hermann, M., & Bock, R. (2002). Loss of the mitochondrial *cox2* intron 1 in a family of monocotyledonous plants and utilization of mitochondrial intron sequences for the construction of a nuclear intron. *Molecular Genetics and Genomics, 267*, 223–230.

Kühn, K., Carrie, C., Giraud, E., Wang, Y., Meyer, E. H., Narsai, R., et al. (2011). The RCC1 family protein RUG3 is required for splicing of *nad2* and complex I biogenesis in mitochondria of *Arabidopsis thaliana*. *The Plant Journal, 67*, 1067–1080.

Kurihara-Yonemoto, S., & Handa, H. (2001). Low temperature affects the processing pattern and RNA editing status of the mitochondrial *cox2* transcripts in wheat. *Current Genetics, 40*, 203–208.

Kurihara-Yonemoto, S., & Kubo, T. (2010). Increased accumulation of intron-containing transcripts in rice mitochondria caused by low temperature: is cold-sensitive RNA editing implicated? *Current Genetics, 56*, 529–541.

Lambowitz, A. M., & Zimmerly, S. (2011). Group II introns: mobile ribozymes that invade DNA. *Cold Spring Harbor Perspectives in Biology, 3*. a003616.

Leblanc, C., Kloareg, B., Loiseaux-de Goër, S., & Boyen, C. (1995). DNA sequence, structure, and phylogenetic relationship of the mitochondrial small-subunit rRNA from the red alga *Chondrus crispus* (Gigartinales rhodophytes). *Journal of Molecular Evolution, 41*, 196–202.

Leister, D., & Kleine, T. (2011). Role of intercompartmental DNA transfer in producing genetic diversity. *International Review of Cell and Molecular Biology, 291*, 73–114.

Li-Pook-Than, J., & Bonen, L. (2006). Multiple physical forms of excised group II intron RNAs in wheat mitochondria. *Nucleic Acids Research, 34*, 2782–2790.

Li-Pook-Than, J., Carrillo, C., Niknejad, N., Calixte, S., Crosthwait, J., & Bonen, L. (2007). Relationship between RNA splicing and exon editing near intron junctions in wheat mitochondria. *Physiologica Plantarum, 129*, 23–33.

Liu, Y., He, J., Chen, Z., Ren, X., Hong, X., & Gong, Z. (2010). ABA overly-sensitive 5 (ABO5), encoding a pentatricopeptide repeat protein required for cis-splicing of mitochondrial *nad2* intron 3, is involved in the abscisic acid response in Arabidopsis. *The Plant Journal, 63*, 749–765.

Liu, Y., Xue, J. Y., Wang, B., Li, L., & Qiu, Y. L. (2011). The mitochondrial genomes of the early land plants *Treubia lacunosa* and *Anomodon rugelii*: dynamic and conservative evolution. *PLoS One, 6*, e25836.

Malek, O., & Knoop, V. (1998). *Trans*-splicing group II introns in plant mitochondria: the complete set of *cis*-arranged homologs in ferns, fern allies, and a hornwort. *RNA, 4*, 1599–1609.

Marienfeld, J. R., & Newton, K. J. (1994). The maize NSC2 abnormal growth mutant has a chimeric *nad4-nad7* mitochondrial gene and is associated with reduced complex I function. *Genetics, 138*, 855–863.

Martin, W., & Koonin, E. V. (2006). Introns and the origin of nucleus-cytosol compartmentalization. *Nature, 440*, 41–45.

Matthes, A., Schmidt-Gattung, S., Kohler, D., Forner, J., Wildrum, S., Raabe, M., et al. (2007). Two DEAD-box proteins may be part of RNA-dependent high-molecular-mass protein complexes in *Arabidopsis* mitochondria. *Plant Physiology, 145*, 1637–1646.

Michel, F., & Westhof, E. (1990). Modelling of the three-dimensional architecture of group I catalytic introns based on comparative sequence analysis. *Journal of Molecular Biology, 216*, 585–610.

Michel, F., & Ferat, J. L. (1995). Structure and activities of group II introns. *Annual Reviews of Biochemistry, 64*, 435–461.

Mohr, G., & Lambowitz, A. M. (2003). Putative proteins related to group II intron reverse transcriptase/maturases are encoded by nuclear genes in higher plants. *Nucleic Acids Research, 31*, 647–652.

Nakagawa, N., & Sakurai, N. (2006). A mutation in At-nMat1a, which encodes a nuclear gene having high similarity to group II intron maturase, causes impaired splicing of mitochondrial NAD4 transcript and altered carbon metabolism in *Arabidopsis thaliana*. *Plant Cell Physiology, 47*, 772–783.

Nielsen, H., & Johansen, S. D. (2009). Group I introns: moving in new directions. *RNA Biology, 6*, 1–9.

Oda, K., Yamato, K., Ohta, E., Nakamura, Y., Takemura, M., Nozato, N., et al. (1992). Gene organization deduced from the complete sequence of liverwort *Marchantia polymorpha* mitochondrial DNA: a primitive form of plant mitochondrial genome. *Journal of Molecular Biology, 223*, 1–7.

Ohyama, K., & Takemura, M. (2008). Molecular evolution of mitochondrial introns in the liverwort *Marchantia polymorpha*. *Proceedings of the Japanese Academy Series B, 84*, 17–22.

Oudot-Le Secq, M. P., & Green, B. R. (2011). Complex repeat structures and novel features in the mitochondrial genomes of the diatoms *Phaeodactylum tricornutum* and *Thalassiosira pseudonana*. *Gene, 476*, 20–26.

Oudot-Le Secq, M. P., Fontaine, J. M., Rousvoal, S., Kloareg, B., Loiseaux-De, & Goër, S. (2001). The complete sequence of a brown algal mitochondrial genome, the ectocarpale *Pylaiella littoralis* (L.) Kjellm. *Journal of Molecular Evolution, 53*, 80–88.

Oudot-Le Secq, M. P., Loiseaux-de Goër, S., Stam, W. T., & Olsen, J. L. (2006). Complete mitochondrial genomes of the three brown algae (Heterokonta: Phaeophyceae) *Dictyota dichotoma, Fucus vesiculosus* and *Desmarestia viridis*. *Current Genetics, 49*, 47–58.

Picardi, E., Horner, D. S., Chiara, M., Schiavon, R., Valle, G., & Pesole, G. (2010). Large-scale detection and analysis of RNA editing in grape mtDNA by RNA deep-sequencing. *Nucleic Acids Research, 38*, 4755–4767.

Pla, M., Mathieu, C., de Paepe, R., Chetrit, P., & Vedel, F. (1995). Deletion of the last two exons of the mitochondrial *nad7* gene results in lack of the NAD7 polypeptide in *Nicotiana sylvestris* CMS mutant. *Molecular and General Genetics, 248*, 79–88.

Pombert, J. F., & Keeling, P. J. (2010). The mitochondrial genome of the entomoparasitic green alga *Helicosporidium*. *PLoS One, 5*, e8954.

Pombert, J. F., Otis, C., Lemieux, C., & Turmel, M. (2005). The chloroplast genome sequence of the green alga *Pseudendoclonium akinetum* (Ulvophyceae) reveals unusual structural features and new insights into the branching order of chlorophyte lineages. *Molecular Biology and Evolution, 22*, 1903–1918.

Pyle, A. M. (2010). The tertiary structure of group II introns: implications for biological function and evolution. *Critical Reviews in Biochemistry and Molecular Biology, 45*, 215–232.

Qiu, Y. L., & Palmer, J. D. (2004). Many independent origins of *trans*-splicing of a plant mitochondrial group II intron. *Journal of Molecular Evolution, 59*, 80–89.

Qiu, Y. L., Cho, Y., Cox, J. C., & Palmer, J. D. (1998). The gain of three mitochondrial introns identifies liverworts as the earliest land plants. *Nature, 394*, 671–674.

Quiroga, C., Kronstad, L., Ritlop, C., Filion, A., & Cousineau, B. (2011). Contribution of base-pairing interactions between group II intron fragments during trans-splicing *in vivo*. *RNA, 17*, 2212–2221.

Ran, J. H., Gao, H., & Wang, X. Q. (2010). Fast evolution of the retroprocessed mitochondrial *rps3* gene in Conifer II and further evidence for the phylogeny of gymnosperms. *Molecular Phylogenetics and Evolution, 54*, 136–149.

Regina, T. M., & Quagliariello, C. (2010). Lineage-specific group II intron gains and losses of the mitochondrial *rps3* gene in gymnosperms. *Plant Physiology and Biochemistry, 48*, 646–654.

Rodríguez-Ezpeleta, N., Philippe, H., Brinkmann, H., Becker, B., & Melkonian, M. (2007). Phylogenetic analyses of nuclear, mitochondrial, and plastid multigene data sets support the placement of *Mesostigma* in the Streptophyta. *Molecular Biology and Evolution, 24*, 723–731.

Sanchez-Puerta, M. V., Cho, Y., Mower, J. P., Alverson, A. J., & Palmer, J. D. (2008). Frequent phylogenetically local horizontal transfer of the *cox1* group I intron in flowering plant mitochondria. *Molecular Biology and Evolution, 25*, 1762–7177.

Sanchez-Puerta, M. V., Abbona, C. C., Zhuo, S., Tepe, E. J., Bohs, L., Olmstead, R. G., et al. (2011). Multiple recent horizontal transfers of the cox1 intron in Solanaceae and extended co-conversion of flanking exons. *BMC Evolutionary Biology, 11*, 277.

Schmitz-Linneweber, C., & Small, I. (2008). Pentatricopeptide repeat proteins: a socket set for organelle gene expression. *Trends in Plant Science, 13*, 663–670.

Schock, I., Gregan, J., Steinhauser, S., Schweyen, R., Brennicke, A., & Knoop, V. (2000). A member of a novel *Arabidopsis thaliana* gene family of candidate Mg^{2+} ion transporters complements a yeast mitochondrial group II intron-splicing mutant. *The Plant Journal, 24*, 489–501.

Seif, E., Leigh, J., Liu, Y., Roewer, I., Forget, L., & Lang, B. F. (2005). Comparative mitochondrial genomics in zygomycetes: bacteria-like RNase P RNAs, mobile elements and a close source of the group I intron invasion in angiosperms. *Nucleic Acids Research, 33*, 734–744.

Smith, D. R., & Lee, R. W. (2008). Nucleotide diversity in the mitochondrial and nuclear compartments of *Chlamydomonas reinhardtii*: investigating the origins of genome architecture. *BMC Evolutionary Biology, 8*, 156.

Smith, D. R., Lee, R. W., Cushman, J. C., Magnuson, J. K., Tran, D., & Polle, J. E. (2010). The *Dunaliella salina* organelle genomes: large sequences, inflated with intronic and intergenic DNA. *BMC Plant Biology, 10*, 83.

Takenaka, M., Verbitskiy, D., van der Merwe, J. A., Zehrmann, A., & Brennicke, A. (2008). The process of RNA editing in plant mitochondria. *Mitochondrion, 8*, 35–46.

Terasawa, K., Odahara, M., Kabeya, Y., Kikugawa, T., Sekine, Y., Fujiwara, M., et al. (2007). The mitochondrial genome of the moss *Physcomitrella patens* sheds new light on mitochondrial evolution in land plants. *Molecular Biology and Evolution, 24*, 699–709.

Thomson, M. C., Macfarlane, J. L., Beagley, C. T., & Wolstenholme, D. R. (1994). RNA editing of *mat-r* transcripts in maize and soybean increases similarity of the encoded protein to fungal and bryophyte group II intron maturases: evidence that *mat-r* encodes a functional protein. *Nucleic Acids Research, 22*, 5745–5752.

Turmel, M., Otis, C., & Lemieux, C. (2003). The mitochondrial genome of *Chara vulgaris*: insights into the mitochondrial DNA architecture of the last common ancestor of green algae and land plants. *Plant Cell, 15*, 1888–1903.

Turmel, M., Côté, V., Otis, C., Mercier, J. P., Gray, M. W., Lonergan, K. M., et al. (1995). Evolutionary transfer of ORF-containing group I introns between different subcellular compartments (chloroplast and mitochondrion). *Molecular Biology and Evolution, 12*, 533–545.

Turmel, M., Otis, C., & Lemieux, C. (2002). The complete mitochondrial DNA sequence of *Mesostigma viride* identifies this green alga as the earliest green plant divergence and predicts a highly compact mitochondrial genome in the ancestor of all green plants. *Molecular Biology and Evolution, 19*, 24–38.

Unseld, M., Marienfeld, J. R., Brandt, P., & Brennicke, A. (1997). The mitochondrial genome of *Arabidopsis thaliana* contains 57 genes in 366,924 nucleotides. *Nature Genetics, 15*, 57–61.

Waller, R. F., & Jackson, C. J. (2009). Dinoflagellate mitochondrial genomes: stretching the rules of molecular biology. *Bioessays, 31*, 237–245.

Wolfe, K. H., Li, W. H., & Sharp, P. M. (1987). Rates of nucleotide substitution vary greatly among plant mitochondrial, chloroplast, and nuclear DNAs. *Proceedings of the National Academy of Sciences of the United States of America, 84*, 9054–9058.

Won, H., & Renner, S. S. (2003). Horizontal gene transfer from flowering plants to Gnetum. *Proceedings of the National Academy of Sciences of the United States of America, 100*, 10824–10829.

CHAPTER EIGHT

Green Algae Genomics: A Mitochondrial Perspective

Elizabeth Rodríguez-Salinas*, Claire Remacle[†] and Diego González-Halphen[‡,1]

[*]Instituto de Fisiología Celular, Universidad Nacional Autónoma de México, México D.F., Mexico
[†]Genetics of Microorganisms, Institute of Plant Biology, University of Liège, Liège, Belgium
[‡]Instituto de Fisiología Celular, Universidad Nacional Autónoma de México, México D.F., Mexico
[1]Corresponding author. E-mail: dhalphen@ifc.unam.mx

Contents

Abstract

Organelle genomics has provided a new perspective for studying the evolution of green algae, mainly by allowing high throughput inter- and intra-species analyses. Therefore, the number of fully sequenced nuclear, chloroplast, and mitochondrial genomes for green algae is continuously expanding. Comparative studies of mitochondrial genomes have revealed an enormous diversity in genome organization, structure, and gene content. Analyses have provided clues to gene transfer

Advances in Botanical Research, Volume 63
ISSN 0065-2296,
http://dx.doi.org/10.1016/B978-0-12-394279-1.00008-9

events from the mitochondrion to the nucleus, as well as to gene acquisitions that have taken place during the evolutionary history of green algae. This review focuses on the diversity observed among the mitochondrial genomes of green algae that have been sequenced. This information, along with several chloroplast genomes and an ever-growing number of fully sequenced nuclear genomes, will help us eventually understand the still obscure phylogenetic relationships among green algae, the cross-talk between their organelles (nucleus, chloroplast and mitochondria), and the genetic basis of the extraordinary metabolic plasticity of chlorophytes.

1. INTRODUCTION

According to the endosymbiotic theory, aerobic organisms originated when an unknown organism, most probably a protoeukaryote or an archea, internalized an α-proteobacteria (Martin & Müller, 1998; Poole & Penny, 2006; Sagan, 1967). Some descendants of this lineage underwent a second endosymbiotic event and internalized a cyanobacteria. The resulting organisms were able to respire and to perform oxygenic photosynthesis (Sagan, 1967). Taxonomically, photosynthetic eukaryotes are currently placed under the group Archaeplastida, which is divided into the lineages Glaucophyta, Rhodophyta (i.e. red algae) and Chloroplastida (i.e. green algae and land plants) (Rodríguez-Ezpeleta *et al.*, 2005). Green algae and land plants comprise the Viridiplantae kingdom. Green algae, whose main morphological and biochemical characteristics are double-membrane chloroplasts that contain the photosynthetic pigments chlorophylls *a* and *b* and the accessory pigments carotenoids and xanthophylls (Melkonian,1990), have been assigned the phylum Chlorophyta.

1.1. Habitats

Green algae are found worldwide in a broad range of habitats. Most of them are free-living aquatic organisms that inhabit fresh, brackish or salt waters (Margullis *et al.*, 1990). However, others are epiphytic (Lüttge & Büdel, 2010), epizoic (Garbary, Bourquea, Herman, & McNeil, 2007), parasitic (de Koning & Keeling, 2006) or endosymbiotic of fungi, protozoans (Nishihara *et al.*, 1998) or animals (Kerney *et al.*, 2011; Lewis & Mueller-Parker, 2004). Some are able to withstand extreme environments such as snow (Mosser, Mosser, & Brock, 1977), desert (Cardon, Gray, & Lewis, 2008), or halites (Lowenstein, Schubert, & Timofeeff, 2011), because they can endure long periods of desiccation. A number of green soil algae have been shown to survive 35 years of desiccation under laboratory conditions (Trainor & Gladych, 1995), or even of millions of years suspended in halites (Lowenstein *et al.*, 2011). The ability to inhabit this wide array of niches suggests that green algae have an underlying metabolic plasticity that must be reflected at the genomic level.

1.2. Taxonomy

Green algae taxonomy has been widely disputed since they were described by Linneo in 1753 (Moestrup, 2006). The conundrum lies within the great diversity of their morphological, ultrastructural and molecular characteristics (Lewis & McCourt, 2004; Pröschold, 2001). Nonetheless, it is widely accepted that the phylum Chlorophyta encompasses four main groups or classes designated as Prasinophyceae (Turmel, Lemieux, *et al.*, 1999), Trebouxiophyceae, Ulvophyceae and Chlorophyceae (Lewis & McCourt, 2004; Nakayama *et al.*, 1998; van den Hoek, Mann, & Jahns, 1995). Phylogenetic studies have shown the group Prasinophyceae to be paraphyletic, while the group Trebouxiophyceae is sister to the groups Ulvophyceae and Chlorophyceae (Pombert, Otis, Lemieux, & Turmel, M, 2004; Pröschold & Leliaert, 2007). A fifth group, designated Pedinophyceae, has been proposed, however its class status remains controversial (Lewis & McCourt, 2004). Analysis of the structural genomic features and phylogenetic analysis of chloroplast sequence data has helped untangle the evolutionary relationships between Chlorophyceae (Brouard, Otis, Lemieux, & Turmel, 2010).

2. CHLOROPHYTE GENOMES OVERVIEW

Although genomics is an emerging discipline within green algae phycology, it has been shown to be a powerful tool for single-gene and whole-genome analyses from an evolutionary perspective. Green algae or chlorophytes contain three genomes: nuclear, chloroplast and mitochondrial (Table 8.1). Nuclear and chloroplast genomes are addressed briefly; the focus of this work is mitochondrial genomes.

2.1. Nuclear

The first chlorophyte genome to be fully sequenced and annotated was *Chalmydomonas reinhardtii* (Chlorophyceae) (Merchant *et al.*, 2007). *C. reinhardtii* is considered a model eukaryotic organism because of its short

Table 8.1 Sequenced genomes of green algae

		Nuclear genome	Mitochondrial genome			Chloroplast genome	
Taxa	Class	Consortium	Accession number	Structure	Size (kb)	Accession number	Size (kb)
Micromonas pusilla	P	DOE Joint Genome Institute	—	—	—	NC_012568	41
Micromonas sp.	P	DOE Joint Genome Institute	NC_012643	Circular	47	NC_012575	72
Monomastix sp.	P	—	—	—	—	NC_012101	114
Nephroselmis olivacea	P	TBestDB	NC_0018239	Circular	45	NC_000927	200
Ostreococcus lucimarinus	P	DOE Joint Genome Institute	—	—	—	—	—
Ostreococcus sp.	P	DOE Joint Genome Institute	—	—	—	NC_008289	71
Ostreococcus tauri	P	DOE Joint Genome Institute	NC_008290	Circular	44	—	—
Pycnococcus provasolii	P	—	NC_013935	Circular	24	NC_012097	80
Pyramimonas parkeae	P	—	—	—	—	NC_012099	101
Acetabularia acetabulum	U	TBestDB	—	—	—	—	—
Bryopsis hypnoides	U	—	—	—	—	NC_013359	153
Oltmannsiellopsis viridis	U	—	NC_008256	Circular	56	NC_008099	151
Pseudendoclonium akinetum	U	—	NC_005926	Circular	95	NC_008114	195
Botryococcus braunii	T	DOE Joint Genome Institute	—	—	—	—	—
Coccomyxa sp.	T	DOE Joint Genome Institute	NC_015316	Circular	65	NC_015084	175

Chlorella sp.	T	DOE Joint Genome Institute	—	—	—	—	—
Chlorella variabilis	T	—	—	—	—	NC_015359	124
Chlorella vulgaris	T	—	—	—	—	NC_001865	150
Helicosporidium sp.	T	TBestDB	—	—	—	NC_008100[b]	37
Leptosira terrestris	T	—	—	—	—	NC_009681	195
Oocystis solitaria	T	—	—	—	—	—	—
Parachlorella kessleri	T	—	—	—	—	NC_012978	123
Prototheca wickerhamii	T	TBestDB	NC_001613	Circular	55	—	—
Chlamydomonas eugametos	C	—	NC_001872	Circular	22	—	—
Chlamydomonas incerta	C	TBestDB	—	—	—	—	—
Chlamydomonas nivalis	C	Aberystwyth University[a]	—	—	—	—	—
Chlamydomonas reinhardtii	C	DOE Joint Genome Institute	NC_001638	Linear: 1 chromosome	15	NC_005353	203
Dunaliella salina	C	DOE Joint Genome Institute	NC_012930	Circular	28	—	—
Dunaliella tertiolecta	C	Yale University	—	—	—	—	—
Floydiella terrestris	C	—	—	—	—	NC_014346	521
Oedogonium cardiacum	C	—	—	—	—	NC_011031	196
Polytomella capuana	C	—	NC_010357	Linear: 1 chromosome	12	—	—
Polytomella parva	C	TBestDB	—	Linear: 2 chromosomes	13.5 y 3.5	—	—
Polytomella sp.	C	—	NC_013472[b]		13 y 3	—	—

(Continued)

Table 8.1 Sequenced genomes of green algae—cont'd

		Nuclear genome	Mitochondrial genome			Chloroplast genome	
Taxa	Class	Consortium	Accession number	Structure	Size (kb)	Accession number	Size (kb)
				Linear: 2 chromosomes			
Scenedesmus obliquus	C	TBestDB	NC_002254	Circular	42	NC_008101	161
Stigeoclonium helveticum	C	–	–	–	–	NC_008372	223
Volvox carteri	C	DOE Joint Genome Institute	Smith and Lee, 2009	Linear: >1 chromosome	30	–	420
Pedinomonas minor	Pe	–	NC_000892	Circular	25	–	–

The complete nuclear genome of taxa in the TBestDB have not been sequenced; only expressed sequenced tags are available. All chloroplast genomes have a circular structure. P, *Prasinophyceae*; U, *Ulvophyceae*; T, *Trebouxiophyceae*; C, *Chlorophyceae*; Pe, *Pedinophyceae*. DOE Joint Genome Institute (http://genome.jgi-psf.org/), TBestDB (http://tbestdb.bcm.umontreal.ca/searches/welcome.php), Yale University (http://www.eng.yale.edu/peccialab/microalgae_sequences.html). GenBank accession numbers are provided (http://www.ncbi.nlm.nih.gov/).

[a]Sequence not available.

[b]Incomplete sequence.

generation time, sexual and asexual reproduction, unicellular and haploid nature, easy genetic manipulation (i.e. its three genomes are subject to specific transformation), and metabolic plasticity (i.e. heterotroph and facultative autotroph; aerobe and facultative anaerobe) (Funes, Franzén, & González-Halphen, 2007; Grossman *et al.*, 2003). Genome analysis revealed that the nuclear genome of *C. reinhardtii* consists of 127 Mb distributed in 17 chromosomes. Proteomic analysis revealed that 50% of the encoded proteins are homologous to their counterparts in other eukaryotes, such as humans and *Arabidopsis thaliana*. Of these proteins 10% are related to flagellar and basal body ultrastructures (cilia and centrioles), and 26% are associated with photosynthesis (thylakoid biogenesis and pigment biosynthesis) (Atteia *et al.*, 2009; Merchant *et al.*, 2007; Rolland *et al.*, 2009). Recently, the genome of *Volvox carteri* (Chlorophyceae) has been completed (Prochnik *et al.*, 2010); however, it is still being assembled. *V. carteri* belongs to a family of multicellular algae (Volvocales) and thus constitutes a model for studying the transition from unicellularity to multicellularity (Herron, Hackett, Aylward, & Michod, 2009). The nuclear genome of *V. carteri* consists of 138 Mb distributed in 14 chromosomes. A preliminary analysis suggests the presence of a large number of proteins involved in cell cycle regulation: the cyclins. *V. carteri* contains five types of D cyclins; *C. reinhardtii* only has three orthologues. The larger number of cyclins probably constituted a key factor in the appearance of processes related to cellular proliferation (Prochnik *et al.*, 2010).

Phytoplankton is responsible for fixing 50–60% of the CO_2 in the planet. It includes cyanobacteria and eukaryotic microalgae, mainly several members of the class Prasinophyceae. Three species from the genus *Ostreococcus* have been described based mainly on the depth of the ocean that they inhabit and thus, on the amount of light they receive: *O. lucimarinus*, *O. tauri* (Palenik, 2007) and *Ostreococcus* sp (Robbens *et al.*, 2007). Analysis of these sequenced genomes has provided important insights into adaptation and speciation processes. First, it was noted that *O. tauri* and *O. lucimarinus* have a different genome size and chromosome number: 12.6 Mb and 20 chromosomes, and 13.2 Mb and 21 chromosomes, respectively (the genome of *Ostreococcus* sp. has not yet been annotated). Despite the different chromosome numbers, 18 of them share similar gene content and order. Second, regarding gene content, the following observations were made. Both species have (1) lost genes that encode transcription factors and proteins related to the cell wall and flagella biosynthesis; (2) contain fused genes that are involved in pigment biosynthesis and nitrate metabolism; (3) contain a unique methylation/demethylation system (i.e. bacterial methyltransferases fused to a chromatin domain) potentially involved in exogenous DNA detection. Third, *O. lucimarinus* encodes a large number of selenium proteins that have been suggested to take part in essential metal metabolism (Palenik, 2007). Other prasinophycean algae whose nuclear genomes have been sequenced are *Micromonas* sp. and *Micromonas pusilla*. These species are 90% identical; the main difference relies on the presence or absence of different riboswitches, repetitive elements and transporters (Worden *et al.*, 2009).

Chlorella members can be either free living or endosymbionts. Thus, the genomes of *Coccomyxa* sp. and *Chlorella* sp. (Trebouxiophyceae) will provide important insights into the differences between free-living and endosymbiotic green algae, respectively (DOE Joint Genome Institute, http://www.jgi.doe.gov/). Furthermore, the genus *Chlorella* hosts a family of double-stranded DNA freshwater viruses that infect approximately 20% of *Chlorella* populations. These viruses play a significant role in global carbon and nitrogen cycles (Van Etten & Dunigan, 2012).

2.2. Chloroplast

At present, 22 chloroplast genomes of green algae have been sequenced (Table 8.1). These genomes have been classified according to the organization of the gene and the direction in which they are transcribed. They are divided into four regions: two inverted palindromic sequences that encode RNAs; and two regions, one short and one large, that contain single copy genes designated as SSC (small single copy) and LSC (large single copy).

The average configuration is a quadripartite structure in which the inverted palindromic repeats are separated by the SSC and LSC regions; and the genes encoding the rRNAs are transcribed in the SSC direction. This gene-partitioning pattern is similar to that of chloroplast genomes of land plants (Turmel, Otis, & Lemieux, 1999). Members of the Prasinophyceae, Pedinophyceae and Trebouxiophyceae classes have this configuration (Belanger *et al.*, 2006; Turmel, Otis, *et al.*, 2009). The exception is the chloroplast genome of *C. vulgaris* (Trebouxiophyceae), which presents a tripartite structure that lacks the inverted palindromic repeats (Belanger *et al.*, 2006). In members of the Ulvophyceae class, the quadripartite structure is conserved; however, the encoded rRNA genes are transcribed in the opposite direction (Pombert, Otis, Lemieux, & Turmel, 2005; Pombert, Lemieux, & Turmel, 2006). Chloroplast genomes within the class

Chlorophyceae reveal great structural diversity. For example, the chloroplast genome of *Stigeoclonium helveticum* lacks the inverted palindromic repeats (Belanger *et al.*, 2006); and *S. obliquus*, *C. reinhardtii* and *Oedogonium cardiacum* contain unique regions of single copy genes of similar length but with variable gene content (Brouard, Otis, Lemieux, & Turmel, 2008; de Cambiaire, Otis, Lemieux, & Turmel, 2006; Maul *et al.*, 2002).

2.3. Mitochondrial

Mitochondrial genomes are usually compact and contain a limited set of genes encoding proteins related to oxidative phosphorylation (OXPHOS) (i.e. complexes I–V), tRNAs and rRNAs (Gray, Burger, & Lang, 1999). A total of 16 mitochondrial green algal genomes have been fully sequenced and annotated (Table 8.2). According to their gene content they have been

Table 8.2 Gene content of completely sequenced chlorophycean mitochondrial genomes

		P				U		T		Pe	C				
Gene		No	Ot	Msp	Pp	Ov	Pa	Csp	Pw	Pm	So	Ds	Ce	Cr	Pc
OXPHOS	*nad1*	+	+	+[a]	+	+	+	+	+	+	+	+	+	+	+
	nad2	+	+	+	+	+	+	+	+	+	+	+	+	+	+
	nad3	+	+	+	+	+	+	+	+	+	+	−	−	−	−
	nad4	+	+	+	+	+	+	+	+	+	+	+	+	+	+
	nad4L	+	+[a]	+	+	+	+	+	+	+	+	−	−	−	−
	nad5	+	+	+	+	+	+	+	+	+	+	+	+	+	+
	nad6	+	+	+[a]	+	+	+	+	+	+	+	+	+	+	+
	nad7	+	+	+	−	+	+	+	+	+	−	−	−	−	−
	nad9	+	+	+	−	+	−	+	+	−	−	−	−	−	−
	nad10	+	+	−	−	−	−	−	−	−	−	−	−	−	−
	cob	+	+[a]	+	+	+	+	+	+	+	+	+	+	+	+
	cox1	+	+[a]	+[a]	+	+	+	+	+	+	+	+	+	+	+
	cox2	+	+	+	+	+	+	+	+	−	+[b]	−	−	−	−
	cox3	+	+	+	+	+	+	+	+	−	+	−	−	−	−
	atp1	+	+	−	−	+	+	+	+	−	−	−	−	−	−
	atp4	+	+	+[a]	+	+	+	+	+	−	−	−	−	−	−
	atp6	+	+	+[a]	+	+	+	+	+	+	+	−	−	−	−
	atp8	+	+[a]	+[a]	+	+	+	+	+	+	−	−	−	−	−
	atp9	+	+	+	+	+	+	+	+	−	+	−	−	−	−
Ribosome	*rpl5*	+	+	+	−	−	+	+	+	−	−	−	−	−	−
	rpl6	+	+	+	−	−	+	−	+	−	−	−	−	−	−

(*Continued*)

Table 8.2 Gene content of completely sequenced chlorophycean mitochondrial genomes—cont'd

		P				U		T		Pe	C				
	Gene	**No**	**Ot**	**Msp**	**Pp**	**Ov**	**Pa**	**Csp**	**Pw**	**Pm**	**So**	**Ds**	**Ce**	**Cr**	**Pc**
	rpl14	+	+	+	−	−	−	−	−	−	−	−	−	−	−
	rpl16	+	+	+	−	+	+	+	+	−	−	−	−	−	−
	rps2	+	+	+	−	+	+	+	+	−	−	−	−	−	−
	rps3	+	+	+	+	+	+	+	+	−	−	−	−	−	−
	rps4	+	+	+	+	−	+	+	+	−	−	−	−	−	−
	rps7	+	+	+	−	−	−	+	+	−	−	−	−	−	−
	rps8	+	+	+	−	−	−	−	−	−	−	−	−	−	−
	rps10	+	+	+	−	−	+	+	+	−	−	−	−	−	−
	rps11	+	+	+	−	+	+	−	+	−	−	−	−	−	−
	rps12	+	+	+	+	+	+	+	+	−	−	−	−	−	−
	rps13	+	+	+	−	+	+	+	+	−	−	−	−	−	−
	rps14	+	+	+	−	+	+	+	+	−	−	−	−	−	−
	rps19	+	+	+	−	+	+	+	+	−	−	−	−	−	−
rRNAs	*rnl*	+	−	−	+	+	+	−	+	+[b]	+	−	+[b]	−	−
	rns	+	−	−	+	+	+	−	+	+	+	−	+[b]	−	−
	rrl	−	+	+[a]	−	−	−	−	−	−	−	−	−	−	−
	rrn	+	−	−	−	+	−	+	+	−	−	+	−	+[b]	+
	rrs		+	+[a]	−	−	−	−	−	−	−	−	−	−	−
Total tRNAs		26	26	34	16	24	25	26	26	8	27	3	3	3	1

[a]Duplicated genes.
[b]Spliced genes.
Gene content of each gene was derived from GenBank (http://www.ncbi.nlm.nih.gov/genomes/GenomesGroup.cgi?taxid=33090&opt=organelle). Unique proteins to each genome that are neither OXPHOS nor ribosomal related were excluded. P, *Prasinophyceae*; U, *Ulvophyceae*; T, *Trebouxiophyceae*; Pe, *Pedinophyceae*; C, *Chlorophyceae*. No, *Nephroselmis olivacea*; Ot, *Ostreococcus tauri*; Pw, *Prototheca wickerhamii*; Msp, *Micromonas* sp.; Pp, *Pycnococcus provasolii*; Ov, *Oltmannsiellopsis viridis*; Pa, *Pseudendoclonium akinetum*; Csp, *Coccomyxa* sp.; Pm, *Pedinomonas minor*; So, *Scenedesmus obliquus*; Ds, *Dunaliella salina*; Ce, *Chlamydomonas eugametos*; Cr, *Chlamydomonas reinhardtii*; Pc, *Polytomella capuana*.

classified into ancestral, reduced and intermediate (Nedelcu, Lee, Lemieux, Gray, & Burger, 2000; Turmel, Lemieux *et al.*, 1999) (Fig. 5.1).

Ancestral genomes are between 44 and 95 kb in size and contain genes that encode OXPHOS proteins, the complete set of tRNAs, and rRNAs (Nedelcu *et al.*, 2000). The latter are located continuously in the mitochondrial genome and seem to be evolving much faster than their nuclear counterparts (Popescu & Lee, 2007). The genomes of the prasinophycean (Turmel, Lemieux *et al.*, 1999), ulvophycean (Pombert *et al.*, 2004) and trebouxiophycean (Pombert & Keeling, 2010; Wolff, Plante, Lang, Kück, & Burger, 1994) algae belong to this group.

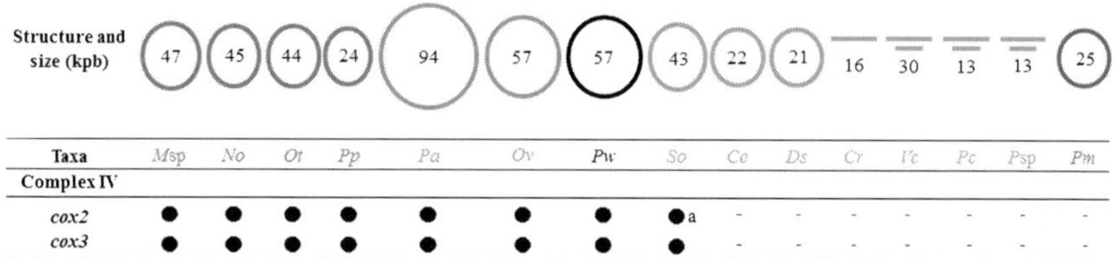

Figure 8.1 *cox2* and *cox3* gene diversity in chlorophyte genomes. The genome structure of chlorophytes. The numbers indicate the size in kilobase pairs. Msp, *Micromonas* sp.; No, *Nephroselmis olivacea*; Ot, *Ostreococcus tauri*; Pp, *Pycnococcus provasolii*; Pa, *Pseudendoclonium akinetum*; Ov, *Oltmannsiellopsis viridis*; Pw, *Prototheca wickerhamii*; So, *Scenedesmus obliquus*; Ce, *Chlamydomonas eugametos*; Ds, *Dunaliella salina*; Cr, *Chlamydomonas reinhardtii*; Vc, *Volvox carteri*; Pc, *Polytomella capuana*; Psp, *Polytomella* sp.; Pm, *Pedinomonas minor*. The circle indicates the presence and the line indicates the absence of the corresponding gene within the mitochondrial genome. cox2a gene. For colour version of this figure, the reader is referred to the online version of this book.

Reduced genomes are between 12 and 28 kb in size and contain genes that encode some OXPHOS proteins, an incomplete set of tRNAs, and some rRNAs. The missing tRNAs must therefore be imported from the cytosol into mitochondria, probably through a mechanism similar to that used by plants (Rubio & Hopper, 2011; Salinas, Duchêne, & Maréchal-Drouard, 2008; Sieber, Placido, El Farouk-Ameqrane, Duchêne, & Maréchal-Drouard, 2011). The ribosomal RNAs may be fragmented and are located intermittently in the mitochondrial genome. The genomes of the chlorophycean and pedinophycean algae belong to this group (Turmel, Lemieux *et al.*, 1999).

The mitochondrial genomes of *P. provasolii* (Prasinophyceae) and *S. obliquus* (Chlorophyceae) share characteristics of both types of genomes. *P. provasolii* has 24 kb, like the reduced type. However, its gene content and organization (except for the rRNA genes) resemble that of the ancestral type genomes (Turmel, Otis, & Lemieux, 2010). The genome of *S. obliquus* is larger, with 43 kb, in the range of the ancestral genomes. Nonetheless, its gene content and arrangement is similar to that of reduced genomes (Nedelcu *et al.*, 2000).

Many green algae use the chlorophycean mitochondrial genetic code (The Genetic Codes-NCBI; http://www.ncbi.nlm.nih.gov/Taxonomy/Utils/wprintgc.cgi), although others have deviant mitochondrial genetic codes (Hayashi-Ishimaru, Ohama, Kawatsu, Nakamura, & Osawa, 1996). Three chlorophytes use an alternate genetic code for certain amino acids: in *Pedinomonas minor* (Pedinophyceae) the stop codon TGA codes tryptophan (Turmel, Lemieux *et al.*, 1999); in *S. obliquus* (Chlorophyceae) the stop codon TAG codes leucine and the triplet TCA codes a stop codon (Nedelcu *et al.*, 2000); and in *P. provasolii* (Prasinophyceae) the stop codon TGA codes tryptophan, and the standard leucine codons TTA and TTG code stop codons (Turmel *et al.*, 2010).

2.4. Mitochondrial Genetics and Mitochondrial Mutants

Transmission of the mitochondrial genome has been studied in *C. reinhardtii*. The pioneer work of Boynton, Harris, Burkhart, Lamerson, and Gillham (1987) demonstrated that sexual zygotes transmit chloroplast and mitochondrial DNA from their opposite mating type, mitochondrial DNA being inherited from the mt^- parent and chloroplast DNA from the mt^+ parent. The fate of mitochondrial DNA was analysed in zygotes and during maturation of the zygospore (Beckers, Munaut, Minet, & Matagne, 1991). The mitochondrial DNA of mt^+ origin is slowly degraded during zygote maturation and light is required for total elimination of mt^+ mitochondrial DNA in the zygospores. Contrary to the situation found for chloroplasts where DNA from mt^+ origin is methylated and transmitted to the progeny (Umen & Goodenough, 2001), no methylation of mitochondrial DNA could be detected. These results were further confirmed by Nakamura's group who correlated the transmission of mt^- mitochondrial DNA to a selective elimination of mt^+ mitochondrial nucleoids after zygote formation (Aoyama, Hagiwara, Misumi, Kuroiwa, & Nakamura, 2006; Nakamura, 2010; Nakamura, Aoyama, & Van Woesik, 2003).

The first mitochondrial mutants in *C. reinhardtii* have been isolated and described by Matagne, Michel-Wolwertz, Munaut, Duyckaerts, and Sluse (1989). These mutants exhibit a terminal deletion of their mitochondrial genome including the *cob* gene. Phenotypically, they are unable to grow in heterotrophic conditions (dark + acetate) contrary to the wild-type strain because they lack the respiratory complex III. These mutants were called *dum* for *d*ark *u*niparental *t*ransmission by the mt^- parent. Other *dum* mutants have since been isolated based on their inability to grow in the dark. They exhibit either long deletions including several genes (such as *cob*, *nad4* and *nad5*) or frameshift mutations in the *nad1*, *nad5* or *nad6* genes (Colin *et al.*, 1995; Dorthu *et al.*, 1992; Duby & Matagne, 1999). Myxothiazol-resistant mutants with substitutions in the *cob* gene have also been characterized (Bennoun, Delosme, & Kück, 1991). *C. reinhardtii* is also the only green organism in which mitochondrial transformation is possible (Randolph-Anderson *et al.*, 1993; Remacle, Cardol, Coosemans, Gaisne, & Bonnefoy,

2006; Yamasaki, Kurokawa, Watanabe, Ikuta, & Ohama, 2005). This technique has been used recently to reconstruct a human mitochondrial complex I mutation in the mitochondrial genome of this alga (Larosa, Coosemans, Motte, Bonnefoy, & Remacle, 2012). Since human mitochondrial mutations are often difficult to study, this system could represent an attractive tool to study the fundamental effects of human mitochondrial complex I mutations (Barbieri *et al.*, 2011). This is of special interest because the other organism in which mitochondrial transformation is feasible, the yeast *Saccharomyces cerevisiae* (Bonnefoy, Remacle, & Fox, 2007), lacks complex I.

3. MITOCHONDRIAL GENOME EVOLUTION

The number of completely sequenced green algae mitochondrial genomes is expanding rapidly. Comparative studies have revealed a great diversity in mitochondrial genome structure, organization and gene content.

3.1. Structure

Most mitochondrial genomes have different molecular structures (Table 8.1), although in most cases they are circular. However, *C. reinhardtii* (Boer & Gray, 1988; Michaelis, Vahrenholz, & Pratje, 1990; Vahrenholz, Riemen, Pratje, Dujon, & Michaelis, 1993), *V. carteri* (Smith & Lee, 2009) and algae from the genus *Polytomella* (Mallet & Lee, 2006), all chlorophycean, have a linear structure. Furthermore, *V. carteri* (Smith & Lee, 2009) and members of the *Polytomella* genus (Mallet & Lee, 2006) have more than one chromosome. Structural differences within the same genus have also been described. For example, the mitochondrial genome of *Polytomella capuana* consists of a single linear chromosome 12 kb in length. In contrast, the genomes of *Polytomella parva* and *Polytomella* sp. consist of two linear chromosomes each: 13 5 kb and 3.5 kb, and 13 kb and 3 kb, respectively (Mallet & Lee, 2006; Smith, Hua *et al.*, 2010).

Telomeric sequences have been described at the ends of linear chromosomes of chlorophycean mitochondrial genomes. These sequences consist of single-stranded inverted repeats of approximately 580 bp (Fan & Lee, 2002). It has been suggested that telomeric sequences, as well as short intergenic repeats that are present throughout the mitochondrial genomes of green algae, act as substrates for recombination events (Nedelcu & Lee, 1998). For example, Smith and Lee (2008) used a segment of an inverted repeat located within the intergenic region of the genes *nad6* and *cob* of *P. capuana* (one linear chromosome) to hypothetically recombine it with the homologous repeats located within the telomeric regions. As a result two products, similar to both linear chromosomes of *P. parva*, were inferred. Therefore, inverted interspersed repeats may be the consequence of the corresponding insertion or deletion events (Nedelcu & Lee, 1998). Short repeats have been described in all members of the chlorophycean class; nonetheless, their abundance within each genome varies. Mitochondrial genomes of the ancestral type are rich in A + T regions with a length between 6 and 17 bp; the reduced type are rich in G + C regions with a length between 9 and 14 bp (Nedelcu & Lee, 1998).

3.2. Organization

Gene content is different among mitochondrial genomes of chlorophytes (Table 8.2). Focussing exclusively on OXPHOS protein coding genes, we have identified contiguous arrangements of three or more consecutive genes (Table 8.3). Usually, gene order (i.e. synteny) is preserved among phylogenetically related organisms, and thus may constitute a molecular tool for assessing evolutionary relationships among green algae (Tamames, 2001). However, as shown in Table 8.3, gene order among chlorophycean algae is highly variable, even among members of the same class.

The first two arrangements were present only in prasinophycean algae. *Micromonas* sp. and *O. tauri* present the arrangement *(atp9)-nad7-nad3*, where the gene *atp9* is located in the complement strand. *Micromonas* sp., *N. olivacea* and *O. tauri* present the arrangement *cox2-cox3-(nad2)-(nad4)-(nad5)* where the *nad* genes are located in the complement strand. In *N. olivacea*, the *nad* genes are oriented in the opposite direction and are located in the coding strand (i.e. *cox2-cox3-nad5-nad4-nad2*). Furthermore, *P. provasolii* presents the *cox3-nad5-nad4-nad2* arrangement. The contiguous *nad5–nad4–nad2* genes are characteristic of eubacterial mitochondrial genomes (Lang *et al.*, 1997). This agrees with the fact that the Prasinophyceae constitute the earliest divergent class within green algae (Turmel, Lemieux *et al.*, 1999). The arrangement *(nad4L)-atp1-nad1* is found in *P. akinetum* (Ulvophyceae) and *P. wickerhamii* (Trebouxiophyceae). Nonetheless, in the latter, all the genes are located in the coding strand. *P. capuana* (Chlorophyceae) and *P. minor* (Pedinophyceae) present the arrangement *cox1-nad4-nad2*.

Table 8.3 OXPHOS Gene arrangements in chlorophycean mitochondrial genomes

Class	Taxa	Gene order
P	*Micromonas* sp.	*(cob)-(nad6[a])-atp6[a]-nad1[a]-(cox1[a])-(atp4[a])-(atp8[a])-(nad4L)-**(atp9)-nad7-nad3**-(nad9)-atp8[a]-atp4[a]-cox1[a]-(nad1[a])-(atp6[a])-nad6[a]-**cox2-cox3-(nad2)-(nad4)-(nad5)***
	Ostreococcus tauri	*cob[a]-(nad1)-(atp6)-nad6-**cox2-cox3-(nad2)-(nad4)-(nad5)**-(cob[a])-(cox1[a])-(atp4[a])-(atp8[a])-(nad4L[a])-**(atp9)-nad7-nad3**-(atp1)-(nad9)-nad4L[a]-atp8[a]-atp4[a]-cox1[a]*
	Nephroselmis olivacea	*(cob)-(nad3)-(nad7)-(atp1)-(atp6)-nad6-**cox2-cox3-nad5-nad4-nad2**-atp9-(cox1)-(nad9)-(nad10)-nad4L-atp8-atp4-(nad1)*
	Pycnococcus provasolii	*cob-nad3-atp4-atp8-atp6-nad4L-nad1-cox2-cox1-atp9-**cox3-nad5-nad4-nad2**-nad6*
U	*Oltmannsiellopsis viridis*	*(cob)-(atp4)-nad9-(nad3)-(nad7)-(atp9)-nad9-(nad3)-(nad7)-(atp9)-(atp1)-nad2-nad5-nad4L-atp6-atp8-cox1-cox2-cox3-nad1-nad6-nad4*
	Pseudendoclonium akinetum	*(cob)-(nad4)-**(nad4L)-atp1-nad1**-nad5-cox2-(cox1)-(nad6)-nad3-atp8-nad2-(cox3)-(atp4)-(atp6)-(nad7)*
T	*Coccomyxa* sp.	*cob-nad5-nad7-nad2-nad4L-cox1-atp1-nad1-nad4-cox2-cox3-atp6*
	Prototheca wickerhamii	*(cob)-(atp6)-(nad3)-nad6-**nad4L-atp1-nad1**-nad2-(cox1)-(nad4)-(nad5)-(nad7)-(cox3)-(cox2)-(atp8)-(nad9)*
C	*Scenedesmus obliquus*	*cob-atp9-cox2-nad5-nad4L-nad3-cox3-nad1-(atp6)-nad2-cox1-nad4-nad6*
	Chlamydomonas eugametos	*cob-cox1-nad1-nad5-nad6-nad4-nad2*
	Chlamydomonas reinhardtii	*(cob)-(nad4)-(nad5)-cox1-nad2-nad6-nad1*
	Dunaliella salina	*cob-nad6-nad5-nad1-nad4-cox1-nad2*
	Polytomella capuana	*(cob)-cox1-nad4-nad2-nad5-nad1-(nad6)*
Pe	*Pedinomonas minor*	*cob-atp8-nad5-atp6-nad4L-cox1-nad4-nad2-nad1-nad6-nad3*

The gene *cob* is set as the start point for analysis, however, the exact gene order is indicated in http://www.ncbi.nlm.nih.gov/genomes/GenomesGroup.cgi?taxid=33090&opt=organelle. Consecutive gene arrangements are highlighted. Parentheses indicate that the gene is located in the complementary DNA strand. P, *Prasinophyceae*; U, *Ulvophyceae*; T, *Trebouxiophyceae*; Pe, *Pedinophyceae*; C, *Chlorophyceae*.
[a]Duplicated genes.

3.3. Gene Dynamics

3.3.1. Gene transfer to the nucleus

Mitochondrial genomes are highly dynamic; they can modify, eliminate or rearrange their genetic material (Gray *et al.*, 1999). After the first endosymbiosis event, a massive gene transfer from the mitochondrial genome to the nuclear genome ensued, which increased the complexity of the nuclear genome and reduced the gene content in mitochondrial genomes (Gray *et al.*, 1999). When relocated to the nuclear genome, mitochondrial genes are usually transferred as a single entity (Adams *et al.*, 1999; Pérez-Martínez *et al.*, 2002). However, nuclear relocation as split genes has been reported (Adams, Ong, & Palmer, 2001; Funes *et al.*, 2002; Gawryluk & Gray, 2009; Gawryluk & Gray, 2010; Pérez-Martínez *et al.*, 2001).

Only two genes are universally located in the mitochondrial genome: *cox1* and *cob*, encoding subunit I of cytochrome *c* oxidase and cytochrome *b*, respectively. In those organisms that contain the whole set of respiratory complexes I–IV, the genes *nad1*, *nad2*, *nad4* and *nad5* are also always present in the mitochondrial genome. The main characteristic of these mitochondria-encoded proteins is that they exhibit a high hydrophobic profile (i.e. they contain from 8 to 16 transmembrane stretches) (Adams & Palmer, 2003). The vast majority of eukaryotic organisms also carry the following genes in their mitochondrial genomes: *cox2*, *cox3*, *nad3*, *nad4L*, *atp6*, and *atp8*. Several of these genes have migrated to the nucleus in some lineages of chlorophycean algae. In *C. reinhardtii*, *Polytomella* sp. and *S. obliquus* (Chlorophyceae), the mitochondrial gene *cox2* is split into two genes: *cox2a* and *cox2b*. The *cox2a* and *cox2b* genes encode a polypeptide that corresponds to a portion of a heterodimeric COX2 subunit; the amino-terminal half and the carboxy-terminal half, respectively (Pérez-Martínez *et al.*, 2001). In *C. reinhardtii* and *Polytomella* sp. both genes are located in the nuclear genome (Pérez-Martínez *et al.*, 2001) but in *S. obliquus* only the gene *cox2b* is located in the nuclear genome, and the gene *cox2a* remains in the mitochondrial genome (Funes *et al.*, 2002). More recently, the distribution of intact and split *cox2* gene sequences was analysed in several algae pertaining to the phylum Chlorophyta. The algae in classes Prasinophyceae, Ulvophyceae, and Trebouxiophyceae all contain orthodox, intact mitochondrial *cox2* genes. In contrast, all the algae in Chlorophyceae that were examined exhibited split *cox2* genes and were separated into two groups: *Scenedesmus*-like algae that have a mitochondrion-localized *cox2a* gene and a nucleus-localized *cox2b* gene, and *Chlamydomonas*-like algae that have both *cox2a* and *cox2b* genes in the nucleus (Rodríguez-Salinas *et al.*, in press).

The *cox3* gene, encoding subunit III of cytochrome *c* oxidase, also migrated to the nucleus in some lineages within Chlorophyceae. This gene is present in the mitochondrial genome of prasinophytes, ulvophytes, trebouxiophytes and some chlorophytes (Table 8.2). However, it is absent in the mitochondrial DNA of the chlorophycean algae *D. salina*, *C. elongatum*, *C. reinhardtii*, *P. capuana* and in the pedinophycean *P. minor*. In these algae, the *cox3* gene may have migrated to the nuclear genome as a single entity. This has been demonstrated in the case of *Polytomella* sp., *C. reinhardtii* and *V. carteri* (Pérez-Martínez *et al.*, 2000; Pérez-Martínez *et al.*, 2002; Prochnik *et al.*, 2010).

Another OXPHOS enzymatic complex that presents subunits of combined origin (i.e. mitochondrion- and nucleus-encoded) is NADH:ubiquinone oxidoreductase (complex I). This is the largest and the most intricate enzyme of the mitochondrial respiratory chain (Efremov & Sazanov, 2011). It is a membrane-bound assembly of approximately 42–43 subunits in *C. reinhardtii* (Cardol *et al.*, 2004). In green algae, genes *nad1*, *nad2*, *nad4*, *nad5*, and *nad6* are present in all the mitochondrial genomes that have been sequenced (Table 8.2). In contrast, the genes *nad3* and *nad4L* are absent in several chlorophycean mitochondrial genomes, except in *S. obliquus*; these genes, which encode subunits essential for the function and assembly of complex I, also migrated to the nucleus in some algal lineages (Cardol *et al.*, 2006) (Table 8.2).

F_1F_O-ATP synthase (complex V) subunits are encoded by *atp* genes. In chlorophytes, the genes *atp1*, *4*, *6*, *8* and *9*, which encode subunits α, b_2, *a*, A6L and *c*, respectively, are typically located in the mitochondrial genome (Table 8.2). Nevertheless, the genes *atp6* and *atp8* are missing from the mitochondrial genomes of several chlorophycean algae. In *C. reinhardtii* and *Polytomella* sp., the gene *atp6* migrated to the nucleus, while *atp8* seems to have been lost. Therefore, these algae do not encode a single *atp* gene in their mitochondrial genome, an exceptional situation that differs from all other mitochondria-bearing eukaryotes. Note that the subunit composition of the chlorophycean mitochondrial ATP synthase is atypical. Opisthokonts and plants contain an orthodox ATP synthase composed of 14–15 conserved subunits that assemble into a rotor containing subunits γ, δ, ε, c_{10}, a catalytic domain formed by three α subunits and three β subunits, and a peripheral stator composed of subunits a, A6L, e, f, g, b_2, OSCP and F6 (Walker and Dickson, 2006). Nonetheless, biochemical and computational analyses have revealed that the ATP synthase of the chlorophycean algae *C. reinhardtii*, *Polytomella* sp. and *V. carteri* lack several of the genes encoding the subunits of the peripheral arm of the enzyme, which exhibits an atypical composition (Cardol *et al.*, 2005; Van Lis, Mendoza-Hernández, Groth, & Atteia, 2007).

Its constituents are nucleus-encoded subunits, which have been named Asa1–9 for ATP synthase-associated proteins (Cano-Estrada *et al.*, 2010; Vázquez-Acevedo *et al.*, 2006). However, their evolutionary origin remains obscure, and to date, few homologues have been identified outside the chlorophycean lineage (Lapaille *et al.*, 2010).

3.3.2. Transfer of other DNA elements

The mitochondrial genome of *Volvox carteri* is highly enriched with short palindromic sequences. It has been proposed that the original palindromic element first appeared in a mitochondrial intron, afterwards spreading to the chloroplast genome and eventually to the nuclear genome (Smith & Lee, 2009). The mitochondrial genome of *Dunaliella salina*, as well as its plastid genome, has extremely high intron and intergenic DNA densities (Smith *et al.*, 2010).

3.3.3. Recent gene acquisitions

The gene *rtl* was identified in the mitochondrial genome of *C. reinhardtii* and was found to be related to the reverse transcriptase-like part of some fungal mitochondrial introns and plasmids (Boer & Gray, 1988). Although the evolutionary origin of the RTL protein (i.e. retro transcriptase-like) is unknown, this gene has been described exclusively in the mitochondrial genomes of *C. reinhardtii* (Lang *et al.*, 1997) and *Chlamydomonas smithii* (Kroymann & Zetsche, 1998). In contrast to the rest of the mitochondrial encoded proteins, the T nucleotide frequency in the third position is low, suggesting it was recently acquired (Boer & Gray, 1988). We have conducted a BLAST analyses using the *rtl* gene of *C. reinhardtii* (NP_042571). The results reveal that a gene that shares 72% identity is present in *Chlamydomonas incerta* (ABC98218), and intriguingly, that a gene that shares 22% of identity is present in the chloroplast genome of *S. obliquus* (YP_636002). Furthermore, the identity shared between the identified genes and the active site of the RTL of *C. reinhardtii* is 54%. The *rtl* gene seems to be absent from the rest of the sequenced green algae genomes (Rodríguez-Salinas & González-Halphen, 2009).

4. COMPARATIVE GENOMICS AND ACQUISITION OF NEW METABOLIC CAPABILITIES

Studying and comparing genome structure and function will help us to understand how evolutionary forces act on genomes; and thus will shed light on green algae adaptive dynamics. However, this field is still in its infancy due to the scarce number of sequenced genomes. An immediate application of comparative genomics includes phylogenetic analysis (i.e. the reconstruction of gene transfer events from one generation to another). These analyses will provide a more solid and reliable reconstruction of the events that led to present-day green algal genomes.

Also, the niches inhabited by chlorophytes are either constantly changing or represent extreme environments. The ability of these algae to adapt to new environmental conditions is just beginning to be studied (Carrera-Martinez, Mateos-Sanz, Lopez-Rodas, & Costas, 2011; Collins, 2011; Li, Lu, Xue, & Xie, 2010). Furthermore, the following genomic characteristics, which have been described in some chlorophytes, may help shed light on the subject.

4.1. Ancestral Organelle Genomes as Evidence of Green Plant Evolution

Green algae and land plants comprise the Viridiplantae kingdom, which consists of two phyla: Chlorophyta (green algae) and Streptophyta (charophyte algae and land plants). A link between both groups has been perceived by taxonomists for centuries (Lewis & McCourt, 2004). The identity of the unicellular flagellate ancestor remains unknown. Through chloroplast genome comparison, it has been estimated that 470 MYA the last common ancestor to both phyla transitioned from an aquatic to a land habitat, thus resulting in the land plant ancestor (Lemieux, Otis, & Turmel, 2000; Lewis & McCourt, 2004; Turmel, Otis, & Lemieux, 2006). This hypothesis has been further supported by single-gene phylogenetic analysis, which suggests that the last common ancestor belonged to the group of present-day charophyte algae (Turmel, Otis, & Lemieux, 2003). In addition, these studies indicate that chloroplast genome architecture has been extremely well conserved (Lemieux *et al.*, 2000; Turmel *et al.*, 2006).

Comparison of chlorophycean mitochondrial genomes accompanied by phylogenetic studies are in agreement with the monophyly and early divergence of the Prasinophyceae class. It is believed that this group contains the most primitive forms of green algae (Bullerwell & Gray, 2004). The analyses of the nuclear genomes of two *Micromonas* species have been used as the start point to infer the genetic composition of the last common ancestor between green algae and land plants. For example, *Micromonas* encodes transcription factors that are also encoded in the modern descendants of the ancestral lineages of land plants (e.g. associated with leaf development), but they are absent in other Prasinophyceae genera (e.g. *Ostreococcus*). Also, genes that encode meiosis-associated proteins (e.g. hydroxyproline-rich glycoproteins, which are expressed exclusively after sexual fusion) have been identified within the *Micromonas* nuclear genome. Nevertheless, none of the members of the Prasinophyceae has been reported to reproduce sexually, as plants do (Worden *et al.*, 2009).

4.2. Evolutionary Mechanisms

Gene and genome duplication have long been considered a major evolutionary force. Since the additional copies are free of selective pressure, they provide a source of evolutionary novelties: new gene functions and expression patterns (Babushok, Ostertag, & Kazazian, 2007). The following alternative outcomes have been suggested for duplicate genes: (1) non-functionalization of one copy (i.e. silencing by degenerative mutations); (2) neo-functionalization of one copy (i.e. acquisition of a novel function and its preservation by natural selection in one copy while the other retains its original function); or (3) sub-functionalization of both copies (i.e. mutation accumulation reduces both gene capacity to the level of the single copy ancestral gene) (Lynch & Conery, 2000). At present, four examples of gene duplication have been described in green algae.

The first two examples are mitochondrial genome duplication within two members of the Prasinophyceae class, *O. tauri* and *Micromonas* sp. The former contains a duplication of a 20-kb segment that encompasses 44% of the full mitochondrial genome. The duplicate segment contains an open reading frame (*orf129*), genes encoding OXPHOS proteins (*cob*, *cox1*, *atp4*, *atp8*, *nad4L*, and *ymf39*), and ribosomal protein coding genes (*rnl*, *rns*, *rrn*) (Robbens *et al.*, 2007). In the latter, the duplicate segment is larger and contains genes encoding OXPHOS proteins (*nad1*, *nad6*, *cox1*, *atp4*, *atp6*, *atp8*) and ribosomal protein coding genes (*rrl* and *rrs*) (NC_012643). Future studies will assess the sequence identity as well as the functionality of each duplicated gene.

The last two examples have been described in members of the *Dunaliella* genus (Chlorophyceae). *D. viridis* contains an inverted duplication of the gene *DvSPT1* within its nuclear genome. This gene encodes a sodium-dependent phosphate transporter. Both genes have the same number of exons and introns and share a sequence identity of 99.7%. When subjected to different salt concentrations, expression levels were higher for the original gene under low salt concentrations, and for the duplicate copy under high salt concentrations (Guan, Meng, Sun, Xu, & Song, 2008). The same scenario has been reported for the duplicated carbonic anhydrase gene *DCA1* in *D. salina* (Li *et al.*, 2010).

5. PERSPECTIVES

The field of genomics has provided new ways to uncover and understand evolutionary mechanisms of mitochondrial genomes, as presented above. Genomics provides insights into different aspects of organism and cell function, i.e. biochemistry, genetics, cellular biology, etc. The different approaches have allowed us to study the remarkable mobility from and to the genomes of green algae; this will help our understanding of their metabolic plasticity. Mitochondria are not alone within green cells; they have coexisted with chloroplasts for more than a billion years. Within this framework, complex interchanges at the genomic level, and interactions at the metabolic level, have been established (Matsuo, Hachisu, Tabata, Fukuzawa, & Obokata, 2011). Adaptive advantages may arise from these successful interactions, representing unique opportunities to gain insight into the transition from single-cell organisms to multicellular photosynthetic organisms. It will thus be interesting in the future to study the metabolism of these two organelles from a co-evolutionary perspective (i.e. events following the endosymbiosis that gave rise to the chloroplast). An interesting example is green algae of the genus *Polytomella*. They lack functional chloroplasts; nonetheless, they may still contain photosynthetic genes (Rodríguez-Salinas and González-Halphen, unpublished observations). The new -omics approaches (May *et al.*, 2008; Wienkoop *et al.*, 2010) and recently developed techniques such as deep sequencing of RNA (Xiong *et al.*, 2012) molecules, and the identification of thousands of proteins by high throughput mass spectrometric analyses will play a key role.

ACKNOWLEDGEMENTS

Research in the laboratory of C.R. is supported by the Fonds National de la Recherche Scientifique (grant numbers 1.5.255.08, 2.4.601.08 and 2.4567.11), by Sunbiopath, an FP7-funded project (GA 245070), Action de la Recherche Concertee ARC07/12-04 and 'Fonds Speciaux du Conseil de la Recherche' from the University of Liege. Research in the laboratory of D.G.-H. is partially supported by grants 128110 (Consejo Nacional de Ciencia y Tecnología, CONACyT, Mexico) and IN203311-3 (Dirección General de Asuntos del Personal Académico, DGAPA-UNAM, Mexico). E.R.-S. received a fellowship (221018) from CONACyT to carry on PhD studies at the program of Biomedical Sciences, UNAM. The authors acknowledge the technical support of Miriam Vázquez-Acevedo on the ongoing research projects.

REFERENCES

Adams, K. L., Ong, H. C., & Palmer, J. D. (2001). Mitochondrial gene transfer in pieces: fission of the ribosomal protein gene rpl2 and partial or complete gene transfer to the nucleus. *Molecular Biology and Evolution, 18*, 2289–2297.

Adams, K. L., & Palmer, J. D. (2003). Evolution of mitochondrial gene content: gene loss and transfer to the nucleus. *Molecular Phylogenetics and Evolution, 29*, 380–395.

Adams, K. L., Song, K., Roessler, P. G., Nugent, J. M., Doyle, J. L., Doyle, J. J., et al. (1999). Intracellular gene transfer in action: dual transcription and multiple silencings of nuclear and mitochondrial cox2 genes in legumes. *Proceedings of the National Academy of Sciences of the United States of America, 96*, 13863–13868.

Atteia, A., Adrait, A., Brugière, S., Tardif, M., van Lis, R., Deusch, O., et al. (2009). A proteomic survey of *Chlamydomonas reinhardtii* mitochondria sheds new light on the metabolic plasticity of the organelle and on the nature of the alpha-proteobacterial mitochondrial ancestor. *Molecular Biology and Evolution, 26*, 1533–1548.

Aoyama, H., Hagiwara, Y., Misumi, O., Kuroiwa, T., & Nakamura, S. (2006). Complete elimination of maternal mitochondrial DNA during meiosis resulting in the paternal inheritance of the mitochondrial genome in *Chlamydomonas* species. *Protoplasma, 228*, 232–242.

Babushok, D. V., Ostertag, E. M., & Kazazian, H. H., Jr. (2007). Current topics in genome evolution: molecular mechanisms of new gene formation. *Cellular and Molecular Life Sciences, 64*, 542–554.

Barbieri, M. R., Larosa, V., Nouet, C., Subrahmanian, N., Remacle, C., & Hamel, P. P. (2011). A forward genetic screen identifies mutants deficient for mitochondrial complex I assembly in *Chlamydomonas reinhardtii. Genetics, 188*, 349–358.

Beckers, M. C., Munaut, C., Minet, A., & Matagne, R. F. (1991). The fate of mtDNAs of mt+ and mt− origin in gametes and zygotes of *Chlamydomonas. Current Genetics, 20*, 239–243.

Belanger, A. S., Brouard, J. S., Charlebois, P., Otis, C., Lemieux, C., & Turmel, M. (2006). Distinctive architecture of the chloroplast genome in the chlorophycean green alga *Stigeoclonium helveticum. Molecular Genetics and Genomics, 276*, 464–477.

Bennoun, P., Delosme, L., & Kück, U. (1991). Mitochondrial genetics of *Chlamydomonas reinhardtii*: resistance mutations marking the cytochrome *b* gene. *Genetics, 127*, 335–343.

Boer, P. H., & Gray, M. W. (1988). Genes encoding a subunit of respiratory NADH dehydrogenase (ND1) and a reverse transcriptase-like protein (RTL) are linked to ribosomal RNA gene pieces in *Chlamydomonas reinhardtii* mitochondrial DNA. *EMBO Journal, 7*, 3501–3508.

Bonnefoy, N., Remacle, C., & Fox, T. D. (2007). Genetic transformation of *Saccharomyces cerevisiae* and *Chlamydomonas reinhardtii* mitochondria. *Methods in Cell Biology, 80*, 525–548.

Boynton, J. E., Harris, E. H., Burkhart, B. D., Lamerson, P. M., & Gillham, N. W. (1987). Transmission of chloroplast and mitochondrial genomes in crosses of *Chlamydomonas. Proceedings of the National Academy of Sciences of the United States of America, 84*, 2391–2395.

Brouard, J. S., Otis, C., Lemieux, C., & Turmel, M. (2008). Chloroplast DNA sequence of the green alga *Oedogonium cardiacum* (Chlorophyceae): unique genome architecture, derived characters shared with the Chaetophorales and novel genes acquired through horizontal transfer. *BMC Genomics, 9*, 290.

Brouard, J. S., Otis, C., Lemieux, C., & Turmel, M. (2010). The exceptionally large chloroplast genome of the green alga *Floydiella terrestris* illuminates the evolutionary history of the Chlorophyceae. *Genome Biology and Evolution, 2*, 240–256.

Bullerwell, D. E., & Gray, M. W. (2004). Evolution of the mitochondrial genome: protist connections to animals, fungi and plants. *Current Opinion in Microbiology, 7*, 528–534.

Cano-Estrada, A., Vázquez-Acevedo, M., Villavicencio-Queijeiro, A., Figueroa-Martínez, F., Miranda-Astudillo, H., Cordeiro, Y., et al. (2010). Subunit-subunit interactions and overall topology of the dimeric mitochondrial ATP synthase of *Polytomella* sp. *Biochimica et Biophysica Acta, 1797*, 1439–1448.

Cardol, P., Gonzalez-Halphen, D., Reyes-Prieto, A., Baurain, D., Matagne, R. F., & Remacle, C. (2005). The mitochondrial oxidative phosphorylation proteome of *Chlamydomonas reinhardtii* deduced from the genome sequencing project. *Plant Physiology, 137*, 447–459.

Cardol, P., Vanrobaeys, F., Devreese, B., Van Beeumen, J., Matagne, R. F., & Remacle, C. (2004). Higher plant-like subunit composition of mitochondrial complex I from *Chlamydomonas reinhardtii*: 31 conserved components among eukaryotes. *Biochimica et Biophysica Acta, 4*, 212–224.

Cardol, P., Lapaille, M., Minet, P., Matagne, R. F., Franck, F., & Cardol, P. (2006). ND3 and ND4L subunits of mitochondrial complex I, both nucleus-encoded in *Chlamydomonas reinhardtii*, are required for the activity and assembly of the enzyme. *Eukaryotic Cell, 5*, 1460–1467.

Cardon, Z. G., Gray, D. W., & Lewis, L. A. (2008). The green algal underground: evolutionary secrets of desert cells. *BioScience, 58*, 114–122.

Carrera-Martinez, D., Mateos-Sanz, A., Lopez-Rodas, V., & Costas, E. (2011). Adaptation of microalgae to a gradient of continuous petroleum contamination. *Aquatic Toxicology, 101*, 342–350.

Colin, M., Dorthu, M. P., Duby, F., Remacle, C., Dinant, M., Wolwertz, M. R., et al. (1995). Mutations affecting the mitochondrial genes encoding cytochrome oxidase subunit I and apocytochrome *b* of *Chlamydomonas reinhardtii*. *Molecular and General Genetics, 249*, 179–184.

Collins, S. (2011). Competition limits adaptation and productivity in a photosynthetic alga at elevated CO_2. *Proceedings of the Royal Society of London B: Biological Sciences, 278*, 247–255.

de Cambiaire, J. C., Otis, C., Lemieux, C., & Turmel, M. (2006). The complete chloroplast genome sequence of the chlorophycean green alga *Scenedesmus obliquus* reveals a compact gene organization and a biased distribution of genes on the two DNA strands. *BMC Evolutionary Biology, 6*, 37.

de Koning, A. P., & Keeling, P. J. (2006). The complete plastid genome sequence of the parasitic green alga *Helicosporidium* sp. is highly reduced and structured. *BMC Biology, 4*, 12.

Dorthu, M. P., Remy, S., Michel-Wolwertz, M. R., Colleaux, L., Breyer, D., Beckers, M. C., et al. (1992). Biochemical, genetic and molecular characterization of new respiratory-deficient mutants in *Chlamydomonas reinhardtii*. *Plant Molecular Biology, 18*, 759–772.

Duby, F., & Matagne, R. F. (1999). Alteration of dark respiration and reduction of phototrophic growth in a mtDNA deletion mutant of *Chlamydomonas* lacking *cob, nd4* and the 3′ end of *nd5*. *The Plant Cell, 11*, 115–125.

Efremov, R. G., & Sazanov, L. A. (2011). Respiratory complex I: 'steam engine' of the cell? *Current Opinion in Structural Biology, 21*, 532–540.

Fan, J., & Lee, R. W. (2002). Mitochondrial genome of the colorless green alga *Polytomella parva*: two linear DNA molecules with homologous inverted repeat termini. *Molecular Biology and Evolution, 19*, 999–1007.

Funes, S., Davidson, E., Reyes-Prieto, A., Magallón, S., Herion, P., King, M. P., et al. (2002). A green algal apicoplast ancestor. *Science, 298*, 2155.

Funes, S., Franzén, L. G., & González-Halphen, D. (2007). *Chlamydomonas reinhardtii* the model of choice to study mitochondria from unicellular photosynthetic organisms. *Methods in Molecular Biology, 372*, 137–149.

Garbary, D. J., Bourquea, G., Herman, T. B., & McNeil, J. A. (2007). Epizoic algae from freshwater turtles in Nova Scotia. *Journal of Freshwater Ecology, 22*, 677–685.

Gawryluk, R. M., & Gray, M. W. (2009). A split and rearranged nuclear gene encoding the iron-sulfur subunit of mitochondrial succinate dehydrogenase in Euglenozoa. *BMC Research Notes, 2*, 16.

Gawryluk, R. M., & Gray, M. W. (2010). An ancient fission of mitochondrial Cox1. *Molecular Biology and Evolution, 27*, 7–10.

Gray, M. W., Burger, G., & Lang, B. F. (1999). Mitochondrial evolution. *Science, 283*, 1476–1481.

Grossman, A. R., Harris, E. E., Hauser, C., Lefebvre, P. A., Martinez, D., Rokhsar, D., et al. (2003). *Chlamydomonas reinhardtii* at the crossroads of genomics. *Eukaryotic Cell, 2*, 1137–1150.

Guan, Z., Meng, X., Sun, Z., Xu, Z., & Song, R. (2008). Characterization of duplicated *Dunaliella viridis* SPT1 genes provides insights into early gene divergence after duplication. *Gene, 423*, 36–42.

Hayashi-Ishimaru, Y., Ohama, T., Kawatsu, Y., Nakamura, K., & Osawa, S. (1996). UAG is a sense codon in several chlorophycean mitochondria. *Current Genetics, 30*, 29–33.

Herron, M. D., Hackett, J. D., Aylward, F. O., & Michod, R. E. (2009). Triassic origin and early radiation of multicellular volvocine algae. *Proceedings of the National Academy of Sciences of the United States of America, 106*, 3254–3258.

Kerney, R., Kim, E., Hangarter, R. P., Heiss, A. A., Bishop, C. D., & Hall, B. K. (2011). Intracellular invasion of green algae in a salamander host. *Proceedings of the National Academy of Sciences of the United States of America, 108*, 6497–6502.

Kroymann, J., & Zetsche, K. (1998). The mitochondrial genome of *Chlorogonium elongatum* inferred from the complete sequence. *Journal of Molecular Evolution, 47*, 431–440.

Lang, B. F., Burger, G., O'Kelly, C. J., Cedergren, R., Golding, G. B., Lemieux, C., et al. (1997). An ancestral mitochondrial DNA resembling a eubacterial genome in miniature. *Nature, 387*, 493–497.

Lapaille, M., Escobar-Ramírez, A., Degand, H., Baurain, D., Rodríguez-Salinas, E., Coosemans, N., et al. (2010). Atypical subunit composition of the chlorophycean mitochondrial FOF1 ATP synthase and the role of Asa7 in stability and oligomycin resistance of the enzyme. *Molecular Biology and Evolution, 27*, 1630–1644.

Larosa, V., Coosemans, N., Motte, P., Bonnefoy, N., & Remacle, C. (2012). Reconstruction of a human mitochondrial complex I mutation in the unicellular green alga *Chlamydomonas*. Plant Journal 70, 759–768.

Lemieux, C., Otis, C., & Turmel, M. (2000). Ancestral chloroplast genome in *Mesostigma viride* reveals an early branch of green plant evolution. *Nature, 403*, 649–652.

Lewis, L. A., & McCourt, R. M. (2004). Green algae and the origin of land plants. *American Journal of Botany, 91*, 1535–1556.

Lewis, L. A., & Muller-Parker, G. (2004). Phylogenetic placement of "zoochlorellae" (Chlorophyta), algal symbiont of the temperate sea anemone *Anthopleura elegantissima*. *The Biological Bulletin, 207*, 87–92.

Li, J., Lu, Y., Xue, L., & Xie, H. (2010). A structurally novel salt-regulated promoter of duplicated carbonic anhydrase gene 1 from *Dunaliella salina*. *Molecular Biology Reports, 37*, 1143–1154.

Lowenstein, T. K., Schubert, B. A., & Timofeeff, M. N. (2011). Microbial communities in fluid inclusions and long-term survival in halite. *GSA Today, 21*, 4–9.

Lüttge, U., & Büdel, B. (2010). Resurrection kinetics of photosynthesis in desiccation-tolerant terrestrial green algae (Chlorophyta) on tree bark. *Plant Biology (Stuttgart, Germany: Online), 12*, 437–444.

Lynch, M., & Conery, J. S. (2000). The evolutionary fate and consequences of duplicate genes. *Science, 290*, 1151–1155.

Mallet, M. A., & Lee, R. W. (2006). Identification of three distinct *Polytomella* lineages based on mitochondrial DNA features. *Journal of Eukaryotic Microbiology, 53*, 79–84.

Martin, W., & Müller, M. (1998). The hydrogen hypothesis for the first eukaryote. *Nature, 392*(6671), 37–41, Mar 5.

Matagne, R. F., Michel-Wolwertz, M. R., Munaut, C., Duyckaerts, C., & Sluse, F. (1989). Induction and characterization of mtDNA mutants in *Chlamydomonas reinhardtii*. *Journal of Cell Biology, 108*, 1221–1226.

Matsuo, M., Hachisu, R., Tabata, S., Fukuzawa, H., & Obokata, J. (2011). Transcriptome analysis of respiration-responsive genes in *Chlamydomonas reinhardtii*: mitochondrial retrograde signaling coordinates the genes for cell proliferation with energy-producing metabolism. *Plant Cell Physiology, 52*, 333–343.

Maul, J. E., Lilly, J. W., Cui, L., de Pamphilis, C. W., Miller, W., Harris, E. H., et al. (2002). The *Chlamydomonas reinhardtii* plastid chromosome: islands of genes in a sea of repeats. *The Plant Cell, 14*, 2659–2679.

May, P., Wienkoop, S., Kempa, S., Usadel, B., Rupprecht, C., Weiss, J., et al. (2008). Metabolomics- and proteomics-assisted genome annotation and analysis of the draft metabolic network of *Chlamydomonas reinhardtii*. *Genetics, 179*, 157–166.

Melkonian, M. (1990). "Phylum Chlorophyta. Introduction to the Chlorophyta". In L. Margulis, J. O. Corliss, M. Melkonian, & D. J. Chapman (Eds.), *Handbook of Protoctista* (pp. 597–599). Boston: Jones & Bartlett.

Merchant, S. S., Prochnik, S. E., Vallon, O., Harris, E. H., Karpowicz, S. J., et al. (2007). The *Chlamydomonas* genome reveals the evolution of key animal and plant functions. *Science, 318*, 245–250.

Michaelis, G., Vahrenholz, C., & Pratje, E. (1990). Mitochondrial DNA of *Chlamydomonas reinhardtii*: the gene for apocytochrome b and the complete functional map of the 15.8 kb DNA. *Molecular and General Genetics, 223*, 211–216.

Moestrup, Ø. 2006. Algal Taxonomy: Historical Overview. eLS. (Wiley Online Library).

Mosser, J. L., Mosser, A. G., & Brock, T. D. (1977). Photosynthesis in the snow: the alga *Chlamydomonas nivalis* (Chlorophyceae). *Journal of Phycology, 13*, 22–27.

Nakamura, S., Aoyama, H., & Van Woesik, R. (2003). Strict paternal transmission of mitochondrial DNA of *Chlamydomonas* species explained by selection against maternal nucleoids. *Protoplasma, 221*, 205–210.

Nakamura, S. (2010). Paternal inheritance of mitochondria in *Chlamydomonas*. *Journal of Plant Research, 123*, 163–170.

Nakayama, T., Marin, B., Kranzc, H. D., Surek, B., Hussc, V. A. R., Inouyea, I., et al. (1998). The basal position of scaly green flagellates among the green algae (Chlorophyta) is revealed by analyses of nuclear-encoded SSU rRNA sequences. *Protist, 149*, 367–380.

Nedelcu, A. M., & Lee, R. W. (1998). Short repetitive sequences in green algal mitochondrial genomes: potential roles in mitochondrial genome evolution. *Molecular Biology and Evolution, 15*, 690–701.

Nedelcu, A. M., Lee, R. W., Lemieux, C., Gray, M. W., & Burger, G. (2000). The complete mitochondrial DNA sequence of *Scenedesmus obliquus* reflects an intermediate stage in the evolution of the green algal mitochondrial genome. *Genome Research, 10*, 819–831.

Nishihara, N., Horiike, S., Takahashi, T., Kosaka, T., Shigenaka, Y., & Hosoya, H. (1998). Cloning and characterization of endosymbiotic algae isolated from *Paramecium bursaria*. *Protoplasma, 203*, 91–99.

Palenik, B., Grimwood, J., Aerts, A., Rouzé, P., Salamov, A., Putnam, N., Dupont, C., Jorgensen, R., Derelle, E., Rombauts, S., Zhou, K., Otillar, R., Merchant, S. S., Podell, S., Gaasterland, T., Napoli, C., Gendler, K., Manuell, A., Tai, V., Vallon, O., Piganeau, G., Jancek, S., Heijde, M., Jabbari, K., Bowler, C., Lohr, M., Robbens, S., Werner, G., Dubchak, I., Pazour, G. J., Ren, Q., Paulsen, I., Delwiche, C., Schmutz, J., Rokhsar, D., Van de Peer, Y., Moreau, H., & Grigoriev, I. V. (2007). The tiny eukaryote *Ostreococcus* provides genomic insights into the paradox of plankton speciation. *Proc Natl Acad Sci U S A, 104*, 7705–7710.

Pérez-Martínez, X., Vazquez-Acevedo, M., Tolkunova, E., Funes, S., Claros, M. G., Davidson, E., et al. (2000). Unusual location of a mitochondrial gene. Subunit III of cytochrome C oxidase is encoded in the nucleus of Chlamydomonad algae. *Journal of Biological Chemistry, 275*, 30144–30152.

Pérez-Martínez, X., Antaramian, A., Vázquez-Acevedo, M., Funes, S., Tolkunova, E., d'Alayer, J., et al. (2001). Subunit II of cytochrome c oxidase in Chlamydomonad algae is a heterodimer encoded by two independent nuclear genes. *Journal of Biological Chemistry, 276*, 11302–11309.

Pérez-Martínez, X., Funes, S., Tolkunova, E., Davidson, E., King, M. P., & González-Halphen, D. (2002). Structure of nuclear-localized cox3 genes in *Chlamydomonas reinhardtii* and in its colorless close relative *Polytomella* sp. *Current Genetics, 40*, 399–404.

Pombert, J. F., Lemieux, C., & Turmel, M. (2006). The complete chloroplast DNA sequence of the green alga *Oltmannsiellopsis viridis* reveals a distinctive quadripartite architecture in the chloroplast genome of early diverging ulvophytes. *BMC Biology, 4*, 3.

Pombert, J. F., Otis, C., Lemieux, C., & Turmel, M. (2004). The complete mitochondrial DNA sequence of the green alga *Pseudendoclonium akinetum* (Ulvophyceae) highlights distinctive evolutionary trends in the chlorophyta and suggests a sister-group relationship between the Ulvophyceae and Chlorophyceae. *Molecular Biology and Evolution, 21*, 922–935.

Pombert, J. F., Otis, C., Lemieux, C., & Turmel, M. (2005). The chloroplast genome sequence of the green alga *Pseudendoclonium akinetum* (Ulvophyceae) reveals unusual structural features and new insights into the branching order of chlorophyte lineages. *Molecular Biology and Evolution, 22*, 1903–1918.

Pombert, J. F., & Keeling, P. J. (2010). The mitochondrial genome of the entomoparasitic green alga *Helicosporidium*. *PLoS ONE, 5*, e8954.

Poole, A. M., & Penny, D. (2006). Evaluating hypotheses for the origin of eukaryotes. *BioEssays, 29*, 74–84.

Popescu, C. E., & Lee, R. W. (2007). Mitochondrial genome sequence evolution in *Chlamydomonas*. *Genetics, 175*, 819–826.

Prochnik, S. E., Umen, J., Nedelcu, A. M., Hallmann, A., Miller, S. M., et al. (2010). Genomic analysis of organismal complexity in the multicellular green alga *Volvox carteri*. *Science, 329*, 223–226.

Pröschold, T., Marin, B., Schlösser, U. G., & Melkonian, M. (2001). Molecular phylogeny and taxonomic revision of Chlamydomonas (Chlorophyta). I. Emendation of Chlamydomonas Ehrenberg and Chloromonas Gobi, and description of Oogamochlamys gen. nov. and Lobochlamys gen. nov. *Protist, 152*, 265–300.

Pröschold, T and Leliaert, F. 2007. 'Systematics of the green algae: conflict of classic and modern approaches'. In Unravelling the Algae: The Past, Present, and Future of the

Algae Systematics, Edited by: Brodie, J and Lewis, J. 123–153. London: Taylor & Francis.

Randolph-Anderson, B. L., Boynton, J. E., Gillham, N. W., Harris, E. H., Johnson, A. M., Dorthu, M. P., et al. (1993). Further characterization of the respiratory deficient *dum1* mutation of *Chlamydomonas reinhardtii* and its use as a recipient for mitochondrial transformation. *Molecular and General Genetics, 236*, 235–244.

Remacle, C., Cardol, P., Coosemans, N., Gaisne, M., & Bonnefoy, N. (2006). High-efficiency biolistic transformation of *Chlamydomonas* mitochondria can be used to insert mutations in complex I genes. *Proceedings of the National Academy of Sciences of the United States of America, 103*, 4771–4776.

Robbens, S., Derelle, E., Ferraz, C., Wuyts, J., Moreau, H., & Van de Peer, Y. (2007). The complete chloroplast and mitochondrial DNA sequence of *Ostreococcus tauri*: organelle genomes of the smallest eukaryote are examples of compaction. *Molecular Biology and Evolution, 24*, 956–968.

Rodríguez-Ezpeleta, N., Brinkmann, H., Burey, S. C., Roure, B., Burger, G., Löffelhardt, W., et al. (2005). Monophyly of primary photosynthetic eukaryotes: green plants, red algae, and glaucophytes. *Current Biology, 15*, 1325–1330.

Rodríguez-Salinas, E., & González-Halphen, D. (2009). Los genomas mitocondriales: ¿Qué nos dicen sobre la evolución de las algas verdes? *Revista Latinoamericana de Microbiología, 52*, 44–57.

Rodríguez-Salinas, E., Riveros-Rosas, H., Li, Z., Fučíková, K., Brand, J. J., Lewis, L. A., & et al. Lineage-specific fragmentation and nuclear relocation of the mitochondrial *cox2* gene in chlorophycean green algae (Chlorophyta). Molecular Phylogenetics and Evolution; *64*, 166–176.

Rolland, N., Atteia, A., Decottignies, P., Garin, J., Hippler, M., Kreimer, G., et al. (2009). Chlamydomonas proteomics. *Current Opinion in Microbiology, 12*, 285–291.

Rubio, M. A., & Hopper, A. K. (2011). Transfer RNA travels from the cytoplasm to organelles. *WIREs RNA, 2*, 802–817.

Sagan, L. (1967). On the origin of mitosing cells. *Journal of Theoretical Biology, 14*, 225–274.

Salinas, T., Duchêne, A. M., & Maréchal-Drouard, L. (2008). Recent advances in tRNA mitochondrial import. *Trends in Biochemical Sciences, 33*, 320–329.

Sieber, F., Placido, A., El Farouk-Ameqrane, S., Duchêne, A. M., & Maréchal-Drouard, L. (2011). A protein shuttle system to target RNA into mitochondria. *Nucleic Acids Research, 39*, e96.

Smith, D. R., & Lee, R. W. (2008). Mitochondrial genome of the colorless green alga *Polytomella capuana*: a linear molecule with an unprecedented GC content. *Molecular Biology and Evolution, 25*, 487–496.

Smith, D. R., & Lee, R. W. (2009). The mitochondrial and plastid genomes of *Volvox carteri*: bloated molecules rich in repetitive DNA. *BMC Genomics, 10*, 132.

Smith, D. R., Hua, J., & Lee, R. W. (2010). Evolution of linear mitochondrial DNA in three known lineages of *Polytomella. Current Genetics, 56*, 427–438.

Smith, D. R., Lee, R. W., Cushman, J. C., Magnuson, J. K., Tran, D., & Polle, J. E. (2010). The *Dunaliella salina* organelle genomes: large sequences, inflated with intronic and intergenic DNA. *BMC Plant Biology, 10*, 83.

Tamames, J. (2001). Evolution of gene order conservation in prokaryotes. *Genome Biology, 2*, 20, 1-20.

The Genetic Codes-NCBI: http://www.ncbi.nlm.nih.gov/Taxonomy/Utils/wprintgc.cgi#SG16.

Trainor, F. R., & Gladych, R. (1995). Survival of algae in a desiccated soil: a 35-year study. *Phycologia, 34*, 191–192.

Turmel, M., Lemieux, C., Burger, G., Lang, B. F., Otis, C., Plante, I., et al. (1999). The complete mitochondrial DNA sequences of *Nephroselmis olivacea* and *Pedinomonas minor.*

Two radically different evolutionary patterns within green algae. *The Plant Cell, 11*, 1717–1730.

Turmel, M., Otis, C., & Lemieux, C. (1999). The complete chloroplast DNA sequence of the green alga *Nephroselmis olivacea*: insights into the architecture of ancestral chloroplast genomes. *Proceedings of the National Academy of Sciences of the United States of America, 96*, 10248–10253.

Turmel, M., Otis, C., & Lemieux, C. (2003). The mitochondrial genome of *Chara vulgaris*: insights into the mitochondrial DNA architecture of the last common ancestor of green algae and land plants. *The Plant Cell, 15*, 1888–1903.

Turmel, M., Otis, C., & Lemieux, C. (2006). The chloroplast genome sequence of *Chara vulgaris* sheds new light into the closest green algal relatives of land plants. *Molecular Biology and Evolution, 23*, 1324–1338.

Turmel, M., Otis, C., & Lemieux, C. (2010). A deviant genetic code in the reduced mitochondrial genome of the picoplanktonic green alga *Pycnococcus provasolii*. *Journal of Molecular Evolution, 70*, 203–214.

Umen, J. G., & Goodenough, U. (2001). Chloroplast DNA methylation and inheritance in *Chlamydomonas*. *Genes and Development, 15*, 2585–2597.

Vahrenholz, C., Riemen, G., Pratje, E., Dujon, B., & Michaelis, G. (1993). Mitochondrial DNA of *Chlamydomonas reinhardtii*: the structure of the ends of the linear 15.8-kb genome suggests mechanisms for DNA replication. *Current Genetics, 24*, 241–247.

van den Hoek, C., Mann, D. G., & Jahns, H. M. (1995). *Algae: An Introduction to Phycology*. Cambridge University Press Cambridge, 623 pp.

van Etten, J. L., & Dunigan, D. D. (2012). Chloroviruses: not your everyday plant virus. *Trends in Plant Sciences, 17*, 1–8.

van Lis, R., Mendoza-Hernández, G., Groth, G., & Atteia, A. (2007). New insights into the unique structure of the F_0F_1-ATP synthase from the chlamydomonad algae *Polytomella* sp. and *Chlamydomonas reinhardtii*. *Plant Physiology, 144*, 1190–1199.

Vázquez-Acevedo, M., Cardol, P., Cano-Estrada, A., Lapaille, M., Remacle, C., & González-Halphen, D. (2006). The mitochondrial ATP synthase of chlorophycean algae contains eight subunits of unknown origin involved in the formation of an atypical stator-stalk and in the dimerization of the complex. *Journal of Bioenergetics and Biomembranes, 38*, 271–282.

Walker, J. E., & Dickson, V. K. (2006). The peripheral stalk of the mitochondrial ATP synthase. *Biochimica et Biophysica Acta, 1757*, 286–296.

Wienkoop, S., Weiss, J., May, P., Kempa, S., , S.Irgang, et al. (2010). Targeted proteomics for *Chlamydomonas reinhardtii* combined with rapid subcellular protein fractionation, metabolomics and metabolic flux analyses. *Molecular Biosystems, 6*, 1018–1031.

Wolff, G., Plante, I., Lang, B. F., Kück, U., & Burger, G. (1994). Complete sequence of the mitochondrial DNA of the chlorophyte alga *Prototheca wickerhamii*. Gene content and genome organization. *Journal of Molecular Biology, 237*, 75–86.

Worden, A. Z., Lee, J. H., Mock, T., Rouze, P., Simmons, M. P., et al. (2009). Green evolution and dynamic adaptations revealed by genomes of the marine picoeukaryotes *Micromonas*. *Science, 324*, 268–272.

Xiong, J., Lu, X., Zhou, Z., Chang, Y., Yuan, D., Tian, M., et al. (2012). Transcriptome analysis of the model Protozoan, *Tetrahymena thermophila*, using deep RNA sequencing. *PLoS ONE, 7*. e30630.

Yamasaki, T., Kurokawa, S., Watanabe, K. I., Ikuta, K., & Ohama, T. (2005). Shared molecular characteristics of successfully transformed mitochondrial genomes in *Chlamydomonas reinhardtii*. *Plant Molecular Biology, 58*, 515–527.

CHAPTER NINE

Recombination in the Stability, Repair and Evolution of the Mitochondrial Genome

Kristina Kühn* and José M. Gualberto[1,†]

*Institut für Biologie, Humboldt-Universität zu Berlin, Germany

†Institute de Biologie Moléculaire des Plantes-CNRS-UPR2357, Université de Strasbourg, Strasbourg, France

[1]Corresponding author. E-mail: jose.gualberto@ibmp-cnrs.unistra.fr

Contents

Advances in Botanical Research, Volume 63
ISSN 0065-2296,
http://dx.doi.org/10.1016/B978-0-12-394279-1.00009-0

Abstract

Mitochondria contain their own genome (mtDNA), whose maintenance and expression is under nuclear control. Plant mtDNAs are considerably larger and more complex than animal or fungal mtDNAs, mostly due to highly active recombination processes that modulate its structure. Frequent homologous, reciprocal recombination involving large sequence repeats results in a multipartite structure of the plant mtDNA. In addition, rare recombination across short repeats can generate new mtDNA configurations that can accumulate and contribute to mtDNA heteroplasmy. Plant mitochondrial heteroplasmy derived from recombination constitutes an important source of genetic diversity, believed to be responsible for the rapid structural evolution of plant mtDNAs. Changes in the stoichiometry of alternative mtDNA configurations can be implemented over only a few plant generations. Recombined mtDNA sequences most probably result from the repair of mtDNA damage by recombination-dependent processes. These repair pathways are only starting to be unravelled. Several genes encoding factors that function in mtDNA recombination have been identified, and preliminary knowledge on mitochondrial recombination pathways has been gained from the study of mutants in which these factors have been inactivated. This review summarizes the present knowledge on plant mtDNA maintenance, including recombination and repair, and on factors that act in these pathways.

1. INTRODUCTION

In eukaryotes the mitochondrion is the cellular site of aerobic energy production. It contains its own genetic information, the mitochondrial (mt) genome (mtDNA), which codes for proteins essential for the assembly of the respiratory complexes. The mtDNA is transmitted maternally in most higher eukaryotes. Because mitochondria are essential to cellular function, defects in the mtDNA or in the machinery for its maintenance and expression are associated with a vast range of dysfunctions. In humans, spontaneous deletions and inherited point mutations have been recognized as causing a wide spectrum of diseases associated with the tissues that are most dependent on mitochondrial oxidative phosphorylation (Smeitink, Zeviani, Turnbull, & Jacobs, 2006). In plants, the mitochondrial genetic information influences several traits of agronomic importance including fertility, plant vigour, chloroplast function and cross-compatibility (Mackenzie & McIntosh, 1999). Mutations that affect plant mtDNA maintenance often result in morphological defects, such as leaf distortion, or even in lethality at the embryonic stage. Changes in the mtDNA are also the cause of many examples of cytoplasmic male sterility (CMS), which is an important agronomic trait used by breeders to produce high-yield hybrids (Budar, Touzet, & De Paepe, 2003; Hanson & Bentolila, 2004).

Our understanding of the processes controlling mtDNA inheritance remains poor. Nevertheless, several features of mtDNA maintenance seem to be common to most higher eukaryotes: recombination processes seem to have central roles in mtDNA repair and replication, in animals, fungi and plants (Backert & Börner, 2000; Kajander, Karhunen, Holt, & Jacobs, 2001; Ling, Makishima, Morishima, & Shibata, 1995). In plants, recombination is also largely responsible for the heteroplasmy of the mtDNA, which is the occurrence of alternative mtDNA configurations (mitotypes) that coexist with the predominant mtDNA at substoichiometric levels. These alternative mitotypes can contain large duplications, deletions and complex rearrangements of mtDNA sequences, but also insertions of foreign sequences. mtDNA rearrangements can in many cases be explained by the action of recombination pathways involving either frequent recombination via large repeated sequences, ectopic recombination involving intermediate-size repeats (IRs) or illegitimate recombination involving sequence microhomologies. Recombination pathways have important functions in the repair of lesions in the mtDNA, in particular DNA breaks that compromise the genome integrity. The heteroplasmy of the mtDNA is dynamic, and the relative proportions of mtDNA mitotypes can vary very rapidly. This genomic shifting can trigger the emergence of a new predominant mitotype and is largely responsible for the evolution of the mtDNA between closely related species.

2. STRUCTURE AND VARIATION OF PLANT MITOCHONDRIAL GENOMES

2.1. High Variability of the mtDNA Structure Among Plant Species

Plant mtDNAs have remarkable features that distinguish them from their animal and fungal counterparts. In particular, higher plants harbour mtDNAs that are highly variable in size and structural organization (Kubo & Newton, 2008). The mtDNAs of animals range in size from around 15 to 18 kb (Boore, 1999) and those of fungi from 18 to more than 100 kb (Cummings & Domenico, 1988). In plants, the mitochondrial genomes are much larger in size. Early studies using physical mapping and reassociation kinetics experiments established sizes for plant mitochondrial genomes ranging between 200 kb and 2400 kb in *Brassica hirta* and muskmelon, respectively (Palmer & Herbon, 1987; Ward, Anderson, & Bendich, 1981). mtDNA sequencing confirmed very large sizes in Cucurbitaceae species (1685 kb in cucumber, *Cucumis sativus*) (Alverson, Rice, Dickinson, Barry, & Palmer, 2011) and even larger sizes in the genus *Silene*, with the mtDNA of *Silene noctiflora* (6.7 Mb) or *Silene conica* (11.3 Mb) being larger than many bacterial genomes (Sloan, Alverson, Chuckalovcak *et al.*, 2012). However, despite their large size, land plant mtDNAs do not encode a proportionately greater number of proteins than mtDNAs in other lineages (Bullerwell & Gray, 2004). The *Arabidopsis* mtDNA (367 kb) codes for 31 mitochondrial proteins and the 17-kb human mtDNA codes for 13 proteins. Differences in gene content among plant species are minor and not linked to genome size (Sloan, Alverson, Chuckalovcak *et al.*, 2012). These differences reflect recent events of gene transfer to the nucleus (Adams & Palmer, 2003).

Despite the structural variability of mtDNAs, gene sequences are generally highly conserved and show very low base substitution rates (Palmer & Herbon, 1988; Wolfe, Li, & Sharp, 1987). However, recent evidence showed high mtDNA substitution rates in certain genera, and even within-species polymorphism (Barr, Keller, Ingvarsson, Sloan, & Taylor, 2007; Cho, Mower, Qiu, & Palmer, 2004; Sloan, Alverson, Chuckalovcak *et al.*, 2012).

2.2. Rapid Evolution of the Plant mtDNA by Recombination

Contrasting with the slow rate of sequence evolution, plant mtDNAs evolve rapidly in structure. Due to high recombination activities in mitochondria, the gene distribution in plant mtDNAs varies even between closely related species, with just a few examples of conserved gene organization. The existence of an active recombination system in plant mitochondria was revealed early by physical mapping of plant mtDNAs. This led to an mtDNA structural model of a circular chromosome (master circle) comprising all mtDNA sequences and subgenomic molecules generated from the master circle by recombination events between large repeated sequences (Fauron, Havlik, & Brettell, 1990; Lonsdale, Hodge, & Fauron, 1984; Palmer & Shields, 1984). Recombination involving inverted repeats inverts the intervening sequences, whereas recombination involving directly oriented repeats generates a pair of subgenomic molecules.

However, observation of the mtDNA from dividing plant cells by electron microscopy, pulse-field gel electrophoresis and moving pictures of

DNA migrating in gels revealed mtDNA structures that were far more complex (Backert, Lurz, Oyarzabal, & Börner *et al.*, 1997; Bendich, 1993; Oldenburg & Bendich, 1998). Circular master circles could not be detected in preparations of plant mtDNA that avoided shearing. Instead, mostly small circular molecules, linear molecules and more complex branched molecules of different sizes were found (Backert & Börner, 2000; Backert, Lurz, Oyarzabal, & Börner, 1996; Backert, Nielsen, & Börner, 1997; Oldenburg & Bendich, 1998). These complex arrays of different structures most likely reflect recombination intermediates and suggest that plant mitochondria utilize recombination-dependent mechanisms for the replication of their genome, as seen for certain bacteriophages (Backert & Börner, 2000; Bendich, 1993; Oldenburg & Bendich, 1996; Oldenburg & Bendich, 2001). A predominance of recombination processes was additionally supported by a high content of DNA in a single-stranded form in plant mitochondria (Backert, Lurz *et al.*, 1997).

In the yeast *Candida albicans*, recent analyses of mtDNA replication intermediates supported that the predominant mechanism of replication initiation involved recombination via a large inverted repeat (Gerhold, Aun, Sedman, Joers, & Sedman, 2010). As most plant mtDNAs similarly contain large repeated sequences, it is possible that, as in *Candida albicans*, plant mitochondria initiate the replication of their genome by repeat-mediated recombination. In addition to branched molecules representing possible recombination intermediates, examination of the mtDNA isolated from replicating cells also revealed circular molecules with tails, also known as sigma-like structures, indicative of rolling-circle replication (Backert, 2002; Backert, Dorfel, Lurz, & Börner, 1996). Rolling-circle replication of subgenomic circular molecules could explain how certain recombination intermediates can be selectively amplified during processes of genomic shifting.

The recombination activity of plant mitochondria plays a major role in the evolution of mitochondrial genomes. This was clearly revealed by the comparison of mtDNA sequences. To date, about 20 seed plant mitochondrial genomes have been sequenced, several from the same species. Intraspecific sequence comparisons showed that intergenic regions change rapidly compared with coding regions, through sequence duplications, inversions, deletions and insertions resulting from recombination (Allen *et al.*, 2007). As an example, the mtDNA sequences of five different maize accessions (two fertile and three CMS cytotypes) revealed big variations in genome size, resulting from large (0.5–120 kb) sequence duplications that

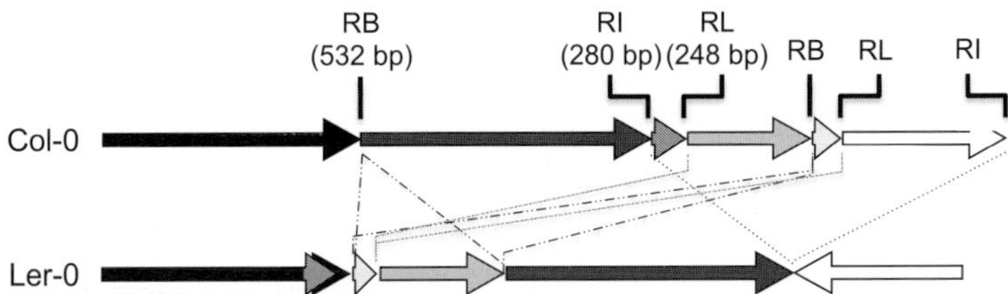

Figure 9.1 ***Rapid evolution of the mtDNA between different* Arabidopsis *accessions.*** The physical map of *Arabidopsis* Col-0 and *Landsberg erecta* (Ler-0) mitochondrial genomes (accession nos. JF729201 and JF729202) (Davila *et al.*, 2011) are represented in linear form, according to syntenic regions. Break points of sequence homology coincide in most cases with the presence of intermediate-size repeats (repeats RB, RL and RI, as defined by Arrieta-Montiel *et al.* (2009)), indicating that recombination across IRs was a major driving force in the diversion of the two genomes.

were cytotype-specific (Allen *et al.*, 2007). Similarly, in *Beta vulgaris*, the sequences of six mitochondrial genomes showed important intraspecific variations in mtDNA size and gene order (Darracq *et al.*, 2011; Satoh *et al.*, 2004). These variations were partially due to duplications representing 5–30% of the mtDNA (Darracq *et al.*, 2011; Satoh *et al.*, 2004). Significant differences between *Arabidopsis thaliana* accessions were also observed (Arrieta-Montiel, Shedge, Davila, Christensen, & Mackenzie, 2009; Davila *et al.*, 2011). Sequence comparisons between accessions Ler-0, Col-0 and C24 showed the loss in Ler-0 of one copy of a large sequence repeat present in Col-0 and C24 (Fig. 9.1). Other events of sequence reshuffling involved IRs. A hypothetical scenario of mtDNA evolution from an ancestral genome through tandem duplication, followed by deletions and inversions has been proposed from the analysis of six maize mtDNA sequences (Darracq, Varre, & Touzet, 2010).

2.3. Heteroplasmy in the Evolution of the mtDNA

The processes driving the rapid evolution of plant mitochondrial genomes are directly linked to the intrinsic heteroplasmy of the mtDNA (Kmiec, Woloszynska, & Janska, 2006). Low copy number alternative mitotypes coexist with the dominant mtDNA genome (Kajander *et al.*, 2000; Small, Suffolk, & Leaver, 1989). These molecules can derive from infrequent homologous recombination (HR) via IRs, which leads to low-abundance crossover products, or from microhomology-mediated recombination yielding complex rearranged mtDNA sequences (see below). Most likely, these products of ectopic and illegitimate recombination result from the repair of mtDNA single-strand and/or double-strand breaks (DSBs) (Cappadocia *et al.*, 2010; Davila *et al.*, 2011; Miller-Messmer *et al.*, in press). Whereas sequences generated by ectopic HR can be created continuously in somatic tissues and might be just 20–100 times less abundant than the parental sequences that are part of the predominant mtDNA, sequences generated by illegitimate recombination involving microhomologies are extremely rare, three orders of magnitude less abundant, less than one per 100 cells (Woloszynska & Trojanowski, 2009).

An important consequence of mtDNA heteroplasmy is that from heteroplasmic parents, individuals can segregate in which substoichiometric mitotypes have been amplified and become the predominant mtDNA. This process is called substoichiometric shifting (SSS) (Kajander *et al.*, 2000; Small *et al.*, 1989). Many examples of rapid shifts between mitotypes have been reported in plants. These shifts can occur during propagation of plants in tissue cultures (Bartoszewski, Malepszy, & Havey, 2004; Kanazawa, Tsutsumi, & Hirai, 1994; Vitart, de Paepe, Mathieu, Chetrit, & Vedel, 1992) or result from mutations in genes involved in recombination surveillance (Abdelnoor *et al.*, 2003; Kuzmin, Duvick, & Newton, 2005; Zaegel *et al.*, 2006). Notably, they have also been observed under normal growth conditions, both in plant cultivars (Chen *et al.*, 2011; Janska, Sarria, Woloszynska, Arrieta-Montiel, & Mackenzie, 1998) and in natural populations (Arrieta-Montiel *et al.*, 2001).

As in plants, mtDNA sequence variants in animals segregate rapidly between generations (Shoubridge, 2000). The ploidy of the mtDNA in animals is high and can reach more than 10,000 copies per cell. To explain SSS, it has been postulated that during gametogenesis, a genetic bottleneck insures the reduction of the mtDNA copy number (Shoubridge, 2000). However, determination of mtDNA copy numbers in germline cells of mice showed that there is no decline in mtDNA copies (Cao *et al.*, 2009). Mitotype reduction could alternatively be achieved through selective replication of mtDNAs during oocyte maturation (Cao *et al.*, 2009; Wai, Teoli, & Shoubridge, 2008). In plants as in animals, rapid mtDNA shifts between alternative mitotypes imply a reduction in segregating units. Compared with animal mitochondria, the ploidy of plant mitochondrial genomes is much lower, reaching only 40–60 copies per cell in young *Arabidopsis* leaves (Preuten *et al.*, 2010; Wang *et al.*, 2010), which equals less than one copy per mitochondrion. As in animals, mtDNA quantification in the egg cells revealed values equivalent to those in somatic tissues, without evidence of a genetic bottleneck by copy number reduction in the female gametophyte (Wang *et al.*, 2010). Nevertheless, such a reduction might occur at earlier stages of gamete maturation.

The process of SSS is under nuclear control, and several genetic loci that influence SSS have been identified in different plant species. In the bean CMS system, the dominant *Fr* restorer gene drastically reduces the copy number of a single mitochondrial subgenome that confers male sterility (Janska *et al.*, 1998). In maize, a recessive nuclear mutation was described that alters the control over mtDNA subgenomes and disrupts their stoichiometric transfer to the progeny, resulting in differences in mtDNA profiles among siblings and between parents and progeny (Kuzmin *et al.*, 2005). In *Arabidopsis*, several factors have been described that are involved in recombination surveillance and affect the steady-state levels of recombined mtDNA sequences. These factors include the MutS-like MSH1, the ssDNA-binding protein OSB1 and the RecA-like protein RECA3 (Fig. 9.2) (Abdelnoor *et al.*, 2003; Shedge, Arrieta-Montiel, Christensen, & Mackenzie, 2007; Zaegel *et al.*, 2006). Disruption of MSH1 or OSB1 will

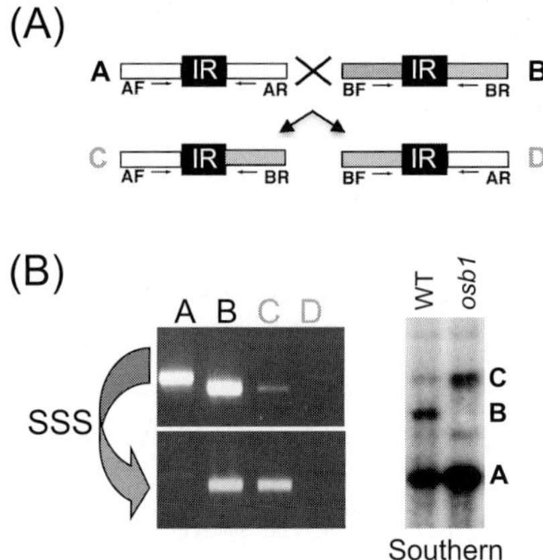

Figure 9.2 ***Recombination between repeated sequences in substoichiometric shifting (SSS) of the* Arabidopsis *mtDNA.*** (A) Representation of the different mtDNA environments that share the same intermediate-size repeated sequence (IR). Forms A and B represent the two major sequence environments present in the main mtDNA; C and D are the reciprocal crossover products. Oligonucleotide primers allowing specific amplification of each sequence are represented by arrows. (B) SSS in mutants of the ssDNA-binding protein OSB1. The left panel shows a polymerase chain reaction experiment, as described in (A), the right panel a Southern blot hybridized with a probe corresponding to *Arabidopsis* repeat RL (256 bp).

lead to SSS after several plant generations and to sorting of the recombined sequences (Davila *et al.*, 2011; Zaegel *et al.*, 2006).

3. RECOMBINATION IN MTDNA MAINTENANCE

3.1. Involvement of Different HR-Dependent Pathways in mtDNA Recombination

Much evidence for the occurrence of HR in plant mitochondria is based on the analysis and comparison of mtDNA sequences (Kubo and Newton, 2008). Direct observation of the mtDNA of dividing cells suggested that replication of plant mitochondrial genomes involves HR-dependent processes (Backert & Börner, 2000; Oldenburg & Bendich, 1996). Biochemical proof for HR in plant mitochondria has been provided by a study that purified a RecA-like strand-invasion activity from soybean mitochondria (Manchekar *et al.*, 2006). HR activities are used in the repair of DNA breaks and gaps (Sekiguchi & Ferguson, 2006; Wyman & Kanaar, 2006). The best studied mechanisms of HR are those induced by DSBs. DSBs are generated by site-specific nucleases, by topoisomerases or by genotoxic agents that directly (gamma radiation, bleomycin) or indirectly (cross-linking agents such as mitomycin C and cisplatin) induce DSBs during replication (Friedberg, 2003). At a DSB, the lost sequence information cannot be recovered from the same DNA molecule and has to be repaired using information from an undamaged DNA copy (San Filippo, Sung, & Klein, 2008). In mitochondria, this copy can be an additional mtDNA molecule or a repeated sequence on the same molecule. In addition to DSBs, other types of DNA lesions, such as single-strand DNA (ssDNA) gaps, can initiate spontaneous recombination. When the replication fork encounters ssDNA gaps it can either pause or dissociate, resulting in a DSB. Re-establishment of the replication fork also requires HR activities (Michel, Boubakri, Baharoglu, LeMasson, & Lestini, 2007).

Pathways of DSB repair by HR have been studied mainly in bacterial and eukaryotic nuclear systems, and have been the subject of many excellent reviews (Aguilera & Gomez-Gonzalez, 2008; Hanawalt, 2007; Hastings, Lupski, Rosenberg, & Ira, 2009; Mannuss, Trapp, & Puchta, 2012; Wyman & Kanaar, 2006). Which repair pathway is mobilized for DSB repair depends on several factors, including the structure of the DNA ends after cleavage and the availability of factors for DSB repair. Fig. 9.3 provides an overview of possible pathways. Detailed information on recombination pathways is

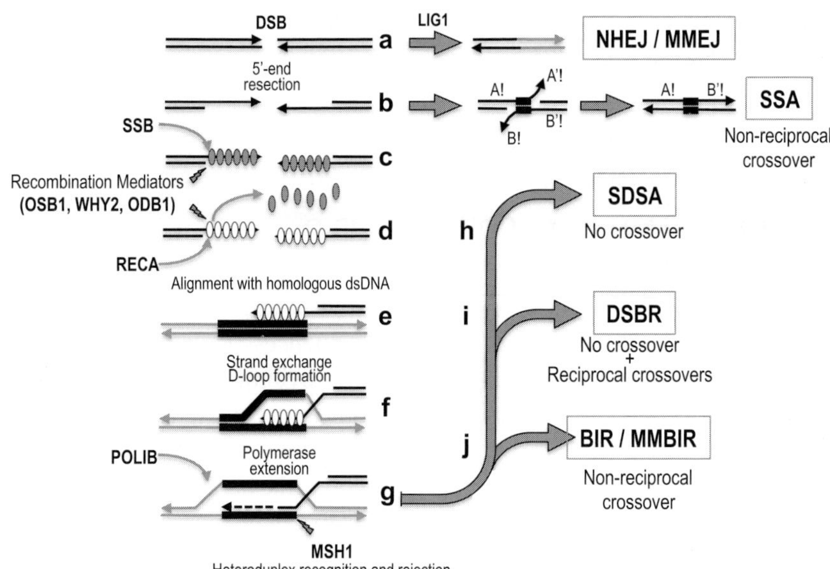

Figure 9.3 ***Different pathways for the repair of double-strand DNA breaks (DSBs).*** (a) Two-ended DNA breaks can be repaired by non-homologous end joining (NHEJ) through the action of a ligase. By microhomology-mediated end joining (MMEJ), broken ends are aligned by pairing a few nucleotides. (b) DSBs are processed to create 3′ free ends required for strand exchange. The ssDNA regions can base-pair at repeated sequences and recombine by the single-strand annealing (SSA) pathway. SSA results in deletions and non-reciprocal crossover. (c) SSB binds with high affinity to ssDNA. It protects ssDNA from degradation and removes secondary structures, but also inhibits recombination. (d) Binding of SSB can be competed by other ssDNA-binding proteins that act as recombination mediators. They promote SSB dissociation and binding of RecA. (e, f) RecA aligns ssDNA with homologous dsDNA (repeated sequence as a thick black line) and promotes strand exchange to form a D-loop. (g) A DNA polymerase (probably PolIB in plant mitochondria) extends the paired 3′ end. From here, several alternative pathways are possible. (h) Synthesis-dependent strand annealing (SDSA): Holliday junctions are not formed and the invading strand is rejected, probably by the action of a helicase, and anneals with the other end of the processed break. This results in faithful repair without crossovers. (i) Double-strand break repair (DSBR). The second end of the broken strand is captured, with formation of a double Holliday junction intermediate. Depending on how junctions are resolved, DSBR results in either gene conversion (without crossover) or in reciprocal crossovers, with both recombination products produced at equimolar amounts. (j) One-ended DNA breaks can be created by replication fork collapse or arrest. Re-establishment of replication forks by recombination is known as break-induced replication (BIR). By microhomology-mediated BIR (MMBIR), single-stranded 3′ tails from the collapsed fork anneal by just a few nucleotides, resulting in complex DNA rearrangements.

available (Hastings, Lupski *et al.*, 2009; Marechal & Brisson, 2010; Wyman & Kanaar, 2006).

3.2. Repeated Sequences Mediating Recombination of Plant Mitochondrial Genomes

In plant mtDNA recombination, repeated sequences play a central role because they are the main sites of intramolecular recombination. For the strand exchange reaction, the length of sequence homology between the invading donor strand and the double-stranded recipient DNA determines the recombination mechanism by which the repeats are mobilized. The minimal length of homology that is required for recombination is an intrinsic characteristic of the recombinase enzymes involved, which in plant mitochondria are eubacterial-type RecAs. The minimal pairing sequence of bacterial RecA can be as small as 25 bp, but there is a strong threshold homology length for recombination. The efficiency increases three orders of magnitude as homology increases from 25 bp to more than 200 bp (Lovett, Hurley, Sutera, Aubuchon, & Lebedeva, 2002). This size range is consistent with the efficiencies of recombination observed in plant mitochondria: large (>1 kb) repeats recombine at high frequency, intermediate-sized (100–1000 bp) repeats recombine sporadically, and short (<50–100 nt) repeats are thought to recombine extremely rarely (Andre, Levy, & Walbot, 1992; Arrieta-Montiel *et al.*, 2009; Sloan, Alverson, Chucklovcak *et al.*, 2012).

Repeats of diverse size and number have been characterized from the 20 or so seed plant mitochondrial genomes sequenced so far. A recent comprehensive analysis of their size and distribution in different species can be found in Alverson, Zhuo *et al.* (2011). The *Cucurbita* mtDNA, which is nearly 1 Mb in size, contains tens of thousands of short (20–40 nt) dispersed repeats that constitute 30% of the mtDNA (Alverson *et al.*, 2010), whereas other genomes contain small numbers of large (1–120 kb) and mostly species-specific segmental duplications (Allen *et al.*, 2007). Major variations in repeat sizes and number are seen among the same species, reflecting their rapid gain and loss following recombination and mtDNA evolution (Fig. 9.4).

3.2.1. Recombination across large repeats

Large repeated sequences (LRs) of sizes that exceed 1 kb are present in most higher plant mtDNAs, and their number varies significantly among plant species (Alverson, Zhuo *et al.*, 2011). Physical mapping and sequencing of mtDNAs showed that in somatic tissues, recombination across LRs appears

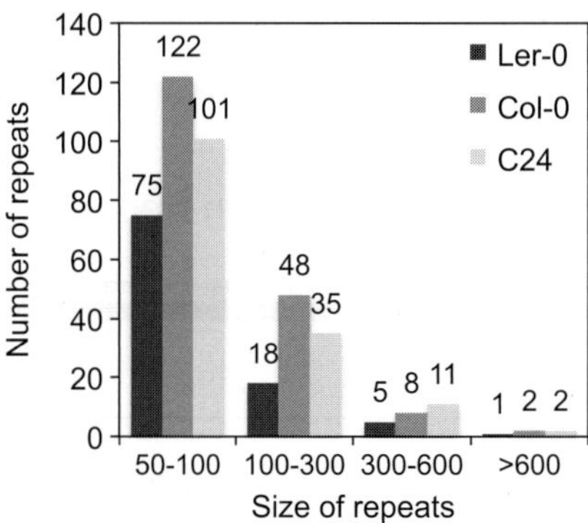

Figure 9.4 ***Number and size of repeats in the sequenced genomes of*** **Arabidopsis.** Repeated sequences found in the published mtDNA sequences of *Arabidopsis thaliana* (accession nos. JF729200, JF729201 and JF729202) have been summarized. Repeats were detected with the REPuter software (http://bibiserv.techfak.uni-bielefeld.de/reputer/submission.html) and classified according to size. A larger number of repeats is found in accessions Col-0 and C24 compared with Ler-0. The Ler-0 mtDNA has lost one copy of one of the two large repeats found in the other accessions.

to be frequent and reciprocal, generating equivalent amounts of both reciprocal crossover products (Fauron *et al.*, 1990; Lonsdale *et al.*, 1984; Ogihara *et al.*, 2005; Palmer & Herbon, 1987). The recombination across LRs is in part responsible for the complex multipartite structure of the mtDNA. It produces multiple genomic and subgenomic molecules that are interconvertible by recombination. The reciprocal nature of the LR-mediated recombination implies that this pathway follows the classic double-stand break repair pathway (DSBR) (Fig. 9.3) (Hastings, Lupski *et al.*, 2009; Marechal & Brisson, 2010; Wyman & Kanaar, 2006).

3.2.2. Recombination involving IRs

IRs ranging between 50 and 500 bp are relatively abundant in plant mtDNAs, but their number can vary significantly between species and within the same species (Alverson, Zhuo *et al.*, 2011) (Fig. 9.4). As expected, IRs are more numerous than LRs. mtDNA analyses indicated that IRs have important roles in modelling genome diversity within individual plants (Andre *et al.*, 1992). The infrequent recombination across this class of repeats generates low copy number alternative mtDNA sequences (also known as sublimons; Small, Isaac, & Leaver, 1987). Recently, recombination across IRs attracted further attention because it has been associated with processes of SSS. SSS was induced following the downregulation of genes involved in recombination surveillance (*OSB1*, *MSH1*, *OSB1*, *RECA2* and *RECA3*) (Miller-Messmer *et al.*, in press; Shedge *et al.*, 2007; Zaegel *et al.*, 2006). *Arabidopsis* mutants disrupted in these genes displayed a plethora of phenotypes and might ultimately be used for the selection of transgene-free CMS varieties (Sandhu, Abdelnoor, & Mackenzie, 2007). IRs were also found to affect SSS under natural conditions. In bean, sequences involved in restoration of fertility in CMS plants by SSS were shown to be maintained by constant recombination across a repeat of 314 bp (Woloszynska & Trojanowski, 2009).

Contrasting with LR-mediated recombination, recombination via IRs is mostly asymmetrical, leading to the accumulation of mainly one of the two reciprocal recombination products. This has been described in detail for the process of SSS induced in *msh1* plants (Arrieta-Montiel *et al.*, 2009; Davila *et al.*, 2011). The increase in the recombined sequence is in some cases, but not always, accompanied by the loss of one of the parental sequences (Davila *et al.*, 2011). In the context of recombination pathways that lead to non-reciprocal crossovers (Fig. 9.3), it has been proposed that recombination across IRs involves mechanisms of replicative repair by the break-induced replication (BIR) pathway (Davila *et al.*, 2011; Shedge *et al.*, 2007). However, in the bean SSS system, quantitative analysis of the four possible environments around the 314-bp repeat suggested that prior to SSS, the steady-state recombination activity in somatic tissues leads to equimolar accumulation of both crossover products, suggesting a DSBR pathway (Woloszynska & Trojanowski, 2009). Studies of *Arabidopsis recA2* and *recA3* mutants indicated that recombination across IRs can be reciprocal or non-reciprocal, depending on the repeat involved (Miller-Messmer *et al.*, in press; Shedge *et al.*, 2007). Therefore, it is not yet clear if it is the recombination pathway that leads to a non-reciprocal crossover, or if it is differential replication that results in a preferential segregation of one of the reciprocal products. A third possibility is that certain events of asymmetric recombination involve a RecA-independent recombination activity (Miller-Messmer *et al.*, in press). The single-strand annealing (SSA) repair pathway (Fig. 9.3) could be responsible for increased recombination activity at certain IRs, rather than the BIR pathway (Davila *et al.*, 2011; Shedge *et al.*, 2007). It is therefore possible that recombination surveillance genes are primarily required for efficient RecA-dependent recombination. Reduced RecA-dependent recombination may de-repress RecA-independent recombination, as described in the bacterial system (Dutra, Sutera, & Lovett, 2007).

There is no clear evidence of the existence of a synthesis-dependent strand annealing (SDSA) pathway in mitochondria. Repair by SDSA results in gene conversion without crossovers that cannot be detected except if polymorphisms exist between the donor and the recipient DNA strands. Like in other genetic systems, SDSA might be selected as the preferred repair pathway in mitochondria because it avoids genomic instability by recombination (Puchta, 2005). SDSA would explain the absence of detectable crossovers across IRs in the repair of the mtDNA of plants treated with the DNA-damaging agent bleomycin (Miller-Messmer *et al.*, in press).

Recombination involving IRs has contributed significantly to the evolution of mtDNA by SSS. Intraspecies comparison of the mtDNA sequences of different *Arabidopsis* accessions confirmed that IRs are often found at breaks of sequence homology (Arrieta-Montiel *et al.*, 2009) (Fig. 9.2). The motor behind these recombination events could be repair processes. Supporting this concept, it was shown that the progeny of *Arabidopsis* plants treated with the gyrase inhibitor ciprofloxacin retained increased levels of recombined mtDNA sequences resulting from replicative DSB repair (Miller-Messmer *et al.*, in press). This induced heteroplasmy of the mtDNA may increase the probability of SSS in following generations.

3.2.3. Recombination involving sequence microhomologies

Small repeated sequences of sizes smaller than 50 bp also influence the evolution of plant mtDNAs. They are often shorter than 10 bp and can more appropriately be considered as sequence microhomologies. In several cases, recombination involving microhomologies generated chimeric sequences that are patchworks composed of mitochondrial sequences and/or sequences of foreign origin (Hanson & Bentolila, 2004). These complex events of recombination often concern open reading frames linked to CMS. As an example, a chimeric gene responsible for CMS-T in maize originated from recombination among portions of the flanking and/or coding regions of mitochondrial genes 26S, ATP6 and of the chloroplastic tRNA-Arg gene (Dewey, Levings, & Timothy, 1986). Many additional examples have been described over the years (Hanson & Bentolila, 2004; Kubo & Newton, 2008). Recombination via microhomologies has also been linked to mitochondrial mutations caused by sequence deletions in mitochondrial genes, such as in maize non-chromosomal stripe (NCS) mutants (Newton & Coe, 1986; Marienfeld & Newton, 1994).

Recombination mediated by microhomologies cannot proceed via HR-dependent pathways because the length of the homologies is often far below the minimal pairing sequence size of RecA, but could result from repair of DSBs by a microhomology-mediated end joining (MMEJ) pathway (Fig. 9.3). MMEJ in plant mitochondria might be independent of Ku proteins and just require DNA polymerase activity and DNA ligase I (Crespan *et al.*, in press). An MMEJ pathway has also been postulated to exist in *Chlamydomonas* chloroplasts to explain rearrangements resulting from DSB repair induced by the *I-CreII* homing endonuclease (Kwon, Huq, & Herrin, 2010). However, the complex recombination events described in plant mitochondria are better explained by the replication-dependent microhomology-mediated BIR (MMBIR) pathway (Hastings, Ira, & Lupski, 2009; Hastings, Lupski *et al.*, 2009). According to this pathway, when strand invasion is limited or not possible because RecA is downregulated, single-strand 3′ tails from the collapsed replication fork will anneal to any available ssDNA through microhomologies, thus priming low-processivity DNA synthesis. The annealing can occur either in front of or behind the position of replication fork collapse, leading to deletions or duplications, respectively, and in either orientation. Successive cycles of template switch will finally generate very complex sequence rearrangements. MMBIR can therefore explain how chimeric mtDNA sequences were assembled from several DNA templates. MMBIR can also explain the upregulation of illegitimate microhomology-mediated recombination in mutants of Whirly ssDNA-binding proteins (Cappadocia *et al.*, 2010; Marechal *et al.*, 2009). These proteins and their possible functions are discussed in Section 4.7.3.

4. GENES INVOLVED IN MTDNA REPLICATION, REPAIR AND RECOMBINATION

Several genes that have important functions in plant mtDNA replication and repair by recombination have been identified (Table 9.1). These comprise genes related to the α-proteobacterial ancestor of mitochondria (Fitzpatrick, Creevey, & McInerney, 2006; Gray, Burger, & Lang, 1999), but also genes that are specific to plants. The mitochondrial and plastidial genome maintenance machineries are related, and components are often encoded by small gene families. The plastid and mitochondrial proteins are either isoforms encoded by different genes or the same protein dually targeted to both organelles (Carrie *et al.*, 2009). Thus, information on the mtDNA maintenance apparatus may also be gained from a recent proteomic study of

Table 9.1 Arabidopsis genes involved in mtDNA replication, repair and recombination

Category	Name	AGI	KO lines	Location	References
DNA polymerase	PolIA	At1g50840	Viable	Mt + Cp	(Carrie *et al.*, 2009; Christensen *et al.*, 2005; Ono *et al.*, 2007; Parent *et al.*, 2011)
	PolIB	At3g20540	Viable	Mt + Cp	(Carrie *et al.*, 2009; Christensen *et al.*, 2005; Ono *et al.*, 2007; Parent *et al.*, 2011)
Helicase/primase	Twinkle	At1g30680	n.d.	Mt + Cp	(Carrie *et al.*, 2009; Shutt & Gray, 2006)
Topoisomerase	GyrA	At3g10690	EL	Mt + Cp	(Wall *et al.*, 2004)
	GyrB	At5g04130	SL	Mt	(Wall *et al.*, 2004)
	TopI	At4g31210	n.d.	Mt + Cp	(Carrie *et al.*, 2009)
Recombinase	RECA2	At2g19490	SL	Mt + Cp	(Miller-Messmer *et al.*, in press; Shedge *et al.*, 2007)
	RECA3	At3g10140	Viable[a]	Mt	(Miller-Messmer *et al.*, in press; Shedge *et al.*, 2007)
ssDNA-binding	SSB1	At4g11060	n.d.	n.d.	(Edmondson *et al.*, 2005; Elo *et al.*, 2003)
	SSB2	At3g18580	SL	Mt	(Edmondson *et al.*, 2005; Elo *et al.*, 2003)
	OSB1	At3g18580	Viable[a]	Mt	(Zaegel *et al.*, 2006)
	OSB3	At5g44785	Viable	Mt + Cp	(Zaegel *et al.*, 2006)
	WHY2	At1g71260	Viable[b]	Mt	(Cappadocia *et al.*, 2010; Krause *et al.*, 2005)
	ODB1	At1g71310	Viable[b]	Mt	(Vermel *et al.*, 2002)
MutS	MSH1	At3g24320	Viable[a]	Mt + Cp	(Xu *et al.*, 2011)
Ligase	LigI	At1g08130	EL	Mt + Nuc	(Sunderland *et al.*, 2006)

Mt, mitochondria; Cp, chloroplast; Nuc, nucleus; EL, embryo lethal; SL, seedling lethal; n.d., not determined.
[a] Increased recombination involving IRs.
[b] Increased recombination involving microhomologies.

nucleoid-enriched plastid fractions prepared from maize (Majeran *et al.*, 2012).

4.1. Type I DNA Polymerases

The single mitochondrial DNA polymerase (DNAP) in yeast and human mitochondria, known as DNAP γ, assumes all DNA synthesis reactions and functions in mitochondrial processes of recombination and repair (Copeland, 2010). A phylogenetic analysis suggested that animal and yeast DNAP γ originated from a T-odd phage DNAP and placed the enzyme in the group of A-type DNAPs, which also contains DNAP I-like enzymes such as the eubacterial DNAP I (Filee, Forterre, Sen-Lin, & Laurent, 2002). In plants, early experiments of DNAP purification from organelles revealed similar characteristics for plastidial and mitochondrial DNAPs (Castroviejo, Tharaud, Tarrago-Litvak, & Litvak, 1979; Heinhorst, Cannon, & Weissbach, 1990; Meissner, Heinhorst, Cannon, & Börner, 1993). The purified proteins had equivalent sizes and were insensitive to inhibitors of nuclear DNAP α. Like animal and yeast mitochondrial DNAP γ, the plant organellar DNAPs contained an intrinsic 3′-5′ exonuclease activity, but in contrast to DNAP γ, they were found to be insensitive to dideoxyribonucleotides. Later, the complete sequencing of plant nuclear genomes revealed the presence of genes potentially encoding organellar DNAP I-like enzymes (Elo, Lyznik, Gonzalez, Kachman, & Mackenzie, 2003; Kimura *et al.*, 2002; Mori *et al.*, 2005; Ono *et al.*, 2007). The two organellar DNAPs from *Arabidopsis*, POLIA and POLIB, are dually targeted to both organelles (Carrie *et al.*, 2009; Christensen *et al.*, 2005; Elo *et al.*, 2003). Accordingly, mutants lacking either enzyme had reduced copy numbers of both mtDNA and cpDNA (Parent, Lepage, & Brisson, 2011). Failure to obtain *polIa polIb* double mutants confirmed essential and redundant functions for the organellar DNAPs, most probably in the replication of the organellar genomes (Parent *et al.*, 2011). *polIB* mutants but not *polIa* mutants were hypersensitive to ciprofloxacin, which specifically induces DNA DSBs in plant organelles, suggesting a function of POLIB in DNA repair (Parent *et al.*, 2011).

4.2. DNA Helicase

In animal mitochondria, Twinkle is the replicative helicase required for unwinding DNA during replication fork progression. Twinkle shares ancestry with the T7 phage protein gp4, which has combined primase–helicase activities (Shutt & Gray, 2006). Homologues of Twinkle are found throughout the eukaryotic lineage including higher plants, algae and mosses (Shutt & Gray, 2006). Although essential residues required for primase activity have been lost from animal Twinkle sequences, these residues are conserved in plant enzymes. Therefore, like gp4, plant Twinkle homologues potentially function as replicative helicases and as primases synthesizing primers required for lagging strand synthesis. *Arabidopsis* TWINKLE is dually targeted to both mitochondria and chloroplasts (Carrie *et al.*, 2009). At present, no functional data are available concerning the possible functions of the plant TWINKLE in mtDNA replication, recombination or repair.

4.3. DNA Ligase

DNA ligase catalyzes the joining of adjacent polynucleotides and thus plays important roles in DNA metabolism. Plants, like all eukaryotes, possess multiple DNA ligases with distinct roles in DNA metabolism. Among these, DNA ligase 1 (LIG1) is present in all eukaryotes. It plays critical roles in both DNA repair and replication and is indispensable for cell viability (Lohman, Tabor, & Nichols, 2011). In *Arabidopsis*, LIG1 accounts for the major DNA ligase activity and has been shown to have essential functions in both DNA replication and excision repair pathways (Cordoba-Canero, Roldan-Arjona, & Ariza *et al.*, 2011). LIG1:green fluorescent protein (GFP) fusion showed that, as in yeast, LIG1 is dually targeted to the nucleus and to mitochondria (Sunderland, West, Waterworth, & Bray, 2006). Although these import experiments failed to detect LIG1 in chloroplasts, the protein was found in the proteome of maize plastidial nucleoids (Majeran *et al.*, 2012). Most probably, LIG1 assumes its function in all three genomic compartments of plant cells. Roles for LIG1 in mtDNA replication, recombination and repair have not yet been studied.

4.4. Topoisomerases

Topoisomerases relieve the supercoiling created by the progression of the DNA replication fork. They catalyze cleavage and religation following the passage of one double-stranded DNA segment across another. Enzymes that cleave only one DNA strand are defined as type I; those that cleave both strands to generate a staggered DSB are in the type II subfamily (Champoux, 2001).

At present there are two different classes of topoisomerases identified in plant mitochondria: a type IA topoisomerase (TopI) (Carrie *et al.*, 2009) and a type IIA topoisomerase, also known as gyrase in prokaryotic systems. Both TopI and gyrase are dually targeted to mitochondria and chloroplasts. Gyrase comprises two subunits, GyrA and GyrB. In *Arabidopsis*, a single GyrA is targeted to both mitochondria and plastids; there are three genes coding for GyrB isoforms, one targeted to mitochondria, one to chloroplasts and one whose subcellular localization has not been determined (Wall, Mitchenall, & Maxwell, 2004). In contrast, in *Nicotiana benthamiana*, both GyrA and GyrB are dually targeted (Cho, Lee *et al.*, 2004). Both TopI and gyrase were identified in *Arabidopsis* and maize plastid proteomes (Majeran *et al.*, 2012; Olinares, Ponnala, & van Wijk, 2010).

In bacteria, DNA gyrase is the only topoisomerase generating negative supercoiling, which is essential for DNA condensation and proper chromosome partitioning during cell division (Holmes & Cozzarelli, 2000). *Arabidopsis* T-DNA *gyrA* mutants are embryo lethal, whereas mutants of *gyrB* lead to seedling-lethal phenotypes (Wall *et al.*, 2004). In *N. benthamiana*, virus-induced gene silencing of GyrA or GyrB induced morphological and physiological abnormalities in both organelles, which contained significantly higher amounts of mtDNA and cpDNA. Flow cytometry and pulse-field gel analysis demonstrated that gyrase plays critical roles in chloroplast nucleoid segregation, probably by regulating DNA topology (Cho, Lee *et al.*, 2004). It probably assumes similar functions in mitochondrial genome partitioning.

Bacterial topoisomerase I is involved in preventing excessive negative supercoiling generated by gyrase and is needed for maintaining DNA at an optimal superhelical density (Tse-Dinh, 1998). There is no information at present regarding the specific functions of Top I in plant organelles.

As in other genetic systems, mitochondrial topoisomerases should be essential for mtDNA recombination activities. RecA recombinases promote strand exchange and join the DNA ends to new partners, whereas topoisomerases catalyze topological changes and restore the original DNA linkage. It is possible that in mitochondria, topoisomerases participate in the dissolution of Holliday junctions, preventing the formation of lethal crossover products (Champoux, 2001).

4.5. MutS-like MSH1

MSH1 encodes a MutS-like protein that has important functions in the surveillance of mtDNA recombination. The gene was identified in *Arabidopsis* as responsible for the cytoplasmically inherited chloroplast mutator

(CHM) phenotype. The *chm* mutation was initially thought to affect plastid functions because of a variegated phenotype. However, the major molecular phenotypes observed in *chm* plants were important rearrangements of the mtDNA (Martinez-Zapater, Gil, Capel, & Somerville, 1992; Sakamoto, Kondo, Murata, & Motoyoshi, 1996). This molecular phenotype revealed *chm* to be a good model to study the molecular basis of SSS. The *chm* gene was found to encode a plant homologue of the yeast mitochondrial protein Msh1, a MutS-like protein (Chi & Kolodner, 1994a; Chi & Kolodner, 1994b) and has been re-named *MSH1* (Abdelnoor *et al.*, 2003). mtDNA rearrangements in *msh1* plants resulted from recombination via IRs ranging from about 50 bp up to 556 bp (Arrieta-Montiel *et al.*, 2009; Davila *et al.*, 2011; Shedge *et al.*, 2007). These IRs represented identical repeats or imperfect repeated sequences. The efficiency of sequence exchange appears to loosely correlate with repeat size and sequence homology (Davila *et al.*, 2011). In most cases, *msh1*-induced recombination is asymmetric, yielding mainly one of the two reciprocal crossover products. The severity of the growth phenotypes and of SSS progression (i.e. replacement of the predominant mitotype by the recombined sequence) evolves over several plant generations: first-generation *msh1* mutants presented low-level recombined mtDNA, whereas advanced generations derived by self-pollination had high copy numbers of crossover products (Arrieta-Montiel *et al.*, 2009). mtDNA recombination was also increased in tobacco and tomato *MSH1* knockdown plants and accompanied by phenotypic defects that included male sterility (Sandhu *et al.*, 2007).

In bacteria, MutS is a protein that participates in mismatch repair and in the suppression of ectopic recombination. MutS binds to mismatches that arise during replication and HR. Together with MutL, it recruits the endonuclease MutH (Schofield & Hsieh, 2003). Plant mitochondrial MSH1 contains a C-terminal GIY-YIG type homing endonuclease domain. A point mutation in this endonuclease domain is sufficient to induce variegation and mtDNA rearrangements (Abdelnoor *et al.*, 2006). It is therefore possible that MSH1 combines mismatch recognition and endonuclease functions.

In recombination, MSH1 is possibly part of a surveillance pathway that recognizes heteroduplexes in strand exchange complexes when there is little sequence homology (Fig. 9.3). MSH1 would then promote heteroduplex rejection (Abdelnoor *et al.*, 2003; Abdelnoor *et al.*, 2006). However, biochemical evidence for this model is still missing.

MSH1 is present in both mitochondria and chloroplasts in several species including *Arabidopsis* (Abdelnoor *et al.*, 2006). Recent results showed that

MSH1 is also active in the maintenance of the plastidial genome, explaining the variegation phenotype of *msh1* plants (Xu *et al.*, 2011). MSH1 constructs containing a mitochondrial-only targeting sequence produced plants that still displayed a variegated phenotype but showed no evidence of altered mitochondrial recombination (Xu *et al.*, 2011). White leaf sectors showed evidence for the accumulation of recombined plastidial DNA, mediated by DNA microhomologies (10–18 bp). MSH1 is therefore involved in the suppression of ectopic and illegitimate recombination in both organellar compartments.

4.6. RecA Recombinases

At the core of organellar HR activities are eubacterial-type RecA proteins. Bacterial RecA catalyzes DNA strand exchange during HR (Chen, Yang, & Pavletich, 2008; Cox, 2007; Muller & Stasiak, 1991; Takahashi & Norden, 1994). Initially, RecA polymerizes on ssDNA in an ATP-dependent manner to form a long and relatively stiff filament called the presynaptic complex, with a stoichiometry of three nucleotides per RecA monomer. The presynaptic filament then binds to dsDNA, forming a synaptic complex that scans for ssDNA–dsDNA homology. RecA is then able to catalyze strand exchange, forming a D-loop, which is a common intermediate in HR pathways. ATP hydrolysis dissociates all DNA intervenients, releasing the heteroduplex and the displaced ssDNA from the donor duplex (Cox, 2007).

RecA-like proteins are present in the mitochondria and/or chloroplasts of land plants. They are encoded by genes inherited from the endosymbiont ancestors of mitochondria and plastids (Lin, Kong, Nei, & Ma, 2006). Mosses possess two RecA genes, one for each organellar compartment (Inouye, Odahara, Fujita, Hasebe, & Sekine, 2008; Odahara, Inouye, Fujita, Hasebe, & Sekine, 2007), whereas *Arabidopsis* has three, named *RECA1*, *RECA2* and *RECA3* (Shedge *et al.*, 2007). RECA1 localizes to plastids and RECA3 to mitochondria, whereas RECA2 is present in both organelles (Shedge *et al.*, 2007). Heterologous expression of RECA2 and RECA3 in *Escherichia coli* revealed that the two proteins are functional RecAs with overlapping but also specific activities (Khazi, Edmondson, & Nielsen, 2003; Miller-Messmer *et al.*, in press). Both RECA2 and RECA3 could complement *E. coli* RecA in the repair of lesions induced by several genotoxic agents, but only RECA2 could complement repair of UV-C induced lesions (Miller-Messmer *et al.*, in press). Accordingly, the *Arabidopsis recA2* and *recA3* mutants displayed different phenotypes. *recA2* plants were unable to develop past the seedling stage but *recA3* plants were phenotypically normal (Miller-Messmer *et al.*, in press, Shedge *et al.*, 2007). Both *recA2* and *recA3* mutants displayed the same molecular phenotypes of increased recombination between IRs, as seen in the recombination surveillance mutants *osb1* and *msh1*, suggesting that RECA2, RECA3, OSB1 and MSH1 act in the same recombination pathways. The increase in HR is exacerbated in *recA2*, where accumulation of crossover products is an order of magnitude higher than in *recA3*. Increased mtDNA reshuffling by recombination in *recA2* is probably the cause of the severe *recA2* phenotype. In *Physcomitrella patens*, disruption of the gene coding for the single mitochondrial RecA (RECA1) also caused mtDNA rearrangements due to ectopic recombination across short (<100 bp) imperfect repeats. Therefore, both in *Arabidopsis* and in *Physcomitrella*, mitochondrial RECAs suppress a recombination pathway that mobilizes short repeats for recombination. This recombination pathway apparently does not depend on RECA and may involve SSA, as discussed in Section 3.2.2 (Fig. 9.3).

RecA-dependent strand exchange is essential for DNA repair by recombination. In chloroplasts, there is extensive evidence for the participation of plastid-targeted RecA in DNA repair (Cerutti, Ibrahim, & Jagendorf, 1993; Cerutti, Johnson, Boynton, & Gillham, 1995; Kwon *et al.*, 2010; Rowan, Oldenburg, & Bendich, 2010). In *Physcomitrella* it was shown that disruption of mitochondrial RECA results in reduced recovery of mtDNA following genotoxic damage (Odahara *et al.*, 2007). Recently, it was also demonstrated that *Arabidopsis* RECA3 has specific functions in mtDNA repair (Miller-Messmer *et al.*, in press). The *recA3* mutants were found to be hypersensitive to several DNA-damaging drugs that induce DNA breaks. It was also shown that in wild-type *Arabidopsis* plants, treatments that induce DNA repair activities result in the accumulation of specific IR-derived crossover products. This induced recombination activity is lost in the *recA3* background, showing that RECA3 is part of an HR-dependent repair activity induced by genotoxic stress (Miller-Messmer *et al.*, in press).

4.7. ssDNA-Binding Proteins

The processes of replication, recombination and repair require unwinding of the DNA strands, exposing ssDNA, which is vulnerable to attack by nucleases and to environmental damages. ssDNA can also form secondary structures that hinder the action of other factors involved in DNA metabolism. This is why ssDNA-binding proteins (SSB) have important roles in almost all DNA-processing reactions (Richard, Bolderson, & Khanna, 2009). They can have multiple functions in binding and protecting ssDNA from degradation, detecting DNA damage, stimulating recombinases and helicases, and mediating the protein–protein interactions necessary for DNA replication, recombination and repair (Richard *et al.*, 2009; Shereda, Kozlov, Lohman, Cox, & Keck, 2008).

In bacteria, SSB is a relatively abundant essential protein that binds ssDNA in a highly cooperative manner, forming protein filaments that saturate long stretches of ssDNA. SSBs act as a sliding platform to recruit and carry its interacting proteins. The high-affinity binding of SSB to ssDNA also limits RecA assembly onto ssDNA (Shereda *et al.*, 2008). To overcome recombination inhibition by SSB, recombination mediators such as bacterial RecO are required. Mediators can be other ssDNA-binding proteins that promote the displacement of SSB and the assembly of RecA on ssDNA (Gasior *et al.*, 2001). Diffusion–displacement of SSB melts secondary structures and stimulates RecA filament elongation (Roy, Kozlov, Lohman, & Ha, 2009).

Plant mitochondria are unusually rich in different types of SSB proteins. These include proteins that are homologues of known bacterial and eukaryotic SSBs, but also plant-specific SSB proteins whose functions are still being unravelled. How these proteins interplay in modulating mitochondrial recombination pathways will be an important subject of future research.

4.7.1. Eubacterial-type SSB

In animals and yeast, a mitochondrial ssDNA-binding protein (mtSSB in animals, RIM1 in yeast) is one of the key components in mtDNA replication and maintenance, essential for mtDNA replication (Maier *et al.*, 2001; Takamatsu *et al.*, 2002; Van Dyck, Foury, Stillman, & Brill, 1992). It is believed to coordinate the functions of DNAP γ and of Twinkle at the mtDNA replication fork (Farr, Wang, & Kaguni, 1999). The primary structure of the mtSSB is homologous to *E. coli* SSB. Eubacterial-type SSBs have also been identified in plant species and are probably central components of the replication and recombination machineries of plant organelles. In *Arabidopsis* and in most flowering plants, there are two SSB genes, *SSB1* and *SSB2* (Edmondson *et al.*, 2005; Elo *et al.*, 2003). Essential functions for SSB2 were confirmed in *Arabidopsis*, where *ssb2* mutants were unable to develop past seedling stage (our unpublished results). *Arabidopsis* SSB1 was shown to have similar properties to *E. coli* SSB: it bound ssDNA but not dsDNA and stimulated RecA-dependent strand invasion (Edmondson *et al.*, 2005).

The *Arabidopsis* SSB1 and SSB2 proteins are predicted to be targeted to mitochondria (our unpublished data). Mitochondrial targeting of *Arabidopsis* SSB1 was confirmed by fusion of the protein N-terminal sequence to GFP (Edmondson *et al.*, 2005). However, contradictory results were obtained for soybean SSB, which was purified from chloroplasts by affinity to a replication origin (oriA) of the chloroplast genome (cpDNA) (Lassen, Kochhar, & Nielsen, 2011). Because chloroplasts and mitochondria have similar DNA replication machineries, it is feasible that at least one SSB may be found in each organelle. SSB1 and SSB2 might be specific for each organellar compartment, or they could be dually targeted to both mitochondria and chloroplasts, as is the case for other components of the replication machinery (Carrie *et al.*, 2009).

4.7.2. Organellar ssDNA-Binding Proteins

The organellar ssDNA-binding proteins (OSB) were initially identified by biochemical purification of mitochondrial proteins with affinity to ssDNA (Vermel *et al.*, 2002). In *Arabidopsis* there are four *OSB* genes that code for proteins targeted to mitochondria (OSB1), plastids (OSB2), or to both organelles (OSB3) (Zaegel *et al.*, 2006). The localization of OSB4 has not been examined. *OSB* sequences are present in all land plants but have not been found in non-plant genomes. The OSB proteins are characterized by the presence of two characteristic domains. A poorly conserved central domain is similar to the OB-fold domain found in other SSB proteins. This domain is followed by one, two or three C-terminal PDF domains. The PDF domain is a plant-specific, well-conserved domain of 50 amino acids. DNA-binding studies showed that *Arabidopsis* OSB1 and OSB2 have a preference for ssDNA and that the PDF motif is required for protein–DNA interaction (Zaegel *et al.*, 2006). The OB-fold domain might be involved in protein oligomerization or in the interaction with other factors.

Arabidopsis osb2 and *osb3* T-DNA insertion mutants did not display any developmental or molecular phenotypes (Zaegel *et al.*, 2006 and our unpublished results). However, *osb1* mutants could develop phenotypes of variegation, leaf distortion and partial sterility, due to the accumulation of aberrant mtDNA sequences resulting from HR between certain IRs (Zaegel *et al.*, 2006). These alternative mitotypes could become predominant following SSS (Zaegel *et al.*, 2006), suggesting that OSB1 is involved in the surveillance of mtDNA HR and prevents unbalanced transmission of the alternative mtDNA configurations. The molecular phenotype of *osb1* is the same as that observed in *recA2* and *recA3* mutants (Miller-Messmer *et al.*, in press, Shedge *et al.*, 2007). Compared with *osb1*, *msh1* plants mobilize the same IRs and additional ones for recombination (Arrieta-Montiel *et al.*, 2009; Davila *et al.*, 2011). Thus, OSB1, RECA2 and RECA3 may function in a different pathway of recombination regulation than MSH1. In agreement with this hypothesis, *recA3 msh1* and *osb1 msh1* double mutants showed more severe phenotypes than the individual mutants (Arrieta-Montiel *et al.*, 2009 and our unpublished results). We propose that OSB1 is a recombination mediator controlling the assembly of RecA-ssDNA presynaptic complexes, whereas MSH1 is responsible for the rejection of heteroduplexes sharing insufficient sequence homology (Fig. 9.3).

Segregation studies of *osb1* plants defined a model for the SSS process. In stage I of SSS (first generations of homozygous *osb1*), HR products are already accumulated. At this stage, the process is reversible if OSB1 activity is restored, presumably because the HR products can be segregated out in the subsequent generations. However, individual lines can segregate in the rearranged mtDNA configurations that are preferentially replicated (stage II). At this stage, a reversion to the wild-type mtDNA is no longer possible, even if the wild-type *OSB1* allele is reintroduced. A similar process was described for *msh1* mutant plants, where prolonged selfing of *msh1* plants resulted in a gradual accumulation of rearranged mtDNA correlating with the severity of morphological phenotypes (Arrieta-Montiel *et al.*, 2009; Davila *et al.*, 2011).

Promoter–β-glucuronidase (GUS) fusion showed that OSB1 is expressed in plant male and female gametophytes (Zaegel *et al.*, 2006). It is therefore tempting to speculate that OSB1 acts as a repressor of recombination during mtDNA replication in gametophytic tissues, in order to reduce the load of aberrant recombined genomes and insure transmission of a functional mtDNA. Promoter studies similarly provided evidence for RECA2 and RECA3 expression in gametophytic cells (Miller-Messmer *et al.*, in press). It is possible that regulation of mtDNA recombination is not only controlled at different steps of the recombination pathway but also at different stages of plant development.

4.7.3. Whirly ssDNA-Binding Proteins

Whirly (WHY) proteins are present mainly in the plant kingdom (reviewed in (Desveaux, Marechal, & Brisson, 2005; Marechal & Brisson, 2010)). They are DNA-binding proteins that have preference for ssDNA and can have multiple functions in the plant cell, both in the nucleus and in plant organelles. All WHY proteins have mitochondria or chloroplast targeting sequences, and their organellar localization has been confirmed both by GFP fusion and by proteomic data (Vermel *et al.*, 2002; Krause *et al.*, 2005; Pfalz, Liere, Kandlbinder, Dietz, & Oelmuller, 2006). In *Arabidopsis*, two *WHY* genes code for plastidial proteins (*WHY1* and *WHY3*) and *WHY2* codes for the mitochondrial counterpart (Krause *et al.*, 2005).

In plastids, WHY proteins have multiple DNA- and RNA-related roles. In maize, the ZmWHY1 protein associates with a subset of chloroplast transcripts and is required for chloroplast transcript maturation (Prikryl, Watkins, Friso, van Wijk, & Barkan, 2008). The protein was also found to bind to the maize chloroplast genome without significant sequence specificity. In barley, although WHY was found associated with chloroplastic intron-containing transcripts, WHY1 knockdown did not affect transcript maturation (Melonek *et al.*, 2010). In mitochondria, roles of WHY proteins in gene expression have not yet been reported. Although overexpression of *Arabidopsis* WHY2 affected the transcription of several mitochondrial genes, this could be a non-specific effect due to masking of the mtDNA by WHY2.

More recently, chloroplast and mitochondrial WHY proteins were found to be involved in the surveillance of illegitimate recombination, preventing error-prone pathways of DSB repair through sequence microhomologies (Cappadocia *et al.*, 2010; Marechal *et al.*, 2009). *Arabidopsis why1 why3* double mutants were shown to develop variegation phenotypes, albeit infrequently. Variegation resulted from cpDNA rearrangements due to recombination involving microhomologies (Marechal *et al.*, 2009). *Arabidopsis* plants deficient in mitochondrial WHY2 were indistinguishable from the wild type, both morphologically and at the molecular level (Marechal *et al.*, 2008). However, under genotoxic stress induced by ciprofloxacin, *why2* plants showed increased mtDNA recombination via sequence microhomologies. Similar rearrangements were also observed in wild-type plants, but only under higher concentrations of ciprofloxacin (Cappadocia *et al.*, 2010). From these observations, it was proposed that WHY proteins are involved in replicative repair by the BIR pathway. Their absence would favour repair by illegitimate recombination according to the MMBIR pathway (Cappadocia *et al.*, 2010; Marechal & Brisson, 2010).

Crosses of the *why1 why3* double mutant with a *polIB* mutant yielded plants with high levels of plastidial DNA rearrangements mediated by sequence microhomologies (Parent *et al.*, 2011). This molecular phenotype was accompanied by severe growth defects (Parent *et al.*, 2011). The aggravated phenotype of *why1 why3 polIB* plants supports roles for plastidial WHY proteins and POLIB in overlapping DNA repair pathways. In contrast, *why2 polIB* mutants had no detectable phenotype, suggesting that in mitochondria, alternative repair mechanisms are in place when WHY2/POLIB-dependent pathways are compromised.

The crystal structure of two potato WHY proteins has been determined, revealing a novel ssDNA-binding domain (Cappadocia *et al.*, 2010; Desveaux, Allard, Brisson, & Sygusch, 2002). Using recombinant potato WHY2, a recent analysis of protein–ssDNA interactions showed that there is cooperative binding of WHY2 to long ssDNA molecules, with the protein forming hexamers of tetramers on the ssDNA. As described for eubacterial SSB, it is possible that cooperative binding of WHY proteins to ssDNA is necessary to condense exposed sequences and protect against spurious recombination (Cappadocia *et al.*, 2011).

4.7.4. RAD52-like ODB1

A novel plant mitochondrial DNA-binding protein has been identified and named ODB1 (organellar DNA-binding protein 1; Janicka *et al.*, in press). The *Arabidopsis ODB1* gene is identical with *RAD52-1* recently described to encode a protein functioning in HR in the nucleus, but also localizing to mitochondria (Samach, Melamed-Bessudo, Avivi-Ragolski, Pietrokovski, & Levy, 2011). In *Arabidopsis*, a second gene (*ODB2*) codes for a plastidial isoform (Janicka *et al.*, in press; Samach *et al.*, 2011). ODB1 was copurified with WHY2 from cauliflower mitochondrial extracts (Janicka *et al.*, in press). In mitochondria, WHY2 and ODB1 were found in large nucleoprotein complexes, and both proteins coimmunoprecipitated in a DNA-dependent manner. *In vitro* assays showed that ODB1 has higher affinity for ssDNA than for dsDNA, but with no apparent sequence specificity (Janicka *et al.*, in press). *Arabidopsis odb1* mutants were morphologically indistinguishable from the wild type and had normal mtDNA levels. However, following genotoxic stress induced by ciprofloxacin, *odb1* plants showed reduced HR activity compared with the wild type. Under the same conditions, *odb1* mutants showed an increase in sequences generated by microhomology-mediated recombination (Janicka *et al.*, in press), as described for *why2* (Cappadocia *et al.*, 2010). These observations identified ODB1 as a further component of HR-dependent DNA repair in plant mitochondria.

ODB proteins are distantly related to the N-terminal SSA domain of the DNA repair protein RAD52. A RAD52-like protein is also present in fungal mitochondria (MGM101p), where it is involved in mtDNA maintenance and repair (Chen, Guan, & Clark-Walker, 1993; Meeusen *et al.*, 1999), but the evolutionary relationship between ODBs and MGM101p is unresolved. In eukaryotes, RAD52 is a recombination mediator with important functions in HR-dependent repair pathways. RAD52 promotes the formation of RAD51-ssDNA filaments by displacing the ssDNA-binding protein RPA (Mortensen, Lisby, & Rothstein, 2009). It is tempting to speculate that ODB1 and ODB2 (named RAD52-1 and RAD52-2 in Samach *et al.*, 2011) act as RAD52 in plant organelles. If we assume that ODB1 performs RAD52/MGM101-like functions in plant mitochondria, ODB1 would be required for efficient loading of RECA onto ssDNA, relieving inhibition by other high-affinity ssDNA-binding proteins.

5. CONCLUSIONS AND FUTURE PERSPECTIVES

The importance of mtDNA and cpDNA repair in the fitness of plants undergoing genotoxic stress has often been underestimated, due to the high ploidy of organelle genomes. Morphological abnormalities described by researchers studying nuclear DNA repair following genotoxic stress have rarely been considered to result from organellar dysfunction (Britt, 1999; Puchta, 2005; Schuermann, Molinier, Fritsch, & Hohn, 2005). This perception has changed with the recent characterization of mutants affected in mtDNA recombination and repair. It is now widely accepted that the processes of mtDNA recombination have central roles in repair, preventing genome instability. But in addition to ensuring genome stability, the processes of mtDNA repair are a major driving force in the evolution of the mtDNA structure (Arrieta-Montiel *et al.*, 2009; Davila *et al.*, 2011; Sloan, Alverson, Wu, Palmer, & Taylor 2012). Recombination may also be responsible for copy correction of point mutations by gene conversion, thus driving the evolution of mtDNA sequences (Davila *et al.*, 2011; Sloan, Alverson, Wu *et al.*, 2012).

The study of plant mtDNA repair processes is hindered by the essential function that mitochondria have in cellular metabolism. Many mutants of genes involved in mtDNA metabolism are unviable and display embryo- or seedling-lethal phenotypes. Proteins specific to plant mtDNA maintenance have often been identified following biochemical approaches. In this context, the recent proteomic analysis of maize plastidial nucleoids is an important new resource for the functional study of genes with putative functions in mtDNA recombination and repair (Majeran *et al.*, 2012). Plastids and mitochondria share many components of their genome maintenance machineries, and many proteins identified in the maize nucleoid have counterparts in *Arabidopsis*. In addition, the recent functional characterization of RECA, WHY, DNAP and MSH1 proteins (Cappadocia *et al.*, 2010; Marechal *et al.*, 2009; Miller-Messmer *et al.*, in press; Parent *et al.*, 2011; Xu *et al.*, 2011) has led to the development of new experimental approaches for the identification of putative repair functions.

In the future it will be important to develop new tools for the study mtDNA recombination that allow a dissociation of mtDNA repair from plastidial and nuclear repair. One possibility is complementation with chloroplast-only targeted proteins, as recently utilized for the study of MSH1 functions (Xu *et al.*, 2011). It will also be important to develop approaches to induce damage specifically in the mtDNA, without affecting the other genetic compartments. Although such an approach is possible for the study of cpDNA maintenance using transplastomic constructs (Kwon *et al.*, 2010; Odom, Baek, Dani, & Herrin, 2008), it is not possible at present to manipulate mtDNA. An alternative approach could be the targeting of specific restriction enzymes into mitochondria. An equivalent approach was used successfully in animals (Bacman, Williams, & Moraes, 2009).

A major challenge in future studies on mtDNA recombination is to define how the numerous ssDNA-binding proteins that are found in mitochondria interplay with each other and with other factors involved in mtDNA maintenance. It is expected that their individual or collective actions promote or restrict access of RECA to its ssDNA substrate. In addition, they might have tissue-specific functions, as hinted by the gametophytic expression of OSB proteins. Mechanisms that regulate mtDNA transmission to the progeny will be a challenging subject of future research.

REFERENCES

Abdelnoor, R. V., Yule, R., Elo, A., Christensen, A. C., Meyer-Gauen, G., & Mackenzie, S. A. (2003). Substoichiometric shifting in the plant mitochondrial genome is influenced by a gene homologous to MutS. *Proceedings of the National Academy of Sciences of the United States of America, 100*, 5968–5973.

Abdelnoor, R. V., Christensen, A. C., Mohammed, S., Munoz-Castillo, B., Moriyama, H., & Mackenzie, S. A. (2006). Mitochondrial genome dynamics in plants and animals: convergent gene fusions of a MutS homologue. *Journal of Molecular Evolution, 63*, 165–173.

Adams, K. L., & Palmer, J. D. (2003). Evolution of mitochondrial gene content: gene loss and transfer to the nucleus. *Molecular Phylogenetics and Evolution, 29*, 380–395.

Aguilera, A., & Gomez-Gonzalez, B. (2008). Genome instability: a mechanistic view of its causes and consequences. *Nature Reviews Genetics, 9*, 204–217.

Allen, J. O., Fauron, C. M., Minx, P., Roark, L., Oddiraju, S., Lin, G. N., et al. (2007). Comparisons among two fertile and three male-sterile mitochondrial genomes of maize. *Genetics, 177*, 1173–1192.

Alverson, A. J., Wei, X., Rice, D. W., Stern, D. B., Barry, K., & Palmer, J. D. (2010). Insights into the evolution of mitochondrial genome size from complete sequences of *Citrullus lanatus* and *Cucurbita pepo* (Cucurbitaceae). *Molecular Biology and Evolution, 27*, 1436–1448.

Alverson, A. J., Rice, D. W., Dickinson, S., Barry, K., & Palmer, J. D. (2011). Origins and recombination of the bacterial-sized multichromosomal mitochondrial genome of cucumber. *The Plant Cell, 23*, 2499–2513.

Alverson, A. J., Zhuo, S., Rice, D. W., Sloan, D. B., & Palmer, J. D. (2011). The mitochondrial genome of the legume *Vigna radiata* and the analysis of recombination across short mitochondrial repeats. *PLoS ONE, 6*. e16404.

Andre, C., Levy, A., & Walbot, V. (1992). Small repeated sequences and the structure of plant mitochondrial genomes. *Trends in Genetics, 8*, 128–132.

Arrieta-Montiel, M., Lyznik, A., Woloszynska, M., Janska, H., Tohme, J., & Mackenzie, S. (2001). Tracing evolutionary and developmental implications of mitochondrial stoichiometric shifting in the common bean. *Genetics, 158*, 851–864.

Arrieta-Montiel, M. P., Shedge, V., Davila, J., Christensen, A. C., & Mackenzie, S. A. (2009). Diversity of the *Arabidopsis* mitochondrial genome occurs via nuclear-controlled recombination activity. *Genetics, 183*, 1261–1268.

Backert, S., Dorfel, P., Lurz, R., & Börner, T. (1996). Rolling-circle replication of mitochondrial DNA in the higher plant *Chenopodium album* (L.). *Molecular and Cellular Biology, 16*, 6285–6294.

Backert, S., Lurz, R., & Börner, T. (1996). Electron microscopic investigation of mitochondrial DNA from *Chenopodium album* (L.). *Current Genetics, 29*, 427–436.

Backert, S., Lurz, R., Oyarzabal, O. A., & Börner, T. (1997). High content, size and distribution of single-stranded DNA in the mitochondria of *Chenopodium album* (L.). *Plant Molecular Biology, 33*, 1037–1050.

Backert, S., Nielsen, B. L., & Börner, T. (1997). The mystery of the rings: structure and replication of mitochondrial genomes from higher plants. *Trends in Plant Science, 2*, 477–483.

Backert, S., & Börner, T. (2000). Phage T4-like intermediates of DNA replication and recombination in the mitochondria of the higher plant *Chenopodium album* (L.). *Current Genetics, 37*, 304–314.

Backert, S. (2002). R-loop-dependent rolling-circle replication and a new model for DNA concatemer resolution by mitochondrial plasmid mp1. *EMBO Journal, 21*, 3128–3136.

Bacman, S. R., Williams, S. L., & Moraes, C. T. (2009). Intra- and inter-molecular recombination of mitochondrial DNA after in vivo induction of multiple double-strand breaks. *Nucleic Acids Research, 37*, 4218–4226.

Barr, C. M., Keller, S. R., Ingvarsson, P. K., Sloan, D. B., & Taylor, D. R. (2007). Variation in mutation rate and polymorphism among mitochondrial genes of *Silene vulgaris*. *Molecular Biology and Evolution, 24*, 1783–1791.

Bartoszewski, G., Malepszy, S., & Havey, M. J. (2004). Mosaic (MSC) cucumbers regenerated from independent cell cultures possess different mitochondrial rearrangements. *Current Genetics, 45*, 45–53.

Bendich, A. J. (1993). Reaching for the ring: the study of mitochondrial genome structure. *Current Genetics, 24*, 564–588.

Boore, J. L. (1999). Animal mitochondrial genomes. *Nucleic Acids Research, 27*, 1767–1780.

Britt, A. B. (1999). Molecular genetics of DNA repair in higher plants. *Trends in Plant Science, 4*, 20–25.

Budar, F., Touzet, P., & De Paepe, R. (2003). The nucleo-mitochondrial conflict in cytoplasmic male sterilities revisited. *Genetica, 117*, 3–16.

Bullerwell, C. E., & Gray, M. W. (2004). Evolution of the mitochondrial genome: protist connections to animals, fungi and plants. *Current Opinion in Microbiology, 7*, 528–534.

Cao, L., Shitara, H., Sugimoto, M., Hayashi, J., Abe, K., & Yonekawa, H. (2009). New evidence confirms that the mitochondrial bottleneck is generated without reduction of mitochondrial DNA content in early primordial germ cells of mice. *PLoS Genetics, 5*. e1000756.

Cappadocia, L., Marechal, A., Parent, J. S., Lepage, E., Sygusch, J., & Brisson, N. (2010). Crystal structures of DNA-Whirly complexes and their role in *Arabidopsis* organelle genome repair. *The Plant Cell, 22*, 1849–1867.

Cappadocia, L., Parent, J. S., Zampini, E., Lepage, E., Sygusch, J., & Brisson, N. (2011). A conserved lysine residue of plant Whirly proteins is necessary for higher order protein assembly and protection against DNA damage. *Nucleic Acids Research, 40*, 258–269.

Carrie, C., Kühn, K., Murcha, M. W., Duncan, O., Small, I. D., O'Toole, N., et al. (2009). Approaches to defining dual-targeted proteins in *Arabidopsis*. *The Plant Journal, 57*, 1128–1139.

Castroviejo, M., Tharaud, D., Tarrago-Litvak, L., & Litvak, S. (1979). Multiple deoxyribonucleic acid polymerases from quiescent wheat embryos. Purification and characterization of three enzymes from the soluble cytoplasm and one from purified mitochondria. *Biochemical Journal, 181*, 183–191.

Cerutti, H., Ibrahim, H. Z., & Jagendorf, A. T. (1993). Treatment of pea (*Pisum sativum* L.) protoplasts with DNA-damaging agents induces a 39-kilodalton chloroplast protein immunologically related to *Escherichia coli* RecA. *Plant Physiology, 102*, 155–163.

Cerutti, H., Johnson, A. M., Boynton, J. E., & Gillham, N. W. (1995). Inhibition of chloroplast DNA recombination and repair by dominant negative mutants of *Escherichia coli* RecA. *Molecular and Cellular Biology, 15*, 3003–3011.

Champoux, J. J. (2001). DNA topoisomerases: structure, function, and mechanism. *Annual Review of Biochemistry, 70*, 369–413.

Chen, J., Guan, R., Chang, S., Du, T., Zhang, H., & Xing, H. (2011). Substoichiometrically different mitotypes coexist in mitochondrial genomes of *Brassica napus* L. *PLoS ONE, 6*. e17662.

Chen, X. J., Guan, M. X., & Clark-Walker, G. D. (1993). MGM101, a nuclear gene involved in maintenance of the mitochondrial genome in *Saccharomyces cerevisiae*. *Nucleic Acids Research, 21*, 3473–3477.

Chen, Z., Yang, H., & Pavletich, N. P. (2008). Mechanism of homologous recombination from the RecA-ssDNA/dsDNA structures. *Nature, 453*, 489–494.

Chi, N. W., & Kolodner, R. D. (1994a). The effect of DNA mismatches on the ATPase activity of MSH1, a protein in yeast mitochondria that recognizes DNA mismatches. *Journal of Biological Chemistry, 269*, 29993–29997.

Chi, N. W., & Kolodner, R. D. (1994b). Purification and characterization of MSH1, a yeast mitochondrial protein that binds to DNA mismatches. *Journal of Biological Chemistry, 269*, 29984–29992.

Cho, H. S., Lee, S. S., Kim, K. D., Hwang, I., Lim, J. S., Park, Y. I., et al. (2004). DNA gyrase is involved in chloroplast nucleoid partitioning. *The Plant Cell, 16*, 2665–2682.

Cho, Y., Mower, J. P., Qiu, Y. L., & Palmer, J. D. (2004). Mitochondrial substitution rates are extraordinarily elevated and variable in a genus of flowering plants. *Proceedings of the National Academy of Sciences of the United States of America, 101*, 17741–17746.

Christensen, A. C., Lyznik, A., Mohammed, S., Elowsky, C. G., Elo, A., Yule, R., et al. (2005). Dual-domain, dual-targeting organellar protein presequences in *Arabidopsis* can use non-AUG start codons. *The Plant Cell, 17*, 2805–2816.

Copeland, W. C. (2010). The mitochondrial DNA polymerase in health and disease. *Subcellular Biochemistry, 50*, 211–222.

Cordoba-Canero, D., Roldan-Arjona, T., & Ariza, R. R. (2011). *Arabidopsis* ARP endonuclease functions in a branched base excision DNA repair pathway completed by LIG1. *The Plant Journal, 68*, 693–702.

Cox, M. M. (2007). Regulation of bacterial RecA protein function. *Critical Reviews in Biochemistry and Molecular Biology, 42*, 41–63.

Crespan, E., Czabany, T., Maga, G., & Hubscher, U. Microhomology-mediated DNA strand annealing and elongation by human DNA polymerases lambda and beta on normal and repetitive DNA sequences. *Nucleic Acids Research*, in press http://dx.doi.org/10.1093/nar/gks186.

Cummings, D. J., & Domenico, J. M. (1988). Sequence analysis of mitochondrial DNA from *Podospora anserina*. *Journal of Molecular Biology, 204*, 815–839.

Darracq, A., Varre, J. S., & Touzet, P. (2010). A scenario of mitochondrial genome evolution in maize based on rearrangement events. *BMC Genomics, 11*, 233.

Darracq, A., Varre, J. S., Marechal-Drouard, L., Courseaux, A., Castric, V., Saumitou-Laprade, P., et al. (2011). Structural and content diversity of mitochondrial genome in beet: a comparative genomic analysis. *Genome Biology and Evolution, 3*, 723–736.

Davila, J. I., Arrieta-Montiel, M. P., Wamboldt, Y., Cao, J., Hagmann, J., Shedge, V., et al. (2011). Double-strand break repair processes drive evolution of the mitochondrial genome in *Arabidopsis*. *BMC Biology, 9*, 64.

Desveaux, D., Allard, J., Brisson, N., & Sygusch, J. (2002). A new family of plant transcription factors displays a novel ssDNA-binding surface. *Nature Structural & Molecular Biology, 9*, 512–517.

Desveaux, D., Marechal, A., & Brisson, N. (2005). Whirly transcription factors: defense gene regulation and beyond. *Trends in Plant Science, 10*, 95–102.

Dewey, R. E., Levings, C. S. I., & Timothy, D. H. (1986). Novel recombinations in the maize mitochondrial genome produce a unique transcriptional unit in the Texas male sterile cytoplasm. *Cell, 44*, 439–449.

Dutra, B. E., Sutera, V. A., Jr., & Lovett, S. T. (2007). RecA-independent recombination is efficient but limited by exonucleases. *Proceedings of the National Academy of Sciences of the United States of America, 104*, 216–221.

Edmondson, A. C., Song, D., Alvarez, L. A., Wall, M. K., Almond, D., McClellan, D. A., et al. (2005). Characterization of a mitochondrially targeted single-stranded DNA-binding protein in *Arabidopsis thaliana*. *Molecular Genetics & Genomics, 273*, 115–122.

Elo, A., Lyznik, A., Gonzalez, D. O., Kachman, S. D., & Mackenzie, S. A. (2003). Nuclear genes that encode mitochondrial proteins for DNA and RNA metabolism are clustered in the *Arabidopsis* genome. *The Plant Cell, 15*, 1619–1631.

Farr, C. L., Wang, Y., & Kaguni, L. S. (1999). Functional interactions of mitochondrial DNA polymerase and single-stranded DNA-binding protein. Template-primer DNA binding and initiation and elongation of DNA strand synthesis. *Journal of Biological Chemistry, 274*, 14779–14785.

Fauron, C. M., Havlik, M., & Brettell, R. I. (1990). The mitochondrial genome organization of a maize fertile cmsT revertant line is generated through recombination between two sets of repeats. *Genetics, 124*, 423–428.

Filee, J., Forterre, P., Sen-Lin, T., & Laurent, J. (2002). Evolution of DNA polymerase families: evidences for multiple gene exchange between cellular and viral proteins. *Journal of Molecular Evolution, 54*, 763–773.

Fitzpatrick, D. A., Creevey, C. J., & McInerney, J. O. (2006). Genome phylogenies indicate a meaningful alpha-proteobacterial phylogeny and support a grouping of the mitochondria with the Rickettsiales. *Molecular Biology and Evolution, 23*, 74–85.

Friedberg, E. C. (2003). DNA damage and repair. *Nature, 421*, 436–440.

Gasior, S. L., Olivares, H., Ear, U., Hari, D. M., Weichselbaum, R., & Bishop, D. K. (2001). Assembly of RecA-like recombinases: distinct roles for mediator proteins in mitosis and meiosis. *Proceedings of the National Academy of Sciences of the United States of America, 98*, 8411–8418.

Gerhold, J. M., Aun, A., Sedman, T., Joers, P., & Sedman, J. (2010). Strand invasion structures in the inverted repeat of *Candida albicans* mitochondrial DNA reveal a role for homologous recombination in replication. *Molecular Cell, 39*, 851–861.

Gray, M. W., Burger, G., & Lang, B. F. (1999). Mitochondrial evolution. *Science, 283*, 1476–1481.

Hanawalt, P. C. (2007). Paradigms for the three rs: DNA replication, recombination, and repair. *Molecular Cell, 28*, 702–707.

Hanson, M. R., & Bentolila, S. (2004). Interactions of mitochondrial and nuclear genes that affect male gametophyte development. *The Plant Cell*(Suppl. 16), S154–169.

Hastings, P. J., Ira, G., & Lupski, J. R. (2009). A microhomology-mediated break-induced replication model for the origin of human copy number variation. *PLoS Genetics, 5*. e1000327.

Hastings, P. J., Lupski, J. R., Rosenberg, S. M., & Ira, G. (2009). Mechanisms of change in gene copy number. *Nature Reviews Genetics, 10*, 551–564.

Heinhorst, S., Cannon, G. C., & Weissbach, A. (1990). Chloroplast and mitochondrial DNA polymerases from cultured soybean cells. *Plant Physiology, 92*, 939–945.

Holmes, V. F., & Cozzarelli, N. R. (2000). Closing the ring: links between SMC proteins and chromosome partitioning, condensation, and supercoiling. *Proceedings of the National Academy of Sciences of the United States of America, 97*, 1322–1324.

Inouye, T., Odahara, M., Fujita, T., Hasebe, M., & Sekine, Y. (2008). Expression and complementation analyses of a chloroplast-localized homolog of bacterial RecA in the moss *Physcomitrella patens*. *Bioscience, Biotechnology, and Biochemistry, 72*, 1340–1347.

Sabina Janicka, S., Kühn, K., Le Ret, M., Bonnard, G., Imbault, P., Augustyniak, H., & Gualberto, J. M. (2012). A RAD52-like single-stranded DNA binding protein affects mitochondrial DNA repair by recombination. *The Plant Journal*, in press.

Janska, H., Sarria, R., Woloszynska, M., Arrieta-Montiel, M., & Mackenzie, S. A. (1998). Stoichiometric shifts in the common bean mitochondrial genome leading to male sterility and spontaneous reversion to fertility. *The Plant Cell, 10*, 1163–1180.

Kajander, O. A., Rovio, A. T., Majamaa, K., Poulton, J., Spelbrink, J. N., Holt, I. J., et al. (2000). Human mtDNA sublimons resemble rearranged mitochondrial genoms found in pathological states. *Human Molecular Genetics, 9*, 2821–2835.

Kajander, O. A., Karhunen, P. J., Holt, I. J., & Jacobs, H. T. (2001). Prominent mitochondrial DNA recombination intermediates in human heart muscle. *EMBO Reports, 2*, 1007–1012.

Kanazawa, A., Tsutsumi, N., & Hirai, A. (1994). Reversible changes in the composition of the population of mtDNAs during dedifferentiation and regeneration in tobacco. *Genetics, 138*, 865–870.

Khazi, F. R., Edmondson, A. C., & Nielsen, B. L. (2003). An *Arabidopsis* homologue of bacterial RecA that complements an *E. coli* recA deletion is targeted to plant mitochondria. *Molecular Genetics & Genomics, 269*, 454–463.

Kimura, S., Uchiyama, Y., Kasai, N., Namekawa, S., Saotome, A., Ueda, T., et al. (2002). A novel DNA polymerase homologous to *Escherichia coli* DNA polymerase I from a higher plant, rice (*Oryza sativa* L.). *Nucleic Acids Research, 30*, 1585–1592.

Kmiec, B., Woloszynska, M., & Janska, H. (2006). Heteroplasmy as a common state of mitochondrial genetic information in plants and animals. *Current Genetics, 50*, 149–159.

Krause, K., Kilbienski, I., Mulisch, M., Rodiger, A., Schafer, A., & Krupinska, K. (2005). DNA-binding proteins of the Whirly family in *Arabidopsis thaliana* are targeted to the organelles. *FEBS Letters, 579*, 3707–3712.

Kubo, T., & Newton, K. J. (2008). Angiosperm mitochondrial genomes and mutations. *Mitochondrion, 8*, 5–14.

Kuzmin, E. V., Duvick, D. N., & Newton, K. J. (2005). A mitochondrial mutator system in maize. *Plant Physiology, 137*, 779–789.

Kwon, T., Huq, E., & Herrin, D. L. (2010). Microhomology-mediated and nonhomologous repair of a double-strand break in the chloroplast genome of *Arabidopsis. Proceedings of the National Academy of Sciences of the United States of America, 107*, 13954–13959.

Lassen, M. G., Kochhar, S., & Nielsen, B. L. (2011). Identification of a soybean chloroplast DNA replication origin-binding protein. *Plant Molecular Biology, 76*, 463–471.

Lin, Z., Kong, H., Nei, M., & Ma, H. (2006). Origins and evolution of the recA/RAD51 gene family: evidence for ancient gene duplication and endosymbiotic gene transfer. *Proceedings of the National Academy of Sciences of the United States of America, 103*, 10328–10333.

Ling, F., Makishima, F., Morishima, N., & Shibata, T. (1995). A nuclear mutation defective in mitochondrial recombination in yeast. *EMBO Journal, 14*, 4090–4101.

Lohman, G. J., Tabor, S., & Nichols, N. M. (2011). DNA ligases. *Current Protocols in Molecular Biology*. Chapter 3, Unit 3 14.

Lonsdale, D. M., Hodge, T. P., & Fauron, C. M. (1984). The physical map and organisation of the mitochondrial genome from the fertile cytoplasm of maize. *Nucleic Acids Research, 12*, 9249–9261.

Lovett, S. T., Hurley, R. L., Sutera, V. A., Jr., Aubuchon, R. H., & Lebedeva, M. A. (2002). Crossing over between regions of limited homology in *Escherichia coli*. RecA-dependent and RecA-independent pathways. *Genetics, 160*, 851–859.

Mackenzie, S., & McIntosh, L. (1999). Higher plant mitochondria. *The Plant Cell, 11*, 571–586.

Maier, D., Farr, C. L., Poeck, B., Alahari, A., Vogel, M., Fischer, S., et al. (2001). Mitochondrial single-stranded DNA-binding protein is required for mitochondrial DNA replication and development in *Drosophila melanogaster. Molecular Biology of the Cell, 12*, 821–830.

Majeran, W., Friso, G., Asakura, Y., Qu, X., Huang, M., Ponnala, L., et al. (2012). Nucleoid-enriched proteomes in developing plastids and chloroplasts from maize leaves: a new conceptual framework for nucleoid functions. *Plant Physiology, 158*, 156–189.

Manchekar, M., Scissum-Gunn, K., Song, D., Khazi, F., McLean, S. L., & Nielsen, B. L. (2006). DNA recombination activity in soybean mitochondria. *Journal of Molecular Biology, 356*, 288–299.

Mannuss, A., Trapp, O., & Puchta, H. (2012). Gene regulation in response to DNA damage. *Biochimica et Biophysica Acta, 1819*, 154–165.

Marechal, A., Parent, J. S., Sabar, M., Veronneau-Lafortune, F., Abou-Rached, C., & Brisson, N. (2008). Overexpression of mtDNA-associated AtWhy2 compromises mitochondrial function. *BMC Plant Biology, 8*, 42.

Marechal, A., Parent, J. S., Veronneau-Lafortune, F., Joyeux, A., Lang, B. F., & Brisson, N. (2009). Whirly proteins maintain plastid genome stability in *Arabidopsis. Proceedings of the National Academy of Sciences of the United States of America, 106*, 14693–14698.

Marechal, A., & Brisson, N. (2010). Recombination and the maintenance of plant organelle genome stability. *New Phytologist, 186*, 299–317.

Marienfeld, J. R., & Newton, K. J. (1994). The maize NCS2 abnormal growth mutant has a chimeric nad4-nad7 mitochondrial gene and is associated with reduced complex I function. *Genetics, 138*, 855–863.

Martinez-Zapater, J. M., Gil, P., Capel, J., & Somerville, C. R. (1992). Mutations at the *Arabidopsis* CHM locus promote rearrangements of the mitochondrial genome. *The Plant Cell, 4*, 889–899.

Meeusen, S., Tieu, Q., Wong, E., Weiss, E., Schieltz, D., Yates, J. R., et al. (1999). Mgm101p is a novel component of the mitochondrial nucleoid that binds DNA and is required for the repair of oxidatively damaged mitochondrial DNA. *Journal of Cell Biology, 145*, 291–304.

Meissner, K., Heinhorst, S., Cannon, G. C., & Börner, T. (1993). Purification and characterization of a gamma-like DNA polymerase from *Chenopodium album* L. *Nucleic Acids Research, 21*, 4893–4899.

Melonek, J., Mulisch, M., Schmitz-Linneweber, C., Grabowski, E., Hensel, G., & Krupinska, K. (2010). Whirly1 in chloroplasts associates with intron containing RNAs and rarely co-localizes with nucleoids. *Planta, 232*, 471–481.

Michel, B., Boubakri, H., Baharoglu, Z., LeMasson, M., & Lestini, R. (2007). Recombination proteins and rescue of arrested replication forks. *DNA Repair (Amst), 6*, 967–980.

Miller-Messmer, M., Kühn, K., Bichara, M., Le Ret, M., Imbault, P., & Gualberto, J.M. RECA-dependent DNA repair results in increased heteroplasmy of the *Arabidopsis* mitochondrial genome. *Plant Physiology*, 159, 211–226.

Mori, Y., Kimura, S., Saotome, A., Kasai, N., Sakaguchi, N., Uchiyama, Y., et al. (2005). Plastid DNA polymerases from higher plants, *Arabidopsis thaliana. Biochemical and Biophysical Research Communications, 334*, 43–50.

Mortensen, U. H., Lisby, M., & Rothstein, R. (2009). Rad52. *Current Biology, 19*, R676–677.

Muller, B., & Stasiak, A. (1991). RecA-mediated annealing of single-stranded DNA and its relation to the mechanism of homologous recombination. *Journal of Molecular Biology, 221*, 131–145.

Newton, K. J., & Coe, E. H. (1986). Mitochondrial DNA changes in abnormal growth (nonchromosomal stripe) mutants of maize. *Proceedings of the National Academy of Sciences of the United States of America, 83*, 7363–7366.

Odahara, M., Inouye, T., Fujita, T., Hasebe, M., & Sekine, Y. (2007). Involvement of mitochondrial-targeted RecA in the repair of mitochondrial DNA in the moss, *Physcomitrella patens. Genes & Genetic Systems, 82*, 43–51.

Odom, O. W., Baek, K. H., Dani, R. N., & Herrin, D. L. (2008). *Chlamydomonas* chloroplasts can use short dispersed repeats and multiple pathways to repair a double-strand break in the genome. *The Plant Journal, 53*, 842–853.

Ogihara, Y., Yamazaki, Y., Murai, K., Kanno, A., Terachi, T., Shiina, T., et al. (2005). Structural dynamics of cereal mitochondrial genomes as revealed by complete nucleotide sequencing of the wheat mitochondrial genome. *Nucleic Acids Research, 33*, 6235–6250.

Oldenburg, D. J., & Bendich, A. J. (1996). Size and structure of replicating mitochondrial DNA in cultured tobacco cells. *The Plant Cell, 8*, 447–461.

Oldenburg, D. J., & Bendich, A. J. (1998). The structure of mitochondrial DNA from the liverwort, *Marchantia polymorpha. Journal of Molecular Biology, 276*, 745–758.

Oldenburg, D. J., & Bendich, A. J. (2001). Mitochondrial DNA from the liverwort *Marchantia polymorpha*: circularly permuted linear molecules, head-to-tail concatemers, and a 5′ protein. *Journal of Molecular Biology, 310*, 549–562.

Olinares, P. D., Ponnala, L., & van Wijk, K. J. (2010). Megadalton complexes in the chloroplast stroma of *Arabidopsis thaliana* characterized by size exclusion chromatography, mass spectrometry, and hierarchical clustering. *Molecular & Cellular Proteomics, 9*, 1594–1615.

Ono, Y., Sakai, A., Takechi, K., Takio, S., Takusagawa, M., & Takano, H. (2007). NtPolI-like1 and NtPolI-like2, bacterial DNA polymerase I homologs isolated from BY-2 cultured tobacco cells, encode DNA polymerases engaged in DNA replication in both plastids and mitochondria. *Plant and Cell Physiology, 48*, 1679–1692.

Palmer, J. D., & Shields, C. R. (1984). Tripartite structure of *Brassica campestris* mitochondrial genome. *Nature, 307*, 436–440.

Palmer, J. D., & Herbon, L. A. (1987). Unicircular structure of the *Brassica hirta* mitochondrial genome. *Current Genetics, 11*, 565–570.

Palmer, J. D., & Herbon, L. A. (1988). Plant mitochondrial DNA evolves rapidly in structure, but slowly in sequence. *Journal of Molecular Evolution, 28*, 87–97.

Parent, J. S., Lepage, E., & Brisson, N. (2011). Divergent roles for the two PolI-like organelle DNA polymerases of *Arabidopsis*. *Plant Physiology, 156*, 254–262.

Pfalz, J., Liere, K., Kandlbinder, A., Dietz, K. J., & Oelmuller, R. (2006). pTAC2, -6, and -12 are components of the transcriptionally active plastid chromosome that are required for plastid gene expression. *The Plant Cell, 18*, 176–197.

Preuten, T., Cincu, E., Fuchs, J., Zoschke, R., Liere, K., & Börner, T. (2010). Fewer genes than organelles: extremely low and variable gene copy numbers in mitochondria of somatic plant cells. *The Plant Journal, 64*, 948–959.

Prikryl, J., Watkins, K. P., Friso, G., van Wijk, K. J., & Barkan, A. (2008). A member of the Whirly family is a multifunctional RNA- and DNA-binding protein that is essential for chloroplast biogenesis. *Nucleic Acids Research, 36*, 5152–5165.

Puchta, H. (2005). The repair of double-strand breaks in plants: mechanisms and consequences for genome evolution. *Journal of Experimental Botany, 56*, 1–14.

Richard, D. J., Bolderson, E., & Khanna, K. K. (2009). Multiple human single-stranded DNA binding proteins function in genome maintenance: structural, biochemical and functional analysis. *Critical Reviews in Biochemistry and Molecular Biology, 44*, 98–116.

Rowan, B. A., Oldenburg, D. J., & Bendich, A. J. (2010). RecA maintains the integrity of chloroplast DNA molecules in *Arabidopsis*. *Journal of Experimental Botany, 61*, 2575–2588.

Roy, R., Kozlov, A. G., Lohman, T. M., & Ha, T. (2009). SSB protein diffusion on single-stranded DNA stimulates RecA filament formation. *Nature, 461*, 1092–1097.

Sakamoto, W., Kondo, H., Murata, M., & Motoyoshi, F. (1996). Altered mitochondrial gene expression in a maternal distorted leaf mutant of *Arabidopsis* induced by chloroplast mutator. *The Plant Cell, 8*, 1377–1390.

Samach, A., Melamed-Bessudo, C., Avivi-Ragolski, N., Pietrokovski, S., & Levy, A. A. (2011). Identification of plant RAD52 homologs and characterization of the *Arabidopsis thaliana* RAD52-like genes. *The Plant Cell.*

San Filippo, J., Sung, P., & Klein, H. (2008). Mechanism of eukaryotic homologous recombination. *Annual Review of Biochemistry, 77*, 229–257.

Sandhu, A. P., Abdelnoor, R. V., & Mackenzie, S. A. (2007). Transgenic induction of mitochondrial rearrangements for cytoplasmic male sterility in crop plants. *Proceedings of the National Academy of Sciences of the United States of America, 104*, 1766–1770.

Satoh, M., Kubo, T., Nishizawa, S., Estiati, A., Itchoda, N., & Mikami, T. (2004). The cytoplasmic male-sterile type and normal type mitochondrial genomes of sugar beet share the same complement of genes of known function but differ in the content of expressed ORFs. *Molecular Genetics & Genomics, 272*, 247–256.

Schofield, M. J., & Hsieh, P. (2003). DNA mismatch repair: molecular mechanisms and biological function. *Annual Review of Microbiology, 57*, 579–608.

Schuermann, D., Molinier, J., Fritsch, O., & Hohn, B. (2005). The dual nature of homologous recombination in plants. *Trends in Genetics, 21*, 172–181.

Sekiguchi, J. M., & Ferguson, D. O. (2006). DNA double-strand break repair: a relentless hunt uncovers new prey. *Cell, 124*, 260–262.

Shedge, V., Arrieta-Montiel, M., Christensen, A. C., & Mackenzie, S. A. (2007). Plant mitochondrial recombination surveillance requires unusual RecA and MutS homologs. *The Plant Cell, 19*, 1251–1264.

Shereda, R. D., Kozlov, A. G., Lohman, T. M., Cox, M. M., & Keck, J. L. (2008). SSB as an organizer/mobilizer of genome maintenance complexes. *Critical Reviews in Biochemistry and Molecular Biology, 43*, 289–318.

Shoubridge, E. A. (2000). Mitochondrial DNA segregation in the developing embryo. *Human Reproduction, 15*(Suppl. 2), 229–234.

Shutt, T. E., & Gray, M. W. (2006). Twinkle, the mitochondrial replicative DNA helicase, is widespread in the eukaryotic radiation and may also be the mitochondrial DNA primase in most eukaryotes. *Journal of Molecular Evolution, 62*, 588–599.

Sloan, D. B., Alverson, A. J., Chuckalovcak, J. P., Wu, M., McCauley, D. E., Palmer, J. D., et al. (2012). Rapid evolution of enormous, multichromosomal genomes in flowering plant mitochondria with exceptionally high mutation rates. *PLoS Biology, 10*. e1001241.

Sloan, D. B., Alverson, A. J., Wu, M., Palmer, J. D., & Taylor, D. R. (2012). Recent acceleration of plastid sequence and structural evolution coincides with extreme mitochondrial divergence in the angiosperm genus *Silene*. *Genome Biology and Evolution, 4*, 294–306.

Small, I., Suffolk, R., & Leaver, C. J. (1989). Evolution of plant mitochondrial genomes via substoichiometric intermediates. *Cell, 58*, 69–76.

Small, I. D., Isaac, P. G., & Leaver, C. J. (1987). Stoichiometric differences in DNA molecules containing the *atpA* gene suggest mechanisms for the generation of mitochondrial genome diversity in maize. *EMBO Journal, 6*, 865–869.

Smeitink, J. A., Zeviani, M., Turnbull, D. M., & Jacobs, H. T. (2006). Mitochondrial medicine: a metabolic perspective on the pathology of oxidative phosphorylation disorders. *Cell Metabolism, 3*, 9–13.

Sunderland, P. A., West, C. E., Waterworth, W. M., & Bray, C. M. (2006). An evolutionarily conserved translation initiation mechanism regulates nuclear or mitochondrial targeting of DNA ligase 1 in *Arabidopsis thaliana*. *The Plant Journal, 47*, 356–367.

Takahashi, M., & Norden, B. (1994). Structure of RecA-DNA complex and mechanism of DNA strand exchange reaction in homologous recombination. *Advances in Biophysics, 30*, 1–35.

Takamatsu, C., Umeda, S., Ohsato, T., Ohno, T., Abe, Y., Fukuoh, A., et al. (2002). Regulation of mitochondrial D-loops by transcription factor A and single-stranded DNA-binding protein. *EMBO Reports, 3*, 451–456.

Tse-Dinh, Y. C. (1998). Bacterial and archeal type I topoisomerases. *Biochimica et Biophysica Acta, 1400*, 19–27.

Van Dyck, E., Foury, F., Stillman, B., & Brill, S. J. (1992). A single-stranded DNA binding protein required for mitochondrial DNA replication in *S. cerevisiae* is homologous to *E. coli* SSB. *EMBO Journal, 11*, 3421–3430.

Vermel, M., Guermann, B., Delage, L., Grienenberger, J. M., Marechal-Drouard, L., & Gualberto, J. M. (2002). A family of RRM-type RNA-binding proteins specific to plant mitochondria. *Proceedings of the National Academy of Sciences of the United States of America, 99*, 5866–5871.

Vitart, V., De paepe, R., Mathieu, C., Chetrit, P., & Vedel, F. (1992). Amplification of substoichiometric recombinant mitochondrial DNA sequences in a nuclear, male sterile mutant regenerated from protoplast culture in *Nicotiana sylvestris*. *Molecular and General Genetics, 233*, 193–200.

Wai, T., Teoli, D., & Shoubridge, E. A. (2008). The mitochondrial DNA genetic bottleneck results from replication of a subpopulation of genomes. *Nature Genetics, 40*, 1484–1488.

Wall, M. K., Mitchenall, L. A., & Maxwell, A. (2004). *Arabidopsis thaliana* DNA gyrase is targeted to chloroplasts and mitochondria. *Proceedings of the National Academy of Sciences of the United States of America, 101*, 7821–7826.

Wang, D. Y., Zhang, Q., Liu, Y., Lin, Z. F., Zhang, S. X., Sun, M. X., & Sodmergen. (2010). The levels of male gametic mitochondrial DNA are highly regulated in angiosperms with regard to mitochondrial inheritance. *The Plant Cell, 22*, 2402–2416.

Ward, B. L., Anderson, R. S., & Bendich, A. J. (1981). Mitochondrial genome is large and variable in a family of plants (Cucurbitaceae). *Cell, 26*, 793–803.

Wolfe, K. H., Li, W. H., & Sharp, P. M. (1987). Rates of nucleotide substitution vary greatly among plant mitochondrial, chloroplast, and nuclear DNAs. *Proceedings of the National Academy of Sciences of the United States of America, 84*, 9054–9058.

Woloszynska, M., & Trojanowski, D. (2009). Counting mtDNA molecules in *Phaseolus vulgaris*: sublimons are constantly produced by recombination via short repeats and undergo rigorous selection during substoichiometric shifting. *Plant Molecular Biology, 70*, 511–521.

Wyman, C., & Kanaar, R. (2006). DNA double-strand break repair: all's well that ends well. *Annual Review of Genetics, 40*, 363–383.

Xu, Y. Z., Arrieta-Montiel, M. P., Virdi, K. S., de Paula, W. B., Widhalm, J. R., Basset, G. J., et al. (2011). MutS HOMOLOG1 is a nucleoid protein that alters mitochondrial and plastid properties and plant response to high light. *The Plant Cell, 23*, 3428–3441.

Zaegel, V., Guermann, B., Le Ret, M., Andres, C., Meyer, D., Erhardt, M., et al. (2006). The plant-specific ssDNA binding protein OSB1 is involved in the stoichiometric transmission of mitochondrial DNA in *Arabidopsis*. *The Plant Cell, 18*, 3548–3563.

CHAPTER TEN

Mitochondrial Genome Evolution and the Emergence of PPR Proteins

Bernard Gutmann, Anthony Gobert and Philippe Giegé[1]
Institut de Biologie Moléculaire des Plantes du CNRS, University of Strasbourg, Strasbourg, France
[1]Corresponding author. E-mail: philippe.giege@ibmp-cnrs.unistra.fr

Contents

Advances in Botanical Research, Volume 63
ISSN 0065-2296,
http://dx.doi.org/10.1016/B978-0-12-394279-1.00010-7

Abstract

The structure of mitochondrial genomes has greatly diverged throughout evolution. They can be very compact in metazoans, whereas they are much larger with considerably lower gene density in higher plants. These changes in structure have occurred in conjunction with the evolution of specific gene expression processes. For example, splicing and RNA editing are typical in plants but never found in animal mitochondria. Most of the organelle-specific gene expression processes rely on the function of nuclear-encoded proteins that do not originate from the organelle bacterial ancestor but have rather evolved during eukaryote history. It has become increasingly evident that pentatricopeptide repeat (PPR) proteins play a major role in mitochondrial gene expression processes. Similar to the mitochondrial genome structure, the number and diversity of PPR proteins have greatly varied during eukaryote evolution. There are, e.g. 80 times more PPR genes in higher plants than in animal genomes. This diversification correlates with the evolution of specific gene expression processes. For instance, the occurrence of PPR genes containing an additional DYW domain, which was proposed to support RNA editing activity, correlates precisely with the occurrence of RNA editing in the respective eukaryote phyla.

1. INTRODUCTION

Mitochondria, together with chloroplasts in plants, are the power stations of eukaryotic cells. They are often described as semi-autonomous organelles because they have retained a genome after the symbiosis of their ancestor with its host and throughout evolution. The finding of single genomes in mitochondria has led to renewed interest in the endosymbiont theory (Gray & Doolittle, 1982; Gray, Burger, & Lang, 1999; Margulis, 1975; Schimper, 1883). The availability of numerous complete mitochondrial genome sequences has contributed to confirm the eubacterial origin of mitochondria. The genome sequences of the eubacterium presenting the highest similarity to mitochondria, *Rickettsia prowazekii* (Andersson *et al.*, 1998), and the most eubacteria-like mitochondrial genome sequence of *Reclinomonas americana* (Lang *et al.*, 1997a) have been determined. These two sequences illustrate the smallest evolutive difference known up to now between a primitive mitochondrial genome and its prokaryotic relative most closely resembling the common ancestor. The general evolutive tendency has been a rapid transfer (or loss) of the information coded in the endosymbiont genome to the nucleus of the host cell. The mitochondrial genome of *Reclinomonas* only contains 11.6% of the genes present on the *Rickettsia* genome. Among the 97 genes in *Reclinomonas* mitochondria, 18 are unique among the mitochondrial sequences investigated to date. Four of these genes code for subunits of a eubacterial-type RNA polymerase.

As a power station, one of the most significant roles of mitochondria is to perform oxidative phosphorylation. This pathway performed by respiratory complexes I, II, III and IV produces a proton gradient between the mitochondrial matrix and the intermembrane space. Protons reenter the matrix dynamically through the ATP synthase complex, also called complex V, thus generating ATP (Millar, Whelan, Soole, & Day, 2011). Nevertheless, alternative pathways also exist in plant mitochondria, with, i.e. the occurrence of alternative dehydrogenases and alternative oxidases (Moore & Siedow, 1992; Rasmusson, Soole, & Elthon, 2004). The mitochondrial genome of plants encodes key components required for mitochondrial energy production. It encodes subunits of the respiratory complexes, proteins involved in cytochrome *c* maturation (ccm), an essential pathway required to obtain functional cytochrome *c* and cytochrome c_1 (Giegé, Grienenberger, & Bonnard, 2008; Hamel, Corvest, Giegé, & Bonnard, 2009) and a subunit of respiratory complex III. In addition, plant mitochondria encode proteins and RNAs required for the translation of mitochondrial genes, i.e. ribosomal proteins, rRNAs and tRNAs (Unseld, Marienfeld, Brandt, & Brennicke, 1997). Mitochondrial gene content is conserved overall between eukaryotes. However, plant mitochondria encode ccm proteins, ribosomal proteins and 5S RNA, which are not found in most other eukaryotes, in particular not in animal mitochondrial genomes (Anderson *et al.*, 1981; Gray, 1982). However, gene content is not correlated with the wide variation of genome sizes. The varying factor is rather gene density; *Arabidopsis* mitochondria encodes only 4% of the proteins present in a eubacterium. The 16.5-kb human mitochondrial genome is 22 times smaller than the *Arabidopsis* mitochondrial genome but encodes only a quarter of the genes present in *Arabidopsis*. The *Marchantia polymorpha* mitochondrial genome, although half as large as the *Arabidopsis* genome, contains approximately the same amount of genes. The coding regions represent about 10% of the *Arabidopsis* genome versus 19% in *Marchantia* (Oda *et al.*, 1992b). These comparisons show that gene density varies in proportion to the differences in genome size.

Analysis of the small numbers of remaining mitochondrial genes indicates that the only direct or indirect purpose of the mitochondrial genome is to express respiratory proteins. However, the mitochondrial genome alone is far from being able to express all the proteins necessary to build up the respective complexes. Of the estimated 75 proteins composing the functional respiratory complexes, only about 25% are encoded in the mitochondrial genome (e.g. 18 proteins for *Arabidopsis*). The remaining proteins are encoded in the nucleus, expressed in the cytosol, imported into mitochondria (Glaser, Sjoling, Tanudji, & Whelan, 1998; Hartl, Pfanner, Nicholson, & Neupert, 1989) and assembled in their respective respiratory complexes. Similarly, the set of tRNAs encoded in plant mitochondria genomes is incomplete and tRNAs have to be imported from the cytosol to ensure translation (Salinas, Duchêne, & Maréchal-Drouard, 2008; Sieber, Duchêne, & Maréchal-Drouard, 2011). Thus, mitochondrial biogenesis completely depends on nuclear-encoded factors, and a number of precise communication and regulatory mechanisms seem to be necessary for mitochondrial biogenesis.

In particular, the expression of the few mitochondrial genes is completely dependent on nuclear-encoded factors. For their expression, mitochondrial genes must be transcribed, their RNA undergoes a number of post-transcriptional maturations and they are translated. In plants, post-transcriptional processes include the maturation of transcript ends, splicing of group II introns and RNA editing (Binder & Brennicke, 2003; Giegé & Brennicke, 2001). All these gene expression processes are specific for higher plant mitochondria and rely entirely on proteins imported from the cytosol. Transcripts are then translated by a mitochondria-specific translation machinery (Giegé & Brennicke, 2001; Kitakawa & Isono, 1991) and can be degraded by a specific degradation process involving polyadenylation (Gagliardi & Leaver, 1999; Holec *et al.*, 2006).

The nature of most proteins involved in these post-transcriptional processes has remained elusive for a long time. The breakthrough came with the discovery of the pentatricopeptide repeat (PPR) protein family (Lurin *et al.*, 2004; Schmitz-Linneweber & Small, 2008). In 2000, as a consequence of the sequencing of *Arabidopsis* nuclear genome, Aubourg and Lecharny found that a novel gene family, unaccounted for until this time, represented up to 1% of the genome (Aubourg, Boudet, Kreis, & Lecharny, 2000). The same year, Small and Peeters (2000) recognized that the genes composing this family contained tandem repeats of motifs resembling and evolutively related to tetratricopeptide repeat (TPR) motifs. Hence, they called the genes in this novel gene family PPR (for pentatricopeptide repeat) because the tandem repeated PPR motifs are composed of 35 amino acids. Small and Peeters (2000) also recognized that PPR proteins were putative RNA-binding proteins most of which were predicted to be localized to organelles. Thus, they proposed that PPR proteins would probably be involved in post-transcriptional processes in plant organelles. A growing number of studies now relate the function of all mitochondria-specific gene expression processes with PPR proteins, as discussed later.

In this chapter, the specific gene expression processes existing in plant mitochondria are reviewed. The evolution of these mitochondrial gene expression processes is analysed and correlations between their occurrence and mitochondrial genome structure are described. The evolutionary history and diversity of PPR proteins are also reviewed and the correlations between the incidence of PPR proteins and the occurrence of specific gene expression processes are described in plants and in different model eukaryotes.

2. SPECIFIC GENE EXPRESSION PROCESSES IN PLANT MITOCHONDRIA

2.1. Transcription

Plant mitochondrial transcription is often polycistronic. Numerous rearrangements that occurred during the evolution of plant mitochondrial genomes have led to the loss of gene organization into functional units (Giegé, Hoffmann, Binder, & Brennicke, 2000; Schuster, 1993). As a consequence, cotranscription does not often involve genes of related function. Contrary to vertebrates in which transcription is initiated at a single point on each DNA strand, plant mitochondrial transcription is initiated at multiple points. Moreover transcription of single genes can be initiated by multiple promoters (Kuhn, Weihe, & Borner, 2005; Lupold, Caoile, & Stern, 1999). Promoter sequences of the *Arabidopsis* mitochondrial genome often contain the consensus motif YRTA, although transcription can also be initiated at non-canonical sites that lack any kind of recognizable consensus motif (Binder & Brennicke, 1993; Fey & Maréchal-Drouard, 1999; Kuhn *et al.*, 2005; Remacle & Maréchal-Drouard, 1996).

Unlike plastids that encode a bacterial-type RNA polymerase (PEP) (Sato, Nakamura, Kaneko, Asamizu, & Tabata, 1999), plant mitochondria rely entirely on nuclear-encoded RNA polymerases for transcription. These enzymes termed NEP (nuclear-encoded polymerase) are expressed in the cytosol and imported into mitochondria. Three genes encode such polymerases in *Arabidopsis*. Among them, RpoTm is targeted to mitochondria, whereas RpoTmp is dual localized in mitochondria and chloroplasts (Chang & Stern, 1999; Hedtke, Borner, & Weihe, 2000). The precise function of RpoTmp remains unclear although recent studies have shown that it could be involved in the transcription of specific genes in mitochondria (Kuhn *et al.*, 2009). Contrary to PEP, NEP enzymes consist of a single polypeptide resembling T3/T7 RNA polymerases (Chang & Stern, 1999) and could require additional protein partners to recognize specific promoter sequences as observed in human and yeast mitochondria (Tracy & Stern, 1995).

Transcription termination in mitochondria is not yet understood. In bacteria and human mitochondria, inverted repeat sequences forming stem loop structures in the 3′ end of transcripts are responsible for transcription arrest. Similar structures are sometimes present in plant mitochondrial transcripts but were found to be involved in RNA maturation (Dombrowski, Brennicke, & Binder, 1997; Hoffmann, Dombrowski, Guha, & Binder, 1999). In Metazoa, proteins of the mTERF family were shown to be localized in mitochondria and involved in transcription termination. These proteins are characterized by a modular architecture based on repetitions of a 30 amino acid module, the mTERF motif, containing leucine zipper-like heptads, which are also encoded in plant genomes and could thus also be involved in transcription termination in plant mitochondria (Roberti *et al.*, 2009). Recently, the *Arabidopsis* mTERF protein RUGOSA2 was shown to be required for mitochondrial and chloroplast biogenesis, although its precise function remains elusive (Quesada *et al.*, 2011).

Studies on plant mitochondrial transcription suggest that transcription is a relaxed and poorly controlled mechanism. The size of precursor transcripts appears to be determined by promoter selection and by the processivity of RNA polymerases (Holec *et al.*, 2006). After transcription, RNA molecules undergo numerous maturation steps. It appears that steady state levels of RNA in plant mitochondria are mainly controlled during these post-transcriptional events rather than during transcription itself (Giegé *et al.*, 2000; Holec *et al.*, 2006).

2.2. RNA Processing

2.2.1. 5′ and 3′ maturation of transcripts

The maturation of plant mitochondrial precursor transcripts involves 5′ and 3′ maturation steps. These maturations could be achieved through direct endoribonuclease activities and/or with 5′ to 3′ exoribonucleases and 3′ to 5′ exoribonucleases. Plant mitochondria do not encode such enzymes. They are thus encoded in the nucleus and must be imported from the cytosol. In higher plant mitochondria, no 5′ to 3′ exoribonuclease has been identified yet. On the contrary, two 3′ to 5′ exoribonucleases were characterized: RNase II is dual localized in mitochondria and plastids and a polynucleotide phosphorylase (PNPase) is found in mitochondria. Studies suggest that the 3′ processing of mitochondrial transcripts is at least a two-step phenomenon, where PNPase first removes large 3′ extensions and RNase II subsequently degrades short nucleotidic extensions to generate the mature 3′ ends (Gagliardi, Perrin, Maréchal-Drouard, Grienenberger, & Leaver, 2001; Perrin *et al.*, 2004b; Perrin, Lange, Grienenberger, & Gagliardi, 2004a). Predicted RNA folds are sometimes present at transcript ends (Forner, Weber, Thuss, Wildum, & Binder, 2007) and can be involved in RNA processing (Dombrowski *et al.*, 1997). However, it is also likely, as initially described in chloroplasts (Pfalz, Bayraktar, Prikryl, & Barkan, 2009; Prikryl, Rojas, Schuster, & Barkan, 2011), that PPR proteins also bind specific sequences in the 5′ and 3′ ends of mitochondrial transcripts to define mRNA termini by serving as barriers for either 5′ or 3′ decay. Plant mitochondrial PPR proteins were found to be required for transcript end maturation *in vivo*, e.g. RPF2 for the 5′ end maturation of *nad9* and *cox3* mRNAs (Jonietz, Forner, Holzle, Thuss, & Binder, 2010) and RPF3 for the maturation of ccmC (Jonietz, Forner, Hildebrandt, & Binder, 2011) in *Arabidopsis* mitochondria.

Similar to other RNAs, tRNAs are transcribed as precursor molecules and have to be matured in 5′ and 3′. These maturation steps are performed by two ubiquitous endoribonuclease activities called RNase P and RNase Z. *Arabidopsis* encodes four RNase Z proteins, two of which are found in mitochondria (Canino *et al.*, 2009). Until recently, all the characterized RNase Ps were found to be ribonucleoproteins composed of an RNA and a variable number of associated proteins, the catalytic activity being held by the RNA component. It was thus believed that RNase P was a universally conserved vestige of the prebiotic RNA world (Bour *et al.*, 2009; Guerrier-Takada, Gardiner, Marsh, Pace, & Altman, 1983). Such ribonucleoprotein RNase P was never identified in plants. A novel type of RNase P composed of a single protein was identified in *Arabidopsis* mitochondria and chloroplasts (Gobert *et al.*, 2010). This protein, termed PRORP1 for proteinaceous RNase P, is a member of the PPR protein family (Aubourg *et al.*, 2000; Hilson *et al.*, 2004). The identification of a PPR protein as a mitochondrial RNase P is consistent with the overall involvement of PPR proteins in plant organelle post-transcriptional processes in plants (Schmitz-Linneweber & Small, 2008). Beyond the maturation of tRNAs, RNase P and RNase Z might also be involved in the maturation of other RNAs in plant mitochondria. A large number of tRNA-like structures, called t-elements, are found in precursor transcripts. Endonucleolytic cleavages that could correspond to RNase P and Z cuts were mapped *in vivo* at the level of the t-elements (Forner *et al.*, 2007) and both recombinant RNase Z and the novel RNase P PRORP1 were able to cleave t-elements *in vitro* (Canino *et al.*, 2009; Gobert *et al.*, 2010). This maturation mechanism based on tRNA-like structures is corroborated by the observation that a CCA extension added by a nucleotidyl transferase is sometimes observed in the 3′ end of mature mitochondrial transcripts (Forner *et al.*, 2007; Kunzmann, Brennicke, & Marchfelder, 1998; Williams, Johzuka, & Mulligan, 2000). Other proteins of the PPR family are proposed to be mitochondrial endoribonucleases. In particular, some of them belonging to the DYW subfamily of PPR proteins were found to have endoribonuclease activity *in vitro* (Nakamura & Sugita, 2008; Okuda *et al.*, 2009).

2.2.2. Splicing of introns

As observed in other eukaryote genomes, plant mitochondrial genomes possess numerous introns that have to be spliced after transcription (see also Chapter 7). Two classes of introns, group I and group II, are typical of organelles, and are distinguished by their characteristic RNA secondary structures first described in fungal mitochondria by Michel, Jacquier, and Dujon (1982). In higher plant mitochondria, group II introns are generally found in several genes (Bonen & Vogel, 2001; Unseld *et al.*, 1997), whereas only one example of a recently acquired group I intron has been found in the *cox1* genes of *Peperomia* and some other higher plants (Cho, Qiu, & Palmer, 1998a; Vaughn, Mason, Sper-Whitis, Kuhlman, & Palmer, 1995). The genes encoded in the *Arabidopsis* mitochondrial genome are interrupted by 23 group II introns varying in size from 485 to about 4000 nucleotides (Unseld *et al.*, 1997). Some genes are interrupted by more than one intron; e.g. *nad7* has four introns. *Trans*-splicing is found in plant mitochondria in several instances where the physically separated exons are flanked by partial group II intron sequences (Chapdelaine & Bonen, 1991; Glanz & Kuck, 2009; Wissinger, Schuster, & Brennicke, 1991). *Trans*-splicing designates splicing events where consecutive exons are not separated linearly by a continuous intron, but are present at distant positions on the genome, some times on the other strand. These *trans*-splicing events occur via the group II intron fragments, which connect the separated exons in two- or three-molecule interactions. In *Arabidopsis* mitochondria, five *trans*-splicing events are detected in *nad1*, *nad2* and in *nad5*; in the latter an exon only 22 nucleotides long is integrated by two *trans*-splicing events (Knoop, Schuster, Wissinger, & Brennicke, 1991). Similar to nuclear introns, group I and II intron splicing occurs by a two-step transesterification reaction. In group I introns, free GTP is needed as a cofactor with its 3′ OH group attacking the 5′ end of the intron. In group II introns, the 5′ end of the intron is attacked by an OH group of a conserved internal A residue (Schmelzer & Schweyen, 1986; van der Veen *et al.*, 1986). The highly conserved structure of the group II introns has been shown to be essential for splicing activity (Fedorova & Zingler, 2007). Contrary to yeast and bacterial group II introns that have autocatalytic activity *in vitro,* the autocatalytic activity of plant organelle introns could never be shown (Michel & Ferat, 1995; Vogel & Borner, 2002). This suggests that protein factors, probably required for proper folding of the intron into its catalytic active form, are required (Lambowitz & Zimmerly, 2004). Some of these factors, called maturases, are encoded by open reading frames (ORFs) present in bacterial group II introns (Meng, Wang, & Liu, 2005; Wank, Sanfilippo, Singh, Matsuura, & Lambowitz, 1999). In plant mitochondria, a similar maturase gene, MatK, is encoded in a *nad1* intron. Moreover, in *Arabidopsis*, four nuclear genes called AtnMat-1a, AtnMat-1b, AtnMat-2a and AtnMat-2b code for maturase-like proteins and have organellar targeting signals (Mohr & Lambowitz, 2003). Genetic studies confirmed that AtnMat-1a and AtnMat-1b were indeed involved in the splicing of the mitochondrial *nad4* and *cox2*, *nad1*, *nad7*, respectively (Keren *et al.*, 2008; Nakagawa & Sakurai, 2006). Other genetic approaches have identified the involvement of CRM and WTF proteins in splicing in chloroplasts. Some proteins of these families appear to have mitochondrial targeting signals, thus suggesting that they could also be involved in splicing in mitochondria (Barkan *et al.*, 2007; Kroeger, Watkins, Friso, Van Wijk, & Barkan, 2009). PPR proteins are also involved in splicing as demonstrated by the involvement of OTP43 for the *trans*-splicing of the mitochondrial *nad1* intron 1 (de Longevialle *et al.*, 2007). A PORR domain protein was recently found to be required for the splicing of introns from the mitochondrial rpl2 and ccmFc transcripts (Colas des Francs-Small *et al.*, 2011).

2.3. RNA Editing

2.3.3. Effects of RNA editing

In plant mitochondria, the term RNA editing designates the specific modification of cytidines into uridines (Covello & Gray, 1989; Gualberto, Lamattina, Bonnard, Weil, & Grienenberger, 1989; Hiesel, Wissinger, Schuster, & Brennicke, 1989). The reverse conversion of uridines into cytidines was also observed, however mostly in mosses and ferns (Knoop, 2004; Malek, Lattig, Hiesel, Brennicke, & Knoop, 1996; Shikanai, 2006; Takenaka, Verbitskiy, Van Der Merwe, Zehrmann, & Brennicke, 2008). In plant mitochondria, RNA editing affects over 500 cytidines (Chateigner-Boutin & Small, 2007; Giegé & Brennicke, 1999; Zehrmann, Van Der Merwe, Verbitskiy, Brennicke, & Takenaka, 2008). Most RNA editing sites are found in mRNAs and were identified by comparison of cDNA and genomic DNA sequences. Some sites are found in tRNA, non-coding RNAs and introns but no site was found in ribosomal RNA. The number of editing sites per gene is highly variable. In *Arabidopsis* mitochondria, complex I and CCM (cytochrome *c* maturation) mRNAs have the highest RNA editing frequency (Giegé & Brennicke, 1999). RNA editing results most of the time in the conversion of the amino acid identity, although in 10% of cases, RNA editing is called silent when the third position of a codon is affected and the amino acid identity is unchanged (Giegé & Brennicke, 1999). In most cases, RNA editing results in the restoration of a residue conserved throughout evolution in orthologue proteins. Thus, RNA editing often affects positions that appear to be essential for the respective protein functions (Bock, Kössel, & Maliga, 1994b). RNA editing can be an essential step to define the translation range. It can create initiation (Chapdelaine & Bonen, 1991) or termination codons in mRNA (Giegé, Rayapuram, Meyer, Grienenberger, & Bonnard, 2004). In maize, the editing of a *nad7* intron is required for its proper folding and thus for efficient splicing (Carrillo & Bonen, 1997). Similarly, RNA editing is required for the proper folding of tRNAs, which is a prerequisite for their maturation (Kunzmann *et al.*, 1998; Maréchal-Drouard, Remacle, Ramamonjisoa, & Dietrich, 1996a; Maréchal-Drouard, Kumar, Remacle, & Small, 1996b). At some sites, RNA editing can be partial, thus leading to heterogeneous populations of transcripts (Bentolila, Elliott, & Hanson, 2008; Chateigner-Boutin & Hanson, 2003).

Consequently, partial editing could lead to the synthesis of different forms of individual proteins. However, some studies have shown that only the proteins resulting from fully edited transcripts are allowed to accumulate in mitochondria (Grohmann, Thieck, Herz, Schroder, & Brennicke, 1994; Lu & Hanson, 1994; Phreaner, Williams, & Mulligan, 1996). This suggests that translation of partially edited transcripts could be inhibited and/or that proteins resulting from partially edited RNA are unstable and rapidly degraded. Thus, it has been hypothesized that instead of a regulatory role, the primary function of RNA editing could have been to correct genomic mutations that appeared during the invasion of land by plants and thus to enable the translation of functional proteins (Shikanai, 2006; Takenaka *et al.*, 2008).

2.3.4. **Cis** *elements and* **trans** *factors involved in RNA editing*

Each RNA editing site is specifically recognized by the editing machinery. Analysis of the hundreds of editing sites present in the plant mitochondrial transcriptome (Bentolila *et al.*, 2008; Giegé & Brennicke, 1999) has not allowed specific consensus signals around editing sites to be defined. However, the distribution of nucleotides around the sites is not random because a strong preference for pyrimidines is observed for the two nucleotides immediately upstream of the sites (Giegé & Brennicke, 1999). However, this is not sufficient to define editing sites and it thus suggests that individual editing sites or groups of editing sites would be recognized independently by *trans*-acting factors. In this case, a large number of *trans* factors can be predicted. A number of *in organello* and *in vitro* RNA editing studies using isolated mitochondria as well as mitochondrial extracts have allowed the minimal nucleotide sequences in 5′ and 3′ of some sites that are required for RNA editing to be determined. For some sites investigated, about 30 nucleotides upstream of the sites are sufficient for site recognition (Farre & Araya, 2001; Neuwirt, Takenaka, Van Der Merwe, & Brennicke, 2005; Staudinger & Kempken, 2003; Takenaka, Neuwirt, & Brennicke, 2004; Takenaka *et al.*, 2007; Verbitskiy, Van Der Merwe, Zehrmann, Brennicke, & Takenaka, 2008). For other sites, an additional short stretch of five or six nucleotides downstream of the editing site is also required for site recognition (Choury, Farre, Jordana, & Araya, 2004; Farre & Araya, 2001). These *cis* elements have to be recognized by *trans* factors responsible for site-specific recognition and for the actual catalytic reaction, a deamination of cytidines into uridines (Blanc, Litvak, & Araya, 1995). RNA editing *trans* factors remained elusive for a long period. The breakthrough came with the identification of the PPR gene family (Small & Peeters, 2000) and with the first identification using a genetic approach of an editing factor, a PPR protein, in chloroplasts (Kotera, Tasaka, & Shikanai, 2005). Since then, a growing number of other genetic studies have also related the function of PPR proteins to RNA editing in mitochondria, i.e. for editing sites in *Arabidopsis* (Zehrmann, Verbitskiy, Van Der Merwe, Brennicke, & Takenaka, 2009b, Verbitskiy *et al.* 2010a), in *Oryza* (Kim *et al.*, 2009) and in *Physcomitrella* mitochondria (Tasaki, Hattori, & Sugita, 2010). In one of the studies, the PPR protein PpPPR_71 was able to bind directly the sequence in *ccmFc* mRNA around the editing site affected in the PpPPR_71 mutant (Tasaki *et al.*, 2010). The identification of PPR proteins as editing factors fits very well with the initial hypothesis that PPR proteins could be sequence-specific factors (Small & Peeters, 2000). Thus, it is now widely accepted that PPR proteins are responsible for site recognition. However, the enzyme(s) performing the catalytic reaction of RNA editing has not been identified. The PPR proteins identified as editing factors all belong to the subfamily of E/E+ and DYW proteins that possess additional domains downstream of the PPR repeats (Lurin *et al.*, 2004). It has been proposed that the DYW domain could hold the actual deaminase activity of RNA editing (Salone *et al.*, 2007) although this has not yet been demonstrated.

2.4. Translation

Plant mitochondria require a fully functional translation machinery to express the 30 or so mRNAs encoded in the mitochondrial genome. Since only a few ribosomal proteins, rRNA and an incomplete set of tRNAs are encoded in plant mitochondria (Unseld *et al.*, 1997), mitochondria must import most of the components of their translational apparatus from the cytosol, e.g. all the required aminoacyl tRNA synthetases (Duchêne *et al.*, 2005) and tRNAs (Michaud, Maréchal-Drouard, & Duchêne, 2010; Salinas *et al.*, 2008). Sequence analysis has shown that translation is usually, but not always, initiated with an AUG codon in higher plant mitochondria. In *Arabidopsis*, GGG, AAU and GUG are possible additional translation initiator triplets (Unseld *et al.*, 1997). Similarly, in *Oenothera* GUG and in radish ACG were found to be potential translation initiation sites (Bock, Brennicke, & Schuster, 1994a; Dong, Wilson, & Makaroff, 1998). Moreover, plant mitochondrial genes can be expressed from mRNAs lacking canonical termination codons with no evidence that alternative termination codons had been created post-transcriptionally by either RNA editing or polyadenylation (Raczynska *et al.*, 2006). In plant mitochondria, sequences resembling Shine and Dalgarno (SD) sequences are very rare and in the absence of an *in vitro* translation system, the function of these sequences in translation initiation could not be determined (Pring, Mullen, & Kempken, 1992). Thus, the mechanism controlling translation initiation remains completely elusive in plant mitochondria. Similarly, potential translation regulation systems are also unknown. However the function of PPR proteins is very likely also connected to plant mitochondrial translation as suggested by the involvement of CRP1 as a chloroplast translation regulator (Schmitz-Linneweber, Williams-Carrier, & Barkan, 2005), by the requirement of PPR proteins for translation in yeast mitochondria (Kuhl, Dujeancourt, Gaisne, Herbert, & Bonnefoy, 2011; Tavares-Carreon *et al.*, 2008), by their association with mitochondrial ribosomes in trypanosomes (Pusnik, Small, Read, Fabbro, & Schneider, 2007), by the association of PTCD3 with the small subunit of mitochondrial ribosomes in human (Davies *et al.*, 2009) and by the association of PPR336 and PNM1 to polysomes in plant mitochondria (Hammani *et al.*, 2011b; Uyttewaal *et al.*, 2008b).

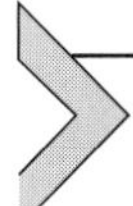

3. EVOLUTION OF MITOCHONDRIAL GENE EXPRESSION PROCESSES AND CORRELATIONS WITH MITOCHONDRIAL GENOMES STRUCTURE

3.1. Evolution of Mitochondrial Transcription

The transcription of mitochondrial genomes has diverged considerably throughout evolution in conjunction with the dramatic structural divergence of mitochondrial genomes in the different eukaryote kingdoms (Tracy & Stern, 1995). This is true in particular for promoter structures. In vertebrates, a single promoter, which can be unidirectional or bidirectional, is present for each strand (Falkenberg, Larsson, & Gustafsson, 2007). For example, in human cells, each strand contains a single promoter region for transcription initiation, the light-strand promoter (LSP) and the heavy-strand promoter (HSP) (Falkenberg *et al.*, 2007). In fungi and plants, multiple promoters are present (Kuhn *et al.*, 2005; Lupold *et al.*, 1999; Shadel & Clayton, 1993). For both phyla, promoter sequences are divergent. Compared with animals, the large quantity of promoters in plants, fungi and trypanosomes correlates with the larger size of mitochondrial genomes and with the decrease in gene density (Tracy & Stern, 1995).

Overall, the transcription of mitochondrial genes appears to be generally catalysed by a single subunit T3/T7 phage-like RNA polymerase, which appears to have been recruited relatively early during the evolution of organelles (Shutt & Gray, 2006). For example, in animals, transcription of the mitochondrial genome is performed by mtRNAP, a protein distantly related to the pol I family of DNA polymerases as well as to all T7-type RNA polymerases, including plant organelle RNA polymerases (Ringel *et al.*, 2011; Tiranti *et al.*, 1997). However, the animal and yeast mitochondrial enzymes are distinguished from their plant counterparts by the occurrence of PPR domains in their N-terminal domains (Kuhl *et al.*, 2011; Lightowlers & Chrzanowska-Lightowlers, 2008). This is paradoxical given the small number of PPR proteins present in animals and yeast compared with the high numbers in plants. In protists, sequences encoding this type of enzymes have also been identified many times (Cermakian *et al.*, 1997; Li, Maga, Cermakian, Cedergren, & Feagin, 2001). They are in particular found in the fully sequenced nuclear genomes of protists (e.g. Fritz-Laylin, Ginger, Walsh, Dawson, & Fulton, 2011). The corresponding RNA polymerase activity has been determined experimentally for some lineages (Grams *et al.*, 2002; Miller, Antes, Qian, & Miller, 2006). However, as pointed out above, the mitochondrial genomes of *R. americana* and of other jakobids still encode a eubacterial-type RNA polymerase (Lang *et al.*, 1997). These enzymes are assumed to be responsible for mitochondrial transcription. However, the presence of an additional phage-type RNA polymerase cannot be ruled out. For instance, a T7-type RNA polymerase is encoded in the mitochondrial genome of the stramenopile *Pylaiella littoralis*, although some of its mitochondrial promoter sequences are typical from bacterial-type RNA polymerases (Oudot-Le Secq, Fontaine, Rousvoal, Kloareg, & Loiseaux-De Goer, 2001). It is thus likely that some protists have retained a dual system involving both functional bacterial and phage-type RNA polymerases for the transcription of their mitochondrial genes. This would be reminiscent of the presence of NEP and PEP RNA polymerases in higher plants chloroplasts (Hedtke, Borner, & Weihe, 1997; Swiatecka-Hagenbruch, Emanuel, Hedtke, Liere, & Borner, 2008).

3.2. Evolution of Splicing in Mitochondria

Similar to transcription, the occurrence of splicing in mitochondrial transcripts has diverged considerably throughout evolution in parallel with the dramatic structural changes in mitochondrial genomes organization in the different eukaryotes (Bonen, 2008; see also Chapter 7). Overall, mitochondrial genes can be interrupted by varying numbers of introns that have been categorized as members of the group I or group II families of mobile genetic elements. These two intron classes are characterized by distinctive structures (Bonen & Vogel, 2001; Lambowitz & Zimmerly, 2004; Michel & Ferat, 1995). Some of these introns encode protein factors that could be involved in their own splicing and/or involved in their dissemination *in vivo*. For instance, it has been shown that group II introns have spread by retrohoming and retrotransposition mechanisms exploiting intron-encoded reverse transcriptase and/or endonuclease activities (Lambowitz & Zimmerly, 2004; Toro, Jimenez-Zurdo, & Garcia-Rodriguez, 2007).

The mitochondrial genomes of higher plants, algae, protists and fungi harbour variable numbers of introns. Land plants distinguish themselves from other eukaryotes because they are particularly rich in group II introns. This increase in intron numbers correlates with the increased mitochondrial genome sizes (Bonen, 2008). In contrast, the small and compact mitochondrial genomes of animals are typically free of introns (i.e. the human genome (Anderson *et al.*, 1981)). However, in a few cases, introns have been identified in animal mitochondrial genes, although these were all restricted to early diverging metazoan lineages such as corals, sponges and sea anemones (Fukami, Chen, Chiou, & Knowlton, 2007). These animal mitochondrial introns apparently all belong to the group I class and could have arisen from a fungal source through recent horizontal genetic transfer. For other eukaryotes, a novel category of introns, group III, has been proposed in *Euglena* genes, although these introns now appear to be group II introns of highly divergent structure (Bonen & Vogel, 2001; Lambowitz & Zimmerly, 2004). Beyond these introns, *Euglena* are also characterized by the presence of so-called twintrons in chloroplasts that correspond to introns localized within other introns. This category of introns was also described in Archaea (Lambowitz & Zimmerly, 2004).

In the precise case of land plants, introns were systematically found in all the mitochondrial genomes sequenced to date, which included both tracheophytes and non-vascular plants (Bonen, 2008). In flowering plants, 23 mitochondrial group II introns have been identified (Unseld *et al.*, 1997). In non-vascular plants, such as the liverwort *M. polymorpha* (Oda, Kohchi, & Ohyama, 1992a) and the moss *Physcomitrella patens* (Terasawa *et al.*, 2007), mitochondrial genes contain 25 and 23 group II introns, respectively as well as seven and two group I introns. Although these intron numbers are similar between vascular and non-vascular plants, little conservation is observed among intron identities. Only three introns are present in both *Physcomitrella* and *Marchantia*, eight in both *Physcomitrella* and *Arabidopsis* and a single one between *Marchantia* and *Arabidopsis* (Bonen, 2008). Beyond group II introns, a group I intron is frequently found in *cox1* genes. These introns are believed to have arisen through multiple events of horizontal gene transfer during evolution (Cho *et al.*, 1998b).

Analysis of the growing amount of land plant mitochondrial genomic data (O'Brien *et al.*, 2003) indicates a complex evolutive history of group II introns, i.e. that introns have apparently been gained and/or lost periodically over time (Bonen, 2008; Dombrovska & Qiu, 2004; Groth-Malonek, Pruchner, Grewe, & Knoop, 2005; Pruchner, Beckert, Muhle, & Knoop, 2002). Intron variation has even been observed among the closely related angiosperms (Gass, Makaroff, & Palmer, 1992; Kudla, Albertazzi, Blazevic, Hermann, & Bock, 2002). The analysis of cDNA sequences has revealed that introns can be lost through a post-transcriptional mechanism because a spliced and edited RNA can be reverse transcribed and reintegrated in the mitochondrial genome by homologous recombination (Geiss, Abbas, & Makaroff, 1994). Similarly, in yeast, the direct involvement of reverse transcriptase has been documented for the deletion of introns from mitochondrial genes (Gargouri, 2005).

In a broader context, the study of mitochondrial introns and mitochondrial splicing mechanisms in general has been of particular interest in terms of evolution, first to understand the origin of introns and second, in the context of the evolutionary study of function of catalytic RNAs, which is helpful to understand the transition from the prebiotic RNA world to the present world. For instance, the two transesterification steps of the splicing mechanism of group II introns are reminiscent of the splicing mechanism of spliceosome introns. It has thus been proposed that the two categories of introns could have evolved from a common ancestor (Jacquier & Michel, 1990; Valadkhan, 2005). It has also been proposed that spliceosomal introns could have arisen after the endosymbiosis of an α-proteobacteria with its host ancestor, i.e. that spliceosome introns could originate from the transfer of group II introns from the symbiont to the eukaryote ancestor host cell (Cavalier-Smith, 1991; Martin & Koonin, 2006).

3.3. Evolution of RNA Editing

RNA editing refers to an ensemble of post-transcriptional modifications that result in the establishment of standard nucleotides in RNA, as opposed to

RNA modifications that typically introduce non-standard nucleotides in RNAs, in particular in tRNAs (Motorin & Helm, 2010). RNA editing was first defined by Benne and colleagues (Benne, 1989; Benne, Van Den Burg, Brakenhoff, Sloof, Van Boom, & Tromp, 1986) when they found that uridines could be inserted or deleted from trypanosome mitochondrial mRNAs in a directional manner (Benne, 1996). In Viridiplantae, RNA editing defines cytidine to uridine (C to U) conversions as well as U to C conversions, called reverse editing, taking place in mitochondria and chloroplasts. C to U conversions are not entirely specific to plants; they are also occasionally found in animals, i.e. in mammals for the editing of a nuclear mRNA (Anant, Macginnitie, & Davidson, 1995) and in marsupials for the editing of a mitochondrial tRNA (Janke & Paabo, 1993). Plant-type mitochondrial RNA editing was also recently described in the protist *Naegleria gruberi* (Rudinger, Fritz-Laylin, Polsakiewicz, & Knoop, 2011a). However, as described in this chapter, the machinery that edits C to U in plants is entirely plant specific. In contrast to the trypanosome RNA editing process, no directionality of editing could be observed in plants. This suggests that RNA editing site recognition and conversion is performed site by site independently. However, just as in kinetoplastids, RNA editing largely seems to serve the purpose of repairing genes because RNA editing mainly restores at the RNA levels codons conserved throughout evolution (Brennicke, Marchfelder, & Binder, 1999).

When plant organelle RNA editing was initially discovered in wheat and *Oenothera* (Covello & Gray, 1989; Gualberto *et al.*, 1989; Hiesel *et al.*, 1989), the wide phylogenetic distribution of RNA editing in plants became immediately evident, i.e. that plant RNA editing would affect at least the majority of angiosperms (flowering plants) because of the deep phylogenetic division between the dicotyledon plant *Oenothera* and the monocotyledon plant wheat. We now know that RNA editing does not only affect angiosperms. It is also found in species of all other land plant clades: in embryophytes, in particular in gymnosperms, ferns and fern allies, mosses, hornworts, and in some liverworts (Sper-Whitis, Russell, & Vaughn, 1994, Hiesel, Von Haeseler, & Brennicke, 1994, Malek et al., 1996, Steinhauser, Beckert, Capesius, Malek, & Knoop, 1999). However, RNA editing has up to now not been identified in green algae including groups closely related in terms of evolution to land plants. RNA editing is also not found in some liverworts such as the *Marchantiopsida* (Oda *et al.*, 1992b). This seemingly unique absence of RNA editing in the liverwort subclade of *Marchantia* has been intriguing for a long time because it suggested that editing might have arisen either through two independent gains or that editing appeared a single time during the emergence of land plants and was subsequently lost in some subclades. The latter proposition was finally retained with the description of abundant RNA editing in *Haplomitrium*, because the clade of this liverwort was found to be basal to both marchantiids where editing is absent and to jungermanniids where RNA editing is present (Groth-Malonek, Wahrmund, Polsakiewicz, & Knoop, 2007). Using the same line of evidence, the evolutive history of U to C reverse editing can now be proposed with increased confidence. Reverse editing is not present in mosses and liverworts and it is present at low frequency in spermatophytes (Gualberto, Weil, & Grienenberger, 1990; Schuster, Hiesel, Wissinger, & Brennicke, 1990), but is, however, very frequent in hornworts, ferns and fern allies (Grewe, Viehoever, Weisshaar, & Knoop, 2009; Steinhauser *et al.*, 1999; Yoshinaga, Iinuma, Mazuzawa, & Uedal, 1996). This suggests that reverse editing might have been acquired in the common ancestor of hornworts and tracheophytes and subsequently progressively lost or decreased in the ancestor of spermatophytes (Groth-Malonek *et al.*, 2005; Qiu *et al.*, 2006).

The occurrence of RNA editing appears to be related to genome contents. Algae that do not perform editing have organelle genomes that are generally AT rich (Smith, 2009). On the other hand, many land plant genomes have a significantly higher GC content. Several studies now indicate that there is a high positive correlation between RNA editing and the GC content of genomes (Malek *et al.*, 1996; Smith, 2009; Yura, Sulaiman, Hatta, Shionyu, & Go, 2009). It has been proposed that during the invasion of land by plants, an evolutive pressure might have existed to reduce the occurrence of adjacent pyrimidines. Ultraviolet (UV) light can damage DNA through the formation of cyclobutane pyrimidine dimers or other deleterious photoproducts (Fujii & Small, 2011). In that sense, RNA editing that restores uridines at the RNA level might have evolved as a mechanism to protect DNA from UV damage. If this is true, RNA editing might have had an important role in the colonization of land by plants (Fujii & Small, 2011).

4. DIVERSITY AND EVOLUTION OF PPR PROTEINS

4.1. Subfamilies of PPR Proteins

In 2000, analysis of the genome sequence of the flowering plant, *Arabidopsis thaliana*, characterized a new class of proteins called PPR (for pentatricopeptide repeat proteins) (Aubourg *et al.*, 2000; Small & Peeters, 2000). A growing number of studies now link the functions of these proteins with the RNA-related processes described above (Table 10.1). These proteins comprise 2–26 motifs repeated in tandem with an average of 12 motifs (Small & Peeters, 2000). The classic PPR motifs are composed of 35 amino acids and have a highly degenerated primary sequence. These motifs resemble and are evolutionarily related to the TPR motifs (Small & Peeters, 2000) involved in protein–protein interaction (Blatch & Lassle, 1999). Each PPR motif is proposed to be composed of two antiparallel alpha helices (Small & Peeters, 2000). The succession of PPR motifs would form a superhelix (Fig. 10.1). Conserved tyrosines in PPR motifs could also be involved in interhelix packing. According to these predictions and unlike TPR motifs, the PPR superhelix could form a positively charged tunnel or inner groove that would be suitable for RNA interaction. Most PPR proteins do not contain any predictable catalytic site. Many PPR proteins are thus predicted to recruit other proteins to carry out catalytic processes (Schmitz-Linneweber & Small, 2008).

The canonical PPR motif, called PPR P, has two variants: PPR-like S (short) and PPR-like L (long) composed of 31 and 35–36 amino acids, respectively (Lurin *et al.*, 2004). PPR proteins are divided into two subfamilies: the P subfamily (P for pure PPR) having only P motifs and the PLS subfamily with a pattern of repetitions of the triple motif P-L-S (Lurin *et al.*, 2004) (Fig. 10.1). Several PPR proteins in the PLS subfamily, previously called the plant combinatorial and modular protein family (Aubourg *et al.*, 2000) can carry other motifs in their C-terminal part (Aubourg *et al.*, 2000). These additional motifs are called E (with an average of 91 amino acids), E+ (33 amino acids) and DYW (106 amino acids). The subfamily of PLS PPR proteins is thus divided into three subgroups: PLS (without additional motifs), E/E+ (with E motifs and for some proteins an additional E+ motif) and DYW (with E, E+ and DYW motifs).

PPR proteins also typically have a predicted organelle-targeting sequence in their N-terminal extremity. Almost three-quarters of the PPR proteins are predicted to be targeted to mitochondria and/or plastids in *Arabidopsis* (Lurin *et al.*, 2004). Moreover, the initial number of PPR proteins predicted to be targeted to organelles is certainly underestimated because of misconstructed gene models, e.g. as already observed for MEF21 and MEF22 (Takenaka, Verbitskiy, Zehrmann, & Brennicke, 2010). However, localization of a high number of PPR proteins is not obviously predictable. Therefore, some PPR proteins are likely to be localized in other

Table 10.1 List of PPR proteins of identified functions and their targets

Species	Accession	Name	Class	Function	Target	Loc	Reference
Neurospora crassa	AF002169	CY A-5	P	Translation	*cox1*	Mt	(Coffin, *et al.*, 1997)
Saccharomyces cerevisiae	YML091C	Rpm2p	P	RNA cleavage	tRNA	Mt	(Morales *et al.*, 1992)
				Unknown		Cy	(Stribinskis & Ramos, 2007)
				Transcription		Nu	(Stribinskis, *et al.*, 2005)
	YGL107C	Rmd9p	P	RNA turnover translation		Mt	(Nouet *et al.*, 2007; Williams, *et al.*, 2007)
	YJL209W	Cbp1p	P	RNA processing	*COB*	Mt	(Dieckmann, *et al.*, 1982; Dieckmann, *et al.*, 1984)
				Translation			(Islas-Osuna, *et al.*, 2002)
	YDR350C	Atp22p	P	Translation	*ATP6*	Mt	(Zeng, *et al.*, 2007)
	YMR257C	Pet111p	P	Translation	*COX2*	Mt	(Mulero & Fox, 1993; Poutre & Fox, 1987)
	YMR282C	Aep2p	P	RNA turnover translation	*ATP9*	Mt	(Ellis, *et al.*, 1999; Finnegan, et al., 1991)
	YMR064W	Aep1p	P	Translation	*ATP9*	Mt	(Payne, *et al.*, 1991; Payne, *et al.*, 1993)
	YPL005W	Aep3p	P	RNA turnover	*ATP8/6*	Mt	(Ellis, *et al.*, 2004)
	YGR150C	Dmr1p	P	Translation		Mt	(Puchta *et al.*, 2010)
	YLR067C	Pet309p	P	RNA turnover translation	*COX1*	Mt	(Manthey & McEwen, 1995; Tavares-Carreon *et al.*, 2008)
	YFL036W	Rpo41p	Other	Transcription		Mt	(Greenleaf, *et al.*, 1986)
Schizosaccharomyces pombe	SPBC1604.02c	Ppr1	P	RNA turnover	*cox2* and *cox3*	Mt	(Kuhl *et al.*, 2011)
	SPBC18H10.11c	Ppr2	P	Translation		Mt	(Kuhl *et al.*, 2011)
	SPBC19G7.07c	Ppr3	P	RNA turnover	*cytb*8.15S	Mt	(Kuhl *et al.*, 2011)

	SPAC8C9.06c	Ppr4	P	Translation	*cox1*	Mt	(Kuhl *et al.*, 2011)
	SPAC1093.01	Ppr5	P	Translation		Mt	(Kuhl *et al.*, 2011)
	SPACC11E1 0.04	Ppr6	P	RNA processing	*atp9*	Mt	(Kuhl *et al.*, 2011)
	SPBC16A3.03c	Ppr7	P	RNA processing	*atp6*	Mt	(Kuhl *et al.*, 2011)
	SPBCI289.06c	Ppr8	P	Translation		Mt	(Kuhl *et al.*, 2011)
	SPAC26H5.12	Rpo41	Other	Transcription		Mt	(Jiang *et al.*, 2011; Kuhl *et al.*, 2011)
Leishmania tarentolae	LmjF15.0410		P	Translation		Mt	(Maslov *et al.*, 2007)
	LmjF24.0830		P	Translation		Mt	(Maslov *et al.*, 2007)
	LmjF29.0430		P	Translation		Mt	(Maslov *et al.*, 2007)
Trypanosoma brucei	Tb10.26.0050		P	Translation		Mt	(Zikova *et al.*, 2008)
	Tb10.389.0260	TbPPR4	P	RNA turnover	9S rRNA and 12S rRNA	Mt	(Pusnik *et al.*, 2007)
	Tb10.70.7960	TbPPR5	P	RNA turnover	9S rRNA and 12S rRNA	Mt	(Pusnik *et al.*, 2007)
				Translation			(Pusnik *et al.*, 2007; Zikova *et al.*, 2008)
	Tb11.01.5980	KPAF2	P	RNA turnover		Mt	(Aphasizheva *et al.*)
	Tb11.01.7930	TbPPR9	P	RNA turnover	COX1 and COX2	Mt	(Pusnik & Schneider, 2012)
	Tb11.02.3180		P	Translation		Mt	(Zikova *et al.*, 2008)
	Tb927.1.1160	TbPPR3	P	RNA turnover	9S rRNA and 12S rRNA	Mt	(Pusnik *et al.* 2007)
				Translation			(Zikova *et al.*, 2008)
	Tb927.1.2990	TbPPR2	P	RNA turnover	9S rRNA and 12S rRNA	Mt	(Pusnik *et al.* 2007)
				Translation			(Zikova *et al.*, 2008)
	Tb927.2.3180	TbPPR1/ KPAF1	P	RNA turnover		Mt	

(Continued)

Table 10.1 List of PPR proteins of identified functions and their targets—cont'd

Species	Accession	Name	Class	Function	Target	Loc	Reference
							(Mingler *et al.*, 2006; Pusnik & Schneider, Weng *et al.*, 2008)
	Tb927.3.4450	TbPPR6	P	RNA turnover	9S rRNA and 12S rRNA	Mt	(Pusnik & *et al.* 2007)
	Tb927.4.4720	TbPPR7	P	RNA turnover	9S rRNA and 12S rRNA	Mt	(Pusnik & *et al.* 2007)
	Tb927.8.3170		P	Translation		Mt	(Zikova *et al.*, 2008)
	Tb927.8.4860		P	Translation		Mt	(Zikova *et al.*, 2008)
Drosophila melanogaster	Q9VJ86	BSF	P	RNA stability	*bcd*	Cy	(Mancebo *et al.*, 2001)
				RNA turnover translation		Mt	(Bratic *et al.*, 2011)
Homo sapiens	NM_133259	LRPPRC/ LRP130	P	Transcription		Nu	(Cooper *et al.*, 2006; Sondheimer *et al.*, 2010)
				transcription		Mt	(Ruzzenente *et al.*, 2011)
				RNA processing			(Sasarman *et al.*; Xu *et al.*, 2004)
	NM_015084	MRPS27	P	Translation		Mt	(Koc *et al.*, 2001)
	NM_015545	PTCD1	P	RNA turnover Translation	Leu-tRNA	Mt	(Rackham *et al.*, 2009)
	NM_024754	PTCD2	P	RNA turnover	*ND5-CYTB*	Mt	(Xu *et al.*, 2008)
	NM_017952	PTCD3	P	Translation		Mt	(Davies *et al.*, 2009)
	NM_014672	MRPP3	Other	RNA cleavage	tRNA	Mt	(Holzmann *et al.*, 2008)
	NM_005035	POLRMT	Other	Transcription		Mt	(Tiranti *et al.*, 1997)
Chlamydomonas reinhardtii	AF330231	MCA1	P	RNA stability	*petA*	Cp	(Loiselay *et al.*, 2008; Raynaud *et al.*, 2007)

		MRL1	P	RNA stability	*rbcL*	Cp	(Johnson *et al.*, 2010)
Physcomitrella patens	AB267806	PpPPR_38	P	RNA cleavage RNA splicing	*clp-5'--rps12* *clpP*	Cp	(Hattori, *et al.*, 2007; Hattori & Sugita, 2009)
	AB647192	PpPPR_43	DYW	RNA splicing	*cox1*	Mt	(Ichinose *et al.*, 2011)
	AB545042	PpPPR_56	DYW	RNA editing	*nad3/nad4*	Mt	(Ohtani *et al.*, 2010)
	AB539865	PpPPR_71	DYW	RNA editing	*ccmFc*	Mt	(Tasaki *et al.*, 2010)
	AB178268	PpPPR_77	DYW	RNA editing	*cox2/cox3*	Mt	(Ohtani *et al.*, 2010)
	AB621361	PpPPR_78	DYW	RNA editing	*cox1/rps14*	Mt	(Uchida, *et al.*, 2011)
	AB621362	PpPPR_79	DYW	RNA editing	*nad5*	Mt	(Uchida *et al.*, 2011)
	AB545044	PpPPR_91	DYW	RNA editing	*nad5*	Mt	(Ohtani *et al.*, 2010)
Arabidopsis thaliana	At1g12700	RPF1	P	RNA cleavage	*nad4*	Mt	(Holzle *et al.*, 2011)
	At1g51965	ABO5	P	RNA splicing	*nad2*	Mt	(Liu *et al.*, 2010)
	At1g61870	PPR336	P	translation		Mt	(Uyttewaal *et al.*, 2008b)
	At1g62670	RPF2	P	RNA cleavage	*nad9/cox3*	Mt	(Jonietz *et al.*, 2010)
	At1g62930	RPF3	P	RNA cleavage	*ccmC*	Mt	(Jonietz *et al.*, 2011)
	At1g74850	PTAC2	P	Transcription		Cp	(Pfalz *et al.*, 2006)
	At1g74900	OTP43	P	RNA splicing	*nad1*	Mt	(de Longevialle *et al.*, 2007)
	At1g80270	PPR596	P	RNA editing	*rps3*	Mt	(Doniwa *et al.*, 2010)
	At2g15820	OTP51	P	RNA splicing	*ycf3*	Cp	(de Longevialle *et al.*, 2008)
	At3g06430	PPR2	P	Translation		Cp	(Lu *et al.*, 2011)
	At3g09650	HCF152	P	RNA cleavage	*psbH-petB*	Cp	(Meierhoff, *et al.*, 2003)
	At3g48250	BIR6	P	RNA splicing	*nad7*	Mt	(Koprivova *et al.*, 2010)
	At4g31850	PGR3	P	RNA stability	*petL* operon	Cp	(Yamazaki, *et al.*, 2004)
	At4g34830	MRL1	P	RNA stability	*rbcL*	Cp	(Johnson *et al.*, 2010)
	At5g60960	PNM1	P	Translation Transcription	 *pnm1*	Mt Nu	(Hammani *et al.*, 2011b)

(Continued)

Table 10.1 List of PPR proteins of identified functions and their targets—cont'd

Species	Accession	Name	Class	Function	Target	Loc	Reference
	At5g67570	DG1	P	Transcription	*pep*	Cp	(Chi *et al.*, 2008, 2010)
	At1g05750	CLB19	E	RNA editing	*rpoA/clpP*	Cp	(Chateigner-Boutin *et al.*, 2008)
	At1g62260	MEF9	E	RNA editing	*nad7*	Mt	(Takenaka, 2010)
	At1g74600	OTP87	E	RNA editing	*nad7(Mt)/atp1(Mt)*	Mt/Cp	(Hammani *et al.*, 2011a)
	At2g20540	MEF21	E	RNA editing	*cox3*	Mt	(Takenaka *et al.*, 2010)
	At2g22410	SLO1	E	RNA editing	*nad4/nad9*	Mt	(Sung *et al.*, 2010)
	At2g45350	CRR4	E	RNA editing	*ndhD*	Cp	(Kotera *et al.*, 2005; Okuda *et al.*, 2006)
	At3g05240	MEF19	E	RNA editing	*cbb206*	Mt	(Takenaka *et al.*, 2010)
	At3g18970	MEF20	E	RNA editing	*rps4*	Mt	(Takenaka *et al.*, 2010)
	At4g25270	OTP70	E	RNA splicing	*rpoc1*	Cp	(Chateigner-Boutin *et al.*, 2011)
	At5g19020	MEF18	E	RNA editing	*nad4*	Mt	(Takenaka *et al.*, 2010)
	At5g55740	CRR21	E	RNA editing	*ndhD*	Cp	(Okuda *et al.*, 2007)
	At5g59200	OTP80	E	RNA editing	*rpl23*	Cp	(Hammani *et al.*, 2009)
	At1g08070	OTP82	DYW	RNA editing	*ndhB* and *ndhG*	Cp	(Okuda *et al.*, 2010)
	At1g11290	CRR22	DYW	RNA editing	*ndhB, ndhD* and *rpoB*	Cp	(Okuda *et al.*, 2009)
	At1g15510	AtECB2VAC1	DYW	RNA editing	*accD/ndhF*	Cp	(Tseng *et al.*, 2010; Yu *et al.*, 2009)
	At1g59720	CRR28	DYW	RNA editing	*ndhB* and *ndhD*	Cp	(Okuda *et al.*, 2009)
	At2g02980	OTP85	DYW	RNA editing	*ndhD*	Cp	(Hammani *et al.*, 2009)
	At2g03880	REME1	DYW	RNA editing	*nad2* and *orfX*	Mt	(Bentolila, *et al.*, 2010)
	At2g25580	MEF8	DYW	RNA editing	*nad5*	Mt	(Takenaka *et al.*, 2010)
	At2g29760	OTP81	DYW	RNA editing	*rps12* intron	Cp	(Hammani *et al.*, 2009)
	At3g12770	MEF22	DYW	RNA editing	*nad3*	Mt	(Takenaka *et al.*, 2010)
	At3g22690	YS1	DYW	RNA editing	*rpoB*	Cp	(Zhou *et al.*, 2008)

	At3g26780	MEF14	DYW	RNA editing	*matR*	Mt	(Verbitskiy, *et al.*, 2011)
	At3g57430	OTP84	DYW	RNA editing	*ndhB, ndhF* and *psbZ*	Cp	(Hammani *et al.*, 2009)
	At3g63370	OTP86	DYW	RNA editing	*rps14*	Cp	(Hammani *et al.*, 2009)
	At4g14850	LOI1 MEF11	DYW	RNA editing	*cox3, nad4* and *ccb 203*	Mt	(Tang *et al.*, 2010; Verbitskiy *et al.*, 2011)
	At5g13270	RARE1	DYW	RNA editing	*accD*	Cp	(Robbins, *et al.*, 2009)
	At5g48910	LPA66	DYW	RNA editing	*psbF*	Cp	(Cai *et al.*, 2009)
	At5g52630	MEF1	DYW	RNA editing	*rps4, nad7 and nad2*	Mt	(Zehrmann, *et al.*, 2009a)
	At1g10270	GRP23	Other	Transcription		Nu	(Ding, *et al.*, 2006)
	At2g16650	PRORP2	Other	RNA cleavage	tRNA	Nu	(Gobert *et al.*, 2010)
	At2g32230	PRORP1	Other	RNA cleavage	tRNA/tRNA-like	Cp/Mt	(Gobert *et al.*, 2010)
	At4g21900	PRORP3	Other	RNA cleavage	tRNA	Nu	(Gobert *et al.*, 2010)
Nicotiana	AAA17740	CRR4	E	RNA editing	*ndhD*	Mt	(Okuda, *et al.* 2008)
Oryza sativa	DQ311052	Rf1a	P	RNA cleavage	B-*atp6-orf79*	Mt	(Wang *et al.*, 2006)
	DQ311054	Rf1b	P	RNA degradation	B-*atp6-orf79*	Mt	(Wang *et al.*, 2006)
	FJ527826	OGR1	DYW	RNA editing	*ccmC, cox2, cox3, nad2,* and *nad4*	Mt	(Kim *et al.*, 2009)
Raphanus sativus	AJ535624	RFK1	P	Translation	*orf125*	Mt	(Koizuka *et al.*, 2003)
	AJ550021	Rfo	P	Translation	*orf138*	Mt	(Uyttewaal *et al.*, 2008a)
Sorghum bicolor	AY661659	Rf1/PPR13	E	Unknown		Mt	(Klein *et al.*, 2005)
Ttiticum aestivum	AAF32491	p63	P	Transcription	*cox2*	Mt	(Ikeda & Gray, 1999)
Zea mays	AF073522	CRP1	P	RNA turnover Translation	*petB/petD* *petA/psaC*	Cp	(Barkan, *et al.*, 1994; Fisk *et al.*, 1999; Schmitz-Linneweber *et al.*, 2005)
	DQ291135	EMP4	P	RNA turnover	*rps2A, rps2B, rps3 rpl16, mttb (orfX)*	Mt	(Gutierrez-Marcos *et al.*, 2007)
	FJ490677	PPR10	P	RNA stability	*atpH/psaJ*	Cp	(Pfalz *et al.*, 2009)
	AY278988	PPR2	P	Translation		Cp	(Williams & Barkan, 2003)

(Continued)

Table 10.1 List of PPR proteins of identified functions and their targets—cont'd

Species	Accession	Name	Class	Function	Target	Loc	Reference
	DQ508419	PPR4	P	RNA splicing	*rps12*	Cp	(Schmitz-Linneweber *et al.*, 2006)
	EU037901	PPR5	P	RNA splicing RNA stabilization	*trnG*	Cp	(Williams-Carrier, *et al.*, 2008) (Beick, *et al.*, 2008)

Among the functional categories indicated, transcription refers to PPR proteins affecting the transcription and/or PPR proteins interacting with components of the transcriptional apparatus or with DNA; RNA turnover refers to PPR protein functions affecting the steady state levels of defined transcripts; RNA cleavage refers to PPR proteins directly involved in the endonucleolytic cleavage of RNA; RNA editing refers to PPR proteins required for cytidine to uridine conversions; RNA splicing refers to PPR protein functions related directly or not to intron splicing; and translation refers to PPR protein functions affecting protein steady state levels and/or to PPR proteins interacting with components of the translation apparatus. Loc refers to localizations deduced from localization experiments and/or from location specific activities. Cp stands for chloroplast, Mt for mitochondria and Nu for the nucleus.

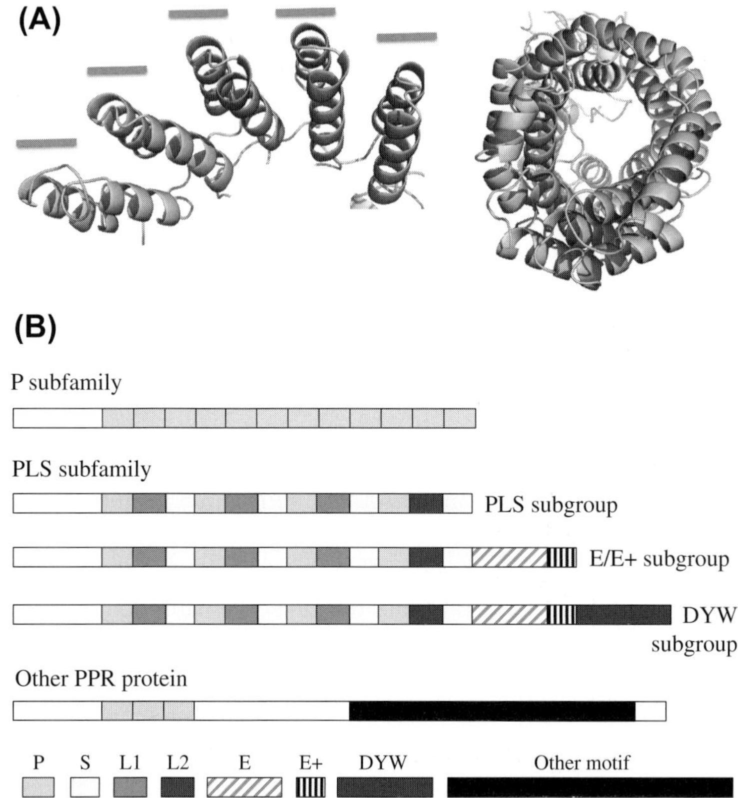

Figure 10.1 Organization and structure of PPR proteins. (A) Alternative tridimensional structure models for PPR domains. Grey bars represent individual PPR repeats that are predicted to be each composed of two anti parallel alpha helices. The succession of PPR repeats could form a superhelix that could be shaped as a crescent and/or a tunnel. In these models, the inner groove or concave surface of the solenoid structure would be very basic and could thus form a suitable platform for interaction with RNA. (B) PPR proteins can be subdivided into subfamilies according to the occurrence of specific domains. The major subfamilies and subgroups as defined by Lurin *et al.* (2004) are represented. The number of PPR motifs are variable in each subfamily and subgroup and the additional motifs can be truncated in the PLS subfamily.

cell compartments. This is the case for PRORP2 and PRORP3 whose localization is strictly restricted to *Arabidopsis* nuclei (Gobert *et al.* 2010). In addition, some proteins can be dual localized in two organelles or in an organelle and the nucleus. In humans, LRPPRC, responsible for the French Canadian variant of the Leigh syndrome, interacts in the nucleus with

PGC1-α to regulate the expression of mitochondrial-encoded genes (Cooper *et al.*, 2006). It is also required in human mitochondria for proper transcription. Its absence in mice mitochondria led to aberrant polyA tails of mRNA encoded on the mitochondrial H strand (Ruzzenente *et al.* 2011). In *Arabidopsis*, PNM1 interacts with polysomes in mitochondria and could thus play a role in mitochondrial translation. However, PNM1 is also localized in the nucleus where it interacts with proteins involved in gene expression regulation, in particular a TCP transcription factor (Hammani *et al.* 2011b). Other PPR proteins such as PRORP1 were found to be dual localized to both mitochondria and chloroplasts (Gobert *et al.* 2010).

4.2. The P Subfamily

PPR proteins belonging to the P subfamily were identified in the three compartments where gene expression takes place as well as in the cytosol. However, possible cytosolic functions of PPR protein remain poorly understood. For example, Rpm2p, a subunit of the mitochondrial ribonucleoproteic RNase P and a transcriptional activator in the nucleus of *Saccharomyces cerevisiae*, was also localized in P-bodies although its potential function at this location is totally unknown (Morales, Dang, Lou, Sulo, & Martin, 1992; Stribinskis & Ramos, 2007).

The P subfamily is the only class of PPR protein represented in animals and fungi. In these organisms, PPR proteins are present in small numbers and were mainly described as involved in translation and RNA turnover.

The involvement of PPR proteins in translation can be proposed for two reasons. PPR proteins can either be directly in interaction or part of ribosomes or can act as translation specificity factors, e.g. through the interaction with transcripts 5′ untranslated regions (UTRs). A decrease in mitochondrial protein PTCD3 in human osteosarcoma cells led to a decrease in protein synthesis (Davies *et al.*, 2009). This protein is associated with MRSP15, a protein of the small subunit of mitochondrial ribosome. On the other hand, DMR1 might be involved, in addition to RNA splicing (Moreno, Buie, Price, & Piva, 2009), in the stability of the small subunit of the ribosome in mitochondria of *S. cerevisiae* by binding with the 15S rRNA (Puchta *et al.*, 2010). PET309 is involved in the turnover of COX1 RNA in *S. cerevisiae* but also in translation through the recognition of the 5′ region of the messenger (Manthey & McEwen, 1995). In *pet309* mutant lines, although the mature COX1 mRNA accumulates at low levels, there is no accumulation of COX1 protein. In a revertant where the accumulation of

COX1 protein is detected, the upstream portion of COX1 5′ UTR has been replaced by the COB 5′ UTR.

Mutations in a large number of PPR protein sequences in both animals and yeast will ultimately affect the accumulation of specific mRNA species. However, the precise functions of the respective PPR proteins often remain elusive. The molecular effects observed could reflect PPR functions related to many processes, e.g. transcription, RNA maturation or RNA degradation. LRPPRC could be involved in the transcription of mitochondrial genes. It is attached to the mitochondrial DNA terminator H2 and the absence of this PPR protein leads to a decrease in transcription but not in the destabilization of RNA (Sondheimer, Fang, Polyak, Falk, & Avadhani, 2010). On the contrary, the BSF protein (Mancebo, Zhou, Shillinglaw, Henzel, & Macdonald, 2001), present in the cytoplasm of *Drosophila* cells, is involved in the stability of bicoid mRNA by binding specifically to the IV/V region present in the transcript 3′ UTR. A mutation in the recognition site IV/V, or a decreased of the BSF protein, lead to decreased levels of IV/V RNA.

In plants, PPR-P proteins appear to be involved in a wide variety of gene expression processes (Fig. 10.2). For the time being, PPR proteins involved in transcription have only been identified in chloroplasts. One of them, DG1, appears to be involved in transcription during the first stage of chloroplast development in *A. thaliana* (Chi *et al.*, 2008). A mutation in the gene (At5g67570) encoding DG1 results in a chlorotic phenotype in young leaves, which disappears after 3 weeks. Furthermore, *in vivo*, DG1 binds SIG6, a sigma factor that plays a role in the regulation of the plastid-encoded RNA polymerase-dependent chloroplast gene transcription (Chi *et al.*, 2010).

Plant PPR-P proteins are involved in splicing. The mitochondrial PPR protein OTP43 (At1g74900) is essential for RNA splicing of the *nad1* gene that encodes the ND1 subunit of complex I of the respiratory chain (de Longevialle *et al.*, 2007). The *nad1* transcript comprises five exons among which the first intron is removed by *trans* splicing. At the macroscopic level, a mutation in OTP43 results in inability of seeds to germinate. At the molecular level, the phenotype is explained by a splicing failure of the first intron of the nad1 transcript.

Plant PPR-P proteins can also be involved in RNA processing, e.g. the PPR-P protein RPF1 (At1g12700) appears to be implicated in the 5′ processing of the *nad4* transcript (Holzle *et al.*, 2011). This protein is part of a subgroup of PPR proteins, typically present at a particular locus of the genome, that was proposed to be involved in cytoplasmic male sterility

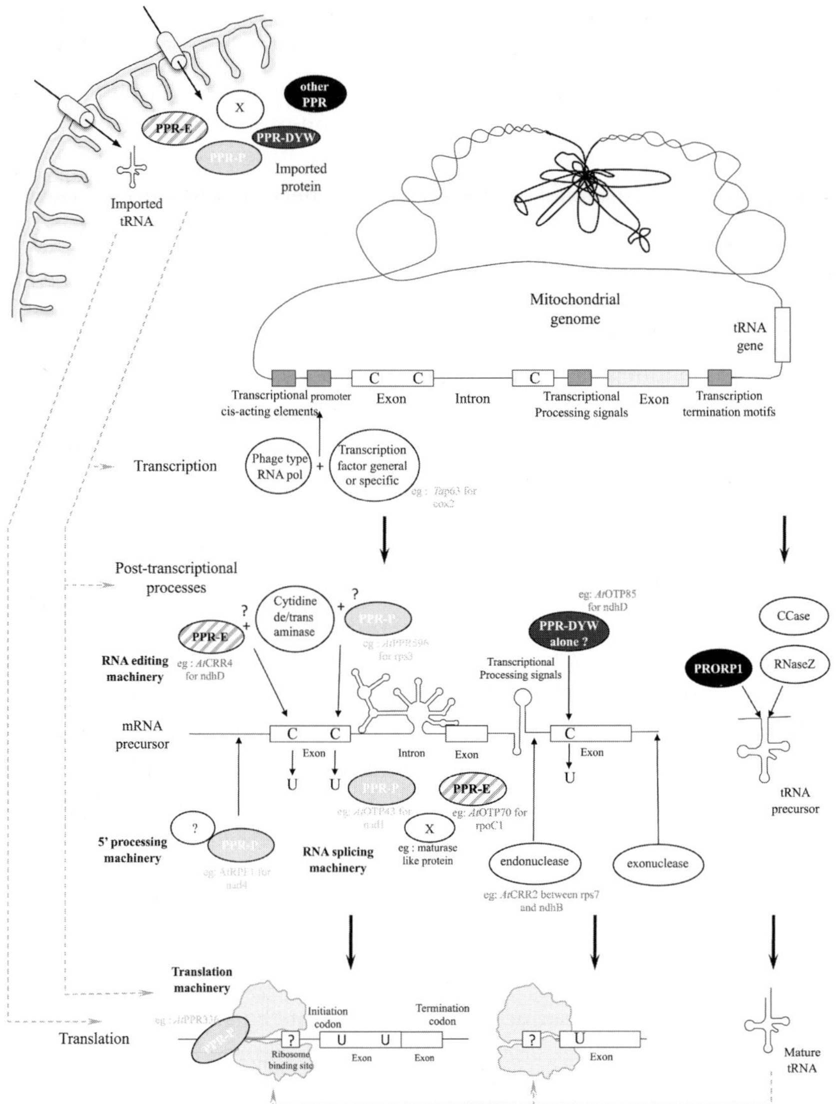

Figure 10.2 Model for the involvement of PPR proteins in plant mitochondrial gene expression processes. Black arrows represent activities performed through PPR protein functions, grey dashed arrows represent the movement and localization of proteins and tRNAs not encoded in mitochondria. Bold black arrows separate the major gene expression stages, i.e. transcription, post-transcriptional modifications and translation.

(CMS) restoration of fertility (Rf) (see also Chapter 5). CMS is a mitochondrial maternally inherited trait caused by aberrant proteins and leading to the absence of fertile pollen. RF1a and RF1b, whose genes are also present in the Rf locus, restore fertility by blocking ORF79 protein synthesis (Wang *et al.*, 2006). The ORF79 gene is present in an operon containing a duplicated ATP6 (B-atp6). RF1a restores fertility by cleaving the transcript B-atp6/org79 while RF1b degrades this transcript.

Concerning RNA editing, PPR proteins involved in this process are typically found to be members of the PLS subfamily (see below). However, PPR596 (At1g80270), a representative of the P subfamily is an exception to this rule as it was found to influence the editing of the partial editing site rps3eU1344SS in the mitochondria of *A. thaliana* (Doniwa *et al.*, 2010).

Plant PPR-P proteins were also recently found to define transcript ends. PPR10, present in chloroplasts of *Zea mays*, binds a region of at least 17 nt in the intergenic region of *atpI-atpH* and *psaJ-rpl33* polycistronic transcripts. Its presence blocks the action of 3′ to 5′ and 5′ to 3′ exonucleases and thus delineates transcript termini (Prikryl *et al.*, 2011).

PPR-P proteins can also be involved in translation regulation as, e.g. PPR10 binds to the *atpI-atpH* transcript and rearranges the secondary structure containing a ribosome binding site, thus allowing the recruitment of ribosomes for *atpH* translation. In addition, as observed in animals and yeast, regulation of translation could also involve PPR proteins associated with ribosomes, such as PPR336 (At1g61870), which was found to interact with mitochondrial polysomes (Uyttewaal *et al.*, 2008b).

4.3. The PLS Subfamily

PPR proteins of the PLS subfamily are only present in land plants (Lurin *et al.*, 2004), although the E subgroup is absent in moss genomes (O'Toole *et al.*, 2008). As described above, the PLS subfamily is divided into three subgroups: PLS, E/E+ and DYW. The PLS subgroup has relatively few representatives and their functions have not been described until now. In contrast, the E/E+ and DYW subgroups have many more representatives and their functions have been the subject of several studies.

4.4. The E/E+ Subgroup

PPR proteins of the E/E+ family were initially found to be involved in RNA editing in chloroplasts. CRR4 was identified by a genetic screen designed to identify activity mutants of the NA(P)DH dehydrogenase complex in chloroplasts (Kotera *et al.*, 2005). CRR4 encoded by At2g45350 is involved in the editing of the first site of ndhD located in its initiation codon. For that, CRR4 binds 25 nucleotides upstream and ten downstream of the editing site allowing the specific recognition of the editing site (Okuda, Nakamura, Sugita, Shimizu, & Shikanai, 2006). Beyond CRR4, the MEF9 gene (At1g62260) codes for a PPR protein with an E domain (Takenaka, 2010). An identified EMS homozygous non-lethal mutant results in the disruption of the E domain. RNA editing at site nad7–200 is affected in this mutant. Thus, the E domain appears to be essential for editing although it does not seem to hold any catalytic activity. Therefore, it was proposed that this domain could be involved in the recruitment of other proteins (Chateigner-Boutin & Small, 2010; Takenaka, 2010).

However, E/E+ PPR proteins cannot be accountable alone for RNA editing site recognition. The number of editing sites largely exceeds that of E/E+ proteins and PPR-DYW proteins were also found to be involved in editing (see below). Moreover, some PPR proteins were shown to be able to recognize a number of different RNA targets. This is the case for SLO1, a mitochondrial-targeted E/E+ PPR protein encoded by At2g22410 in *A. thaliana* (Sung, Tseng, & Hsieh, 2010). The homozygous T-DNA insertion mutant presents late-germination, slow-growth and delayed development phenotypes. Moreover, it affects the editing of two sites: nad4–449 and nad9–328. The region immediately upstream of the editing site of nad4 and nad9 has a high sequence similarity, as 9 of 15 nucleotides are identical. Thus, this similar sequence should be the specific *cis*-element recognized by SLO1.

However, all E/E+ PPR proteins do not seem to be involved in RNA editing. A T-DNA insertion in the *OTP70* gene (At4g25270) (Chateigner-Boutin *et al.*, 2011) results in decreased levels of mRNA molecules transcribed by the plastid-encoded polymerase (PEP). OTP70 is involved indirectly in transcription because it is necessary for the splicing of rpoC1, a subunit of the PEP polymerase. Unlike MEF9, the E domain of OTP70 is not essential for its function as the mutant could be complemented with OTP70 lacking the E domain. However, the OTP70 E domain is shorter by about ten amino acids compared with the E domains of proteins such as MEF9 (Chateigner-Boutin *et al.*, 2011). Thus; it was proposed that the ten amino acid sequence could be involved in the recruitment of specific factors required for editing (Chateigner-Boutin & Small, 2010).

The fertility restorer genes found in land plants do not seem to be restricted to the P subfamily of PPR proteins. PPR13 encodes a protein with

14 PPR repeats and an E domain. This mitochondrial-targeted protein was proposed to function as the *Sorghum* Rf1 gene (Klein *et al.*, 2005) (see also Chapter 5).

4.5. The DYW Subgroup

While E/E+ PPR proteins could recognize editing sites and recruit proteins necessary for the editing activity, the DYW proteins might be directly involved in the C to U deamination reaction that defines RNA editing. The DYW domain has an HxEx_nCxxC motif presenting some sequence identity with the active site of cytidine deaminases and mammalian editing enzymes (Salone *et al.*, 2007). In cytidine deaminases, the conserved cysteines and histidine bind a zinc atom in the enzyme catalytic site and the glutamate is involved in the deamination reaction.

Similar to E/E+ PPR proteins, DYW PPR proteins were mostly described as involved in RNA editing. Some proteins have a large number of targets in mitochondria, e.g. MEF11 in *A. thaliana* and OGR1 in *Oryza sativa.* Knock out or knock down mutations of the *MEF11* gene (At4g14850) (Verbitskiy, Zehrmann, Van Der Merwe, Brennicke, & Takenaka, 2010b) (also called *LOI1* (Tang, Kobayashi, Suzuki, Matsumoto, & Muranaka, 2010)) confers an enhanced lovastatin and clomazone tolerance (Tang *et al.*, 2010). At the molecular level, the LOI1 protein is involved in editing of three mRNAs encoding proteins of the respiratory chain: cox3, nad4 and ccb203. The comparison between the molecular phenotypes of the mutants and the response from wild-type plants to treatments with inhibitors of the respiratory chain suggests that the resistance to lovastatin and clomazone in the silenced line results indirectly from defects in the cytochrome pathway and complex I, respectively (Tang *et al.*, 2010). Similar to E/E+ PPR proteins, the common RNA targets of DYW PPR proteins share sequence similarities. Common substrates of MEF11 share seven or eight identical nucleotides in the immediate vicinity of editing sites (Verbitskiy *et al.*, 2010b). In *O. sativa, OGR1* is a single-exon gene (FJ527826) that encodes a PPR protein essential for normal growth and development (Kim *et al.*, 2009). Similar to MEF11, this protein is essential for the editing of different sites, i.e. for seven editing sites in five different transcripts.

For the time being, only two DYW PPR proteins have been described as not involved in RNA editing: CRR2 in *A. thaliana* and PpPPR_43 in *P. patens.* A mutation in the *CRR2* gene (At3g46790), which encodes a protein targeted to the chloroplast, leads to a defect in NDH activity caused by a defect in the accumulation of the NDH complex and, more precisely, in the accumulation of NDHB (Hashimoto, Endo, Peltier, Tasaka, & Shikanai, 2003). The ndhB mRNA is present in a tricistronic precursor also containing rps12 and rps7. In the intergenic region between rps7 and ndhB, two processing sites are present at positions 12 and 180 upstream of ndhB. Cleavage at position −180 appears to be independent from CRR2 function, whereas cleavage at position −12 is affected in CRR2 mutants, which led to the hypothesis that CRR2 might be directly involved in ndhB transcript processing (Hashimoto *et al.*, 2003). On the other hand, PpPPR_43, a protein localized in *P. patens* mitochondria, appears to be involved in the splicing of the third intron of *cox1* transcript (Ichinose, Tasaki, Sugita, & Sugita, 2011). The motif proposed to be involved in the deaminase activity of DYW PPR protein (HxEx$_n$CxxC) is absent in PpPPR_43. Similar to OTP70 (Chateigner-Boutin *et al.*, 2011), the E domain of PpPPR_43 is not required for splicing and neither is the DYW domain of PpPPR_43.

4.6. Other PPR Proteins

A number of other proteins containing PPR domains cannot be assigned to one of the particular subfamilies defined by Lurin *et al.* (2004). These proteins contain a PPR motif in association with additional domains carrying catalytic activities that do not belong to the E/E+ or DYW subfamilies. Among them, the human mitochondrial RNA polymerase and RNase P proteins were first defined.

Similar to other transcripts, tRNAs are transcribed as precursor molecules and undergo a number of post-transcriptional modifications. An essential step of their maturation is the cleavage of the 5′ leader sequence of pre-tRNAs. This activity is achieved by an endonuclease called RNase P. Until recently, RNase P enzymes were always found to be ribonucleoproteins with a ribozyme holding the catalytic activity (Evans, Marquez, & Pace, 2006). However, protein-only RNase P enzymes were recently described, initially in human mitochondria (Holzmann *et al.*, 2008). Human mitochondrial RNase P is composed of three proteins: a tRNA methyltransferase, a short chain dehydrogenase/reductase and a protein with PPR motif as well as an NYN metallonuclease domain (Holzmann *et al.*, 2008). This PPR protein has orthologues in plants, including three in *A. thaliana* (PRORP1 to 3). PRORP2 and 3 are both localized in the nucleus, while PRORP1 is present in mitochondria and chloroplasts. PRORP1 was shown to perform as a single protein in the endonucleolytic release of pre-tRNA 5′ leaders, which defines RNase P activity (Gobert *et al.* 2010). Plant mitochondrial genomes encode both monocistronic and polycistronic transcripts. In some cases, transcripts are matured by endonucleolytic cleavage at the level of tRNA-like structures called t-elements. PRORP1 is able to cleave some of these tRNA-like structures *in vitro*, thus suggesting that PROPR enzymes are also involved in the maturation of mRNAs in mitochondria (Gobert *et al.* 2010).

The mitochondrial RNA polymerase in humans (POLRMT or mtRNAP) and in yeast (rpo41) is composed of a single subunit. However, it acts in conjunction with two transcription factors: TFAM and TFB2M (Tiranti *et al.*, 1997; Wang & Shadel, 1999). mtRNAP has in its N-terminal part a PPR domain containing two PPR repeats and a T7-like catalytic carboxy-terminal domain (Ringel *et al.*, 2011). PPR motifs are also present in rpo41 (Lipinski, Puchta, Surendranath, Kudla, & Golik, 2011). A deletion of the 185 first nucleotides, which contain two PPR motifs, does not directly affect the initiation of transcription but leads to instability or a loss of the mitochondrial genome (Wang & Shadel, 1999). The PPR domain was thus proposed to be involved in protein–protein interaction (Rodeheffer, Boone, Bryan, & Shadel, 2001). However, determination of the crystal structure of mtRNAP has suggested that the PPR domain might be indirectly required for transcription initiation as it sequesters the tip of the AT-rich recognition loop (Ringel *et al.*, 2011). The PPR domain might also be involved in the stabilization of nascent mitochondrial transcripts.

4.7. Evolution of PPR Proteins

4.7.1. Distribution of PPR genes

The growing availability of genomic sequences has enabled a large number of PPR genes to be identified. This has helped to provide a pattern for the distribution and expansion of this novel class of genes during the evolution of eukaryotes. Since their initial discovery, it has been evident that PPR proteins are eukaryote specific (Lurin *et al.*, 2004). There are hardly any PPR coding genes in prokaryotic genomes. A few exceptions are found; e.g. in the genome of the plant pathogen *Ralstonia solanacearum* (Salanoubat *et al.*, 2002), although these genes have most probably arisen from horizontal gene transfer (see below) (Lurin *et al.*, 2004). When PPR gene numbers are compared in different eukaryote phyla, the most obvious observation is that an exponential increase in the number of PPR genes has taken place during the evolution of land plants.

In unicellular eukaryotes and animals, the PPR proteins coding genes are restricted to the P subclass and are only present in small numbers. For example, there are 15 members in *S. cerevisiae* (Lipinski *et al.*, 2011), about 30 in *Trypanosoma brucei* (Mingler *et al.*, 2006; Pusnik *et al.*, 2007) and seven in humans (Rackham & Filipovska). Similar to non-plant eukaryotes, green algae do not encode many PPR proteins (e.g. 11 in *Chlamydomonas reinhardtii*) (Johnson *et al.*, 2010) and no additional domains are found in the C-terminal part of these proteins (Lurin *et al.*, 2004). In contrast, the number of PPR proteins coding genes is much higher in land plants (e.g. 450 in *A. thaliana* and 477 in *O. sativa*), although these gene numbers remain relatively smaller for mosses (e.g. 103 in *P. patens*) (O'Toole *et al.*, 2008). Moreover, mosses distinguish themselves from other land plants by the relative low number of PPR proteins belonging to the PLS subclass. For instance, *P. patens* only encodes 10 proteins of the DYW subgroup and none from the E/E+ subgroup (O'Toole *et al.*, 2008).

4.7.2. Origin of PPR genes

The wide distribution of PPR genes in eukaryotes from very divergent phyla suggests that PPR genes appeared early in the evolution of eukaryotes. However, a bacterial ancestor for early eukaryote PPR genes has not yet been identified. This could be due to the very low primary sequence conservation of PPR motifs. Because of rapid sequence divergence, the ancestor of PPR motifs might be barely recognizable in bacteria. Nevertheless, phylogenetic studies clearly show that PPR proteins found in eukaryotes do not originate from the few PPR genes identified in bacterial genomes. These bacterial PPR genes appear to originate from modern plant to prokaryote horizontal gene transfer events (Lurin *et al.*, 2004).

4.7.3. How and when the PPR gene family expanded

In contrast with moss PPR genes, which often contain introns, higher plants PPR genes are only rarely interrupted by introns. This has suggested from the start that the exponential increase in the number of PPR genes in higher plants could have arisen through one or more waves of retrotransposition events (Lurin *et al.*, 2004). Two lines of evidence now suggest that this is indeed the case. First, phylogenetic analysis of the few higher plant PPR genes containing introns has revealed that these genes cluster with moss PPR genes (O'Toole *et al.*, 2008). This further suggests that the intron-containing PPR genes could correspond to ancient PPR genes that were used as templates for the PPR gene family expansion (O'Toole *et al.*, 2008).

Second, the positions of introns in the higher plant PPR genes containing few introns also provide evidence to support the occurrence of expansion by retrotransposition. In these genes, introns are restricted to the 5′ end of PPR sequences. These positions are coherent with the occurrence of 3′ to 5′ retrotranscription events that would have favoured the conservation of some 5′ introns when retrotranscription was not complete (O'Toole *et al.*, 2008).

The precise timing of the spectacular increase in PPR gene numbers remains elusive. However, as supported by strong phylogenetic data (O'Toole *et al.*, 2008), it is now evident that it happened before the separation between monocotyledon and dicotyledon plants. Since then, very little gene loss or gain has seemingly occurred, thus suggesting a large conservation of PPR functions among angiosperm species (O'Toole *et al.*, 2008).

5. CORRELATIONS BETWEEN THE EMERGENCE OF PPR PROTEINS AND SPECIFIC GENE EXPRESSION PROCESSES

Mitochondria have retained a genome except in a few anaerobic organisms (van der Giezen, 2009). However, these genomes vary considerably in size and gene numbers between organisms. In addition, genes coded by the mitochondrial genomes can support extra features such as introns of types I and II and mutations reversed by C to U and U to C RNA editing (described in Section 3). In order to produce correct proteins, numerous post-transcriptional mechanisms are needed. PPR proteins are major players among these processes occurring in mitochondria and have coevolved with them.

5.1. Emergence of PPR Proteins and Occurrence of Specific Processes in Different Phyla

5.1.1. PPR proteins in metazoan

In animals, the number of PPR proteins is limited to six or seven members in humans (POLRMT, PTCD1-3, LRPPRC, MRPS27 and MRPP3/PRORP) and two members in *Drosophila* (DS3, BSF). However, these PPR proteins, with the exception of PRORP, were discovered using an algorithm specifically designed for plants. The use of a fungus-specific program allowed the discovery of 12 new PPR proteins in yeast (Lipinski *et al.*, 2011). Similarly, Sterky *et al.* (2010) used an animal-specific program in order to verify the number of pentatricopeptide repeats in LRPPRC (LRP103).

They found ten additional PPR repeats in the human protein compared with those identified with the plant-specific algorithm. In conclusion, specific computational algorithms might be necessary in order to detect the full range of PPR proteins occurring in the different eukaryotic phyla.

The metazoan mitochondrial genomes are compact (less than 20 kb), they do not contain any introns (with few exceptions) and no RNA editing is occurring. Transcription produces a few main RNA precursors. Post-transcriptional processes leading to mature transcripts are mainly mediated by nucleases (RNase P, RNase Z/Elac2) at tRNA locations (Sanchez *et al.*, 2011). The overall processing is quite straightforward compared with other eukaryote mitochondria. In humans, PPR proteins are involved in mitochondrial transcription. A PPR-containing RNA polymerase (POLRMT) is responsible for the transcription of the three RNA precursors (H1, H2, L). The structure of this protein was recently resolved (Ringel *et al.*, 2011). In addition, PPR proteins are involved in the endonuclease activities leading to mature transcripts. MRPP3/PRORP is part of the human mitochondrial RNase P composed of a complex of three proteins (Holzmann *et al.*, 2008). PDTC1 interacts with Elac2 and therefore might be implicated in the processing of the 3′ end of tRNAs (Sanchez *et al.*, 2011). It was previously shown that PDTC1 could interact specifically with $tRNA^{Leu}$ in unprocessed transcripts (Rackham *et al.*, 2009) and that PDTC1 was a negative regulator for the correct maturation of $tRNA^{Leu}$. PTCD2 and LRPPRC are involved in RNA maturation as well. In mice, PTCD2 was shown to regulate the processing of cytochrome *b*. Downregulation of PTCD2 leads to an accumulation of RNA precursors and a reduced quantity of mature cytochrome *b* mRNA (Xu *et al.*, 2008). Recent studies on LRPPRC have shown a specific mitochondrial localization (Sterky, Ruzzenente, Gustafsson, Samuelsson, & Larsson, 2010) and a stabilization effect on RNA precursors (Sondheimer *et al.*, 2010). However, other studies implicate this protein in nuclear functions and it has been hypothesized that LRPPRC is involved in the coordination of transcription between the nucleus and mitochondria (Mili & Pinol-Roma, 2003). MRPS27 and PTCD3 are part of the translational apparatus. MRPS27 is a mitochondrial ribosomal protein of the small subunit 27 and PTCD3 associates with the mitochondrial small ribosomal subunit and was shown to be important for protein synthesis (Davies *et al.*, 2009).

5.1.2. PPR proteins in fungi

In yeast, 15 PPR gene members have been recognized (Lipinski *et al.*, 2011). As in metazoan mitochondria, yeast PPRs are involved in transcription

(Rpo41p, the catalytic subunit of the mitochondrial RNA polymerase), post-transcriptional mechanism (Rpm2p, the protein component of mitochondrial RNase P) and translation (translational activators and protein interacting with rRNA). However, several PPR proteins have an unknown function. Yeast PPR proteins bind COB, COX1, COX2, ATP6, ATP8 and ATP9 mRNA. They were described as stabilizing RNA and as translation activators. However, the binding of Pet111p to COX3 transcripts has a negative effect on its translation.

5.1.3. PPR proteins in chlorophyta

In green alga, the mitochondrial genome is compact and linear. It is 15.7-kb long for *C. reinhardtii*. The amount of PPR genes in *C. reinhardtii* (12 PPR genes) is reminiscent of metazoan and yeast although chlorophyta are characterized by the occurrence of an additional organelle, i.e. the chloroplast. The only two PPR proteins studied in *Chlamydomonas* (MRL1 and MCA1) are targeted to the chloroplast and function in post-transcriptional processes and translation activation (Johnson *et al.*, 2010; Loiselay *et al.*, 2008). However, the *C. reinhardtii* mitochondrial genome only encodes 13 genes and it could be expected to require only a limited number of PPR proteins for its expression. To date, no PPR protein has been shown to function in *Chlamydomonas* mitochondria. In addition to PPR proteins, a family of 38–40 repeats seems to perform several functions in chloroplasts (Eberhard *et al.*, 2011). The members of this family perform identical functions in organelles (Eberhard *et al.*, 2011) and therefore might replace the function of bona fide PPR proteins. In *Ostreococcus tauri*, a PPR-containing PRORP protein has RNase P activity but the localization of this enzyme in the cell is unknown (Lai *et al.*, 2011). This PRORP enzyme is also encoded in the nuclear genome of *C. reinhardtii* (Merchant *et al.*, 2007).

5.1.4. PPR proteins in viridiplantae

In land plants, the mitochondrial genomes are large (200–2000 bp) and contain an average of 50–70 genes. The presence of group II introns and mutated nucleotides in the genomes has led to the occurrence of new post-transcriptional mechanisms in the organelles, i.e. splicing and RNA editing. Plant nuclear genomes encode a large amount of PPR proteins in comparison with other organisms (103 in *P. patens*, 450 in *A. thaliana*, 477 in *O. sativa*). Most PPR proteins are predicted to be targeted to mitochondria and chloroplasts (Lurin *et al.*, 2004) but some also have nuclear functions (Hammani *et al.*, 2011b).

The increase in PPR protein numbers in land plants is correlated with the incidence of C to U RNA editing, especially for the PLS PPR proteins (Fujii & Small, 2011). The PLS class usually accounts for half of all PPR genes (~225 of 450 in *Arabidopsis*) but represents four times the amount of P class PPR genes in *Selaginella* (>800 of 1050) (Fujii & Small, 2011). In *Arabidopsis*, ~525 editing sites are found in mitochondria and 34 in the chloroplast (Chateigner-Boutin & Small, 2010), however, *Selaginella* has hyperedited organellar genomes with more than 2150 editing sites in the mitochondrial genome (Hecht, Grewe, & Knoop, 2011) and 104 sites in the chloroplast genome (Fujii & Small, 2011). Thus, there is a direct link between the number of PLS class PPRs and the number of editing sites. In addition to RNA editing, the plant organellar genomes contain group II introns (and a few species with group I introns). In *Arabidopsis* mitochondrial genome, 23 introns are present and need to be spliced from primary transcripts. PPR proteins are involved in the splicing of several introns. In plants, the organellar RNA polymerases do not contain PPR repeats. However, PPR proteins interacting with these polymerases were identified and could thus have a role in the transcription in plants (Pfalz *et al.*, 2009; Suzuki *et al.*, 2004). pTAC2 was found in the purified chloroplastic PEP fraction of *Arabidopsis* and tobacco. In *Arabidopsis*, the *ptac2* mutant has the same phenotype as the mutants of the core components of PEP. The mutants have a pale green phenotype and the PEP-dependent transcripts are less abundant (Pfalz *et al.*, 2009). In plant organelles, the tRNA 5′ maturation is mediated by the PPR-containing PRORP1. This protein does not require additional proteins for its function, unlike the human PRORP operating in mitochondria. A subset of P class PPR proteins are associated with polysomes (Uyttewaal *et al.*, 2008b). However, up to now, plant PPR proteins were not found to be part of the core structure of mitoribosomes and chlororibosomes.

5.1.5. PPR proteins in trypanosomids

Trypanosomids have the largest set of PPR proteins after land plants. The nuclear genome of *T. brucei*, the agent of sleeping sickness, codes for 28 PPR proteins (Pusnik *et al.*, 2007). These proteins could be specifically involved in mitochondrial translation and/or in the biogenesis of ribosomal RNA in trypanosome mitochondria because of the eight PPR proteins investigated, seven are present in this organelle; they are membrane bound, similar to mitoribosomes and are required for the accumulation of rRNAs. Knock out mutations of single PPR proteins induce degradation of other PPR proteins.

This could indicate collaborative functions and possibly a specific role for PPR proteins in trypanosomids (Pusnik *et al.*, 2007).

5.2. Correlations between the Occurrence of PPR Proteins and Gene Expression Processes

5.2.1. PPR proteins and transcription

PPR proteins were found to be involved in mitochondrial transcription in related eukaryote groups such as fungi and metazoans. In yeast and humans, mitochondrial RNA polymerases are T7-phage related and possess a PPR domain (Ringel *et al.*, 2011). These proteins have a limited number of PPR repeats in the respective PPR domains, e.g. two for the human RNA polymerase. However, not all phage-type RNA polymerases contain PPR domains; for instance, plant mitochondrial RNA polymerases are also phage related but do not contain PPR repeats. Moreover, some plant RNA polymerases, e.g. the ones encoded in plastid genomes, are not of phage type but rather resemble bacterial RNA polymerases. This distribution suggests that PPR domains in phage-type RNA polymerases might have been acquired after the separation of opisthokonts (which contain both fungi and metazoans) from other eukaryote groups. This acquisition does not correlate with a particular mitochondrial genome structure, e.g. metazoans have evolved a very compact mitochondrial DNA, whereas fungi have retained a relatively larger mitochondrial genome. Similarly, the occurrence of PPR-containing RNA polymerase does not correlate with the incidence of most post-transcriptional processes in mitochondria (Table 10.2). In all instances, the occurrence of PPR-containing RNA polymerase in fungi and metazoans and not in, e.g. Viridiplantae is paradoxical given the very small numbers of PPR genes in these groups compared with higher plants and the considerable diversity of functions that PPR proteins have acquired in plants.

5.2.2. PPR proteins and RNA splicing

RNA splicing is mainly found in plant organelles. Taking *A. thaliana* as an example, its mitochondrial genome contains 23 group II introns (Unseld *et al.*, 1997) that need to be spliced at the RNA level. Among the 50 or so *Arabidopsis* PPR proteins studied so far, at least five are involved in RNA splicing (four P-PPR and one E-PPR, Table 10.1). In mitochondria, OTP43 is essential for the splicing of the intron 1 of *nad1* (de Longevialle *et al.*, 2007), ABO5 for the splicing of the intron 3 of *nad2* (Liu *et al.*, 2010) and BIR6 for the splicing of the intron 1 of *nad7* (Koprivova *et al.*, 2010).

Table 10.2 Correlations between mitochondrial genome features, the incidence of gene expression processes and the occurrence of PPR proteins in representative species

	Mitochondrial genomes					PPR genes		Gene expression processes					
	Accession	**Size (kb)**	**Gene numbers**	**GC content**	**C→U editing sites**	**#PPR**	**#PLS**	**TransC**	**RNA editing**	**Splicing**	**RNA turnover**	**RNA processing**	**TransL**
Homo sapiens	NC 012920	16,569	37	44.4	–	7	–	+P	–	–	+P	+P	+P
Saccharomyces cerevisiae	NC_001224	85,779	35	17.1	–	15	–	+P	–	+P?	+P	+P	+P
Chlamydomonas reinhardtii	NC_001638	15,758	13	45.2	–	12	–	+	–	–	+	+	+
Physcomitrella patens	NC_007945	105,340	69	40.6	11	103	10	+	+PLS	+P/PLS	+	+P	+
Selaginella moellendorfii	JF338143-7	261,212	20	68.1	2,139	>1000	~144	+	+	+	+	+	+
Arabidopsis thaliana	NC_001284	366,924	57	44.8	525	450	87	+	+P/PLS	+P	+	+P	+P
Oryza sativa japonica	NC_011033	490,520	55	43.9	>400	477	90	+	+PLS	+	+P	+P	+
Trypanosoma brucei	M94286.1	23,016★	14	23.3	–	28	–	+	–	–	+P	+	+P

+ and – indicate whether gene expression processes take place or not in the respective species. P and PLS show whether PPR proteins of the P or the PLS subclasses were shown to be involved or not in the respective mechanisms in mitochondria of the different organisms. ★ refers to the maxicircle only in *Trypanosoma*. It does not take into account the numerous minicircles whose total size approximates 1 kb. Gene expression processes are as defined in Table I. TransC stands for transcription and TransL for translation.

PPR proteins are also implicated in the splicing of introns present in the plastid genome (OTP70 is involved in the splicing of rpoC1 (Chateigner-Boutin *et al.*, 2011) and OTP51 in the splicing of intron 2 of *ycf3* (de Longevialle *et al.*, 2008)). However, splicing also occurs in non-plant phyla, i.e. in fungi. To our knowledge, the only non-plant PPR protein that was proposed to be involved in splicing is Ccm1 (Dmr1) from *S. cerevisiae* (Moreno *et al.*, 2009). DMR1 is necessary for the removal of intron 4, a group I intron, in Cox1 and Cob (Moreno *et al.*, 2009). However, Puchta *et al.* (2010) argue DMR1 is implicated in the stability of 15S rRNA and therefore the splicing effect would be an indirect effect resulting from compromised translation. In all instances, the involvement of PPR proteins in splicing correlates with the increase of intron numbers in genomes. Thus, the occurrence of PPR proteins involved in splicing also correlates with the increase of mitochondrial genome sizes in eukaryotes (Table 10.2).

5.2.3. PPR proteins and RNA editing

As stated earlier, organellar C to U RNA editing is almost exclusively found in plants with the exception of the heterolobosean protest, *N. gruberi*. At this stage, it is not known whether the RNA editing mechanism in *N. gruberi* arose from horizontal gene transfer or if the plant-type RNA editing mechanism is more ancestral than previously expected (Rudinger, Szovenyi, Rensing, & Knoop, 2011b). In Viridiplantae, RNA editing seems to be present only in the embryophyta clade (no editing occurs in chlorophyta and streptophyta alga). However, subsequent loss of this mechanism of RNA editing has occurred in the marchantiid liverworts in which no editing site is present in organellar genomes (Rudinger, Polsakiewicz, & Knoop, 2008). In contrast, some species have hyperedited organellar genomes, e.g. the tracheophyta *Selaginella moellendorffii* with more than 2000 sites in the mitochondrial genome (Hecht *et al.*, 2011) and the hornwort *Anthoceros formosae* with more than 500 sites in the chloroplast genome (Kugita, Yamamoto, Fujikawa, Matsumoto, & Yoshinaga, 2003). The presence of RNA hyperediting and the great number of PLS-type PPR proteins correlate with the high GC content of *S. moellendorfii* mitochondrial genome (68.1%, Table 10.2) and in general a positive correlation exists between GC content and RNA editing (Malek *et al.*, 1996). The current hypothesis is that the expansion of the PPR family allowed the editing of organellar transcripts and genome sequence drift. This sequence drift might have been advantageous for the first land plants to come out of the aqueous environment and be resistant to terrestrial environmental conditions (Fujii & Small, 2011). In all instances, the huge expansion of the PPR gene family in plants strictly correlates with the occurrence of RNA editing in eukaryotes and with the incidence of large mitochondrial genomes (Table 10.2). Moreover, the number of editing sites does seem to correlate with the number of PLS PPR genes (O'Toole *et al.*, 2008). Among the different subcategories of PPR proteins, it appears that the occurrence of DYW PPR proteins strictly correlates with the occurrence of RNA editing. The DYW domain is only found in embryophyta. More precisely, for example, among liverworts, both RNA editing and DYW PPR proteins are found in jungermanniid species, whereas both editing and DYW proteins are absent from the closely related Marchantiidae (Salone *et al.*, 2007).

5.3. Concluding Remarks

Two main features of higher plant mitochondria are the occurrence of very large mitochondrial genomes and very high numbers of nuclear-encoded mitochondrial PPR proteins compared with other eukaryotes. The incidence of these two features should not be considered as independent but should rather be regarded as evolutionarily connected phenomena. It becomes evident that the spectacular size increase of mitochondrial genomes in land plants has taken place in conjunction with the remarkable expansion of the PPR gene family, most probably through waves of retrotransposition events. The increase of mitochondrial genome sizes has been accompanied by sequence drifts that have resulted in the introduction of, e.g. high intron numbers compared with mitochondria in other eukaryotes and in the introduction of hundreds of T/C mutations in coding sequences. These features of plant mitochondrial DNA have only been conserved because they could be reversed at the RNA level through the function of members of the expanded PPR family. Moreover, the resulting diversification of post-transcriptional processes required in higher plants, e.g. the appearance of RNA editing, has occurred in conjunction with the diversification of PPR protein organization in plants, i.e. with the appearance of the specific subgroup of PLS PPR proteins. Thus, the small numbers of P class PPR proteins found in distantly related eukaryotes such as fungi, metazoan or trypanosomids and their orthologues in plants are believed to hold the ancestral PPR proteins functions, whereas specific groups of PPR proteins only found in some phyla hold functions that have specifically evolved in conjunction with the evolution of mitochondrial genomes in the respective phyla.

ACKNOWLEDGEMENTS

This work was supported by the French Centre National de la Recherche Scientifique. A.G. was supported by an ANR Blanc research grant (PRO-RNase P, ANR 11 BSV8 008 01) to P.G. B.G. was supported by a PhD grant from the University of Strasbourg.

REFERENCES

Anant, S., Macginnitie, A. J., & Davidson, N. O. (1995). Apobec-1, the catalytic subunit of the mammalian apolipoprotein b mRNA editing enzyme, is a novel RNA-binding protein. *Journal of Biological Chemistry, 270*, 14762–14767.

Anderson, S., Bankier, A. T., Barrell, B. G., De Bruijn, M. H., Coulson, A. R., Drouin, J., et al. (1981). Sequence and organization of the human mitochondrial genome. *Nature, 290*, 457–465.

Andersson, S. G., Zomorodipour, A., Andersson, J. O., Sicheritz-Ponten, T., Alsmark, U. C., Podowski, R. M., et al. (1998). The genome sequence of *Rickettsia prowazekii* and the origin of mitochondria. *Nature, 396*, 133–140.

Aphasizheva, I., Maslov, D., Wang, X., Huang, L. & Aphasizhev, R. Pentatricopeptide repeat proteins stimulate mRNA adenylation/uridylation to activate mitochondrial translation in trypanosomes. *Molecular Cell, 42*, 106–117.

Aubourg, S., Boudet, N., Kreis, M., & Lecharny, A. (2000). *Arabidopsis thaliana*, 1% of the genome codes for a novel protein family unique to plants. *Plant Molecular Biology, 42*, 603–613.

Barkan, A., Walker, M., Nolasco, M., & Johnson, D. (1994). A nuclear mutation in maize blocks the processing and translation of several chloroplast mRNAs and provides evidence for the differential translation of alternative mRNA forms. *EMBO Journal, 13*, 3170–3181.

Barkan, A., Klipcan, L., Ostersetzer, O., Kawamura, T., Asakura, Y., & Watkins, K. P. (2007). The CRM domain: an RNA binding module derived from an ancient ribosome-associated protein. *RNA, 13*, 55–64.

Beick, S., Schmitz-Linneweber, C., Williams-Carrier, R., Jensen, B., & Barkan, A. (2008). The pentatricopeptide repeat protein PPR5 stabilizes a specific tRNA precursor in maize chloroplasts. *Molecular and Cellular Biology, 28*, 5337–5347.

Benne, R., Van Den Burg, J., Brakenhoff, J., Sloof, P., Van Boom, J. H., & Tromp, M. C. (1986). Major transcript of the frameshift *coxII* from trypanosome mitochondria contains four nucleotides that are not encoded in the DNA. *Cell, 46*, 819–826.

Benne, R. (1989). RNA-editing in trypanosome mitochondria. *Biochimica et Biophysica Acta, 1007*, 131–139.

Benne, R. (1996). RNA editing: how a message is changed. *Current Opinion in Genetics & Development, 6*, 221–231.

Bentolila, S., Elliott, L. E., & Hanson, M. R. (2008). Genetic architecture of mitochondrial editing in *Arabidopsis thaliana*. *Genetics, 178*, 1693–1708.

Bentolila, S., Knight, W., & Hanson, M. (2010). Natural variation in *Arabidopsis* leads to the identification of REME1, a pentatricopeptide repeat-DYW protein controlling the editing of mitochondrial transcripts. *Plant Physiology, 154*, 1966–1982.

Binder, S., & Brennicke, A. (1993). A transfer RNA gene transcription initiation site is similar to messenger RNA and rRNA promoters in plant mitochondria. *Nucleic Acids Research, 21*, 5012–5019.

Binder, S., & Brennicke, A. (2003). Gene expression in plant mitochondria: transcriptional and post-transcriptional control. *Proceedings of the Royal Society of London B: Biological Sciences, 358*, 181–188, discussion 188–189.

Blanc, V., Litvak, S., & Araya, A. (1995). RNA editing in wheat mitochondria proceeds by a deamination mechanism. *FEBS Letters, 373*, 56–60.

Blatch, G. L., & Lassle, M. (1999). The tetratricopeptide repeat: a structural motif mediating protein-protein interactions. *Bioessays, 21*, 932–939.

Bock, H., Brennicke, A., & Schuster, W. (1994a). *Rps3* and *rpl16* genes do not overlap in *Oenothera* mitochondria: GTG as a potential translation initiation codon in plant mitochondria? *Plant Molecular Biology, 24*, 811–818.

Bock, R., Kössel, H., & Maliga, P. (1994b). Introduction of a heterologous editing site into the tobacco plastid genome: the lack of RNA editing leads to a mutant phenotype. *EMBO Journal, 13*, 4623–4628.

Bonen, L., & Vogel, J. (2001). The ins and outs of group II introns. *Trends in Genetics, 17*, 322–331.

Bonen, L. (2008). Cis- and *trans*-splicing of group II introns in plant mitochondria. *Mitochondrion, 8*, 26–34.

Bour, T., Akaddar, A., Lorber, B., Blais, S., Balg, C., Candolfi, E., et al. (2009). Plasmodial aspartyl-tRNA synthetases and peculiarities in *Plasmodium falciparum*. *Journal of Biological Chemistry, 284*, 18893–18903.

Bratic, A., Wredenberg, A., Gronke, S., Stewart, J. B., Mourier, A., Ruzzenente, B., et al. (2011). The bicoid stability factor controls polyadenylation and expression of specific mitochondrial mRNAs in *Drosophila melanogaster*. *PLoS Genetics*, 7. e1002324.

Brennicke, A., Marchfelder, A., & Binder, S. (1999). RNA editing. *FEMS Microbiology Reviews, 23*, 297–316.

Cai, W., Ji, D., Peng, L., Guo, J., Ma, J., Zou, M., et al. (2009). LPA66 is required for editing psbF chloroplast transcripts in *Arabidopsis*. *Plant Physiology, 150*, 1260–1271.

Canino, G., Bocian, E., Barbezier, N., Echeverria, M., Forner, J., Binder, S., et al. (2009). *Arabidopsis* encodes four tRNase Z enzymes. *Plant Physiology, 150*, 1494–1502.

Carrillo, C., & Bonen, L. (1997). RNA editing status of nad7 intron domains in wheat mitochondria. *Nucleic Acids Research, 25*, 403–409.

Cavalier-Smith, T. (1991). Intron phylogeny: a new hypothesis. *Trends in Genetics*, 7, 145–148.

Cermakian, N., Ikeda, T. M., Miramontes, P., Lang, B. F., Gray, M. W., & Cedergren, R. (1997). On the evolution of the single-subunit RNA polymerases. *Journal of Molecular Evolution, 45*, 671–681.

Chang, C. C., & Stern, D. B. (1999). DNA-binding factors assemble in a sequence-specific manner on the maize mitochondrial atpA promoter. *Current Genetics, 35*, 506–511.

Chapdelaine, Y., & Bonen, L. (1991). The wheat mitochondrial gene for subunit-I of the NADH dehydrogenase complex - a *trans*-splicing model for this gene-in-pieces. *Cell, 65*, 465–472.

Chateigner-Boutin, A. L., & Hanson, M. R. (2003). Developmental co-variation of RNA editing extent of plastid editing sites exhibiting similar cis-elements. *Nucleic Acids Research, 31*, 2586–2594.

Chateigner-Boutin, A. L., & Small, I. (2007). A rapid high-throughput method for the detection and quantification of RNA editing based on high-resolution melting of amplicons. *Nucleic Acids Research, 35*, e114.

Chateigner-Boutin, A. L., & Small, I. (2010). Plant RNA editing. *RNA Biology*, 7, 213–219.

Chateigner-Boutin, A. L., Ramos-Vega, M., Guevara-Garcia, A., Andres, C., De La Luz Gutierrez-Nava, M., Cantero, A., et al. (2008). CLB19, a pentatricopeptide repeat protein required for editing of rpoA and clpP chloroplast transcripts. *The Plant Journal, 56*, 590–602.

Chateigner-Boutin, A. L., Des Francs-Small, C. C., Delannoy, E., Kahlau, S., Tanz, S. K., De Longevialle, A. F., et al. (2011). OTP70 is a pentatricopeptide repeat protein of the

E subgroup involved in splicing of the plastid transcript rpoC1. *The Plant Journal, 65*, 532–542.

Chi, W., Ma, J., Zhang, D., Guo, J., Chen, F., Lu, C., et al. (2008). The pentratricopeptide repeat protein DELAYED GREENING1 is involved in the regulation of early chloroplast development and chloroplast gene expression in *Arabidopsis*. *Plant Physiology, 147*, 573–584.

Chi, W., Mao, J., Li, Q., Ji, D., Zou, M., Lu, C., et al. (2010). Interaction of the pentatricopeptide-repeat protein DELAYED GREENING 1 with sigma factor SIG6 in the regulation of chloroplast gene expression in *Arabidopsis* cotyledons. *The Plant Journal, 64*, 14–25.

Cho, Y., Qiu, Y.-L., & Palmer, J. D. (1998a). Explosive invasion of plant mitochondria by a group I intron. *Proceedings of the National Academy of Sciences of the United States of America, 95*, 14244–14249.

Choury, D., Farre, J. C., Jordana, X., & Araya, A. (2004). Different patterns in the recognition of editing sites in plant mitochondria. *Nucleic Acids Research, 32*, 6397–6406.

Coffin, J. W., Dhillon, R., Ritzel, R. G., & Nargang, F. E. (1997). The *Neurospora crassa* cya-5 nuclear gene encodes a protein with a region of homology to the *Saccharomyces cerevisiae* PET309 protein and is required in a post-transcriptional step for the expression of the mitochondrially encoded COXI protein. *Current Genetics, 32*, 273–280.

Colas Des Francs-Small, C., Kroeger, T., Zmudjak, M., Ostersetzer-Biran, O., Rahimi, N., Small, I., et al. (2011). A PORR domain protein required for rpl2 and ccmF(C) intron splicing and for the biogenesis of c-type cytochromes in *Arabidopsis* mitochondria. *The Plant Journal.*

Cooper, M. P., Qu, L., Rohas, L. M., Lin, J., Yang, W., Erdjument-Bromage, H., et al. (2006). Defects in energy homeostasis in Leigh syndrome French Canadian variant through PGC-1alpha/LRP130 complex. *Genes & Development, 20*, 2996–3009.

Covello, P. S., & Gray, M. W. (1989). RNA editing in plant mitochondria. *Nature, 341*, 662–666.

Davies, S. M., Rackham, O., Shearwood, A. M., Hamilton, K. L., Narsai, R., Whelan, J., et al. (2009). Pentatricopeptide repeat domain protein 3 associates with the mitochondrial small ribosomal subunit and regulates translation. *FEBS Letters, 583*, 1853–1858.

De Longevialle, A. F., Meyer, E. H., Andres, C., Taylor, N. L., Lurin, C., Millar, A. H., et al. (2007). The pentatricopeptide repeat gene OTP43 is required for *trans*-splicing of the mitochondrial nad1 Intron 1 in *Arabidopsis thaliana*. *The Plant Cell, 19*, 3256–3265.

De Longevialle, A. F., Hendrickson, L., Taylor, N. L., Delannoy, E., Lurin, C., Badger, M., et al. (2008). The pentatricopeptide repeat gene OTP51 with two LAGLIDADG motifs is required for the cis-splicing of plastid ycf3 intron 2 in *Arabidopsis thaliana*. *The Plant Journal, 56*, 157–168.

Dieckmann, C. L., Pape, L. K., & Tzagoloff, A. (1982). Identification and cloning of a yeast nuclear gene (CBP1) involved in expression of mitochondrial cytochrome *b*. *Proceedings of the National Academy of Sciences of the United States of America, 79*, 1805–1809.

Dieckmann, C. L., Koerner, T. J., & Tzagoloff, A. (1984). Assembly of the mitochondrial membrane system. CBP1, a yeast nuclear gene involved in 5′ end processing of cytochrome *b* pre-mRNA. *Journal of Biological Chemistry, 259*, 4722–4731.

Ding, Y. H., Liu, N. Y., Tang, Z. S., Liu, J., & Yang, W. C. (2006). *Arabidopsis* GLUTAMINE-RICH PROTEIN23 is essential for early embryogenesis and encodes a novel nuclear PPR motif protein that interacts with RNA polymerase II subunit III. *The Plant Cell, 18*, 815–830.

Dombrovska, O., & Qiu, Y. L. (2004). Distribution of introns in the mitochondrial gene nad1 in land plants: phylogenetic and molecular evolutionary implications. *Molecular Phylogenetics and Evolution, 32*, 246–263.

Dombrowski, S., Brennicke, A., & Binder, S. (1997). 3′-Inverted repeats in plant mitochondrial mRNAs are processing signals rather than transcription terminators. *EMBO Journal, 16*, 5069–5076.

Dong, F. G., Wilson, K. G., & Makaroff, C. A. (1998). The radish (*Raphanus sativus* L.) mitochondrial cox2 gene contains an ACG at the predicted translation initiation site. *Current Genetics, 34*, 79–87.

Doniwa, Y., Ueda, M., Ueta, M., Wada, A., Kadowaki, K., & Tsutsumi, N. (2010). The involvement of a PPR protein of the P subfamily in partial RNA editing of an *Arabidopsis* mitochondrial transcript. *Gene, 454*, 39–46.

Duchêne, A. M., Giritch, A., Hoffmann, B., Cognat, V., Lancelin, D., Peeters, N. M., et al. (2005). Dual targeting is the rule for organellar aminoacyl-tRNA synthetases in *Arabidopsis thaliana*. *Proceedings of the National Academy of Sciences of the United States of America, 102*, 16484–16489.

Eberhard, S., Loiselay, C., Drapier, D., Bujaldon, S., Girard-Bascou, J., Kuras, R., et al. (2011). Dual functions of the nucleus-encoded factor TDA1 in trapping and translation activation of atpA transcripts in *Chlamydomonas reinhardtii* chloroplasts. *The Plant Journal, 67*, 1055–1066.

Ellis, T. P., Lukins, H. B., Nagley, P., & Corner, B. E. (1999). Suppression of a nuclear aep2 mutation in *Saccharomyces cerevisiae* by a base substitution in the 5′-untranslated region of the mitochondrial oli1 gene encoding subunit 9 of ATP synthase. *Genetics, 151*, 1353–1363.

Ellis, T. P., Helfenbein, K. G., Tzagoloff, A., & Dieckmann, C. L. (2004). Aep3p stabilizes the mitochondrial bicistronic mRNA encoding subunits 6 and 8 of the H^+-translocating ATP synthase of *Saccharomyces cerevisiae*. *Journal of Biological Chemistry, 279*, 15728–15733.

Evans, D., Marquez, S. M., & Pace, N. R. (2006). RNase P: interface of the RNA and protein worlds. *Trends in Biochemical Sciences, 31*, 333–341.

Falkenberg, M., Larsson, N. G., & Gustafsson, C. M. (2007). DNA replication and transcription in mammalian mitochondria. *Annual Review of Biochemistry, 76*, 679–699.

Farre, J. C., & Araya, A. (2001). Gene expression in isolated plant mitochondria: high fidelity of transcription, splicing and editing of a transgene product in electroporated organelles. *Nucleic Acids Research, 29*, 2484–2491.

Fedorova, O., & Zingler, N. (2007). Group II introns: structure, folding and splicing mechanism. *Biological Chemistry, 388*, 665–678.

Fey, J., & Maréchal-Drouard, L. (1999). Compilation and analysis of plant mitochondrial promoter sequences: An illustration of a divergent evolution between monocot and dicot mitochondria. *Biochemical and Biophysical Research Communications, 256*, 409–414.

Finnegan, P. M., Payne, M. J., Keramidaris, E., & Lukins, H. B. (1991). Characterization of a yeast nuclear gene, AEP2, required for accumulation of mitochondrial mRNA encoding subunit 9 of the ATP synthase. *Current Genetics, 20*, 53–61.

Fisk, D. G., Walker, M. B., & Barkan, A. (1999). Molecular cloning of the maize gene crp1 reveals similarity between regulators of mitochondrial and chloroplast gene expression. *EMBO Journal, 18*, 2621–2630.

Forner, J., Weber, B., Thuss, S., Wildum, S., & Binder, S. (2007). Mapping of mitochondrial mRNA termini in *Arabidopsis thaliana*: t-elements contribute to 5′ and 3′ end formation. *Nucleic Acids Research*.

Fritz-Laylin, L. K., Ginger, M. L., Walsh, C., Dawson, S. C., & Fulton, C. (2011). The *Naegleria* genome: a free-living microbial eukaryote lends unique insights into core eukaryotic cell biology. *Research in Microbiology, 162*, 607–618.

Fujii, S., & Small, I. (2011). The evolution of RNA editing and pentatricopeptide repeat genes. *New Phytology, 191*, 37–47.

Fukami, H., Chen, C. A., Chiou, C. Y., & Knowlton, N. (2007). Novel group I introns encoding a putative homing endonuclease in the mitochondrial cox1 gene of Scleractinian corals. *Journal of Molecular Evolution, 64*, 591–600.

Gagliardi, D., & Leaver, C. J. (1999). Polyadenylation accelerates the degradation of the mitochondrial mRNA associated with cytoplasmic male sterility in sunflower. *EMBO Journal, 18*, 3757–3766.

Gagliardi, D., Perrin, R., Maréchal-Drouard, L., Grienenberger, J. M., & Leaver, C. J. (2001). Plant mitochondrial polyadenylated mRNAs are degraded by a 3′- to 5′-exoribonuclease activity, which proceeds unimpeded by stable secondary structures. *Journal of Biological Chemistry, 276*, 43541–43547.

Gargouri, A. (2005). The reverse transcriptase encoded by ai1 intron is active in trans in the retro-deletion of yeast mitochondrial introns. *FEMS Yeast Research, 5*, 813–822.

Gass, D. A., Makaroff, C. A., & Palmer, J. D. (1992). Variable intron content of the NADH dehydrogenase subunit-4 gene of plant mitochondria. *Current Genetics, 21*, 423–430.

Geiss, K. T., Abbas, G. M., & Makaroff, C. A. (1994). Intron loss from the NADH dehydrogenase subunit 4 gene of lettuce mitochondrial DNA - evidence for homologous recombination of a cDNA intermediate. *Molecular and General Genetics, 243*, 97–105.

Giegé, P., & Brennicke, A. (1999). RNA editing in *Arabidopsis* mitochondria effects 441 C to U changes in ORFs. *Proceedings of the National Academy of Sciences of the United States of America, 96*, 15324–15329.

Giegé, P., & Brennicke, A. (2001). From gene to protein in higher plant mitochondria. *Comptes Rendus de l'Académie des Sciences – Series III, 324*, 209–217.

Giegé, P., Hoffmann, M., Binder, S., & Brennicke, A. (2000). RNA degradation buffers asymmetries of transcription in *Arabidopsis* mitochondria. *EMBO Reports, 1*, 164–170.

Giegé, P., Rayapuram, N., Meyer, E. H., Grienenberger, J. M., & Bonnard, G. (2004). CcmF(C) involved in cytochrome *c* maturation is present in a large sized complex in wheat mitochondria. *FEBS Letters, 563*, 165–169.

Giegé, P., Grienenberger, J. M., & Bonnard, G. (2008). Cytochrome *c* biogenesis in mitochondria. *Mitochondrion, 8*, 61–73.

Glanz, S., & Kuck, U. (2009). *Trans*-splicing of organelle introns–a detour to continuous RNAs. *Bioessays, 31*, 921–934.

Glaser, E., Sjoling, S., Tanudji, M., & Whelan, J. (1998). Mitochondrial protein import in plants. Signals, sorting, targeting, processing and regulation. *Plant Molecular Biology, 38*, 311–338.

Gobert, A., Gutmann, B., Taschner, A., Gößringer, M., Holzmann, J., Hartmann, R. K., et al. (2010). A single *Arabidopsis* organellar protein has RNase P activity. *Nature Structural & Molecular Biology, 17*, 740–744.

Grams, J., Morris, J. C., Drew, M. E., Wang, Z., Englund, P. T., & Hajduk, S. L. (2002). A trypanosome mitochondrial RNA polymerase is required for transcription and replication. *Journal of Biological Chemistry, 277*, 16952–16959.

Gray, M. W., & Doolittle, W. F. (1982). Has the endosymbiot hypothesis been proven? *Microbiological Reviews, 46*, 1–42.

Gray, M. W., Burger, G., & Lang, B. F. (1999). Mitochondrial evolution. *Science, 283*, 1476–1481.

Gray, M. W. (1982). Mitochondrial genome diversity and the evolution of mitochondrial DNA. *Canadian Journal of Biochemistry, 60*, 157–171.

Greenleaf, A. L., Kelly, J. L., & Lehman, I. R. (1986). Yeast RPO41 gene product is required for transcription and maintenance of the mitochondrial genome. *Proceedings of the National Academy of Sciences of the United States of America, 83*, 3391–3394.

Grewe, F., Viehoever, P., Weisshaar, B., & Knoop, V. (2009). A *trans*-splicing group I intron and tRNA-hyperediting in the mitochondrial genome of the lycophyte *Isoetes engelmannii*. *Nucleic Acids Research, 37*, 5093–5104.

Grohmann, L., Thieck, O., Herz, U., Schroder, W., & Brennicke, A. (1994). Translation of *nad9* mRNAs in mitochondria from *Solanum tuberosum* is restricted to completely edited transcripts. *Nucleic Acids Research, 22*, 3304–3311.

Groth-Malonek, M., Pruchner, D., Grewe, F., & Knoop, V. (2005). Ancestors of *trans*-splicing mitochondrial introns support serial sister group relationships of hornworts and mosses with vascular plants. *Molecular Biology and Evolution, 22*, 117–125.

Groth-Malonek, M., Wahrmund, U., Polsakiewicz, M., & Knoop, V. (2007). Evolution of a pseudogene: exclusive survival of a functional mitochondrial nad7 gene supports *Haplomitrium* as the earliest liverwort lineage and proposes a secondary loss of RNA editing in Marchantiidae. *Molecular Biology and Evolution, 24*, 1068–1074.

Gualberto, J. M., Lamattina, L., Bonnard, G., Weil, J. H., & Grienenberger, J. M. (1989). RNA editing in wheat mitochondria results in the conservation of protein sequences. *Nature, 341*, 660–662.

Gualberto, J. M., Weil, J. H., & Grienenberger, J. M. (1990). Editing of the wheat *coxIII* transcript: evidence for twelve C to U and one U to C conversions and for sequence similarities around editing sites. *Nucleic Acids Research, 18*, 3771–3776.

Guerrier-Takada, C., Gardiner, K., Marsh, T., Pace, N., & Altman, S. (1983). The RNA moiety of ribonuclease P is the catalytic subunit of the enzyme. *Cell, 35*, 849–857.

Gutierrez-Marcos, J. F., Dal Pra, M., Giulini, A., Costa, L. M., Gavazzi, G., Cordelier, S., et al. (2007). Empty pericarp4 encodes a mitochondrion-targeted pentatricopeptide repeat protein necessary for seed development and plant growth in maize. *The Plant Cell, 19*, 196–210.

Hamel, P., Corvest, V., Giegé, P., & Bonnard, G. (2009). Biochemical requirements for the maturation of mitochondrial c-type cytochromes. *Biochimica et Biophysica Acta, 1793*, 125–138.

Hammani, K., Okuda, K., Tanz, S. K., Chateigner-Boutin, A. L., Shikanai, T., & Small, I. (2009). A study of new *Arabidopsis* chloroplast RNA editing mutants reveals general features of editing factors and their target sites. *The Plant Cell, 21*, 3686–3699.

Hammani, K., Des Francs-Small, C. C., Takenaka, M., Tanz, S. K., Okuda, K., Shikanai, T., et al. (2011a). The pentatricopeptide repeat protein OTP87 is essential for RNA editing of nad7 and atp1 transcripts in *Arabidopsis* mitochondria. *Journal of Biological Chemistry, 286*, 21361–21371.

Hammani, K., Gobert, A., Hleibieh, K., Choulier, L., Small, I., & Giegé, P. (2011b). An *Arabidopsis* dual-localized pentatricopeptide repeat protein interacts with nuclear proteins involved in gene expression regulation. *The Plant Cell, 23*, 730–740.

Hartl, F. U., Pfanner, N., Nicholson, D. W., & Neupert, W. (1989). Mitochondrial protein import. *Biochimica et Biophysica Acta, 988*, 1–46.

Hashimoto, M., Endo, T., Peltier, G., Tasaka, M., & Shikanai, T. (2003). A nucleus-encoded factor, CRR2, is essential for the expression of chloroplast ndhB in *Arabidopsis*. *The Plant Journal, 36*, 541–549.

Hattori, M., & Sugita, M. (2009). A moss pentatricopeptide repeat protein binds to the 3′ end of plastid clpP pre-mRNA and assists with mRNA maturation. *FEBS Journal, 276*, 5860–5869.

Hattori, M., Miyake, H., & Sugita, M. (2007). A pentatricopeptide repeat protein is required for RNA processing of clpP Pre-mRNA in moss chloroplasts. *Journal of Biological Chemistry, 282*, 10773–10782.

Hecht, J., Grewe, F., & Knoop, V. (2011). Extreme RNA editing in coding islands and abundant microsatellites in repeat sequences of *Selaginella moellendorffii* mitochondria: the root of frequent plant mtDNA recombination in early tracheophytes. *Genome Biology and Evolution, 3*, 344–358.

Hedtke, B., Borner, T., & Weihe, A. (1997). Mitochondrial and chloroplast phage-type RNA polymerases in *Arabidopsis*. *Science, 277*, 809–811.

Hedtke, B., Borner, T., & Weihe, A. (2000). One RNA polymerase serving two genomes. *EMBO Reports, 1*, 435–440.

Hiesel, R., Wissinger, B., Schuster, W., & Brennicke, A. (1989). RNA editing in plant mitochondria. *Science, 246*, 1632–1634.

Hiesel, R., Von Haeseler, A., & Brennicke, A. (1994). Plant mitochondrial nucleic acid sequences as a tool for phylogenetic analysis. *Proceedings of the National Academy of Sciences of the United States of America, 91*, 634–638.

Hilson, P., Allemeersch, J., Altmann, T., Aubourg, S., Avon, A., Beynon, J., et al. (2004). Versatile gene-specific sequence tags for *Arabidopsis* functional genomics: transcript profiling and reverse genetics applications. *Genome Research, 14*, 2176–2189.

Hoffmann, M., Dombrowski, S., Guha, C., & Binder, S. (1999). Cotranscription of the rpl5-rps14-cob gene cluster in pea mitochondria. *Molecular and General Genetics, 261*, 537–545.

Holec, S., Lange, H., Kuhn, K., Alioua, M., Borner, T., & Gagliardi, D. (2006). Relaxed transcription in *Arabidopsis* mitochondria is counterbalanced by RNA stability control mediated by polyadenylation and polynucleotide phosphorylase. *Molecular and Cellular Biology, 26*, 2869–2876.

Holzle, A., Jonietz, C., Torjek, O., Altmann, T., Binder, S., & Forner, J. (2011). A restorer of fertility-like PPR gene is required for 5′-end processing of the nad4 mRNA in mitochondria of *Arabidopsis thaliana*. *The Plant Journal, 65*, 737–744.

Holzmann, J., Frank, P., Loffler, E., Bennett, K. L., Gerner, C., & Rossmanith, W. (2008). RNase P without RNA: identification and functional reconstitution of the human mitochondrial tRNA processing enzyme. *Cell, 135*, 462–474.

Ichinose, M., Tasaki, E., Sugita, C., & Sugita, M. (2011). A PPR-DYW protein is required for splicing of a group II intron of cox1 pre-mRNA in *Physcomitrella patens*. *The Plant Journal*.

Ikeda, T. M., & Gray, M. W. (1999). Characterization of a DNA-binding protein implicated in transcription in wheat mitochondria. *Molecular and Cellular Biology, 19*, 8113–8122.

Islas-Osuna, M. A., Ellis, T. P., Marnell, L. L., Mittelmeier, T. M., & Dieckmann, C. L. (2002). Cbp1 is required for translation of the mitochondrial cytochrome *b* mRNA of *Saccharomyces cerevisiae*. *J Biol Chem, 277*, 37987–37990.

Jacquier, A., & Michel, F. (1990). Base pairing interactions involving the 5′ and 3′ terminal nucleotides of group II self splicing introns. *Journal of Molecular Biology, 213*, 437–447.

Janke, A., & Paabo, S. (1993). Editing of a transfer RNA anticodon in marsupial mitochondria changes its codon recognition. *Nucleic Acids Research., 21*, 1523–1525.

Jiang, H., Sun, W., Wang, Z., Zhang, J., Chen, D., & Murchie, A. I. (2011). Identification and characterization of the mitochondrial RNA polymerase and transcription factor in the fission yeast *Schizosaccharomyces pombe*. *Nucleic Acids Research, 39*, 5119–5130.

Johnson, X., Wostrikoff, K., Finazzi, G., Kuras, R., Schwarz, C., Bujaldon, S., et al. (2010). MRL1, a conserved pentatricopeptide repeat protein, is required for stabilization of rbcL mRNA in *Chlamydomonas* and *Arabidopsis*. *The Plant Cell, 22*, 234–248.

Jonietz, C., Forner, J., Holzle, A., Thuss, S., & Binder, S. (2010). RNA PROCESSING FACTOR2 is required for 5′ end processing of nad9 and cox3 mRNAs in mitochondria of *Arabidopsis thaliana*. *The Plant Cell, 22*, 443–453.

Jonietz, C., Forner, J., Hildebrandt, T., & Binder, S. (2011). RNA PROCESSING FACTOR3 is crucial for the accumulation of mature ccmC transcripts in mitochondria of *Arabidopsis* accession Columbia. *Plant Physiology, 157*, 1430–1439.

Keren, I., Klipcan, L., Bezawork-Geleta, A., Kolton, M., Shaya, F., & Ostersetzer-Biran, O. (2008). Characterization of the molecular basis of group II intron RNA recognition by CRS1-CRM domains. *Journal of Biological Chemistry, 283*, 23333–23342.

Kim, S. R., Yang, J. I., Moon, S., Ryu, C. H., An, K., Kim, K. M., et al. (2009). Rice OGR1 encodes a pentatricopeptide repeat-DYW protein and is essential for RNA editing in mitochondria. *The Plant Journal, 59*, 738–749.

Kitakawa, M., & Isono, K. (1991). The mitochondrial ribosomes. *Biochimie, 73*, 813–825.

Klein, R. R., Klein, P. E., Mullet, J. E., Minx, P., Rooney, W. L., & Schertz, K. F. (2005). Fertility restorer locus Rf1 [corrected] of sorghum (*Sorghum bicolor* L.) encodes a pentatricopeptide repeat protein not present in the colinear region of rice chromosome 12. *Theoretical and Applied Genetics, 111*, 994–1012.

Knoop, V., Schuster, W., Wissinger, B., & Brennicke, A. (1991). *Trans*-splicing integrates an exon of 22 nucleotides into the *nad5* messenger RNA in higher plant mitochondria. *EMBO Journal, 10*, 3483–3493.

Knoop, V. (2004). The mitochondrial DNA of land plants: peculiarities in phylogenetic perspective. *Current Genetics, 46*, 123–139.

Koc, E. C., Burkhart, W., Blackburn, K., Koc, H., Moseley, A., & Spremulli, L. L. (2001). Identification of four proteins from the small subunit of the mammalian mitochondrial ribosome using a proteomics approach. *Protein Science, 10*, 471–481.

Koizuka, N., Imai, R., Fujimoto, H., Hayakawa, T., Kimura, Y., Kohno-Murase, J., et al. (2003). Genetic characterization of a pentatricopeptide repeat protein gene, orf687, that restores fertility in the cytoplasmic male-sterile Kosena radish. *The Plant Journal, 34*, 407–415.

Koprivova, A., Des Francs-Small, C. C., Calder, G., Mugford, S. T., Tanz, S., Lee, B. R., et al. (2010). Identification of a pentatricopeptide repeat protein implicated in splicing of intron 1 of mitochondrial nad7 transcripts. *Journal of Biological Chemistry, 285*, 32192–32199.

Kotera, E., Tasaka, M., & Shikanai, T. (2005). A pentatricopeptide repeat protein is essential for RNA editing in chloroplasts. *Nature, 433*, 326–330.

Kroeger, T. S., Watkins, K. P., Friso, G., Van Wijk, K. J., & Barkan, A. (2009). A plant-specific RNA-binding domain revealed through analysis of chloroplast group II intron splicing. *Proceedings of the National Academy of Sciences of the United States of America, 106*, 4537–4542.

Kudla, J., Albertazzi, F. J., Blazevic, D., Hermann, M., & Bock, R. (2002). Loss of the mitochondrial cox2 intron 1 in a family of monocotyledonous plants and utilization of mitochondrial intron sequences for the construction of a nuclear intron. *Molecular Genetics and Genomics, 267*, 223–230.

Kugita, M., Yamamoto, Y., Fujikawa, T., Matsumoto, T., & Yoshinaga, K. (2003). RNA editing in hornwort chloroplasts makes more than half the genes functional. *Nucleic Acids Research, 31*, 2417–2423.

Kuhl, I., Dujeancourt, L., Gaisne, M., Herbert, C. J., & Bonnefoy, N. (2011). A genome wide study in fission yeast reveals nine PPR proteins that regulate mitochondrial gene expression. *Nucleic Acids Research, 39*, 8029–8041.

Kuhn, K., Weihe, A., & Borner, T. (2005). Multiple promoters are a common feature of mitochondrial genes in *Arabidopsis*. *Nucleic Acids Research, 33*, 337–346.

Kuhn, K., Richter, U., Meyer, E. H., Delannoy, E., De Longevialle, A. F., O'Toole, N., et al. (2009). Phage-type RNA polymerase RPOTmp performs gene-specific transcription in mitochondria of *Arabidopsis thaliana*. *The Plant Cell, 21*, 2762–2779.

Kunzmann, A., Brennicke, A., & Marchfelder, A. (1998). 5′ end maturation and RNA editing have to precede tRNA 3′ processing in plant mitochondria. *Proceedings of the National Academy of Sciences of the United States of America, 95*, 108–113.

Lai, L. B., Bernal-Bayard, P., Mohannath, G., Lai, S. M., Gopalan, V., & Vioque, A. (2011). A functional RNase P protein subunit of bacterial origin in some eukaryotes. *Molecular Genetics and Genomics, 286*, 359–369.

Lambowitz, A. M., & Zimmerly, S. (2004). Mobile group II introns. *Annual Review of Genetics, 38*, 1–35.

Lang, B. F., Burger, G., O'Kelly, C. J., Cedergren, R., Golding, G. B., Lemieux, C., et al. (1997). An ancestral mitochondrial DNA resembling a eubacterial genome in miniature. *Nature, 387*, 493–496.

Li, J., Maga, J. A., Cermakian, N., Cedergren, R., & Feagin, J. E. (2001). Identification and characterization of a *Plasmodium falciparum* RNA polymerase gene with similarity to mitochondrial RNA polymerases. *Molecular and Biochemical Parasitology, 113*, 261–269.

Lightowlers, R. N., & Chrzanowska-Lightowlers, Z. M. (2008). PPR (pentatricopeptide repeat) proteins in mammals: important aids to mitochondrial gene expression. *Biochemical Journal, 416*, e5–e6.

Lipinski, K. A., Puchta, O., Surendranath, V., Kudla, M., & Golik, P. (2011). Revisiting the yeast PPR proteins–application of an Iterative Hidden Markov Model algorithm reveals new members of the rapidly evolving family. *Molecular Biology and Evolution, 28*, 2935–2948.

Liu, Y., He, J., Chen, Z., Ren, X., Hong, X., & Gong, Z. (2010). ABA overly-sensitive 5 (ABO5), encoding a pentatricopeptide repeat protein required for cis-splicing of mitochondrial nad2 intron 3, is involved in the abscisic acid response in *Arabidopsis*. *The Plant Journal, 63*, 749–765.

Loiselay, C., Gumpel, N. J., Girard-Bascou, J., Watson, A. T., Purton, S., Wollman, F. A., et al. (2008). Molecular identification and function of cis- and trans-acting determinants for petA transcript stability in *Chlamydomonas reinhardtii* chloroplasts. *Molecular and Cellular Biology, 28*, 5529–5542.

Lu, B., & Hanson, M. R. (1994). A single homogeneous form of ATP6 protein accumulates in petunia mitochondria despite the presence of differentially edited atp6 transcripts. *The Plant Cell, 6*, 1955–1968.

Lu, Y., Li, C., Wang, H., Chen, H., Berg, H., & Xia, Y. (2011). AtPPR2, an *Arabidopsis* pentatricopeptide repeat protein, binds to plastid 23S rRNA and plays an important role in the first mitotic division during gametogenesis and in cell proliferation during embryogenesis. *The Plant Journal, 67*, 13–25.

Lupold, D. S., Caoile, A. G., & Stern, D. B. (1999). Genomic context influences the activity of maize mitochondrial cox2 promoters. *Proceedings of the National Academy of Sciences of the United States of America, 96*, 11670–11675.

Lurin, C., Andres, C., Aubourg, S., Bellaoui, M., Bitton, F., Bruyere, C., et al. (2004). Genome-wide analysis of *Arabidopsis* pentatricopeptide repeat proteins reveals their essential role in organelle biogenesis. *The Plant Cell, 16*, 2089–2103.

Malek, O., Lattig, K., Hiesel, R., Brennicke, A., & Knoop, V. (1996). RNA editing in bryophytes and a molecular phylogeny of land plants. *EMBO Journal, 15*, 1403–1411.

Mancebo, R., Zhou, X., Shillinglaw, W., Henzel, W., & Macdonald, P. M. (2001). BSF binds specifically to the bicoid mRNA 3′ untranslated region and contributes to stabilization of bicoid mRNA. *Molecular and Cellular Biology, 21*, 3462–3471.

Manthey, G. M., & McEwen, J. E. (1995). The product of the nuclear gene PET309 is required for translation of mature mRNA and stability or production of intron-containing RNAs derived from the mitochondrial COX1 locus of *Saccharomyces cerevisiae*. *EMBO Journal, 14*, 4031–4043.

Maréchal-Drouard, L., Cosset, A., Remacle, C., Ramamonjisoa, D., & Dietrich, A. (1996a). A single editing event is a prerequisite for efficient processing of potato mitochondrial phenylalanine tRNA. *Molecular and Cellular Biology, 16*, 3504–3510.

Maréchal-Drouard, L., Kumar, R., Remacle, C., & Small, I. (1996b). RNA editing of larch mitochondrial tRNA(His) precursors is a prerequisite for processing. *Nucleic Acids Research, 24*, 3229–3234.

Margulis, L. (1975). Symbiotic theory of the origin of eukaryotic organelles: criteria for proof. *Symposia of the Society for Experimental Biology, 29*, 21–27.

Martin, W., & Koonin, E. V. (2006). Introns and the origin of nucleus-cytosol compartmentalization. *Nature, 440*, 41–45.

Maslov, D. A., Spremulli, L. L., Sharma, M. R., Bhargava, K., Grasso, D., Falick, A. M., et al. (2007). Proteomics and electron microscopic characterization of the unusual mitochondrial ribosome-related 45S complex in *Leishmania tarentolae*. *Molecular and Biochemical Parasitology, 152*, 203–212.

Meierhoff, K., Felder, S., Nakamura, T., Bechtold, N., & Schuster, G. (2003). HCF152, an *Arabidopsis* RNA binding pentatricopeptide repeat protein involved in the processing of chloroplast psbB-psbT-psbH-petB-petD RNAs. *The Plant Cell, 15*, 1480–1495.

Meng, Q., Wang, Y., & Liu, X. Q. (2005). An intron-encoded protein assists RNA splicing of multiple similar introns of different bacterial genes. *Journal of Biological Chemistry, 280*, 35085–35088.

Merchant, S., et al. (2007). The *Chlamydomonas* genome reveals the evolution of key animal and plant functions. *Science, 318*, 245–250.

Michaud, M., Maréchal-Drouard, L., & Duchêne, A. M. (2010). RNA trafficking in plant cells: targeting of cytosolic mRNAs to the mitochondrial surface. *Plant Molecular Biology, 73*, 697–704.

Michel, F., & Ferat, J. L. (1995). Structure and activities of group II introns. *Annual Review of Biochemistry, 64*, 435–461.

Michel, F., Jacquier, A., & Dujon, B. (1982). Comparison of fungal mitochondrial introns reveals extensible homologies in RNA secondary structure. *Biochimie, 64*, 867–881.

Mili, S., & Pinol-Roma, S. (2003). LRP130, a pentatricopeptide motif protein with a noncanonical RNA-binding domain, is bound *in vivo* to mitochondrial and nuclear RNAs. *Molecular and Cellular Biology, 23*, 4972–4982.

Millar, A. H., Whelan, J., Soole, K. L., & Day, D. A. (2011). Organization and regulation of mitochondrial respiration in plants. *Annual Review of Plant Biology, 62*, 79–104.

Miller, M. L., Antes, T. J., Qian, F., & Miller, D. L. (2006). Identification of a putative mitochondrial RNA polymerase from *Physarum polycephalum*: characterization, expression, purification, and transcription *in vitro*. *Current Genetics, 49*, 259–271.

Mingler, M. K., Hingst, A. M., Clement, S. L., Yu, L. E., Reifur, L., & Koslowsky, D. J. (2006). Identification of pentatricopeptide repeat proteins in *Trypanosoma brucei*. *Molecular and Biochemical Parasitology, 150*, 37–45.

Mohr, G., & Lambowitz, A. M. (2003). Putative proteins related to group II intron reverse transcriptase/maturases are encoded by nuclear genes in higher plants. *Nucleic Acids Research, 31*, 647–652.

Moore, A. L., & Siedow, J. N. (1992). The nature and regulation of the alternative oxidase of plant mitochondria. *Biochemical Society Transactions, 20*, 361–363.

Morales, M. J., Dang, Y. L., Lou, Y. C., Sulo, P., & Martin, N. C. (1992). A 105-kDa protein is required for yeast mitochondrial RNase P activity. *Proceedings of the National Academy of Sciences of the United States of America, 89*, 9875–9879.

Moreno, J. I., Buie, K. S., Price, R. E., & Piva, M. A. (2009). Ccm1p/Ygr150cp, a pentatricopeptide repeat protein, is essential to remove the fourth intron of both COB and COX1 pre-mRNAs in *Saccharomyces cerevisiae*. *Current Genetics, 55*, 475–484.

Motorin, Y., & Helm, M. (2010). tRNA stabilization by modified nucleotides. *Biochemistry, 49*, 4934–4944.

Mulero, J. J., & Fox, T. D. (1993). PET111 acts in the 5′-leader of the *Saccharomyces cerevisiae* mitochondrial COX2 mRNA to promote its translation. *Genetics, 133*, 509–516.

Nakagawa, N., & Sakurai, N. (2006). A mutation in At-nMat1a, which encodes a nuclear gene having high similarity to group II intron maturase, causes impaired splicing of mitochondrial NAD4 transcript and altered carbon metabolism in *Arabidopsis thaliana. Plant and Cell Physiology, 47*, 772–783.

Nakamura, T., & Sugita, M. (2008). A conserved DYW domain of the pentatricopeptide repeat protein possesses a novel endoribonuclease activity. *FEBS Letters, 582*, 4163–4168.

Neuwirt, J., Takenaka, M., Van Der Merwe, J. A., & Brennicke, A. (2005). An *in vitro* RNA editing system from cauliflower mitochondria: editing site recognition parameters can vary in different plant species. *RNA, 11*, 1563–1570.

Nouet, C., Bourens, M., Hlavacek, O., Marsy, S., Lemaire, C., & Dujardin, G. (2007). Rmd9p controls the processing/stability of mitochondrial mRNAs and its over-expression compensates for a partial deficiency of oxa1p in *Saccharomyces cerevisiae. Genetics, 175*, 1105–1115.

O'Brien, E. A., Badidi, E., Barbasiewicz, A., Desousa, C., Lang, B. F., & Burger, G. (2003). GOBASE–a database of mitochondrial and chloroplast information. *Nucleic Acids Research, 31*, 176–178.

Oda, K., Kohchi, T., & Ohyama, K. (1992a). Mitochondrial DNA of *Marchantia-polymorpha* as a single circular form with no incorporation of foreign DNA. *Bioscience, Biotechnology and Biochemistry, 56*, 132–135.

Oda, K., Yamato, K., Ohta, E., Nakamura, Y., Takemura, M., Nozato, N., et al. (1992b). Gene organization deduced from the complete sequence of liverwort *Marchantia polymorpha* mitochondrial DNA. A primitive form of plant mitochondrial genome. *Journal of Molecular Biology, 223*, 1–7.

Ohtani, S., Ichinose, M., Tasaki, E., Aoki, Y., Komura, Y., & Sugita, M. (2010). Targeted gene disruption identifies three PPR-DYW proteins involved in RNA editing for five editing sites of the moss mitochondrial transcripts. *Plant and Cell Physiology, 51*, 1942–1949.

Okuda, K., Nakamura, T., Sugita, M., Shimizu, T., & Shikanai, T. (2006). A pentatricopeptide repeat protein is a site recognition factor in chloroplast RNA editing. *Journal of Biological Chemistry, 281*, 37661–37667.

Okuda, K., Myouga, F., Motohashi, R., Shinozaki, K., & Shikanai, T. (2007). Conserved domain structure of pentatricopeptide repeat proteins involved in chloroplast RNA editing. *Proceedings of the National Academy of Sciences of the United States of America, 104*, 8178–8183.

Okuda, K., Habata, Y., Kobayashi, Y., & Shikanai, T. (2008). Amino acid sequence variations in *Nicotiana* CRR4 orthologs determine the species-specific efficiency of RNA editing in plastids. *Nucleic Acids Research, 36*, 6155–6164.

Okuda, K., Chateigner-Boutin, A. L., Nakamura, T., Delannoy, E., Sugita, M., Myouga, F., et al. (2009). Pentatricopeptide repeat proteins with the DYW motif have distinct molecular functions in RNA editing and RNA cleavage in *Arabidopsis* chloroplasts. *The Plant Cell, 21*, 146–156.

Okuda, K., Hammani, K., Tanz, S. K., Peng, L., Fukao, Y., Myouga, F., et al. (2010). The pentatricopeptide repeat protein OTP82 is required for RNA editing of plastid ndhB and ndhG transcripts. *The Plant Journal, 61*, 339–349.

O'Toole, N., Hattori, M., Andres, C., Iida, K., Lurin, C., Schmitz-Linneweber, C., et al. (2008). On the expansion of the pentatricopeptide repeat gene family in plants. *Molecular Biology and Evolution, 25*, 1120–1128.

Oudot-Le Secq, M. P., Fontaine, J. M., Rousvoal, S., Kloareg, B., & Loiseaux-De Goer, S. (2001). The complete sequence of a brown algal mitochondrial genome, the ectocarpale *Pylaiella littoralis* (L.) Kjellm. *Journal of Molecular Evolution, 53*, 80–88.

Payne, M. J., Schweizer, E., & Lukins, H. B. (1991). Properties of two nuclear pet mutants affecting expression of the mitochondrial oli1 gene of *Saccharomyces cerevisiae. Current Genetics, 19*, 343–351.

Payne, M. J., Finnegan, P. M., Smooker, P. M., & Lukins, H. B. (1993). Characterization of a second nuclear gene, AEP1, required for expression of the mitochondrial OLI1 gene in *Saccharomyces cerevisiae*. *Current Genetics, 24*, 126–135.

Perrin, R., Lange, H., Grienenberger, J. M., & Gagliardi, D. (2004a). AtmtPNPase is required for multiple aspects of the 18S rRNA metabolism in *Arabidopsis thaliana* mitochondria. *Nucleic Acids Research, 32*, 5174–5182.

Perrin, R., Meyer, E. H., Zaepfel, M., Kim, Y. J., Mache, R., Grienenberger, J. M., et al. (2004b). Two exoribonucleases act sequentially to process mature 3′-ends of atp9 mRNAs in *Arabidopsis* mitochondria. *Journal of Biological Chemistry, 279*, 25440–25446.

Pfalz, J., Liere, K., Kandlbinder, A., Dietz, K. J., & Oelmuller, R. (2006). pTAC2, -6, and -12 are components of the transcriptionally active plastid chromosome that are required for plastid gene expression. *The Plant Cell, 18*, 176–197.

Pfalz, J., Bayraktar, O. A., Prikryl, J., & Barkan, A. (2009). Site-specific binding of a PPR protein defines and stabilizes 5′ and 3′ mRNA termini in chloroplasts. *EMBO Journal, 28*, 2042–2052.

Phreaner, C. G., Williams, M. A., & Mulligan, R. M. (1996). Incomplete editing of rps12 transcripts results in the synthesis of polymorphic polypeptides in plant mitochondria. *The Plant Cell, 8*, 107–117.

Poutre, C. G., & Fox, T. D. (1987). PET111, a *Saccharomyces cerevisiae* nuclear gene required for translation of the mitochondrial mRNA encoding cytochrome *c* oxidase subunit II. *Genetics, 115*, 637–647.

Prikryl, J., Rojas, M., Schuster, G., & Barkan, A. (2011). Mechanism of RNA stabilization and translational activation by a pentatricopeptide repeat protein. *Proceedings of the National Academy of Sciences of the United States of America, 108*, 415–420.

Pring, D. R., Mullen, J. A., & Kempken, F. (1992). Conserved sequence blocks 5′ to start codons of plant mitochondrial genes. *Plant Molecular Biology, 19*, 313–317.

Pruchner, D., Beckert, S., Muhle, H., & Knoop, V. (2002). Divergent intron conservation in the mitochondrial nad2 gene: signatures for the three bryophyte classes (mosses, liverworts, and hornworts) and the lycophytes. *Journal of Molecular Evolution, 55*, 265–271.

Puchta, O., Lubas, M., Lipinski, K. A., Piatkowski, J., Malecki, M., & Golik, P. (2010). DMR1 (CCM1/YGR150C) of *Saccharomyces cerevisiae* encodes an RNA-binding protein from the pentatricopeptide repeat family required for the maintenance of the mitochondrial 15S ribosomal RNA. *Genetics, 184*, 959–973.

Pusnik, M., & Schneider, A. A. trypanosomal pentatricopeptide repeat protein stabilizes the mitochondrial mRNAs of cytochrome oxidase subunit 1 and 2. *Eukaryotic Cell.*

Pusnik, M., Small, I., Read, L. K., Fabbro, T., & Schneider, A. (2007). Pentatricopeptide repeat proteins in *Trypanosoma brucei* function in mitochondrial ribosomes. *Molecular and Cellular Biology, 27*, 6876–6888.

Qiu, Y. L., Li, L., Wang, B., Chen, Z., Knoop, V., Groth-Malonek, M., et al. (2006). The deepest divergences in land plants inferred from phylogenomic evidence. *Proceedings of the National Academy of Sciences of the United States of America, 103*, 15511–15516.

Quesada, V., Sarmiento-Manus, R., Gonzalez-Bayon, R., Hricova, A., Perez-Marcos, R., Gracia-Martinez, E., et al. (2011). *Arabidopsis* RUGOSA2 encodes an mTERF family member required for mitochondrion, chloroplast and leaf development. *The Plant Journal.*

Rackham, O., Davies, S. M., Shearwood, A. M., Hamilton, K. L., Whelan, J., & Filipovska, A. (2009). Pentatricopeptide repeat domain protein 1 lowers the levels of mitochondrial leucine tRNAs in cells. *Nucleic Acids Research, 37*, 5859–5867.

Raczynska, K. D., Le Ret, M., Rurek, M., Bonnard, G., Augustyniak, H., & Gualberto, J. M. (2006). Plant mitochondrial genes can be expressed from mRNAs lacking stop codons. *FEBS Letters, 580*, 5641–5646.

Rasmusson, A. G., Soole, K. L., & Elthon, T. E. (2004). Alternative NAD(P)H dehydrogenases of plant mitochondria. *Annual Review of Plant Biology, 55*, 23–39.

Raynaud, C., Loiselay, C., Wostrikoff, K., Kuras, R., Girard-Bascou, J., Wollman, F. A., et al. (2007). Evidence for regulatory function of nucleus-encoded factors on mRNA stabilization and translation in the chloroplast. *Proceedings of the National Academy of Sciences of the United States of America, 104*, 9093–9098.

Remacle, C., & Maréchal-Drouard, L. (1996). Characterization of the potato mitochondrial transcription unit containing 'native' trnS (GCU), trnF (GAA) and trnP (UGG). *Plant Molecular Biology, 30*, 553–563.

Ringel, R., Sologub, M., Morozov, Y. I., Litonin, D., Cramer, P., & Temiakov, D. (2011). Structure of human mitochondrial RNA polymerase. *Nature*.

Robbins, J. C., Heller, W. P., & Hanson, M. R. (2009). A comparative genomics approach identifies a PPR-DYW protein that is essential for C-to-U editing of the *Arabidopsis* chloroplast accD transcript. *RNA, 15*, 1142–1153.

Roberti, M., Polosa, P. L., Bruni, F., Manzari, C., Deceglie, S., Gadaleta, M. N., et al. (2009). The MTERF family proteins: mitochondrial transcription regulators and beyond. *Biochimica et Biophysica Acta, 1787*, 303–311.

Rodeheffer, M. S., Boone, B. E., Bryan, A. C., & Shadel, G. S. (2001). Nam1p, a protein involved in RNA processing and translation, is coupled to transcription through an interaction with yeast mitochondrial RNA polymerase. *Journal of Biological Chemistry, 276*, 8616–8622.

Rudinger, M., Polsakiewicz, M., & Knoop, V. (2008). Organellar RNA editing and plant-specific extensions of pentatricopeptide repeat proteins in jungermanniid but not in marchantiid liverworts. *Molecular Biology and Evolution, 25*, 1405–1414.

Rudinger, M., Fritz-Laylin, L., Polsakiewicz, M., & Knoop, V. (2011a). Plant-type mitochondrial RNA editing in the protist *Naegleria gruberi*. *RNA*.

Rudinger, M., Szovenyi, P., Rensing, S. A., & Knoop, V. (2011b). Assigning DYW-type PPR proteins to RNA editing sites in the funariid mosses *Physcomitrella patens* and *Funaria hygrometrica*. *The Plant Journal*.

Ruzzenente, B., Metodiev, M. D., Wredenberg, A., Bratic, A., Park, C. B., Camara, Y., et al. (2011). LRPPRC is necessary for polyadenylation and coordination of translation of mitochondrial mRNAs. *EMBO Journal*.

Salanoubat, M., Genin, S., Artiguenave, F., Gouzy, J., Mangenot, S., Arlat, M., et al. (2002). Genome sequence of the plant pathogen *Ralstonia solanacearum*. *Nature, 415*, 497–502.

Salinas, T., Duchêne, A. M., & Maréchal-Drouard, L. (2008). Recent advances in tRNA mitochondrial import. *Trends in Biochemical Sciences, 33*, 320–329.

Salone, V., Rudinger, M., Polsakiewicz, M., Hoffmann, B., Groth-Malonek, M., Szurek, B., et al. (2007). A hypothesis on the identification of the editing enzyme in plant organelles. *FEBS Letters, 581*, 4132–4138.

Sanchez, M. I., Mercer, T. R., Davies, S. M., Shearwood, A. M., Nygard, K. K., Richman, T. R., et al. (2011). RNA processing in human mitochondria. *Cell Cycle, 10*, 2904–2916.

Sasarman, F., Brunel-Guitton, C., Antonicka, H., Wai, T., & Shoubridge, E. A. LRPPRC and SLIRP interact in a ribonucleoprotein complex that regulates posttranscriptional gene expression in mitochondria. *Molecular Biology of the Cell, 21*, 1315–1323.

Sato, S., Nakamura, Y., Kaneko, T., Asamizu, E., & Tabata, S. (1999). Complete structure of the chloroplast genome of *Arabidopsis thaliana*. *DNA Research, 6*, 283–290.

Schimper, A. F. W. (1883). Uber die Entwicklung der Chlorophyllkorner und Farbkörner (Teil 1). *Botanische Zeitung, 41*, 105–114.

Schmelzer, C., & Schweyen, R. J. (1986). Self-splicing of group II introns *in vitro*: mapping of the branch point and mutational inhibition of lariat formation. *Cell, 46*, 557–565.

Schmitz-Linneweber, C., & Small, I. (2008). Pentatricopeptide repeat proteins: a socket set for organelle gene expression. *Trends in Plant Science, 13*, 663–670.

Schmitz-Linneweber, C., Williams-Carrier, R., & Barkan, A. (2005). RNA immunoprecipitation and microarray analysis show a chloroplast Pentatricopeptide repeat protein to be associated with the 5′ region of mRNAs whose translation it activates. *The Plant Cell, 17*, 2791–2804.

Schmitz-Linneweber, C., Williams-Carrier, R. E., Williams-Voelker, P. M., Kroeger, T. S., Vichas, A., & Barkan, A. (2006). A pentatricopeptide repeat protein facilitates the *trans*-splicing of the maize chloroplast rps12 pre-mRNA. *The Plant Cell, 18*, 2650–2663.

Schuster, W., Hiesel, R., Wissinger, B., & Brennicke, A. (1990). RNA editing in the cytochrome b locus of the higher plant *Oenothera berteriana* includes a U to C transition. *Molecular and Cellular Biology, 10*, 2428–2431.

Schuster, W. (1993). Ribosomal protein gene rpl5 is cotranscribed with the nad3 gene in *Oenothera* mitochondria. *Molecular and General Genetics, 240*, 445–449.

Shadel, G. S., & Clayton, D. A. (1993). Mitochondrial transcription initiation. Variation and conservation. *Journal of Biological Chemistry, 268*, 16083–16086.

Shikanai, T. (2006). RNA editing in plant organelles: machinery, physiological function and evolution. *Cellular and Molecular Life Sciences, 63*, 698–708.

Shutt, T. E., & Gray, M. W. (2006). Bacteriophage origins of mitochondrial replication and transcription proteins. *Trends in Genetics, 22*, 90–95.

Sieber, F., Duchêne, A. M., & Maréchal-Drouard, L. (2011). Mitochondrial RNA import: from diversity of natural mechanisms to potential applications. *International Review of Cell and Molecular Biology, 287*, 145–190.

Small, I. D., & Peeters, N. (2000). The PPR motif - a TPR-related motif prevalent in plant organellar proteins. *Trends in Biochemical Sciences, 25*, 46–47.

Smith, D. R. (2009). Unparalleled GC content in the plastid DNA of Selaginella. *Plant Molecular Biology, 71*, 627–639.

Sondheimer, N., Fang, J. K., Polyak, E., Falk, M. J., & Avadhani, N. G. (2010). Leucine-rich pentatricopeptide-repeat containing protein regulates mitochondrial transcription. *Biochemistry, 49*, 7467–7473.

Sper-Whitis, G. L., Russell, A. L., & Vaughn, J. C. (1994). Mitochondrial RNA editing of cytochrome c oxidase subunit II (coxII) in the primitive vascular plant *Psilotum nudum*. *Biochimica et Biophysica Acta, 1218*, 218–220.

Staudinger, M., & Kempken, F. (2003). Electroporation of isolated higher-plant mitochondria: transcripts of an introduced cox2 gene, but not an atp6 gene, are edited in organello. *Molecular Genetics and Genomics, 269*, 553–561.

Steinhauser, S., Beckert, S., Capesius, I., Malek, O., & Knoop, V. (1999). Plant mitochondrial RNA editing: extreme in hornworts and dividing the liverworts? *Journal of Molecular Evolution, 48*, 303–312.

Sterky, F. H., Ruzzenente, B., Gustafsson, C. M., Samuelsson, T., & Larsson, N. G. (2010). LRPPRC is a mitochondrial matrix protein that is conserved in metazoans. *Biochemical and Biophysical Research Communications, 398*, 759–764.

Stribinskis, V., & Ramos, K. S. (2007). Rpm2p, a protein subunit of mitochondrial RNase P, physically and genetically interacts with cytoplasmic processing bodies. *Nucleic Acids Research, 35*, 1301–1311.

Stribinskis, V., Heyman, H. C., Ellis, S. R., Steffen, M. C., & Martin, N. C. (2005). Rpm2p, a component of yeast mitochondrial RNase P, acts as a transcriptional activator in the nucleus. *Molecular and Cellular Biology, 25*, 6546–6558.

Sung, T. Y., Tseng, C. C., & Hsieh, M. H. (2010). The SLO1 PPR protein is required for RNA editing at multiple sites with similar upstream sequences in *Arabidopsis* mitochondria. *The Plant Journal.*

Suzuki, J. Y., Ytterberg, A. J., Beardslee, T. A., Allison, L. A., Wijk, K. J., & Maliga, P. (2004). Affinity purification of the tobacco plastid RNA polymerase and *in vitro* reconstitution of the holoenzyme. *The Plant Journal, 40*, 164–172.

Swiatecka-Hagenbruch, M., Emanuel, C., Hedtke, B., Liere, K., & Borner, T. (2008). Impaired function of the phage-type RNA polymerase RpoTp in transcription of chloroplast genes is compensated by a second phage-type RNA polymerase. *Nucleic Acids Research, 36*, 785–792.

Takenaka, M., Neuwirt, J., & Brennicke, A. (2004). Complex cis-elements determine an RNA editing site in pea mitochondria. *Nucleic Acids Research, 32*, 4137–4144.

Takenaka, M., Verbitskiy, D., Van Der Merwe, J. A., Zehrmann, A., Plessmann, U., Urlaub, H., et al. (2007). *In vitro* RNA editing in plant mitochondria does not require added energy. *FEBS Letters, 581*, 2743–2747.

Takenaka, M., Verbitskiy, D., Van Der Merwe, J. A., Zehrmann, A., & Brennicke, A. (2008). The process of RNA editing in plant mitochondria. *Mitochondrion, 8*, 35–46.

Takenaka, M., Verbitskiy, D., Zehrmann, A., & Brennicke, A. (2010). Reverse genetic screening identifies five E-class PPR proteins involved in RNA editing in mitochondria of *Arabidopsis thaliana*. *Journal of Biological Chemistry, 285*, 27122–27129.

Takenaka, M. (2010). MEF9, an E-subclass pentatricopeptide repeat protein, is required for an RNA editing event in the nad7 transcript in mitochondria of *Arabidopsis*. *Plant Physiology, 152*, 939–947.

Tang, J., Kobayashi, K., Suzuki, M., Matsumoto, S., & Muranaka, T. (2010). The mitochondrial PPR protein LOVASTATIN INSENSITIVE 1 plays regulatory roles in cytosolic and plastidial isoprenoid biosynthesis through RNA editing. *The Plant Journal, 61*, 456–466.

Tasaki, E., Hattori, M., & Sugita, M. (2010). The moss pentatricopeptide repeat protein with a DYW domain is responsible for RNA editing of mitochondrial ccmFc transcript. *The Plant Journal, 62*, 560–570.

Tavares-Carreon, F., Camacho-Villasana, Y., Zamudio-Ochoa, A., Shingu-Vazquez, M., Torres-Larios, A., & Perez-Martinez, X. (2008). The pentatricopeptide repeats present in Pet309 are necessary for translation but not for stability of the mitochondrial COX1 mRNA in yeast. *Journal of Biological Chemistry, 283*, 1472–1479.

Terasawa, K., Odahara, M., Kabeya, Y., Kikugawa, T., Sekine, Y., Fujiwara, M., et al. (2007). The mitochondrial genome of the moss *Physcomitrella patens* sheds new light on mitochondrial evolution in land plants. *Molecular Biology and Evolution, 24*, 699–709.

Tiranti, V., Savoia, A., Forti, F., D'Apolito, M. F., Centra, M., Rocchi, M., et al. (1997). Identification of the gene encoding the human mitochondrial RNA polymerase (h-mtRPOL) by cyberscreening of the expressed sequence tags database. *Human Molecular Genetics, 6*, 615–625.

Toro, N., Jimenez-Zurdo, J. I., & Garcia-Rodriguez, F. M. (2007). Bacterial group II introns: not just splicing. *FEMS Microbiology Reviews, 31*, 342–358.

Tracy, R. L., & Stern, D. B. (1995). Mitochondrial transcription initiation: promoter structures and RNA polymerases. *Current Genetics, 28*, 205–216.

Tseng, C. C., Sung, T. Y., Li, Y. C., Hsu, S. J., Lin, C. L., & Hsieh, M. H. Editing of accD and ndhF chloroplast transcripts is partially affected in the *Arabidopsis* vanilla cream1 mutant. *Plant Molecular Biology, 73*, 309–323.

Uchida, M., Ohtani, S., Ichinose, M., Sugita, C., & Sugita, M. (2011). The PPR-DYW proteins are required for RNA editing of rps14, cox1 and nad5 transcripts in *Physcomitrella patens* mitochondria. *FEBS Letters, 585*, 2367–2371.

Unseld, M., Marienfeld, J. R., Brandt, P., & Brennicke, A. (1997). The mitochondrial genome of *Arabidopsis thaliana* contains 57 genes in 366,924 nucleotides. *Nature Genet., 15*, 57–61.

Uyttewaal, M., Arnal, N., Quadrado, M., Martin-Canadell, A., Vrielynck, N., Hiard, S., et al. (2008a). Characterization of *Raphanus sativus* pentatricopeptide repeat proteins encoded by the fertility restorer locus for Ogura cytoplasmic male sterility. *The Plant Cell, 20*, 3331–3345.

Uyttewaal, M., Mireau, H., Rurek, M., Hammani, K., Arnal, N., Quadrado, M., et al. (2008b). PPR336 is associated with polysomes in plant mitochondria. *Journal of Molecular Biology, 375*, 626–636.

Valadkhan, S. (2005). snRNAs as the catalysts of pre-mRNA splicing. *Current Opinion in Chemical Biology, 9*, 603–608.

Van Der Giezen, M. (2009). Hydrogenosomes and mitosomes: conservation and evolution of functions. *Journal of Eukaryotic Microbiology, 56*, 221–231.

Van Der Veen, R., Arnberg, A. C., Van Der Horst, G., Bonen, L., Tabak, H. F., & Grivell, L. A. (1986). Excised group II introns in yeast mitochondria are lariats and can be formed by self-splicing *in vitro*. *Cell, 44*, 225–234.

Vaughn, J. C., Mason, M. T., Sper-Whitis, G. L., Kuhlman, P., & Palmer, J. D. (1995). Fungal origin by horizontal transfer of a plant mitochondrial group I intron in the chimeric CoxI gene of *Peperomia*. *Journal of Molecular Evolution, 41*, 563–572.

Verbitskiy, D., Van Der Merwe, J. A., Zehrmann, A., Brennicke, A., & Takenaka, M. (2008). Multiple specificity recognition motifs enhance plant mitochondrial RNA editing *in vitro*. *Journal of Biological Chemistry, 283*, 24374–24381.

Verbitskiy, D., Zehrmann, A., Brennicke, A., & Takenaka, M. (2010a). A truncated MEF11 protein shows site-specific effects on mitochondrial RNA editing. *Plant Signaling & Behavior, 5*, 558–560.

Verbitskiy, D., Zehrmann, A., Van Der Merwe, J. A., Brennicke, A., & Takenaka, M. (2010b). The PPR protein encoded by the LOVASTATIN INSENSITIVE 1 gene is involved in RNA editing at three sites in mitochondria of *Arabidopsis thaliana*. *The Plant Journal, 61*, 446–455.

Verbitskiy, D., Hartel, B., Zehrmann, A., Brennicke, A., & Takenaka, M. (2011). The DYW-E-PPR protein MEF14 is required for RNA editing at site matR-1895 in mitochondria of *Arabidopsis thaliana*. *FEBS Letters, 585*, 700–704.

Vogel, J., & Borner, T. (2002). Lariat formation and a hydrolytic pathway in plant chloroplast group II intron splicing. *EMBO Journal, 21*, 3794–3803.

Wang, Y., & Shadel, G. S. (1999). Stability of the mitochondrial genome requires an amino-terminal domain of yeast mitochondrial RNA polymerase. *Proceedings of the National Academy of Sciences of the United States of America, 96*, 8046–8051.

Wang, Z., Zou, Y., Li, X., Zhang, Q., Chen, L., Wu, H., et al. (2006). Cytoplasmic male sterility of rice with boro II cytoplasm is caused by a cytotoxic peptide and is restored by two related PPR motif genes via distinct modes of mRNA silencing. *The Plant Cell, 18*, 676–687.

Wank, H., Sanfilippo, J., Singh, R. N., Matsuura, M., & Lambowitz, A. M. (1999). A reverse transcriptase/maturase promotes splicing by binding at its own coding segment in a group II intron RNA. *Molecular Cell, 4*, 239–250.

Weng, J., Aphasizheva, I., Etheridge, R. D., Huang, L., Wang, X., Falick, A. M., et al. (2008). Guide RNA-binding complex from mitochondria of trypanosomatids. *Molecular Cell, 32*, 198–209.

Williams, P. M., & Barkan, A. (2003). A chloroplast-localized PPR protein required for plastid ribosome accumulation. *The Plant Journal, 36*, 675–686.

Williams, M. A., Johzuka, Y., & Mulligan, R. M. (2000). Addition of non-genomically encoded nucleotides to the 3′-terminus of maize mitochondrial mRNAs: truncated rps12 mRNAs frequently terminate with CCA. *Nucleic Acids Research, 28*, 4444–4451.

Williams, E. H., Butler, C. A., Bonnefoy, N., & Fox, T. D. (2007). Translation initiation in *Saccharomyces cerevisiae* mitochondria: functional interactions among mitochondrial ribosomal protein Rsm28p, initiation factor 2, methionyl-tRNA-formyltransferase and novel protein Rmd9p. *Genetics, 175*, 1117–1126.

Williams-Carrier, R., Kroeger, T., & Barkan, A. (2008). Sequence-specific binding of a chloroplast pentatricopeptide repeat protein to its native group II intron ligand. *RNA, 14*, 1930–1941.

Wissinger, B., Schuster, W., & Brennicke, A. (1991). Trans splicing in *Oenothera* mitochondria: *nad1* mRNAs are edited in exon and *trans*-splicing group-II intron sequences. *Cell, 65*, 473–482.

Xu, F., Morin, C., Mitchell, G., Ackerley, C., & Robinson, B. H. (2004). The role of the LRPPRC (leucine-rich pentatricopeptide repeat cassette) gene in cytochrome oxidase assembly: mutation causes lowered levels of COX (cytochrome *c* oxidase) I and COX III mRNA. *Biochemical Journal, 382*, 331–336.

Xu, F., Ackerley, C., Maj, M. C., Addis, J. B., Levandovskiy, V., Lee, J., et al. (2008). Disruption of a mitochondrial RNA-binding protein gene results in decreased cytochrome *b* expression and a marked reduction in ubiquinol-cytochrome *c* reductase activity in mouse heart mitochondria. *Biochemical Journal, 416*, 15–26.

Yamazaki, H., Tasaka, M., & Shikanai, T. (2004). PPR motifs of the nucleus-encoded factor, PGR3, function in the selective and distinct steps of chloroplast gene expression in *Arabidopsis*. *The Plant Journal, 38*, 152–163.

Yoshinaga, K., Iinuma, H., Mazuzawa, T., & Uedal, K. (1996). Extensive RNA editing of U to C in addition to C to U substitutions in the *Rbcl* transcript of hornwort chloroplasts and the origin of RNA editing in green plants. *Nucleic Acids Research., 24*, 1008–1014.

Yu, Q. B., Jiang, Y., Chong, K., & Yang, Z. N. (2009). AtECB2, a pentatricopeptide repeat protein, is required for chloroplast transcript accD RNA editing and early chloroplast biogenesis in *Arabidopsis thaliana*. *The Plant Journal, 59*, 1011–1023.

Yura, K., Sulaiman, S., Hatta, Y., Shionyu, M., & Go, M. (2009). RESOPS: a database for analyzing the correspondence of RNA editing sites to protein three-dimensional structures. *Plant and Cell Physiology, 50*, 1865–1873.

Zehrmann, A., Van Der Merwe, J. A., Verbitskiy, D., Brennicke, A., & Takenaka, M. (2008). Seven large variations in the extent of RNA editing in plant mitochondria between three ecotypes of *Arabidopsis thaliana*. *Mitochondrion, 8*, 319–327.

Zehrmann, A., Verbitskiy, D., Hartel, B., Brennicke, A., & Takenaka, M. (2009a). RNA editing competence of trans-factor MEF1 is modulated by ecotype-specific differences but requires the DYW domain. *FEBS Letters, 584*, 4181–4186.

Zehrmann, A., Verbitskiy, D., Van Der Merwe, J. A., Brennicke, A., & Takenaka, M. (2009b). A DYW domain-containing pentatricopeptide repeat protein is required for RNA editing at multiple sites in mitochondria of *Arabidopsis thaliana*. *The Plant Cell, 21*, 558–567.

Zeng, X., Hourset, A., & Tzagoloff, A. (2007). The *Saccharomyces cerevisiae* ATP22 gene codes for the mitochondrial ATPase subunit 6-specific translation factor. *Genetics, 175*, 55–63.

Zhou, W., Cheng, Y., Yap, A., Chateigner-Boutin, A. L., Delannoy, E., Hammani, K., et al. (2008). The *Arabidopsis* gene YS1 encoding a DYW protein is required for editing of rpoB transcripts and the rapid development of chloroplasts during early growth. *The Plant Journal.*

Zikova, A., Panigrahi, A. K., Dalley, R. A., Acestor, N., Anupama, A., Ogata, Y., et al. (2008). *Trypanosoma brucei* mitochondrial ribosomes: affinity purification and component identification by mass spectrometry. *Molecular & Cellular Proteomics, 7*, 1286–1296.

CHAPTER ELEVEN

Evolution of Protein Import Pathways

Beata Kmiec*, Elzbieta Glaser*, Owen Duncan†, James Whelan† and Monika W. Murcha†,1

*Department of Biochemistry and Biophysics, Arrhenius Laboratories for Natural Science, Stockholm University, Stockholm, Sweden
†Australian Research Council Centre of Excellence in Plant Energy Biology, University of Western Australia, Crawley, WA, Australia
[1]Corresponding author. monika.murcha@uwa.edu.au

Contents

Abstract

Although a single endosymbiotic event is accepted to have led to the evolution of mitochondria, the machinery that is necessary to import the hundreds of cytosol-synthesized precursor proteins has diverged between various lineages. Plants have the additional requirement to sort protein between mitochondria and plastids, and yet there are also many cases of dual-targeted proteins. The machinery that achieved this process, the mitochondrial protein import apparatus, is composed of a variety of multi-subunit protein complexes on the outer and inner mitochondrial membranes. In plant mitochondria, there are differences in protein composition and/or function in the

Advances in Botanical Research, Volume 63
ISSN 0065-2296,
http://dx.doi.org/10.1016/B978-0-12-394279-1.00011-9

translocase of the outer membrane and in the intermembrane space. Furthermore, whereas the inner membrane translocases appear more conserved, there is an expansion in the preprotein and amino acid transporter (PRAT) family of proteins that suggest that neofunctionalization has occurred within this family. Our understanding of the processing and degradation of mitochondrial targeting signals in all systems is based on the intensive studies in plants.

1. EVOLUTION OF MITOCHONDRIA: FROM BACTERIUM TO ENDOSYMBIONT

It is widely accepted that mitochondria in all eukaryotic organisms arose from a single endosymbiosis (Gray, Burger, & Lang, 2001). The nature of this remarkable endosymbiosis that is the driving force behind the evolution of the eukaryotic cell is much studied and debated (Esser & Martin, 2007; Pisani, Cotton, & McInerney, 2007). Although the progenitor to mitochondria is accepted to be of alpha-proteobacterial lineage (Thrash *et al.*, 2011), the nature of the host remains unknown. What is clear is that over the intervening time period of at least 1.5 billion years (van der Giezen & Tovar, 2005) following the establishment of this relationship, the genome of the original alpha-proteobacterium has been greatly reduced, leading to a strict dependence of the mitochondria on the nucleus. This is shown by the existence of an estimated 1500 (Millar, Whelan, Soole, & Day, 2011) nuclear-encoded mitochondrial proteins that are synthesized on cytosolic ribosomes before being transported into mitochondria. The transfer of this genetic material is likely to be the result of a continuous selective pressure to optimize and coordinate the metabolic and biosynthetic processes occurring in the endosymbiote (Kurland & Andersson, 2000). This gene transfer is common to all organelles (see also Chapter 2), however, it relies on the establishment of transport mechanisms to return the protein products of these genes across the organelles membranes to their sites of action (Gross & Bhattacharya, 2009).

How this transport machinery was established is the subject of continued debate. The outsiders hypothesis (Gross & Bhattacharya, 2009) suggests that evolution of the import apparatus was driven by the host (Fig. 11.1). In this model, the host assumed control starting at the outer membrane following the transfer of two essential genes encoding β-barrel outer membrane proteins. These evolved in the host genome, allowing the host-encoded proteins access to the intermembrane space, then following the evolution of new components, the inner membrane and subsequently, the matrix. On

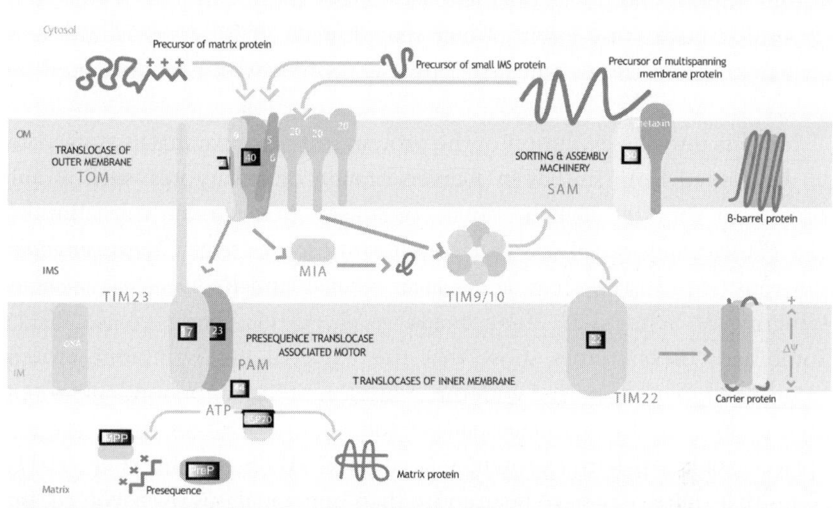

Figure 11.1 ***An overview of the protein import pathways into mitochondria.*** The import of precursor proteins via the general, carrier or mitochondrial intermembrane space assembly (MIA) pathways are shown. The numbers relate to the designation of the protein in each complex, e.g. Tom9 etc. The numbers within boxes indicate proteins in which orthologous protein in yeast can be identified. See the colour plate.

the other hand, the insiders view suggests that the endosymbiont played an active role in the establishment of the protein import apparatus, using preexisting components allowing access of host cell proteins to each compartment of the organelle (Alcock, Clements, Webb, Lithgow, 2010). Each of the core subunits of the extant protein import apparatus is likely to be related to preexisting bacterial proteins lending weight to the insiders perspective. These theories may help us understand the earliest steps in establishing protein import into mitochondria, however, over the proceeding 1.5 billion years; great diversity has arisen in the process of returning proteins to the mitochondria across the eukaryotic kingdoms. These mechanistic differences exist at the numerous steps of getting a protein translated in the cytosol to a functional complex within the mitochondrion. These differences are important to define and understand as they will give a greater understanding of plant cell biology, and how two endosymbiotic organelles (i.e. mitochondria and plastids) can coexist in the same cell environment with proteins targeted to each or both organelles, that is, location specific and dual-targeted proteins. The definition of the plant mitochondrial protein import apparatus will also give insights into how protein import, and hence organelle biogenesis and function, is regulated in a complex eukaryotic environment that displays cellular specialization to form organs and tissues and development profiles with a multifaceted life cycle.

Insights into the evolution of the protein import apparatus in eukaryotes can be gained from studies in a diverse range of eukaryotes, such as apicomplexan parasites, which contain organisms that contain mitochondria (e.g. *Plasmodium*), organisms that contain mitosomes (e.g. *Cryptosporidium*), and parasitic protists such as *Trichomonas vaginalis* and *Reclinomonas americana* (Lithgow & Schneider, 2010; Rada *et al.*, 2011; Tong *et al.*, 2011). Combined, these studies show that the bacterial endosymbiont protein export machinery coexisted with the evolution of the mitochondrial protein import apparatus. In *Trypanosoma brucei,* the absence of a Tom40 orthologue, replaced with a β-barrel protein related to OMP85 (Pusnik *et al.*, 2011), means that there may have been more than one apparatus that evolved, but that the machinery most prevalent today either coevolved or replaced previous components. If the protein import apparatus of the model system, yeast (*Saccharomyces cerevisiae*), is used as a comparison with plants, two points are evident; namely, that yeast appears to contain components that are apparently absent in plants and plants contain components that do not have an obvious yeast homologue (Carrie, Murcha, & Whelan, 2010a). However, in this context, it is important to define the terms present or absent. As given in detail below, no orthologous proteins to yeast Tom20 or Tom70 are present in plants (Carrie *et al.*, 2010a; Chan, Likic, Waller, Mulhern, & Lithgow, 2006; Lister, Murcha, & Whelan, 2003; Perry, Hulett, Likic, Lithgow, & Gooley, 2006a); instead, functionally equivalent proteins of 20 kDa and 64 kDa in size, termed Tom20 and OM64, are present and function as preprotein receptors (Lister *et al.*, 2007). These cases likely represent an example of convergent evolution of unrelated proteins resulting in a similar function in both yeast and plants.

2. TARGETING SIGNALS

Most mitochondrial proteins are synthesized in the cytosol on free polyribosomes as precursor proteins equipped with N-terminal extensions called mitochondrial targeting peptides (mTPs) or presequences. mTPs function as targeting signals and are necessary and sufficient for targeting of precursor proteins to the organelle (for reviews see Neupert & Herrmann, 2007; Schleiff & Becker, 2011). Membrane proteins usually lack the N-terminal extensions and contain non-cleavable targeting and sorting signals within the protein itself, that is, internal targeting signals (Brix, Rudiger, Bukau, Schneider-Mergener, & Pfanner, 1999; Rapaport, 2003). The targeting signals are recognized by receptors at the mitochondrial outer membrane and the precursor protein is translocated into the mitochondria through translocases of the outer and the inner mitochondrial membranes (TOM and TIM, respectively). Precursor proteins equipped with mTPs are translocated into the mitochondrial matrix where they are processed by the general mitochondrial processing peptidase (MPP), which, in plants, is integrated into the bc1 complex of the respiratory chain, MPP/bc1 (Braun, Emmermann, Kruft, & Schmitz, 1992; Glaser, Eriksson, & Sjoling, 1994). Some of the intermembrane space (IMS) proteins and some membrane proteins contain a so-called bipartite targeting peptide consisting of mTP followed by a peptide located directly downstream of the mTP that is cleaved off by an inner membrane-bound processing peptidase (IMP) on transport of the protein from the matrix to the IMS (Bomer *et al.*, 1997). Moreover, many IMS proteins contain a characteristic cysteine motif as a targeting signal forming disulphide bonds on import to trap the protein in the IMS (Milenkovic *et al.*, 2009; Tokatlidis, 2005). The cleaved free targeting peptides are degraded in the matrix by a newly identified presequence protease (PreP) (Stahl *et al.*, 2002). Also other proteases might be involved in additional trimming of the precursor and degradation of presequence peptides.

2.1. Specificity of Targeting, Mistargeting and Dual Targeting

In contrast with other organisms, plants contain, in the same cell, two organelles involved in energy metabolism mitochondria and chloroplasts, which share many similarities in biogenetic processes. Most chloroplast proteins, like the mitochondrial proteins, are nuclear encoded and synthesized in the cytosol as precursor proteins with N-terminal chloroplast-targeting peptides (cTPs), also called transit peptides. This raises an important question of intracellular protein sorting between these organelles. It has been demonstrated that, generally, targeting of proteins to mitochondria and chloroplasts is organelle specific. A number of *in vivo* studies using transgenic approaches as well as *in vitro* import studies showed high targeting specificity into mitochondria and chloroplasts that was dependent on respective targeting peptides (Boutry, Nagy, Poulsen, Aoyagi, & Chua, 1987; Silva-Filho, Wieers, Flugge, Chaumont, & Boutry, 1997). Mis-sorting of mitochondrial proteins into chloroplasts has not been reported, however, mis-sorting of chloroplast proteins *in vitro* into mitochondria has been documented (Bhushan, Kuhn, Berglund, Roth, & Glaser, 2006; Cleary *et al.*, 2002; Huang, Hack, Thornburg, & Myers, 1990; Lister, Chew, Rudhe, Lee, & Whelan, 2001). Development of a so-called dual import system for the simultaneous *in vitro* import of precursor proteins into isolated mitochondria and chloroplasts was important for *in vitro* protein import studies as it abolished the mistargeting of chloroplast precursors into mitochondria observed in a single organelle import system (Rudhe, Chew, Whelan, & Glaser, 2002b). *In vitro* mis-sorting of chloroplast proteins to mitochondria may be interpreted to indicate involvement of cytosolic factors in protein sorting. Several such factors have been reported over the years to be involved in binding to the mTPs (Dessi *et al.*, 2003; Hachiya *et al.*, 1993; Murakami & Mori, 1990; Ono & Tuboi, 1990; Yano, Terada, & Mori, 2003) and cTPs (May & Soll, 2000; Schleiff, Motzkus, & Soll, 2002), however, their role in protein sorting is not understood. Mistargeting studies also show higher targeting specificity of the chloroplast translocase for native organellar precursors than the mitochondrial protein translocation machinery (Rudhe, Chew, Whelan, & Glaser, 2002b). The mature protein may also play a role in the import of precursors into mitochondria and chloroplasts (Schleiff & Soll, 2000). Regions in the mature portion of the small subunit of Rubisco (SSU) have been shown to enhance the interaction with components of the chloroplast import apparatus (Dabney-Smith, van Den Wijngaard, Treece, Vredenberg, & Bruce, 1999).

However, there exists a group of proteins that are encoded by a single gene in the nucleus, translated in the cytosol and targeted to both chloroplasts and mitochondria (see Chapter 12). These proteins contain an N-terminal signal peptide referred to as a dual-targeting peptide (dTP) that is recognized by both mitochondrial and chloroplast receptors and has the capacity to target the precursor protein to both the mitochondrial matrix and chloroplast stroma. Additionally, dual-targeted proteins may contain twin presequences containing two distinct targeting domains, one for each organelle (Peeters & Small, 2001; Silva-Filho, 2003). To date, this group includes 51 proteins including enzymes related to translation (e.g. 18 aminoacyl-tRNA synthetases, the three subunits of the tRNA-dependent amidotransferase) protection against oxidative stress, DNA synthesis and processing, cellular protein folding and turnover as well as energy conversion

(Berglund, Pujol, Duchene, & Glaser, 2009a; Carrie *et al.*, 2009; Duchene *et al.*, 2005; Pujol *et al.*, 2008). The number of dual-targeted proteins is constantly growing and based on a prediction program (ATP, ambiguous targeting predictor), there might be as many as 523 dual-targeted proteins in *Arabidopsis* thaliana (Mitschke *et al.*, 2009).

2.2. Properties of Targeting Peptides

Despite the fact that plant mTPs vary greatly in length, from 19 to 109 amino acid residues in Arabidopsis and from 18 to 117 in rice, most of them are in the range of 20–70 residues with an average length of 50 and 45 amino acid residues in Arabidopsis and rice, respectively (Bhushan *et al.*, 2006; Heazlewood *et al.*, 2004; Huang, Taylor, Whelan, & Millar, 2009; Zhang & Glaser, 2002). They have a high content of hydroxylated, hydrophobic and positively charged amino acid residues, and are almost completely deficient in acidic amino acids. Furthermore, proline and glycine are well represented. Comparison of plant mTPs with cTPs showed a striking similarity in the overall amino acid composition (Bhushan *et al.*, 2006; Huang *et al.*, 2009; Zhang & Glaser, 2002). Some differences have been reported; cTPs are about 9–20 amino acid residues longer and contain more serine residues than plant mTPs. Furthermore, there is a decreased amount of arginine resulting in a substantially lower hydrophilicity as well as an increased amount of proline and serine and in the N-terminal portion of 16–20 amino acids of the cTPs in comparison with mTPs (Bhushan *et al.*, 2006; Huang *et al.*, 2009). However, it has been suggested that it is not specific sequence properties but rather certain structural or physicochemical features that are recognized by organellar import receptors (Bhushan *et al.*, 2006); for a review see Schleiff and Becker (2011).

An important property of mTPs for targeting and interactions with import receptors is that they have the potential to form amphiphilic α-helices (von Heijne, 1986). Eighty-four percent and 90% of Arabidopsis and rice mTPs, respectively, form an α-helix within the first ten amino acid residues (Huang *et al.*, 2009). Chloroplast transit peptides were initially thought to be random coils but, as more sequences of transit peptides have become available, experimental studies and computer predictions have shown that chloroplast transit peptides might also form helical structures (Bruce, 2000; Lancelin *et al.*, 1994).

The amphiphilic α-helical character of mTPs has been shown to be important for interaction of mTPs with the Tom20 receptor both in yeast

(Abe *et al.*, 2000) and in Arabidopsis (Perry *et al.*, 2006b; Rimmer *et al.*, 2011). The structure of the cytosolic domain of yeast Tom20 in a complex with the mTP of aldehyde dehydrogenase showed that the hydrophobic surface of the amphiphilic α-helix of the mTP interacts with the hydrophobic binding groove of Tom20 (Abe *et al.*, 2000). AtTom20 possesses a discontinuous bidentate hydrophobic binding site for presequences, at which hydrophobic binding regions of amphiphilic α-helical plant mTPs bind (Rimmer *et al.*, 2011).

Nearly all mTPs and cTPs contain Hsp70 binding motifs (Zhang & Glaser, 2002). Comparison of yeast mTPs with plant mTPs revealed that yeast mTPs are shorter (about ten residues on average), less hydrophilic and show lower propensity to form amphiphilic α-helices, which may be explained by the fact that due to the lack of chloroplasts in yeast, the targeting signal may be less specific (Mossmann, Meisinger, & Vogtle, 2011; Vogtle *et al.*, 2009).

Analysis of dual-targeting peptides (dTPs) revealed that they are very similar to mTPs and cTPs in the overall amino acid composition and it is difficult to identify specific features for dTPs. dTPs are about 50–60 amino acid residues long (Berglund *et al.*, 2009a). It has been suggested that the dTPs might be intermediary in character in comparison with mTPs and cTPs (Berglund *et al.*, 2009a; Peeters & Small, 2001). There is an overall significant increase in phenylalanines, leucines and serines and a decrease in acidic amino acids and glycine in dTPs in comparison with single mTPs and cTPs (Berglund *et al.*, 2009a; Pujol, Marechal-Drouard, & Duchene, 2007). The N-terminal portion of dTPs has significantly more serines than mTPs. The amount of arginines is similar to that in mTPs but significantly higher than in cTPs. The lack of an amphiphilic α-helix has been observed (Peeters & Small, 2001), however, in some cases the amphiphilic α-helix is present with a hydrophilic side formed by a polar residue, serine (Berglund *et al.*, 2009a; Pujol *et al.*, 2007). Circular dichroism spectroscopy of the dual-targeting peptide of threonyl-tRNA synthetase (ThrRS-dTP(2–60)) indicated that the peptide has the propensity to form an α-helical structure in membrane mimetic environments (Berglund *et al.*, 2009b). Furthermore, purified ThrRS-dTP(2–60) inhibited the import of the F1β precursor into mitochondria and of the SSU precursor into chloroplasts at micromolar concentrations showing that dual and organelle-specific proteins use the same organellar import pathways (Berglund *et al.*, 2009b).

Studies in which dTPs of aminoacyl-tRNA synthetases fused to green fluorescent protein (GFP) were imported *in vitro* in single and dual import systems (Rudhe *et al.*, 2002b) as well as *in vivo* revealed that, for most constructs, determinants for import into both organelles were localized in the N-terminal portion of dTPs (Berglund *et al.*, 2009b). However, for a few dTPs, the targeting information has been shown to be organized in domains as, for example, in RNA polymerase RpoT:2 (Hedtke, Borner, & Weihe, 2000), glutathione reductase (Rudhe *et al.*, 2002a), PreP (Bhushan *et al.*, 2003) and aminoacyl-tRNA synthetases (Berglund *et al.*, 2009a). For some dTPs of aminoacyl-tRNA synthetases, deletion of the N-terminal portion of dTPs did not affect organellar import and the dual-targeting capacity has also been shown to be affected by the nature of the mature protein (Chew & Whelan, 2003). This implies that determinants for dualn-targeting are not yet fully understood and it will be very interesting to study interaction of dTPs with organellar import receptors.

3. TRANSLOCASE OF THE OUTER MEMBRANE AND SORTING AND ASSEMBLY MACHINERY

The translocase of the outer membrane (TOM) and sorting and assembly machinery (SAM) multi-subunit protein complexes are the gateway to mitochondria and are responsible for the insertion of proteins into the four mitochondrial compartments. All nuclear-encoded mitochondrial proteins, with the exception of some signal-anchored proteins (Ahting, Waizenegger, Neupert, & Rapaport, 2005), pass through the TOM complex; they are then passed to other multi-subunit protein complexes on the outer membrane (SAM), the IMS or inner membrane (TIM17:23 or TIM22) (Chacinska, Koehler, Milenkovic, Lithgow, & Pfanner, 2009). The TOM complex is responsible for a number of functions: first, it is responsible for the recognition of precursor proteins via cytosolic-faced receptor components; second, it is responsible for translocation of protein across the outer membrane via the Tom40 channel; and third, it mediates the transfer of proteins from the *cis* side of the Tom40 channel to one of the other protein import complexes. However, it has been observed that the plant TOM complex is not highly conserved in terms of orthology with regard to the yeast TOM complex. In plants, orthologs to Tom40 and Tom7 can be identified, whereas orthologues to the other TOM components, Tom20, Tom9, Tom5 and Tom6, are either plant specific or too small in size to define their orthology with confidence (Lister *et al.*, 2003). The notable differences between the TOM complex in plants (Arabidopsis rabidopsis) and yeast are:

- The receptor components of the plant TOM complex, Tom20 and OM64 are not orthologs to the yeast receptor components Tom20 and Tom70 (Carrie *et al.*, 2010a).
- In plant mitochondria, Tom9 is the ortholog to yeast Tom22; the latter has a cytosolic receptor domain that acts as a secondary receptor and a *trans* domain that interacts with Tim components. Plant Tom9 lacks this receptor domain and is likely an ortholog of yeast Tom22 that has lost the receptor domain (Macasev *et al.*, 2004).
- Tom20 appears to be more tightly bound to the core TOM complex in plants than in yeast, as shown by the fact that Tom20 co-migrates with the TOM complex in plants on blue native-polyacrylamide gel electrophoresis, whereas in yeast it is dissociated (Dekker *et al.*, 1998; Lister *et al.*, 2007; Werhahn *et al.*, 2001).

Overall, this means that the primary receptor components of plants differ from those of yeast (and mammalian) mitochondria. Closer analysis of the mechanism of recognition between precursor proteins and plant Tom20 revealed two preprotein binding sites compared with a single binding with yeast Tom20 (Rimmer *et al.*, 2011). Yeast Tom70 has nine tetratricopeptide repeat (TPR) domains (Chan *et al.*, 2006) and receptor binding for precursor proteins is via an HSP90 and/or an HSP70 associated mechanism (Young, Hoogenraad, & Hartl, 2003). Although the mechanism of binding of precursor proteins to OM64 is not known, modelling and comparisons with the plastid import receptor Toc64 suggests that both are predicted to bind HSP90 (which has been experimentally shown for Toc64) (Qbadou *et al.*, 2006), although the mechanisms of how Toc64 and OM64 interact with HSP90 are likely to be different (Mirus, Bionda, von Haeseler, & Schleiff, 2009). Despite these differences, it is notable that inactivation of both Tom20 and Tom70 in yeast and Tom20 and OM64 in plants results in a severe growth phenotype in the former (viability is likely maintained by the receptor function of Tom22) and is lethal in the latter (Duncan, Murcha, & Whelan, 2012). Thus, although yeast Tom20 and Tom70 are not orthologous to plant Tom20 and OM64, they essentially carry out the same functions.

There is little functional information on the SAM complex of plant mitochondria. Orthologues of SAM50 are present in the Arabidopsis genome (Carrie *et al.*, 2010a; Lister *et al.*, 2003), although there is no evidence of their function. The only putative component of the SAM complex that has been studied so far in plants is metaxin. Arabidopsis plants display severe growth abnormalities when metaxin is inactivated (Lister *et al.*, 2007). It has been proposed to be involved in the import of β-barrel proteins into plant mitochondria as *metaxin* mutant plants accumulate porin in the cytosol. Metaxin from plants and mammals and Sam37 from yeast have all been implicated in the import of β-barrel proteins; structurally all three contain two glutathione-*S*-transferase (GST) domains and a single transmembrane domain, although Sam37 is unique in that the transmembrane domain is located at the N-terminus (Carrie *et al.*, 2010a). However, there is no sequence similarity between metaxin in plants and Sam37 in yeast to suggest orthology and rather it may be similar to what has been observed with Tom20, where a common problem has been solved by convergent evolution that produces a similar, but evolutionary unrelated protein.

4. THE INTERMEMBRANE SPACE

The intermembrane space contains two sets of protein complexes involved in protein import, the tiny Tims, involved in the import of carrier pathway precursors from the outer membrane to the inner membrane and the mitochondrial intermembrane space assembly (MIA) complex, involved in the oxidative folding and maturation of IMS proteins including the tiny Tims themselves (Chacinska *et al.*, 2009).

4.1. The Tiny Tims

The tiny Tims are a highly conserved protein family consisting of Tim8, 9, 10 and 13. They are found in a wide range of eukaryotic lineages ranging from yeast, algae, moss, plants and mammals (Carrie *et al.*, 2010a). Following translocation from the TOM channel, carrier precursor proteins first bind to a Tim9–Tim10 hexameric complex located in the IMS. This complex acts as a chaperone, transporting the hydrophobic precursor protein through the IMS to the TIM22 complex of the inner membrane (Koehler *et al.*, 1998; Senapin, Chen, & Clark-Walker, 2003). An additional hexameric complex formed by Tim8–Tim13 also exists and is involved in the transfer of carrier proteins in a similar manner, although its function has only so far been described with the import of Tim23 and, unlike Tim9 and Tim10, these tiny Tims are not essential for yeast viability (Davis, Alder, Jensen, & Johnson, 2007). In addition to proteins containing internal targeting signals, the tiny Tims have been shown to import β-barrel proteins back into the outer membrane from the IMS (Chacinska *et al.*, 2009). The tiny Tims most likely arose from a single ancestral protein (Hewitt, Alcock, & Lithgow *et al.*, 2011) and some eukaryotes that contain mitosomes such as *Trichomonas* or *Crytosporidium* can lack either Tim9 and 10 or both, even with the presence of carrier-like proteins, suggesting that although the tiny Tims are not absolutely essential, various other mechanisms of transport may have been acquired to compensate (Gentle *et al.*, 2007; Waller *et al.*, 2009). Yeast also contain an additional component, Tim12, thought to be peripherally attached to TIM22 and involved in the transfer of soluble precursor proteins bound to the TIM9:10 complex to the inner membrane translocase (Lionaki *et al.*, 2008). Tim12 cannot be identified in any other organisms via homology-based searches and it is proposed that Tim12 most likely evolved as a result of a single gene duplication event exclusively in yeast (Carrie *et al.*, 2010a).

Although the carrier import pathway and the tiny Tims are present in all mitochondria, biochemical reconstitution of the tiny Tims has only been carried out in plants (Lister, Mowday, Whelan, & Millar, 2002). In these assays, mitochondria were initially depleted of IMS proteins by osmotic shock and salt wash. The wash fractions were fractionated by anion-exchange chromatography before being added back into an *in vitro* protein import assay containing the depleted mitochondria. The *in vitro* import of two carrier proteins could be directly correlated to the presence of Tim9 and Tim10 within these fractions, thus showing a functional role for these proteins in the carrier import pathway (Lister *et al.*, 2002).

A unique feature of some plant carrier proteins is the presence of cleavable presequences (Laloi, 1999; Mozo, Fischer, Flugge, & Schmitz, 1995). Removal of the presequence does not affect its import ability nor does the presequence alone have the ability to target proteins to the mitochondria (Murcha, Elhafez, Millar, & Whelan, 2004). It was shown that IMS fractions could stimulate the import of N-terminal presequence carrier proteins to a greater extent than those with only internal targeting signals (Murcha, Millar, & Whelan, 2005a). Thus, plants may contain additional components in the IMS that can specifically recognize the N-terminal cleavable presequence of plant carrier proteins. In addition, presequence processing was shown to be a multistep process as the initial processing was carried out by the MPP on the matrix side of the inner membrane followed by additional processing to the mature protein by a yet undetermined serine protease located within the IMS (Murcha *et al.*, 2004).

4.2. The MIA Pathway

The MIA import pathway is the most recently characterized import pathway in mitochondria, responsible for the import of the IMS proteome of cysteine-rich proteins such as the tiny Tims. The MIA pathway consists of only two components, Mia40 and Erv1, both of which are essential for yeast viability and form disulphide bonds within precursor proteins prior to release into the IMS (Chacinska *et al.*, 2004; Mesecke *et al.*, 2005; Rissler *et al.*, 2005; Terziyska *et al.*, 2005).

MIA precursor proteins contain conserved cysteine residues in twin CX(9)C or CX(3)C motifs (Gabriel *et al.*, 2007). Mia40 acts as a receptor for the precursor transferring from the outer membrane to the IMS, where the conserved cysteine residues are oxidized by Mia40, which is itself subsequently oxidized by Erv1 (Hell, 2008). The MIA Erv1 pathway differs in plants in that Mia40 has been determined to be co-localized to both the mitochondria and the peroxisome (Carrie *et al.*, 2010c). In addition, Mia40 is not an essential import component in plants as deletion of this protein does not affect the import pathway or result in any observable plant phenotype, suggesting that the plant disulphide relay system is functional in the presence of Erv1 alone (Carrie *et al.*, 2010b). Furthermore, some organisms such as trypanosomes, lack Mia40, but contain an Erv1 homologue and the substrates of the MIA pathway (Hewitt *et al.*, 2011). No prokaryotic homologues to Mia40 can be identified, which suggests that Mia40 has evolved specifically for the eukaryotic lineage, and the dual location of Mia40 suggests that additional features may have been acquired in higher eukaryotes such as plants.

In yeast, the MIA import pathway has most recently been associated with an additional complex, MINOS1 (mitochondrial inner membrane organizing system) or mitofilin, a large complex of the inner membrane responsible for cristae formation and inner membrane organization (John *et al.*, 2005; Rabl *et al.*, 2009). Yeast deletion strains of MINOS1 components exhibited a reduced import ability via the MIA pathway (von der Malsburg *et al.*, 2011). Although the effects on other import pathways could not be determined due to the MINOS1 mutants also exhibiting a reduced membrane potential, a direct association of MINOS1 and Mia40 was confirmed via affinity chromatography (von der Malsburg *et al.*, 2011). In addition, the outer membrane import components, Sam50 and metaxin, were also found to co-isolate with the MINOS1 complex (Xie *et al.*, 2007) suggesting that this complex linking the inner and outer membrane may have a prominent role in protein import. The diverse functions and mechanisms in both morphology and biogenesis are being uncovered, but gene homology searches could not identify any homologues to this highly conserved protein complex in plants (Murcha, unpublished data).

5. THE INNER MEMBRANE

Unlike the outer membrane, the inner membrane is impermeable to most solutes and maintains a tight electrochemical gradient, and as such, protein translocation through or into the inner membrane needs to be carried out with the aid of specific inner membrane translocases. There are two translocases of the inner membrane, the TIM22 complex and the TIM17:23 complex, responsible for the translocation of carrier protein and general pathway proteins, respectively. These complexes consist of channel-forming pores and a variety of associated proteins involved in the recognition, translocation and assembly of precursor proteins (Chacinska *et al.*, 2009).

5.1. The TIM17:23 Complex

The TIM17:23 complex contains the major translocation pore responsible for the import of the majority of mitochondrial proteins. This complex contains the channel-forming Tim23 and Tim17, and several additional subunits (Chacinska *et al.*, 2009). Extensively characterized in yeast, this translocase was shown to exist as two functionally and structurally distinct forms termed TIM23-SORT and TIM23-PAM (presequence translocase associated motor) (Chacinska *et al.*, 2009; Wiedemann, Van der Laan, Hutu, Rehling,& Pfanner, 2007). TIM23-SORT contains Tim17 and Tim23 along with Tim50 and Tim21, and is involved in the direct translocation of precursor proteins from the TOM complex through the inner membrane (Mokranjac & Neupert, 2010; van der Laan, Hutu, & Rehling, 2010). This complex forms super complex structures with both the TOM complex and complex III of the mitochondrial respiratory chain, which is thought to promote efficient tranlsocation by utilizing the cristae junction contact sites and the proton motive force of the inner membrane (Chacinska *et al.*, 2005; Saddar, Dienhart, & Stuart, 2008; Wiedemann *et al.*, 2007).

The second form of the TIM17:23 complex, termed TIM23-PAM, functions to transfer proteins through the inner membrane to the matrix.

This complex similarly consists of Tim17, Tim23, Tim50 but also the PAM complex, mainly comprising mtHSP70, TIM44 and its associated co-chaperones, PAM16, PAM17 and PAM18 (Chacinska *et al.*, 2010).

Orthologs exist for all these components in plants although several unique features have been uncovered. For instance, plant Tim17-2 contains a C-terminal extension of up to 143 amino acids. This C-terminal extension is observed in all higher plant species and must first be removed to complement a yeast deletion strain of Tim17 (Murcha, Elhafez, Millar, & Whelan, 2005b). The C-terminal extension exhibits presequence binding properties and can link both the outer and inner membranes (Carrie *et al.*, 2010a; Murcha, Lister, Ho, & Whelan, 2003; Murcha *et al.*, 2005b). In contrast, the N-terminus of yeast Tim23 spans the IMS linking the inner and outer membrane and similarly can bind precursor proteins directing them to the TIM17:23 translocation pore (Donzeau *et al.*, 2000). Plant Tim23 lacks any extension on either terminus and only contains the four transmembrane spanning regions characteristic of a translocation pore (Murcha *et al.*, 2003), and thus Tim23 and Tim17 from plants and yeast may be an example of convergent evolution.

5.2. The TIM22 Complex

The TIM22 complex is responsible for the insertion of carrier import pathway proteins that are multi-spanning in the inner membrane (Sirrenberg et al., 1996). So far, this complex, which has only been characterized in yeast, consists of the TIM22 pore and accessory proteins TIM54, 18 and 12. Only Tim22 and Tim12 have been shown to be essential for yeast viability (Kerscher, Holder, Srinivasan, Leung, & Jensen, 1997; Kerscher, Sepuri, & Jensen, 2000; Sirrenberg *et al.*, 1998). In Arabidopsis, TIM22 is encoded by two identical genes that are located on different chromosomes, and this protein has been shown to have the ability to functionally complement a yeast deletion strain of Tim22 (Murcha *et al.*, 2007). It is surprising that homologues to TIM54, 18 and 12 cannot be identified in any plant genome database (Carrie *et al.*, 2010a), and this raises the possibility that, in plants, the TIM22 channel may function alone or in conjunction with additional plant-specific components. Recent experiments characterizing the mitochondrial complex proteome by systemic liquid chromatography and mass spectrometry (LC/MS) may have uncovered a novel feature of Arabidopsis Tim22. Tim22 was identified to migrate exclusively in the same complex as respiratory complex I subunits at 1500 kDa (Klodmann,

Senkler, Rode, & Braun, 2011). In the same study, complex I subunits encoded by At1g68680, At1g72170 and At2g27730 were also seen to migrate with known import components such as Tim23, Tim44 and Tim17 at ~100 kDa (Klodmann *et al.*, 2011). The suggestion that the mitochondrial respiratory complexes and import components may be interlinked is an emerging area. Several independent studies on both yeast and plants have provided such evidence, the first of which determined that plant MPP was integrated into the cytochrome bc1 complex in plants (Braun *et al.*, 1992; Braun, Emmermann, Kruft, Bodicker, & Schmitz, 1995; Emmermann, Braun, & Schmitz, 1994; Eriksson, Sjoling, Glaser, 1996). In yeast, Tim23 has been shown to interact with complex III (Gebert *et al.*, 2011b; Saddar *et al.*, 2008; Wiedemann *et al.*, 2007) and most recently, sdh3, a subunit of complex II in yeast is also a subunit of the TIM22 complex (Gebert *et al.*, 2011b).

5.3. Multigene Families

A distinctive feature of higher eukaryotes is the existence of large multigene families and this remains accurate for many proteins of the protein import pathway. One such example is the Tim17, 23 and 22 proteins, thought to have derived as a result of gene duplication. This gene family of three members in yeast has expanded to 16 members in Arabidopsis (Murcha *et al.*, 2007). The preprotein and amino acid transporter (PRAT) proteins are characterized by four transmembrane spanning regions and a [G/A]X2[F/Y}X10RX3DX6[G/A/S]GX3G, motif where X is any amino acid, originating from a single bacterial amino acid transporter LivH (Rassow, Dekker, van Wilpe, Meijer, & Soll, 1999). The Arabidopsis members of the PRAT family have been shown to be located in the mitochondria, chloroplast or both. Of the ten mitochondrial PRAT proteins, three encode for Tim17, three encode for Tim23, two encode for Tim22 and three encode for unique plant-specific components of yet undetermined function (Murcha *et al.*, 2007). One such protein encoded by At2g42210 (an orthologue to the bovine complex I subunit termed B14.7 (NDUFA11) has been identified to be a component of complex I of the respiratory chain by two independent studies (Klodmann *et al.*, 2011; Meyer, Solheim, Tanz, Bonnard, & Millar, 2011). Another PRAT protein encoded by At4g26670 may also be associated with complex I, as it is also orthologous to B14.7 (NDUFA11) (Murcha, unpublished data). Although the location of this PRAT protein was determined to be predominantly chloroplastic by immunodetection, *in vitro* and *in vivo* protein targeting was inconclusive (Murcha *et al.*, 2007), and thus further investigation of this protein may reveal a mitochondrial location and an association with complex I.

5.4. OXA Import Pathway

An additional import pathway located termed OXA (OXidase Assembly) involves several complexes of the inner membrane and is involved in the transfer of mitochondrial-encoded proteins from the matrix into the inner membrane (Wang & Dalbey, 2011). First identified by characterization of an Oxa1 yeast mutant that was defective in cytochrome *c* assembly (Bauer, Behrens, Esser, Michaelis, & Pratje, 1994; Bonnefoy, Chalvet, Hamel, Slonimski, & Dujardin, 1994; Meyer, Bomer, & Pratje, 1997), Oxa1 has been shown to be involved in the insertion of mitochondrial-encoded respiratory subunits in a cotranslational manner (Wang & Dalbey, 2011). This initially involves Oxa1 and Mba1 interacting with ribosomes and cotranslationally inserting only the N-terminus of the precursor into the inner membrane. Further modifications by proteases such as Imp1/2 (inner membrane protease) cleave the N-terminal targeting signal followed by insertion of the C terminus of the precursor by the Oxa2 complex, consisting of Cox18, Pnt1 and Mss2 (Wang & Dalbey, 2011).

Homologues to Oxa1 can also be found in both chloroplasts and bacteria, and thus Oxa1 is thought to have resulted from three independent duplication events, followed by specialization of function within the mitochondria, chloroplast and bacteria, an example of divergent evolution (Funes, Kauff, van der Sluis, Ott, & Herrmann, 2011). In Arabidopsis, homologues to Oxa1 and Imp1/2 have been identified (Lister *et al.*, 2003) although very little functional characterization has been carried out to determine if its function has specialized to meet the unique demands of plant mitochondria.

6. MITOCHONDRIAL PRECURSOR PROCESSING AND PRESEQUENCE DEGRADATION

As described above, the proteins synthesized in the cytosol, but destined to reside in mitochondria, contain mitochondrial targeting signals. In the case of matrix-localized proteins, the signal is localized at the N-terminus of the precursor sequence as an extension, referred to as the presequence. The presequence is cleaved off after reaching the mitochondrial matrix in a reaction termed processing. Global identification of semi-tryptic peptides revealed that over 77%, 67% and 69% of mitochondrial proteins identified in Arabidopsis, rice and yeast, respectively, contain cleavable presequences (Huang *et al.*, 2009; Vogtle *et al.*, 2009). Most of our knowledge about processing peptidases comes from studies performed in yeast, in which seven different proteases were shown to be involved in mitochondrial precursor processing. In addition to MPP, responsible for processing of most of the mitochondrial precursors, processing activity was also shown for other proteases in yeast: m-AAA and Atp23 (reviewed by Gerdes, Tatsuta, & Langer, 2012; Mossmann *et al.*, 2011). In addition, it was shown for some precursors, that after MPP or m-AAA processing, they are cleaved again by other proteases. To date, four yeast proteases were shown to perform secondary cleavage steps and they can be divided into two groups. Processing mediated by IMP and Pcp1 (processing of cytochrome *c* peroxidase 1) results in release of their substrates into the mitochondrial intermembrane space or the inner membrane. Cleavage mediated by Oct1 (octapeptidyl aminopeptidase 1) and Icp55 (intermediate cleaving protease) occurs in the matrix. Typical substrates of Oct1 and Icp55 are MPP-generated intermediates, which have a destabilizing amino acid at their N-termini. In these cases, the secondary processing event increases proteins stability (Mossmann *et al.*, 2011).

To date, MPP is the only plant mitochondrial protease with well-characterized processing activity. *In silico* analyses show that Arabidopsis contains homologues of all the yeast processing proteases listed above (Kwasniak, Pogorzelec, Migdal, Smakowska, & Janska, 2011), however their function is still unknown. Studies on Arabidopsis showed that homologues of m-AAA (AtFtsH3 and AtFtsH10) and Pcp1 (AtRbl12) proteases are localized in mitochondria, however their processing activity remains to be confirmed (Kmiec-Wisniewska *et al.*, 2008; Kolodziejczak *et al.*, 2002).

Processing of mitochondrial protein precursors gives rise to mature proteins and leads to release of their presequences to the mitochondrial matrix. Due to their hydrophobic nature, the presequence peptides can interact with the mitochondrial inner membrane. Several research groups have reported harmful effects of these peptides on the structural and functional integrity of mitochondrial membranes (Hugosson, Andreu, Boman, & Glaser, 1994; Maduke & Roise, 1993; Wieprecht, Apostolov, Beyermann, & Seelig, 2000; Zardeneta & Horowitz, 1992). Addition of presequences to mitochondria results in membrane lysis and dissipation of the membrane potential (Nicolay, Laterveer, & van Heerde, 1994; Roise, Horvath, Tomich, Richards, & Schatz, 1986). Therefore presequences should be removed, either through export or degradation. Analyses of the fate of mitochondrial presequences led to identification of a peptide-degrading protease, termed presequence protease (PreP) (Stahl *et al.*, 2002). PreP is present in the mitochondrial matrix and has been shown to degrade not only presequences but also other unstructured peptides, including toxic amyloid β peptide accumulating in brain mitochondria during the development of Alzheimer disease (Alikhani *et al.*, 2011a; Falkevall *et al.*, 2006). In yeast, another peptidase, called Prd1, has been proposed to be involved in the degradation of unstructured peptides (Kambacheld, Augustin, Tatsuta, Muller, & Langer, 2005).

6.1. MPP

MPP is a general mitochondrial processing peptidase, responsible for cleaving off presequences of most of the mitochondrial preproteins (reviewed by Gakh, Cavadini, & Isaya, 2002; Glaser & Dessi, 1999). MPP is a metallopeptidase that belongs to the pitrilysin family (MEROPS M16 protease family). The active form of MPP is a heterodimer formed by the catalytic β-MPP subunit, carrying an inverted Zn-binding motif (HXXEHX74-76E), and proteolytically inactive α-MPP subunit, responsible for substrate binding (Yang *et al.*, 1991). The crystal structure of yeast MPP showed that the active site of the enzyme is located in the large central cavity situated between β-MPP and α-MPP subunits (Taylor *et al.*, 2001). The cavity is negatively charged, which facilitates binding of positively charged presequences. Although MPP is a general peptidase and acts on several hundred different substrates, it still recognizes and cleaves precursors at specific sites. MPP substrates in plants share three conserved MPP recognition motives: −3R motif, which is the most common one, RX(F/Y/L)↓(S/A)(S/T), where ↓ depicts MPP cleavage; −2R motif, RX↓XST; and non-R motif, (F/Y)↓(S/A) (Huang *et al.*, 2009). These three MPP motifs were also identified in yeast presequences (Vogtle *et al.*, 2009), however the existence of the −3R motif in yeast was questioned due to the identification of Icp55 protease, removing one amino acid from MPP-generated intermediates. It was suggested that the −3R motif does not refer to MPP cleavage alone, but is a part of a motif for two-step cleavage, first by MPP and then by Icp55 (Vogtle *et al.*, 2009). Since −3R is the most common MPP motif in plants, it would be interesting to analyse the activity of the plant homologue of Icp55 protease and its function in the processing of mitochondrial preproteins in plants. Yeast presequences possess an additional motif −10R, RX↓(F/L/I)XX(T/S/G)XXXX↓, where the first ↓ depicts MPP cleavage, and the second ↓ depicts cleavage by Oct1. The latest analysis of plant proteome does not support the existence of this motif in plant presequences (Huang *et al.*, 2009). In addition to MPP recognition motifs, processing can also be influenced by residues located upstream of the cleavage site and in the mature portion of the protein (reviewed by Glaser & Dessi, 1999).

In plants, the MPP complex is integrated into the inner mitochondrial membrane as α-MPP and β-MPP are integral core subunits of the cytochrome *c* oxidoreductase complex (bc1 complex). The bc1 complex in plants is thus bifunctional, involved in both electron transfer and protein processing, but inhibiting one of these activities does not affect the other (Eriksson, Sjoling, & Glaser, 1994). In contrast to the MPP localization in plants, in yeast and mammals, protein processing is spatially separated from electron transfer. Core subunits of the bc1 complex do not possess proteolytic activity and MPP complex resides in the mitochondrial matrix (Ou, Itu, Okazaki, & Omura, 1989; Yang, Jensen, Yaffe, Oppliger, & Schatz, 1988). An intermediate situation was observed in *Neurospora crassa*, in which a fraction of the β-MPP subunit is incorporated in the bc1 complex, but α-MPP resides in the matrix (reviewed by Glaser and Dessi, 1999). Analysis of the evolutionary connection between MPP and bc1 complexes in lower plants and algae led to a hypothesis that, in early mitochondria such as in plants, MPP was incorporated into the bc1 complex (Braun *et al.*, 1995; Brumme, Kruft, Schmitz, & Braun, 1998). According to this hypothesis, in the course of evolution of fungi and animals, due to increasing energetic demands of the cell, α-MPP and β-MPP duplicated and the MPP complex detached from the mitochondrial membrane. The core subunits of the bc1 complex lost the processing activity, which allowed independent regulation of protein processing and respiration (Braun *et al.*, 1995). In plants, that kind of detachment was not required, probably due to the presence of chloroplasts, a second type of organelles involved in the bioenergetics of the cell (Brumme *et al.*, 1998).

6.2. PreP

PreP is the only characterized mitochondrial peptidase involved in degradation of presequences in plants. PreP protease was initially identified in

Solanum tuberosum and in Arabidopsis (Moberg *et al.*, 2003; Stahl *et al.*, 2002). Arabidopsis possesses two PreP isoforms, namely AtPreP1 and AtPreP2, with sequence identity of 86.7% (Bhushan *et al.*, 2005; Moberg *et al.*, 2003). AtPreP, like MPP, belongs to the pitrilysin protease family and contains an inverted Zn-binding motif in the active centre. The crystal structure of AtPreP1 revealed that the active chamber is formed by two halves connected by a hinge region (Johnson *et al.*, 2006). The chamber of 10,000 $Å^3$ is big enough to fit peptides, but too small to accommodate larger folded proteins. This is in agreement with degradation studies, which showed that both AtPreP isoforms degrade unstructured peptides in the range of 10–65 amino acids in length (Stahl *et al.*, 2005). Using peptides derived from the F1β presequence, it was shown that AtPreP1 cleaves its substrate into smaller fragments than AtPreP2 (10–16 amino acids and 10–23 amino acids, respectively). AtPreP1 displays greater preference for positively charged amino acid residues in the C-terminal part of the cleaved peptide (P1′position) than AtPreP2 (Stahl *et al.*, 2005). AtPreP1 and AtPreP2 can degrade both mitochondrial presequences and chloroplast transit peptides. Localization studies confirmed that both isoforms are dual-targeted. Furthermore, precursors of both isoforms can be dual targeted to both mitochondria and chloroplasts using an ambiguous targeting peptide (Bhushan *et al.*, 2003; Bhushan *et al.*, 2005). Transcript and protein level analyses showed that AtPreP isoforms differ in expression level; AtPreP1 is the dominant isoform. Although AtPreP1 and AtPreP2 differ slightly in substrate cleavage preference and expression levels, they functionally complement each other, since single knockouts do not show a strong phenotype (Nilsson Cederholm, Backman, Pesaresi, Leister, & Glaser, 2009), reviewed by Kmiec and Glaser, (2011). Deletion of both isoforms on the other hand, results in decreased size and growth rate of plants. Double knockout plants display a phenotype connected with both mitochondrial and chloroplastic abnormalities: chlorosis, altered starch content, partial loss of integrity of the inner mitochondrial membrane and an affect on mitochondrial respiration. The *prep1prep2* plant phenotype is completely rescued by overexpression of AtPreP1 (Nilsson Cederholm *et al.*, 2009). These abnormalities are mostly seen early in plant development. Lack of a more severe phenotype might imply the existence of additional proteases involved in degradation of harmful peptides.

The PreP gene is highly conserved in all kingdoms of life (T. Eneqvist, personal communication). After identification of AtPreP, its homologues were also found in yeast and mammals (Falkevall *et al.*, 2006; Jonson,

Rehfeld, & Johnsen, 2004). AtPreP1 complemented yeast PreP homologue (Mop112/Cym1) deletion mutants indicating functional conservation of this protein in different species (Alikhani *et al.*, 2011b). Presequences of AtPreP and human PreP (hPreP) are predicted to consist of 85 and 29 amino acids, respectively, whereas the *S. cerevisiae* Cym1/Mop112 presequence contains only seven residues. Whereas the AtPreP presequence has the capacity to target GFP in the mitochondrial matrix of plants, yeast and mammals, the hPreP presequence only targets GFP in the matrix of mammalian and yeast mitochondria. The Cym1/Mop112 presequence has a much weaker targeting capacity overall and only ensures mitochondrial sorting in its host species yeast.

A human homologue of PreP, hPreP, apart from degrading presequences, is also involved in the degradation of amyloid β peptide associated with Alzheimer disease (Alikhani *et al.*, 2011a; Falkevall *et al.*, 2006). The activity of hPreP is decreased in the brain mitochondria of patients suffering from Alzheimer disease (Alikhani *et al.*, 2011a). These two facts link hPreP to the pathology of Alzheimer disease.

Mitochondrial presequence peptide degradation in yeast was also proposed to be conducted by the Prd1 peptidase localized in the intermembrane space (Kambacheld *et al.*, 2005). Analysis of the double yeast knockout of Prd1 and Mop112/Cym1 revealed increased release of peptides from mitochondria and suggests that these two peptidases cooperate in the degradation of peptides in yeast mitochondria (Kambacheld *et al.*, 2005). Analysis of the Arabidopsis genome shows the presence of two Prd1 homologues, however their function remains unknown (Kwasniak *et al.*, 2011).

7. CONCLUSIONS

Our knowledge of the mitochondrial import process in plants has progressed since the initial reports of *in vivo* and *in vitro* import of protein into mitochondria 25 year ago (Boutry, Nagy, Poulsen, Aoyagi, & Chua, 1987; Whelan, Dolan, & Harmey, 1988; Whelan, Knorpp, Glaser, 1990). The unique aspects of protein import into plant mitochondria associated with preprotein receptors and processing and degradation of targeting signals indicate that the plant machinery is different to that in the model yeast system. However, there are many aspects of protein targeting to mitochondria that remain to be understood. The realization that dual targeting of protein to mitochondria and plastids (and peroxisomes) is quite common (Carrie *et al.*, 2009) means that it is even more critical to elucidate the features that define specific targeting of precursor protein to mitochondria, and any cytosolic factors associated with this process. What roles, if any, do the variety of processing peptidases and proteases play in the regulation of turnover of mitochondrial proteins (Li, Nelson, Solheim, Whelan, & Millar, 2012), and can the peptides derived from these processes act as retrograde signals to integrate mitochondrial and cellular function (Moller & Sweetlove, 2010)? The expansion of the PRAT family of proteins is intriguing, raising the possibility that they play a role in mitochondria in addition to protein import. The characterization of many of the translocases associated with protein import is incomplete, especially for the SAM and TIM22. The recent findings that the import translocases and respiratory chain complexes interact with each other in yeast (Gebert *et al.*, 2011a; van der Laan *et al.*, 2006) raises the possibility of a variety of novel interactions in plant mitochondria.

REFERENCES

Abe, Y., Shodai, T., Muto, T., Mihara, K., Torii, H., Nishikawa, S., et al. (2000). Structural basis of presequence recognition by the mitochondrial protein import receptor Tom20. *Cell, 100*, 551–560.

Ahting, U., Waizenegger, T., Neupert, W., & Rapaport, D. (2005). Signal-anchored proteins follow a unique insertion pathway into the outer membrane of mitochondria. *Journal of Biological Chemistry, 280*, 48–53.

Alcock, F., Clements, A., Webb, C., & Lithgow, T. (2010). Evolution. Tinkering inside the organelle. *Science, 327*, 649–650.

Alikhani, N., Guo, L., Yan, S., Du, H., Pinho, C. M., Chen, J. X., et al. (2011a). Decreased proteolytic activity of the mitochondrial amyloid-beta degrading enzyme, PreP peptidasome, in Alzheimer's disease brain mitochondria. *Journal of Alzheimer's Disorders, 27*, 75–87.

Alikhani, N., Berglund, A. K., Engmann, T., Spanning, E., Vogtle, F. N., Pavlov, P., et al. (2011b). Targeting capacity and conservation of PreP homologues localization in mitochondria of different species. *Journal of Molecular Biology, 410*, 400–410.

Bauer, M., Behrens, M., Esser, K., Michaelis, G., & Pratje, E. (1994). Pet1402, a nuclear gene required for proteolytic processing of cytochrome-oxidase subunit-2 in yeast. *Molecular and General Genetics, 245*, 272–278.

Berglund, A. K., Pujol, C., Duchene, A. M., & Glaser, E. (2009a). Defining the determinants for dual targeting of amino acyl-tRNA synthetases to mitochondria and chloroplasts. *Journal of Molecular Biology, 393*, 803–814.

Berglund, A. K., Spanning, E., Biverstahl, H., Maddalo, G., Tellgren-Roth, C., Maler, L., et al. (2009b). Dual targeting to mitochondria and chloroplasts: characterization of Thr-tRNA synthetase targeting peptide. *Molecular Plant, 2*, 1298–1309.

Bhushan, S., Kuhn, C., Berglund, A. K., Roth, C., & Glaser, E. (2006). The role of the N-terminal domain of chloroplast targeting peptides in organellar protein import and miss-sorting. *FEBS Letters, 580*, 3966–3972.

Bhushan, S., Lefebvre, B., Stahl, A., Wright, S. J., Bruce, B. D., Boutry, M., et al. (2003). Dual targeting and function of a protease in mitochondria and chloroplasts. *EMBO Reports, 4*, 1073–1078.

Bhushan, S., Stahl, A., Nilsson, S., Lefebvre, B., Seki, M., Roth, C., et al. (2005). Catalysis, subcellular localization, expression and evolution of the targeting peptides degrading protease, AtPreP2. *Plant Cell Physiology, 46*, 985–996.

Bomer, U., Meijer, M., Guiard, B., Dietmeier, K., Pfanner, N., & Rassow, J. (1997). The sorting route of cytochrome b2 branches from the general mitochondrial import pathway at the preprotein translocase of the inner membrane. *Journal of Biological Chemistry, 272*, 30439–30446.

Bonnefoy, N., Chalvet, F., Hamel, P., Slonimski, P. P., & Dujardin, G. (1994). Oxa1, a *Saccharomyces cerevisiae* nuclear gene whose sequence is conserved from prokaryotes to eukaryotes controls cytochrome oxidase biogenesis. *Journal of Molecular Biology, 239*, 201–212.

Boutry, M., Nagy, F., Poulsen, C., Aoyagi, K., & Chua, N. H. (1987). Targeting of bacterial chloramphenicol acetyltransferase to mitochondria in transgenic plants. *Nature, 328*, 340–342.

Braun, H. P., Emmermann, M., Kruft, V., & Schmitz, U. K. (1992). The general mitochondrial processing peptidase from potato is an integral part of cytochrome *c* reductase of the respiratory chain. *EMBO Journal, 11*, 3219–3227.

Braun, H. P., Emmermann, M., Kruft, V., Bodicker, M., & Schmitz, U. K. (1995). The general mitochondrial processing peptidase from wheat is integrated into the cytochrome bc1-complex of the respiratory chain. *Planta, 195*, 396–402.

Brix, J., Rudiger, S., Bukau, B., Schneider-Mergener, J., & Pfanner, N. (1999). Distribution of binding sequences for the mitochondrial import receptors Tom20, Tom22, and Tom70 in a presequence-carrying preprotein and a non-cleavable preprotein. *Journal of Biological Chemistry, 274*, 16522–16530.

Bruce, B. D. (2000). Chloroplast transit peptides: structure, function and evolution. *Trends in Cell Biology, 10*, 440–447.

Brumme, S., Kruft, V., Schmitz, U. K., & Braun, H. P. (1998). New insights into the coevolution of cytochrome *c* reductase and the mitochondrial processing peptidase. *Journal of Biological Chemistry, 273*, 13143–13149.

Carrie, C., Giraud, E., & Whelan, J. (2009). Protein transport in organelles: dual targeting of proteins to mitochondria and chloroplasts. *FEBS Journal, 276*, 1187–1195.

Carrie, C., Murcha, M. W., & Whelan, J. (2010a). An in silico analysis of the mitochondrial protein import apparatus of plants. *BMC Plant Biology, 10*, 249.

Carrie, C., Giraud, E., Duncan, O., Xu, L., Wang, Y., Huang, S., et al. (2010b). Conserved and novel functions for *Arabidopsis thaliana* MIA40 in assembly of proteins in mitochondria and peroxisomes. *Journal of Biological Chemistry, 285*, 36138–36148.

Chacinska, A., Koehler, C. M., Milenkovic, D., Lithgow, T., & Pfanner, N. (2009). Importing mitochondrial proteins: machineries and mechanisms. *Cell, 138*, 628–644.

Chacinska, A., Pfannschmidt, S., Wiedemann, N., Kozjak, V., Sanjuan Szklarz, L. K., Schulze-Specking, A., et al. (2004). Essential role of Mia40 in import and assembly of mitochondrial intermembrane space proteins. *EMBO Journal, 23*, 3735–3746.

Chacinska, A., van der Laan, M., Mehnert, C. S., Guiard, B., Mick, D. U., Hutu, D. P., et al. (2010). Distinct forms of mitochondrial TOM–TIM supercomplexes define signal-dependent states of preprotein sorting. *Molecular Cell Biology, 30*, 307–318.

Chacinska, A., Lind, M., Frazier, A. E., Dudek, J., Meisinger, C., Geissler, A., et al. (2005). Mitochondrial presequence translocase: switching between TOM tethering and motor recruitment involves Tim21 and Tim17. *Cell, 120*, 817–829.

Chan, N. C., Likic, V. A., Waller, R. F., Mulhern, T. D., & Lithgow, T. (2006). The C-terminal TPR domain of Tom70 defines a family of mitochondrial protein import receptors found only in animals and fungi. *Journal of Molecular Biology, 358*, 1010–1022.

Chew, O., & Whelan, J. (2003). Dual targeting ability of targeting signals is dependent on the nature of the mature protein. *Functional Plant Biology, 30*, 805–812.

Cleary, S. P., Tan, F. C., Nakrieko, K. A., Thompson, S. J., Mullineaux, P. M., Creissen, G. P., et al. (2002). Isolated plant mitochondria import chloroplast precursor proteins in vitro with the same efficiency as chloroplasts. *Journal of Biological Chemistry, 277*, 5562–5569.

Dabney-Smith, C., van Den Wijngaard, P. W., Treece, Y., Vredenberg, W. J., & Bruce, B. D. (1999). The C terminus of a chloroplast precursor modulates its interaction with the translocation apparatus and PIRAC. *Journal of Biological Chemistry, 274*, 32351–32359.

Davis, A. J., Alder, N. N., Jensen, R. E., & Johnson, A. E. (2007). The Tim9p/10p and Tim8p/13p complexes bind to specific sites on Tim23p during mitochondrial protein import. *Molecular Biology of the Cell, 18*, 475–486.

Dekker, P. J., Ryan, M. T., Brix, J., Muller, H., Honlinger, A., & Pfanner, N. (1998). Preprotein translocase of the outer mitochondrial membrane: molecular dissection and assembly of the general import pore complex. *Molecular Cell Biology, 18*, 6515–6524.

Dessi, P., Pavlov, P. F., Wallberg, F., Rudhe, C., Brack, S., Whelan, J., et al. (2003). Investigations on the in vitro import ability of mitochondrial precursor proteins synthesized in wheat germ transcription–translation extract. *Plant Molecular Biology, 52*, 259–271.

Donzeau, M., Kaldi, K., Adam, A., Paschen, S., Wanner, G., Guiard, B., et al. (2000). Tim23 links the inner and outer mitochondrial membranes. *Cell, 101*, 401–412.

Duchene, A. M., Giritch, A., Hoffmann, B., Cognat, V., Lancelin, D., Peeters, N. M., et al. (2005). Dual targeting is the rule for organellar aminoacyl-tRNA synthetases in *Arabidopsis thaliana. Proceedings of the National Academy of Sciences of the United States of America, 102*, 16484–16489.

Duncan, O., Murcha, M. W., & Whelan, J. (2012). Unique components of the plant mitochondrial protein import apparatus. *Biochimica et Biophysica Acta.* PMID: 22406071.

Emmermann, M., Braun, H. P., & Schmitz, U. K. (1994). The mitochondrial processing peptidase from potato: a self-processing enzyme encoded by two differentially expressed genes. *Molecular and General Genetics, 245*, 237–245.

Eriksson, A. C., Sjoling, S., & Glaser, E. (1994). The ubiquinol cytochrome c oxidoreductase complex of spinach leaf mitochondria is involved in both respiration and protein processing. *Biochimica et Biophysica Acta, 1186*, 221–231.

Eriksson, A. C., Sjoling, S., & Glaser, E. (1996). Characterization of the bifunctional mitochondrial processing peptidase (MPP)/bc1 complex in *Spinacia oleracea. Journal of Bioenergetics and Biomembranes, 28*, 285–292.

Esser, C., & Martin, W. (2007). Supertrees and symbiosis in eukaryote genome evolution. *Trends in Microbiology, 15*, 435–437.

Falkevall, A., Alikhani, N., Bhushan, S., Pavlov, P. F., Busch, K., Johnson, K. A., et al. (2006). Degradation of the amyloid beta-protein by the novel mitochondrial peptidasome, PreP. *Journal of Biological Chemistry, 281*, 29096–29104.

Funes, S., Kauff, F., van der Sluis, E. O., Ott, M., & Herrmann, J. M. (2011). Evolution of YidC/Oxa1/Alb3 insertases: three independent gene duplications followed by functional specialization in bacteria, mitochondria and chloroplasts. *Biological Chemistry, 392*, 13–19.

Gabriel, K., Milenkovic, D., Chacinska, A., Muller, J., Guiard, B., Pfanner, N., et al. (2007). Novel mitochondrial intermembrane space proteins as substrates of the MIA import pathway. *Journal of Molecular Biology, 365*, 612–620.

Gakh, O., Cavadini, P., & Isaya, G. (2002). Mitochondrial processing peptidases. *Biochimica et Biophysica Acta, 1592*, 63–77.

Gebert, N., Gebert, M., Oeljeklaus, S., von der Malsburg, K., Stroud, D. A., Kulawiak, B., et al. (2011a). Dual function of Sdh3 in the respiratory chain and TIM22 protein translocase of the mitochondrial inner membrane. *Molecular Cell, 44*, 811–818.

Gentle, I. E., Perry, A. J., Alcock, F. H., Likic, V. A., Dolezal, P., Ng, E. T., et al. (2007). Conserved motifs reveal details of ancestry and structure in the small TIM chaperones of the mitochondrial intermembrane space. *Molecular Biology of Evolution, 24*, 1149–1160.

Gerdes, F., Tatsuta, T., & Langer, T. (2012). Mitochondrial AAA proteases – towards a molecular understanding of membrane-bound proteolytic machines. *Biochimica et Biophysica Acta, 1823*, 49–55.

Glaser, E., & Dessi, P. (1999). Integration of the mitochondrial-processing peptidase into the cytochrome bc1 complex in plants. *Journal of Bioenergetics and Biomembranes, 31*, 259–274.

Glaser, E., Eriksson, A., & Sjoling, S. (1994). Bifunctional role of the bc1 complex in plants. Mitochondrial bc1 complex catalyses both electron transport and protein processing. *FEBS Letters, 346*, 83–87.

Gray, M. W., Burger, G., & Lang, B. F. (2001). The origin and early evolution of mitochondria. *Genome Biology, 2*. REVIEWS1018.

Gross, J., & Bhattacharya, D. (2009). Mitochondrial and plastid evolution in eukaryotes: an outsiders' perspective. *Nature Reviews: Genetics, 10*, 495–505.

Hachiya, N., Alam, R., Sakasegawa, Y., Sakaguchi, M., Mihara, K., & Omura, T. (1993). A mitochondrial import factor purified from rat liver cytosol is an ATP-dependent conformational modulator for precursor proteins. *EMBO Journal, 12*, 1579–1586.

Heazlewood, J. L., Tonti-Filippini, J. S., Gout, A. M., Day, D. A., Whelan, J., & Millar, A. H. (2004). Experimental analysis of the Arabidopsis mitochondrial proteome highlights signaling and regulatory components, provides assessment of targeting prediction programs, and indicates plant-specific mitochondrial proteins. *Plant Cell, 16*, 241–256.

Hedtke, B., Borner, T., & Weihe, A. (2000). One RNA polymerase serving two genomes. *EMBO Reports, 1*, 435–440.

Hell, K. (2008). The Erv1-Mia40 disulfide relay system in the intermembrane space of mitochondria. *Biochimica et Biophysica Acta, 1783*, 601–609.

Hewitt, V., Alcock, F., & Lithgow, T. (2011). Minor modifications and major adaptations: the evolution of molecular machines driving mitochondrial protein import. *Biochimica et Biophysica Acta, 1808*, 947–954.

Huang, J., Hack, E., Thornburg, R. W., & Myers, A. M. (1990). A yeast mitochondrial leader peptide functions in vivo as a dual targeting signal for both chloroplasts and mitochondria. *Plant Cell, 2*, 1249–1260.

Huang, S., Taylor, N. L., Whelan, J., & Millar, A. H. (2009). Refining the definition of plant mitochondrial presequences through analysis of sorting signals, N-terminal modifications, and cleavage motifs. *Plant Physiology, 150*, 1272–1285.

Hugosson, M., Andreu, D., Boman, H. G., & Glaser, E. (1994). Antibacterial peptides and mitochondrial presequences affect mitochondrial coupling, respiration and protein import. *European Journal of Biochemistry, 223*, 1027–1033.

John, G. B., Shang, Y. L., Li, L., Renken, C., Mannella, C. A., Selker, J. M. L., et al. (2005). The mitochondrial inner membrane protein mitofilin controls cristae morphology. *Molecular Biology of the Cell, 16*, 1543–1554.

Johnson, K. A., Bhushan, S., Stahl, A., Hallberg, B. M., Frohn, A., Glaser, E., et al. (2006). The closed structure of presequence protease PreP forms a unique 10,000 Angstroms3 chamber for proteolysis. *EMBO Journal, 25*, 1977–1986.

Jonson, L., Rehfeld, J. F., & Johnsen, A. H. (2004). Enhanced peptide secretion by gene disruption of CYM1, a novel protease in *Saccharomyces cerevisiae*. *European Journal of Biochemistry, 271*, 4788–4797.

Kambacheld, M., Augustin, S., Tatsuta, T., Muller, S., & Langer, T. (2005). Role of the novel metallopeptidase Mop112 and saccharolysin for the complete degradation of proteins residing in different subcompartments of mitochondria. *Journal of Biological Chemistry, 280*, 20132–20139.

Kerscher, O., Sepuri, N. B., & Jensen, R. E. (2000). Tim18p is a new component of the Tim54p-Tim22p translocon in the mitochondrial inner membrane. *Molecular Biology of the Cell, 11*, 103–116.

Kerscher, O., Holder, J., Srinivasan, M., Leung, R. S., & Jensen, R. E. (1997). The Tim54p-Tim22p complex mediates insertion of proteins into the mitochondrial inner membrane. *Journal of Cell Biology, 139*, 1663–1675.

Klodmann, J., Senkler, M., Rode, C., & Braun, H. P. (2011). Defining the protein complex proteome of plant mitochondria. *Plant Physiology, 157*, 587–598.

Kmiec, B., & Glaser, E. (2011). A novel mitochondrial and chloroplast peptidasome, PreP. *Physiologia Plantarum*. http://dx.doi.org/10.1111/j.1399-3054.2011.01531.x.

Kmiec-Wisniewska, B., Krumpe, K., Urantowka, A., Sakamoto, W., Pratje, E., & Janska, H. (2008). Plant mitochondrial rhomboid, AtRBL12, has different substrate specificity from its yeast counterpart. *Plant Molecular Biology, 68*, 159–171.

Koehler, C. M., Merchant, S., Oppliger, W., Schmid, K., Jarosch, E., Dolfini, L., et al. (1998). Tim9p, an essential partner subunit of Tim10p for the import of mitochondrial carrier proteins. *EMBO Journal, 17*, 6477–6486.

Kolodziejczak, M., Kolaczkowska, A., Szczesny, B., Urantowka, A., Knorpp, C., Kieleczawa, J., et al. (2002). A higher plant mitochondrial homologue of the yeast m-AAA protease. Molecular cloning, localization, and putative function. *Journal of Biological Chemistry, 277*, 43792–43798.

Kurland, C. G., & Andersson, S. G. (2000). Origin and evolution of the mitochondrial proteome. *Microbiology and Molecular Biology Reviews, 64*, 786–820.

Kwasniak, M., Pogorzelec, L., Migdal, I., Smakowska, E., & Janska, H. (2011). Proteolytic system of plant mitochondria. *Physiologia Plantarum*. http://dx.doi.org/10.1111/j.1399-3054.2011.01542.x.

Laloi, M. (1999). Plant mitochondrial carriers: an overview. *Cellular and Molecular Life Science, 56*, 918–944.

Lancelin, J. M., Bally, I., Arlaud, G. J., Blackledge, M., Gans, P., Stein, M., et al. (1994). NMR structures of ferredoxin chloroplastic transit peptide from *Chlamydomonas reinhardtii* promoted by trifluoroethanol in aqueous solution. *FEBS Letters, 343*, 261–266.

Li, L., Nelson, C. J., Solheim, C., Whelan, J., & Millar, A. H. (2012). Determining degradation and synthesis rates of *Arabidopsis* proteins using the kinetics of progressive 15N labeling of 2D gel-separated protein spots. *Molecular & Cellular Proteomics*. PMID: 22215636.

Lionaki, E., de Marcos Lousa, C., Baud, C., Vougioukalaki, M., Panayotou, G., & Tokatlidis, K. (2008). The essential function of Tim12 in vivo is ensured by the assembly interactions of its C-terminal domain. *Journal of Biological Chemistry, 283*, 15747–15753.

Lister, R., Murcha, M. W., & Whelan, J. (2003). The Mitochondrial Protein Import Machinery of Plants (MPIMP) database. *Nucleic Acids Research, 31*, 325–327.

Lister, R., Mowday, B., Whelan, J., & Millar, A. H. (2002). Zinc-dependent intermembrane space proteins stimulate import of carrier proteins into plant mitochondria. *Plant Journal, 30*, 555–566.

Lister, R., Chew, O., Rudhe, C., Lee, M. N., & Whelan, J. (2001). *Arabidopsis thaliana* ferrochelatase-I and -II are not imported into *Arabidopsis* mitochondria. *FEBS Letters, 506*, 291–295.

Lister, R., Carrie, C., Duncan, O., Ho, L. H., Howell, K. A., Murcha, M. W., et al. (2007). Functional definition of outer membrane proteins involved in preprotein import into mitochondria. *Plant Cell, 19*, 3739–3759.

Lithgow, T., & Schneider, A. (2010). Evolution of macromolecular import pathways in mitochondria, hydrogenosomes and mitosomes. *Philosophical Transactions of the Royal Society of London Series B Biological Science, 365*, 799–817.

Macasev, D., Whelan, J., Newbigin, E., Silva-Filho, M. C., Mulhern, T. D., & Lithgow, T. (2004). Tom22′, an 8-kDa trans-site receptor in plants and protozoans, is a conserved feature of the TOM complex that appeared early in the evolution of eukaryotes. *Molecular Biology and Evolution, 21*, 1557–1564.

Maduke, M., & Roise, D. (1993). Import of a mitochondrial presequence into protein-free phospholipid vesicles. *Science, 260*, 364–367.

May, T., & Soll, J. (2000). 14-3-3 proteins form a guidance complex with chloroplast precursor proteins in plants. *Plant Cell, 12*, 53–64.

Mesecke, N., Terziyska, N., Kozany, C., Baumann, F., Neupert, W., Hell, K., et al. (2005). A disulfide relay system in the intermembrane space of mitochondria that mediates protein import. *Cell, 121*, 1059–1069.

Meyer, E. H., Solheim, C., Tanz, S. K., Bonnard, G., & Millar, A. H. (2011). Insights into the composition and assembly of the membrane arm of plant complex I through analysis of subcomplexes in *Arabidopsis* mutant lines. *Journal of Biological Chemistry, 286*, 26081–26092.

Meyer, W., Bomer, U., & Pratje, E. (1997). Mitochondrial inner membrane bound Pet1402 protein is rapidly imported into mitochondria and affects the integrity of the cytochrome oxidase and ubiquinol-cytochrome *c* oxidoreductase complexes. *Biological Chemistry, 378*, 1373–1379.

Milenkovic, D., Ramming, T., Muller, J. M., Wenz, L. S., Gebert, N., Schulze-Specking, A., et al. (2009). Identification of the signal directing Tim9 and Tim10 into the intermembrane space of mitochondria. *Molecular Biology of the Cell, 20*, 2530–2539.

Millar, A. H., Whelan, J., Soole, K. L., & Day, D. A. (2011). Organization and regulation of mitochondrial respiration in plants. *Annual Review of Plant Biology, 62*, 79–104.

Mirus, O., Bionda, T., von Haeseler, A., & Schleiff, E. (2009). Evolutionarily evolved discriminators in the 3-TPR domain of the Toc64 family involved in protein translocation at the outer membrane of chloroplasts and mitochondria. *Journal of Molecular Modelling, 15*, 971–982.

Mitschke, J., Fuss, J., Blum, T., Hoglund, A., Reski, R., Kohlbacher, O., et al. (2009). Prediction of dual protein targeting to plant organelles. *New Phytologist, 183*, 224–235.

Moberg, P., Stahl, A., Bhushan, S., Wright, S. J., Eriksson, A., Bruce, B. D., et al. (2003). Characterization of a novel zinc metalloprotease involved in degrading targeting peptides in mitochondria and chloroplasts. *Plant Journal, 36*, 616–628.

Mokranjac, D., & Neupert, W. (2010). The many faces of the mitochondrial TIM23 complex. *Biochimica et Biophysica Acta, 1797*, 1045–1054.

Moller, I. M., & Sweetlove, L. J. (2010). ROS signalling–specificity is required. *Trends in Plant Science, 15*, 370–374.

Mossmann, D., Meisinger, C., & Vogtle, F. N. (2011). Processing of mitochondrial presequences. *Biochimica et Biophysica Acta*. PMID: 22172993.

Mozo, T., Fischer, K., Flugge, U. I., & Schmitz, U. K. (1995). The N-terminal extension of the ADP/ATP translocator is not involved in targeting to plant mitochondria in vivo. *Plant Journal, 7*, 1015–1020.

Murakami, K., & Mori, M. (1990). Purified presequence binding factor (PBF) forms an import-competent complex with a purified mitochondrial precursor protein. *EMBO Journal, 9*, 3201–3208.

Murcha, M. W., Millar, A. H., & Whelan, J. (2005a). The N-terminal cleavable extension of plant carrier proteins is responsible for efficient insertion into the inner mitochondrial membrane. *Journal of Molecular Biology, 351*, 16–25.

Murcha, M. W., Lister, R., Ho, A. Y., & Whelan, J. (2003). Identification, expression, and import of components 17 and 23 of the inner mitochondrial membrane translocase from *Arabidopsis*. *Plant Physiology, 131*, 1737–1747.

Murcha, M. W., Elhafez, D., Millar, A. H., & Whelan, J. (2004). The N-terminal extension of plant mitochondrial carrier proteins is removed by two-step processing: the first cleavage is by the mitochondrial processing peptidase. *Journal of Molecular Biology, 344*, 443–454.

Murcha, M. W., Elhafez, D., Millar, A. H., & Whelan, J. (2005b). The C-terminal region of TIM17 links the outer and inner mitochondrial membranes in *Arabidopsis* and is essential for protein import. *Journal of Biological Chemistry, 280*, 16476–16483.

Murcha, M. W., Elhafez, D., Lister, R., Tonti-Filippini, J., Baumgartner, M., Philippar, K., et al. (2007). Characterization of the preprotein and amino acid transporter gene family in *Arabidopsis*. *Plant Physiology, 143*, 199–212.

Neupert, W., & Herrmann, J. M. (2007). Translocation of proteins into mitochondria. *Annual Review of Biochemistry, 76*, 723–749.

Nicolay, K., Laterveer, F. D., & van Heerde, W. L. (1994). Effects of amphipathic peptides, including presequences, on the functional integrity of rat liver mitochondrial membranes. *Journal of Bioenergetics and Biomembranes, 26*, 327–334.

Nilsson Cederholm, S., Backman, H. G., Pesaresi, P., Leister, D., & Glaser, E. (2009). Deletion of an organellar peptidasome PreP affects early development in *Arabidopsis thaliana*. *Plant Molecular Biology, 71*, 497–508.

Ono, H., & Tuboi, S. (1990). Purification and identification of a cytosolic factor required for import of precursors of mitochondrial proteins into mitochondria. *Archives of Biochemistry and Biophysics, 280*, 299–304.

Ou, W. J., Ito, A., Okazaki, H., & Omura, T. (1989). Purification and characterization of a processing protease from rat liver mitochondria. *EMBO Journal, 8*, 2605–2612.

Peeters, N., & Small, I. (2001). Dual targeting to mitochondria and chloroplasts. *Biochimica et Biophysica Acta, 1541*, 54–63.

Perry, A. J., Hulett, J. M., Likic, V. A., Lithgow, T., & Gooley, P. R. (2006a). Convergent evolution of receptors for protein import into mitochondria. *Current Biology, 16*, 221–229.

Pisani, D., Cotton, J. A., & McInerney, J. O. (2007). Supertrees disentangle the chimerical origin of eukaryotic genomes. *Molecular Biology and Evolution, 24*, 1752–1760.

Pujol, C., Marechal-Drouard, L., & Duchene, A. M. (2007). How can organellar protein N-terminal sequences be dual targeting signals? In silico analysis and mutagenesis approach. *Journal of Molecular Biology, 369*, 356–367.

Pujol, C., Bailly, M., Kern, D., Maréchal-Drouard, L., Becker, H., & Duchêne, A. M. (2008). Dual-targeted tRNA-dependent amidotransferase ensures both mitochondrial and chloroplastic Gln-tRNAGln synthesis in plants. *Proceedings of the National Academy of Sciences of the United States of America, 105*(17), 6481–6485.

Pusnik, M., Schmidt, O., Perry, A. J., Oeljeklaus, S., Niemann, M., Warscheid, B., et al. (2011). Mitochondrial preprotein translocase of trypanosomatids has a bacterial origin. *Current Biology, 21*, 1738–1743.

Qbadou, S., Becker, T., Mirus, O., Tews, I., Soll, J., & Schleiff, E. (2006). The molecular chaperone Hsp90 delivers precursor proteins to the chloroplast import receptor Toc64. *The EMBO Journal, 25*, 1836–1847.

Rabl, R., Soubannier, V., Scholz, R., Vogel, F., Mendl, N., Vasiljev-Neumeyer, A., et al. (2009). Formation of cristae and crista junctions in mitochondria depends on antagonism between Fcj1 and Su e/g. *Journal of Cell Biology, 185*, 1047–1063.

Rada, P., Dolezal, P., Jedelsky, P. L., Bursac, D., Perry, A. J., Sedinova, M., et al. (2011). The core components of organelle biogenesis and membrane transport in the hydrogenosomes of *Trichomonas vaginalis*. *PLoS ONE, 6*, e24428.

Rapaport, D. (2003). Finding the right organelle. Targeting signals in mitochondrial outer-membrane proteins. *EMBO Reports, 4*, 948–952.

Rassow, J., Dekker, P. J., van Wilpe, S., Meijer, M., & Soll, J. (1999). The preprotein translocase of the mitochondrial inner membrane: function and evolution. *Journal of Molecular Biology, 286*, 105–120.

Rimmer, K. A., Foo, J. H., Ng, A., Petrie, E. J., Shilling, P. J., Perry, A. J., et al. (2011). Recognition of mitochondrial targeting sequences by the import receptors Tom20 and Tom22. *Journal of Molecular Biology, 405*, 804–818.

Rissler, M., Wiedemann, N., Pfannschmidt, S., Gabriel, K., Guiard, B., Pfanner, N., et al. (2005). The essential mitochondrial protein Erv1 cooperates with Mia40 in biogenesis of intermembrane space proteins. *Journal of Molecular Biology, 353*, 485–492.

Roise, D., Horvath, S. J., Tomich, J. M., Richards, J. H., & Schatz, G. (1986). A chemically synthesized pre-sequence of an imported mitochondrial protein can form an amphiphilic helix and perturb natural and artificial phospholipid bilayers. *EMBO Journal, 5*, 1327–1334.

Rudhe, C., Clifton, R., Whelan, J., & Glaser, E. (2002a). N-terminal domain of the dual-targeted pea glutathione reductase signal peptide controls organellar targeting efficiency. *Journal of Molecular Biology, 324*, 577–585.

Rudhe, C., Chew, O., Whelan, J., & Glaser, E. (2002b). A novel in vitro system for simultaneous import of precursor proteins into mitochondria and chloroplasts. *Plant Journal, 30*, 213–220.

Saddar, S., Dienhart, M. K., & Stuart, R. A. (2008). The F1F0-ATP synthase complex influences the assembly state of the cytochrome bc(1)-cytochrome oxidase supercomplex and its association with the TIM23 machinery. *Journal of Biological Chemistry, 283*, 6677–6686.

Schleiff, E., & Soll, J. (2000). Travelling of proteins through membranes: translocation into chloroplasts. *Planta, 211*, 449–456.

Schleiff, E., & Becker, T. (2011). Common ground for protein translocation: access control for mitochondria and chloroplasts. *Nature Reviews: Molecular Cell Biology, 12*, 48–59.

Schleiff, E., Motzkus, M., & Soll, J. (2002). Chloroplast protein import inhibition by a soluble factor from wheat germ lysate. *Plant Molecular Biology, 50*, 177–185.

Senapin, S., Chen, X. J., & Clark-Walker, G. D. (2003). Transcription of TIM9, a new factor required for the petite-positive phenotype of *Saccharomyces cerevisiae*, is defective in spt7 mutants. *Current Genetics, 44*, 202–210.

Silva-Filho, M. C. (2003). One ticket for multiple destinations: dual targeting of proteins to distinct subcellular locations. *Current Opinions in Plant Biology, 6*, 589–595.

Silva-Filho, M. D., Wieers, M. C., Flugge, U. I., Chaumont, F., & Boutry, M. (1997). Different in vitro and in vivo targeting properties of the transit peptide of a chloroplast envelope inner membrane protein. *Journal of Biological Chemistry, 272*, 15264–15269.

Sirrenberg, C., Bauer, M. F., Guiard, B., Neupert, W., & Brunner, M. (1996). Import of carrier proteins into the mitochondrial inner membrane mediated by Tim22. *Nature, 384*, 582–585.

Sirrenberg, C., Endres, M., Folsch, H., Stuart, R. A., Neupert, W., & Brunner, M. (1998). Carrier protein import into mitochondria mediated by the intermembrane proteins Tim10/Mrs11 and Tim12/Mrs5. *Nature, 391*, 912–915.

Stahl, A., Moberg, P., Ytterberg, J., Panfilov, O., Brockenhuus Von Lowenhielm, H., Nilsson, F., et al. (2002). Isolation and identification of a novel mitochondrial metalloprotease (PreP) that degrades targeting presequences in plants. *Journal of Biological Chemistry, 277*, 41931–41939.

Stahl, A., Nilsson, S., Lundberg, P., Bhushan, S., Biverstahl, H., Moberg, P., et al. (2005). Two novel targeting peptide degrading proteases, PrePs, in mitochondria and chloroplasts, so similar and still different. *Journal of Molecular Biology, 349*, 847–860.

Taylor, A. B., Smith, B. S., Kitada, S., Kojima, K., Miyaura, H., Otwinowski, Z., et al. (2001). Crystal structures of mitochondrial processing peptidase reveal the mode for specific cleavage of import signal sequences. *Structure, 9*, 615–625.

Terziyska, N., Lutz, T., Kozany, C., Mokranjac, D., Mesecke, N., Neupert, W., et al. (2005). Mia40, a novel factor for protein import into the intermembrane space of mitochondria is able to bind metal ions. *FEBS Letters, 579*, 179–184.

Thrash, J. C., Boyd, A., Huggett, M. J., Grote, J., Carini, P., Yoder, R. J., et al. (2011). Phylogenomic evidence for a common ancestor of mitochondria and the SAR11 clade. *Scientific Reports, 1*, 13.

Tokatlidis, K. (2005). A disulfide relay system in mitochondria. *Cell, 121*, 965–967.

Tong, J., Dolezal, P., Selkrig, J., Crawford, S., Simpson, A. G., Noinaj, N., et al. (2011). Ancestral and derived protein import pathways in the mitochondrion of *Reclinomonas americana*. *Molecular Biology and Evolution, 28*, 1581–1591.

van der Giezen, M., & Tovar, J. (2005). Degenerate mitochondria. *EMBO Reports, 6*, 525–530.

van der Laan, M., Hutu, D. P., & Rehling, P. (2010). On the mechanism of preprotein import by the mitochondrial presequence translocase. *Biochimica et Biophysica Acta, 1803*, 732–739.

van der Laan, M., Wiedemann, N., Mick, D. U., Guiard, B., Rehling, P., & Pfanner, N. (2006). A role for Tim21 in membrane-potential-dependent preprotein sorting in mitochondria. *Current Biology, 16*, 2271–2276.

Vogtle, F. N., Wortelkamp, S., Zahedi, R. P., Becker, D., Leidhold, C., Gevaert, K., et al. (2009). Global analysis of the mitochondrial N-proteome identifies a processing peptidase critical for protein stability. *Cell, 139*, 428–439.

von der Malsburg, K., Muller, J. M., Bohnert, M., Oeljeklaus, S., Kwiatkowska, P., Becker, T., et al. (2011). Dual role of mitofilin in mitochondrial membrane organization and protein biogenesis. *Development Cell, 21*, 694–707.

von Heijne, G. (1986). Mitochondrial targeting sequences may form amphiphilic helices. *EMBO Journal, 5*, 1335–1342.

Waller, R. F., Jabbour, C., Chan, N. C., Celik, N., Likic, V. A., Mulhern, T. D., et al. (2009). Evidence of a reduced and modified mitochondrial protein import apparatus in microsporidian mitosomes. *Eukaryotic Cell, 8*, 19–26.

Wang, P., & Dalbey, R. E. (2011). Inserting membrane proteins: the YidC/Oxa1/Alb3 machinery in bacteria, mitochondria, and chloroplasts. *Biochimica et Biophysica Acta, 1808*, 866–875.

Werhahn, W., Niemeyer, A., Jansch, L., Kruft, V., Schmitz, U. K., & Braun, H. (2001). Purification and characterization of the preprotein translocase of the outer mitochondrial membrane from *Arabidopsis*. Identification of multiple forms of TOM20. *Plant Physiology, 125*, 943–954.

Whelan, J., Dolan, L., & Harmey, M. A. (1988). Import of precursor proteins into *Viica faba* mitochondria. *FEBS Letters, 236*, 217–220.

Whelan, J., Knorpp, C., & Glaser, E. (1990). Sorting of precursor proteins between isolated spinach leaf mitochondria and chloroplasts. *Plant Molecular Biology, 14*, 977–982.

Wiedemann, N., Van der Laan, M., Hutu, D. P., Rehling, P., & Pfanner, N. (2007). Sorting switch of mitochondrial presequence translocase involves coupling of motor module to respiratory chain. *Journal of Cell Biology, 179*, 1115–1122.

Wieprecht, T., Apostolov, O., Beyermann, M., & Seelig, J. (2000). Interaction of a mitochondrial presequence with lipid membranes: role of helix formation for membrane binding and perturbation. *Biochemistry, 39*, 15297–15305.

Xie, J., Marusich, M. F., Souda, P., Whitelegge, J., & Capaldi, R. A. (2007). The mitochondrial inner membrane protein mitofilin exists as a complex with SAM50, metaxins 1 and 2, coiled-coil-helix coiled-coil-helix domain-containing protein 3 and 6 and DnaJC11. *FEBS Letters, 581*, 3545–3549.

Yang, M., Jensen, R. E., Yaffe, M. P., Oppliger, W., & Schatz, G. (1988). Import of proteins into yeast mitochondria: the purified matrix processing protease contains two subunits which are encoded by the nuclear MAS1 and MAS2 genes. *EMBO Journal, 7*, 3857–3862.

Yang, M. J., Geli, V., Oppliger, W., Suda, K., James, P., & Schatz, G. (1991). The MAS-encoded processing protease of yeast mitochondria. Interaction of the purified enzyme with signal peptides and a purified precursor protein. *Journal of Biological Chemistry, 266*, 6416–6423.

Yano, M., Terada, K., & Mori, M. (2003). AIP is a mitochondrial import mediator that binds to both import receptor Tom20 and preproteins. *Journal of Cell Biology, 163*, 45–56.

Young, J. C., Hoogenraad, N. J., & Hartl, F. U. (2003). Molecular chaperones Hsp90 and Hsp70 deliver preproteins to the mitochondrial import receptor Tom70. *Cell, 112*, 41–50.

Zardeneta, G., & Horowitz, P. M. (1992). Analysis of the perturbation of phospholipid model membranes by rhodanese and its presequence. *Journal of Biological Chemistry, 267*, 24193–24198.

Zhang, X. P., & Glaser, E. (2002). Interaction of plant mitochondrial and chloroplast signal peptides with the Hsp70 molecular chaperone. *Trends in Plant Science, 7*, 14–21.

CHAPTER TWELVE

Macromolecules Trafficking to Plant Mitochondria

Morgane Michaud and Anne-Marie Duchêne[1]

Institut de Biologie Moléculaire des Plantes, UPR 2357 du CNRS, Université de Strasbourg, Strasbourg, France

[1]Corresponding author. E-mail: anne-marie.duchene@ibmp-cnrs.unistra.fr

Contents

Advances in Botanical Research, Volume 63
ISSN 0065-2296,
http://dx.doi.org/10.1016/B978-0-12-394279-1.00012-0

Abstract

The plant mitochondrial genome contains a limited number of genes. Mitochondrial activity is therefore highly dependent on the nuclear genome. Nearly all mitochondrial proteins are nuclear encoded and synthesized in the cytosol. Their subsequent import into mitochondria requires particular signals and machinery depending on the final submitochondrial localization of the proteins. These signals are located at the protein level and/or at the mRNA level. Several mitochondrial proteins are also localized in another subcellular compartment. These dual-targeted proteins have retained a combination of signals allowing their correct sorting to different compartments. Furthermore, many transfer RNAs, essential for mitochondrial translation, are also nuclear encoded and imported into mitochondria. In plants, they represent one-third to one-half of the set of mitochondrial tRNAs. However, the signals and mechanisms involved in their translocation through the mitochondrial membranes have not been completely deciphered.

1. INTRODUCTION

Mitochondria are found in all aerobic eukaryotes. These organelles with two membranes originate from an α-bacterium and have conserved a genome. During evolution, most of the genes were lost or transferred to the nucleus. Present-day mitochondrial genomes contain a limited number of genes. In higher plants, the number of proteins encoded by mitochondrial DNA ranges from 25 to 40, there are three ribosomal RNA genes, and the number of transfer RNA (tRNA) genes varies from 11 to 22. Mitochondrial activity is therefore highly dependent on the nuclear genome. Most mitochondrial proteins and numerous mitochondrial tRNAs are nuclear encoded and imported from the cytosol. The import of these macromolecules requires particular import mechanisms. In plants, the presence of another organelle, the chloroplast, which is also of bacterial origin and imports most of its proteins, leads to a specific differentiation of import mechanisms.

The mechanism of protein import has been analysed in detail in yeast, and the differences in the plant system are summarized in this chapter. In particular, the different protein import pathways and the targeting signals are presented, as well as examples of protein dual targeting to mitochondria and other compartments. Targeting of messenger RNA to the mitochondrial surface and its role in protein import are also discussed. Transfer RNA import and its function in mitochondrial translation is described. DNA uptake into plant mitochondria is summarized.

2. PROTEIN IMPORT INTO MITOCHONDRIA

The different pathways and components of the mitochondrial import machinery are described briefly and the main differences between plants and yeast or mammals are highlighted. For more details about the evolution or functions of these pathways/components, see Chapter 11.

2.1. Protein Import Pathways

Protein import into mitochondria involves the cooperation of different complexes to bring mitochondrial precursors to their correct localization. The outer membrane (OM) is the first gateway to mitochondria. Translocase of the outer membrane of mitochondria (TOM) is the main complex of the OM and nearly all mitochondrial proteins have to pass through this complex before reaching their final destination. Subsequently, four main

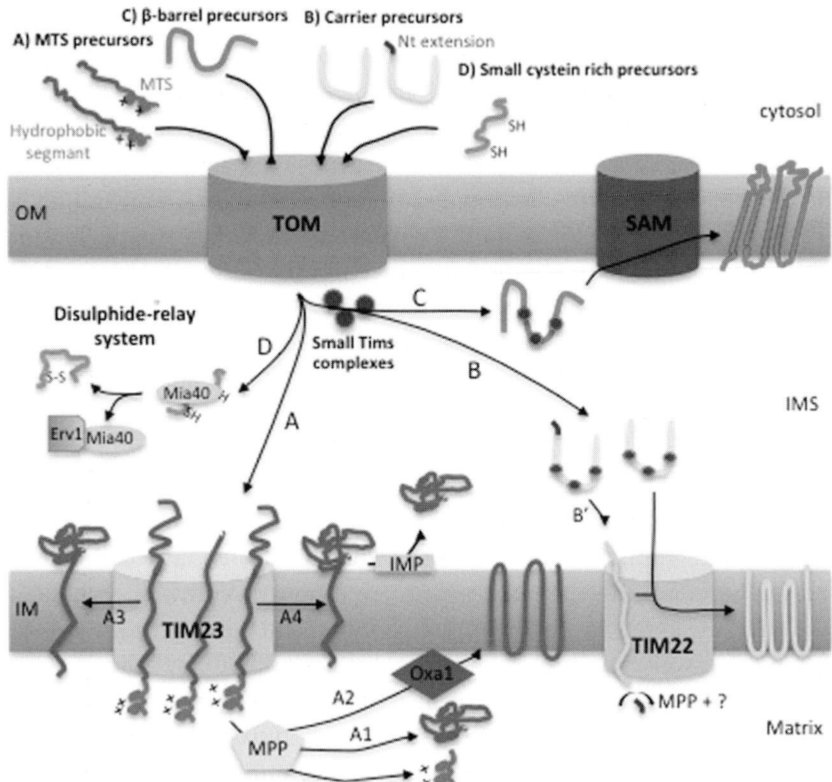

Figure 12.1 Protein import pathways into mitochondria. (A) Import pathways for MTS-containing precursors. After their translocation through the TOM complex, proteins with a cleavable MTS reach the TIM23 complex. MTS are cleaved by the MPP (present in the matrix or the MPP/bc1 complex in yeast and plants, respectively). Matrix proteins and multiple membrane-spanning proteins of the IM are completely translocated into the matrix (A1, A2) and membrane proteins are inserted into the IM with the help of Oxa1 (A2). Transfer of precursors with a hydrophobic segment is stopped in the Tim23 channel and proteins are released into the IM (A3, A4). IMS proteins are released in the IMS by a cleavage catalysed by the IMP (A4). No homologues of IMP have been found in plants and the presence of the A4 pathway is still elusive. (B) Import pathways for carrier proteins, with or without N-terminal extension. N-terminal extension has been found only in plants and mammals. After crossing the TOM complex, carrier precursors bind to the small Tims protein complexes in the IMS. Then they reach the TIM22 complex involved in their insertion into the IM. The N-terminal extension present in some precursors is cleaved by MPP and an unknown endopeptidase (B′). (C) Import pathway of β-barrel precursors. Precursors interact with the small Tims complexes in the IMS after their translocation through the TOM complex. Proteins are inserted into the OM by the SAM complex. (D) Import pathway of small cystein-rich proteins. After translocation in the IMS by the TOM complex, the disulphide relay system catalyses the formation of intramolecular disulphide bonds allowing the folding of the proteins. MTS,

pathways can be distinguished depending on precursor features (Fig. 12.1) (for review see Balsera, Soll, & Bolter, 2009; Bolender, Sickmann, Wagner, Meisinger, & Pfanner et al., 2008; Chacinska, Koehler, Milenkovic, Lithgow & Pfanner, 2009; Mokranjac & Neupert, 2009; Neupert & Herrmann, 2007). Most mitochondrial proteins located in the matrix, the inner membrane (IM) and the intermembrane space (IMS) present a mitochondrial targeting signal (MTS) at their N-terminal end (see Section IIIA). This signal targets precursors to the translocase of the IM of mitochondria 23 (TIM23) complex after passing through the TOM complex (Fig. 12.1A). Translocation through the TIM23 complex is dependent on the membrane potential across the IM ($\Delta\Psi$) and on ATP hydrolysis. Precursors reach their final destination by four different pathways (Fig. 12.1, A1–A4). Cleavable MTSs are removed by the mitochondrial processing peptidase (MPP) (Figs 12.1A and 12.2C) (Eriksson, Sjoling, & Glaser, 1996; Glaser, Eriksson, & Sjoling, 1994). The second pathway is for carrier proteins with an even number of transmembrane segments. Their insertion into the IM involves the TIM22 complex and is $\Delta\Psi$ dependent (for more details, see Fig. 12.1, B). In plants and mammals, some of these carriers contain an N-terminal extension cleaved by MPP and an unidentified peptidase (Fig. 12.1, B′). The third pathway is for β-barrel proteins. Once in the IMS, these proteins are inserted into the OM by the sorting and assembly machinery (SAM) complex (Fig. 12.1, C). Finally, cysteine-rich (cys-rich) proteins of the IMS are trapped by the Mia40/Erv1 disulphide relay system present in the IMS (Fig. 12.1, D). The conserved small Tim proteins in the IMS interact with hydrophobic precursors and prevent their aggregation (Fig. 12.1, B and C).

2.2. Protein Import Complexes

Overall, the subunits of import complexes described in yeast are conversed in plants. The main variations are observed for complexes of the OM (Fig. 12.2) (Balsera *et al.*, 2009; Lister & Whelan, 2006; Perry *et al.*, 2008). Another significant difference compared with yeast is the presence of several isoforms for nearly all import subunits in plants (Carrie, Murcha, & Whelan, 2010b; Lister *et al.*, 2004; Lister, Murcha, & Whelan, 2003).

mitochondrial targeting signal; TOM, translocase of outer membrane of mitochondria; TIM, translocase of inner membrane of mitochondria; SAM, sorting and assembly machinery; IMS, intermembrane space; MPP, mitochondrial processing peptidase; IM, inner membrane; OM, outer membrane; IMP, intermediate peptidase. See the colour plate.

2.2.1. The TOM complex

The conserved core of the TOM complex is composed of the Tom40 translocase and three small Tom subunits (Tom5, Tom6, Tom7) involved in the assembly and dynamics of this complex (Fig. 12.2A) (Carrie *et al.*, 2010b; Chacinska *et al.*, 2009). The TOM complex is also composed of receptors involved in precursor recognition. In yeast, the Tom20 and Tom22 receptors are mainly involved in the recognition of MTS-containing precursors, and Tom70 is mainly involved in the recognition of carrier proteins (Neupert & Herrmann, 2007) (Fig. 12.2A). In plants, no homologues were found for Tom70. The Tom22 subunit, called Tom9, is deleted from it cytosolic acidic domain involved in the binding of precursors (Carrie *et al.*, 2010b; Macasev, Newbigin, Whelan, & Lithgow, 2000; Macasev *et al.*, 2004). Plant TOM complex also contains a Tom20 receptor but it is not evolutionarily related to the Tom20 subunit from yeast and mammals (Lister & Whelan, 2006; Perry *et al.*, 2006). Two other potential receptors are found in plant mitochondria: metaxin and mtOM64 (Fig. 12.2A). In mammals, metaxin acts as a preprotein receptor and interferes with the import of β-barrel proteins (Armstrong, Komiya, Bergman, Mihara & Bornstein, 1997; Kozjak-Pavlovic *et al.*, 2007). Similarly in plants, metaxin interacts with several mitochondrial proteins including β-barrel proteins (Lister *et al.*, 2007). MtOM64 is a plant-specific subunit that is a clear homologue of Toc64, one of the receptors of the translocator of outer membrane of chloroplast (TOC) complex (Chew *et al.*, 2004). MtOM64 plays a role in the mitochondrial import of some precursors and is able to bind several preproteins, suggesting a role as a receptor (Lister *et al.*, 2007).

2.2.2. The SAM complex

The SAM complex is composed of the well-conserved Sam50, a pore-forming subunit promoting the insertion of β-barrel proteins into the OM (Becker, Gebert, Pfanner, & Van Der Laan, 2009; Kutik *et al.*, 2008; Neupert & Herrmann, 2007). The other subunits are not conserved between yeast and other species (Fig. 12.2B). Yeast Sam35 and Sam37 subunits are involved in the recognition of β-barrel proteins and in their release into the OM, respectively (see Section III.B.2) (Chan & Lithgow, 2008; Kutik *et al.*, 2008). Other proteins interacting with the SAM complex in yeast, such as Mdm10 or Mim1, are involved in the insertion of α-helical proteins into the OM and in the assembly of TOM complexes (Becker *et al.*, 2008; Meisinger *et al.*, 2004; Waizenegger, Schmitt, Zivkovic, Neupert, &

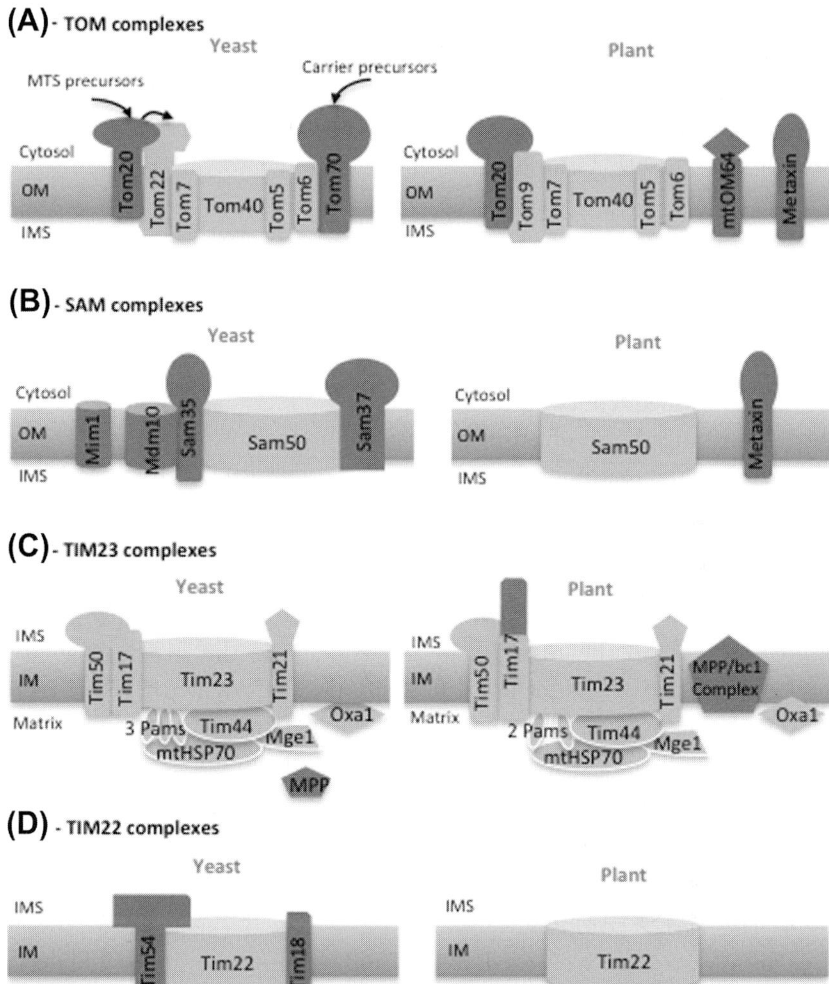

Figure 12.2 Composition of the four main complexes involved in protein import into yeast and plant mitochondria. The evolution-related subunits between yeast and plant are in orange and the non-conserved subunits are in green. (A) Translocase of the outer membrane of mitochondria (TOM) complexes. (B) Sorting and assembly machinery (SAM). Note that the association of metaxin with TOM and/or SAM complexes is still elusive in plants. (C) Translocase of the inner membrane of mitochondria 23 (TIM23) complexes. The mitochondrial processing peptidase (MPP) and Oxa1 are not part of the TIM23 complexes but are involved in the processing of mitochondrial targeting signals (MTS) and the insertion of IM proteins, respectively. Note that in plants, MPP is part of the bc1 complex. (D) Translocase of the inner membrane of mitochondria 22 (TIM22) complexes. IMS, intermembrane space; IM, inner membrane; OM, outer membrane. See the colour plate.

Rapaport, 2005; Yamano, Tanaka-Yamano, & Endo, 2010). Until now, components associated with Sam50 in other organisms remained elusive. Metaxin interferes with the import of β-barrels in mammals and plants but direct interaction with Sam50 has not been shown until now (Kozjak-Pavlovic *et al.*, 2007; Lister *et al.*, 2007).

2.2.3. The TIM23 complex

The overall composition of the TIM23 complex is similar between plants and yeast (Fig. 12.2C) (Carrie *et al.*, 2010b; Lister *et al.*, 2003). Tim23 constitutes the translocation pore and is associated with three other membrane-embedded components: Tim50, Tim17 and Tim21. These subunits are mainly involved in the transfer of precursors from the TOM complex into the matrix or the IM (Chacinska *et al.*, 2005; Mokranjac *et al.*, 2009; Neupert & Herrmann, 2007). Plant Tim17 contains a C-terminal extension of 85 amino acids that crosses the OM. The role of this extension is still elusive (Murcha, Elhafez, Millar, & Whelan, 2005a; Murcha, Lister, Ho, & Whelan, 2003). The TIM23 complex is also composed of an import motor, located on the matrix side and containing the matrix chaperon mtHSP70. MtHSP70 binds to preproteins after sorting across Tim23 and assists with their folding (Bolender *et al.*, 2008; Neupert & Herrmann, 2007). Five subunits that regulate the ATPase activity of mtHSP70 and its association with the translocation pore are also part of this mechanism (Fig. 12.2C).

2.2.4. The TIM22 complex

In addition to the well-conserved Tim22 translocase, two additional subunits, Tim54 and Tim18, are found only in yeast (Carrie *et al.*, 2010b) and might be involved in the assembly and stability of the TIM22 complex (Becker *et al.*, 2009; Wagner *et al.*, 2008). The others components of TIM22 complexes are not known in plants or mammals (Fig. 12.2D).

2.2.5. The Mia40/Erv1 disulphide relay system

Mia40 and Erv1 are present in yeast and in plants. In plants, Mia40 is not essential for the function of the disulphide relay system and is also targeted to peroxisomes. Thus Evr1, alone or with the help of unidentified components, seems to be mainly involved in the formation of disulphide bonds (Carrie *et al.*, 2010a).

3. TARGETING SIGNALS

Signals and pathways of protein targeting and import into mitochondria depend on the final submitochondrial localization and on the topology of the mature protein. This last feature is mostly true for membrane proteins. These proteins can be anchored in the IM or the OM by a single membrane-spanning segment present at the N- or C-terminal end or by multiple membrane-spanning segments. Mitochondrial OM also contains a group of proteins with a particular structure, the β-barrel proteins, which form a pore constituted of several β-strands. Consequently, a multitude of targeting signals have been developed by mitochondria during evolution (Fig. 12.3). Some, like MTS, are common to the vast majority of mitochondrial precursors. Others are specific to a subset of proteins. Only MTS is cleavable. In this chapter, the term precursor is used for all proteins that have not reached their final destination even if no maturation occurs.

3.1. The Cleavable N-Terminal Targeting Signal: MTS

MTSs are signals present at the N-terminal end of matrix proteins and of some IM proteins in plants. They are involved in the recognition of precursors by the TOM complex and in their subsequent import through the TOM and TIM23 complexes. These signals are removed in the matrix by MPP to obtain the mature form of the mitochondrial proteins (Fig. 12.1, A). In general, MTSs are sufficient to import precursors into mitochondria but in some cases, sequences in the mature protein are also necessary (Kimura *et al.*, 1993; Lee & Whelan, 2004).

3.1.1. Features of plant MTS

Plant MTSs present important variations in length, ranging from 13 to 117 amino acids with an average of 40–50 residues (Huang, Taylor, Whelan, & Millar, 2009; Sjoling & Glaser, 1998; Zhang & Glaser, 2002; Zhang *et al.*, 2001). They are enriched in basic (Arg, Lys), hydroxylated (Ser, Thr) and hydrophobic (Ala, Leu) residues and have a low content in acidic and aromatic residues compared with mature sequences (Fig. 12.3) (Glaser, Sjoling, Tanudji, & Whelan, 1998; von Heijne, 1986; Zhang & Glaser, 2002; Zhang *et al.*, 2001). Plant MTSs are longer than in yeast (7–9 amino acids longer) and have a higher Ser content (17% in plants, 7% in yeast) (Sjoling & Glaser, 1998). No consensus motifs can be defined. Analyses of the amino acid distribution along plant MTSs have revealed some common

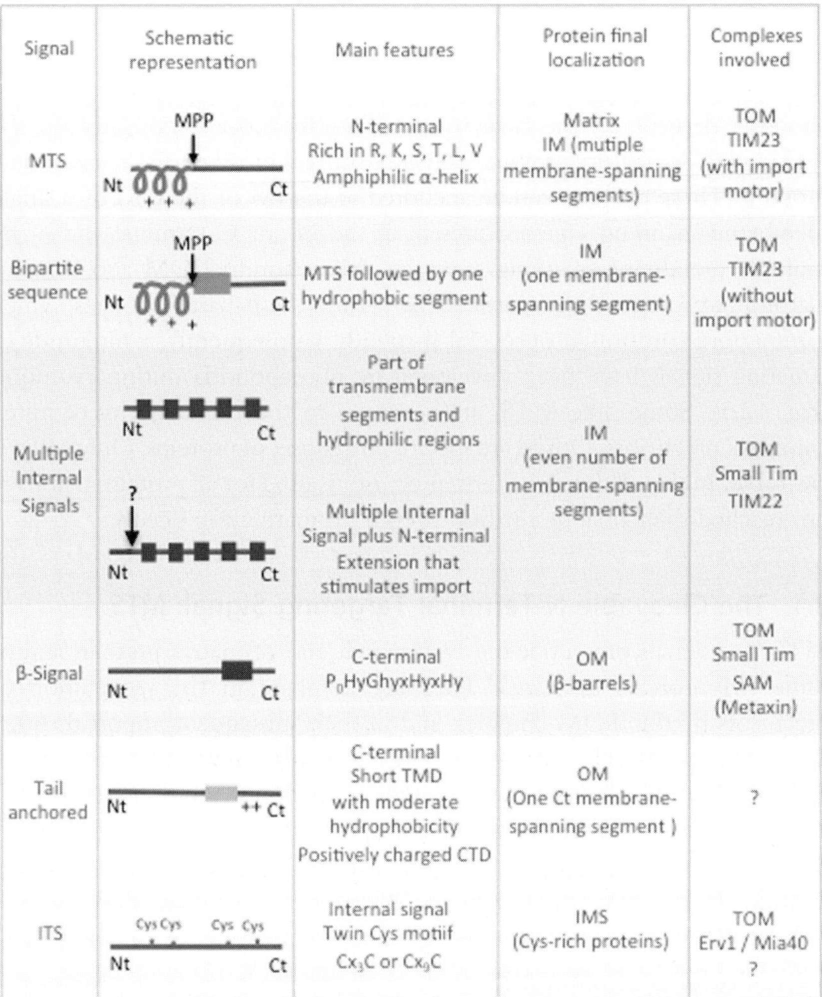

Signal	Schematic representation	Main features	Protein final localization	Complexes involved
MTS	MPP; Nt; Ct; + + +	N-terminal Rich in R, K, S, T, L, V Amphiphilic α-helix	Matrix IM (mutiple membrane-spanning segments)	TOM TIM23 (with import motor)
Bipartite sequence	MPP; Nt; Ct; + + +	MTS followed by one hydrophobic segment	IM (one membrane-spanning segment)	TOM TIM23 (without import motor)
Multiple Internal Signals	Nt; Ct ?; Nt; Ct	Part of transmembrane segments and hydrophilic regions Multiple Internal Signal plus N-terminal Extension that stimulates import	IM (even number of membrane-spanning segments)	TOM Small Tim TIM22
β-Signal	Nt; Ct	C-terminal P_0HyGhyxHyxHy	OM (β-barrels)	TOM Small Tim SAM (Metaxin)
Tail anchored	Nt; ++ Ct	C-terminal Short TMD with moderate hydrophobicity Positively charged CTD	OM (One Ct membrane-spanning segment)	?
ITS	Cys Cys Cys Cys; Nt; Ct	Internal signal Twin Cys motiif Cx_3C or Cx_9C	IMS (Cys-rich proteins)	TOM Erv1 / Mia40 ?

Figure 12.3 Summary of the main protein signals involved in mitochondrial targeting in plants. The main features of the signals, the final mitochondrial localization of proteins and the import machineries involved are shown. ITS, internal targeting sequence; TMD, transmembrane domain; CTD, C-terminal domain; P_0, polar residue; Hy, large hydrophobic residue; hy, hydrophobic residue. *Adapted from Chacinska et al. (2009).* For colour version of this figure, the reader is referred to the online version of this book.

features and conserved positions. For instance, Arg residues are enriched in the N-terminal part (residues 1–20) of plant MTS (Bhushan, Kuhn, Berglund, Roth, & Glaser, 2006; Pujol, Marechal-Drouard, & Duchene, 2007) and hydrophobic amino acids are overrepresented in position 2 (Pujol *et al.*, 2007). As in mammals and yeast, plant MTSs have a propensity to adopt α-helical amphiphilic structures, with a positively charged side and a hydrophobic side, when they are in hydrophobic environments (Moberg *et al.*, 2004; Rimmer *et al.*, 2011; Roise *et al.*, 1988; von Heijne, Steppuhn, & Herrmann, 1989). The hydrophobic side is involved in the interaction of MTS with Tom20 receptor in both plants and mammals (Abe *et al.*, 2000; Rimmer *et al.*, 2011).

3.1.2. Model of import of MTSs containing precursors into mitochondria: the binding chain hypothesis

In yeast and mammals, a model called the binding chain hypothesis has been proposed for translocation of MTS-containing precursors from the cytosol into mitochondria (Komiya, Rospert, Koehler, Looser, Schatz, & Mihara, 1998; Pfanner & Geissler, 2001). This model arises from the observation that different subunits of the TOM complex (Tom20, Tom22, Tom40) and of the TIM23 complex (Tim23, Tim50) are able to interact directly with MTS (Pfanner & Geissler, 2001). Thus, after an initial binding of precursors to the Tom20 receptor, they are transferred from one binding site to another, allowing their progression into the TOM and TIM23 complexes. Some binding steps are mediated by ionic interactions between positively charged MTSs and acidic components of the mitochondrial import machinery (Moczko *et al.*, 1997). Positively charged residues are also important for $\Delta\psi$-dependent translocation of the precursor through the TIM23 complex (Fig. 12.1, A) (Neupert & Herrmann, 2007). Such a mechanism has not yet been demonstrated in plants but is likely to exist. The IMS side of the TOM complex and the components of the TIM23 complex are well conserved between yeast and plant mitochondria (Fig. 12.2C) and the overall features of the presequences are also conserved.

3.1.3. The bipartite presequences

IM proteins with one transmembrane segment are targeted and imported to their final mitochondrial subcompartment by a stop–transfer mechanism that involves bipartite presequences (Fig. 12.1, A3 and Fig. 12.3) (Braun, Emmermann, Kruft, & Schmitz, 1992; Neupert & Herrmann, 2007). These presequences are composed of an MTS followed by a hydrophobic segment. The MTS allows the transfer of the precursor from the cytosol to the TIM23 complex. Once in the Tim23 channel, the hydrophobic segment stops the transfer and the precursor is released into the IM (Fig. 12.1, A3). In yeast, some IMS proteins also follow this pathway. After transfer of precursor into the IM, a cleavage after the hydrophobic segment, catalysed by the intermediate peptidase (IMP), releases the protein in the IMS (Fig. 12.1, A4) (Gakh, Cavadini, & Isaya, 2002; Neupert & Herrmann, 2007).

3.2. Non-Cleavable targeting signals

3.2.1. Multiple internal signals

Multiple internal signals (MIS) are found in a group of IM proteins containing several membrane-spanning segments. This group is composed of some Tim subunits (Tim17, Tim23 and Tim22) and of IM carrier proteins containing an even number of transmembrane segments. The insertion of these proteins into the IM is dependent on $\Delta\Psi$ and involves the TIM22 complex (Fig. 12.1, B). In yeast, MISs are composed of signals of about ten residues that are distributed along the protein (Chacinska *et al.*, 2009). These signals contain portions of transmembrane segments and flanking hydrophilic regions that are able to interact with the Tom70 receptor and with small Tims once in the IMS (Neupert & Herrmann, 2007; Vasiljev *et al.*, 2004). The precursor/small Tims complex binds to the TIM22 complex. This binding is mediated by the small Tims and/or by the direct binding of the C-terminal end of precursors to Tim22 (Brandner, Rehling, & Truscott, 2005; Chacinska *et al.*, 2009). Precursors are subsequently inserted into the IM. Two main differences can be noted in plants: the absence of homologues for the Tom70 receptor (Fig. 12.2A) and the presence of a cleavable N-terminal extension for some carriers compared with yeast orthologues (Fig. 12.3) (Murcha, Elhafez, Millar, & Whelan, 2004; Winning, Sarah, Purdue, Day, & Leaver, 1992). The last feature has also been described for some carriers of mammalian mitochondria (Zara, Ferramosca, Palmisano, Palmieri, & Rassow, 2003; Zara, Palmieri, Mahlke, & Pfanner, 1992). Receptor(s) involved in the binding of MISs containing precursors at the surface of plant mitochondria has not been clearly identified. However, Tom20 could be involved in such binding for two reasons. First, in yeast, Tom20 can be used as a receptor instead of Tom70 by some precursors of the carrier import pathway (Kurz, Martin, Rassow, Pfanner & Ryan, 1999). Second, the import efficiency of the phosphate translocator (PiC) carrier is dramatically decreased in plant mitochondria lacking the three Tom20 isoforms (Lister *et al.*, 2007). The roles of the N-terminal extension in IM import in plants have been studied for two carriers: the adenine nucleotide transporter (ANT) from potato and the PiC from maize (Mozo, Fischer, Flugge, & Schmitz, 1995; Murcha *et al.*, 2004; Murcha, Millar, & Whelan, 2005b; Winning *et al.*, 1992). The overall charge of these extensions is positive and they are composed of more than 50% of hydrophobic residues with a high Ser content. N-terminal extensions are not essential for the import into mitochondria. They are proposed to stimulate the import of carrier proteins by favouring the direct or indirect interaction (through small Tims complexes) of the preprotein with the TIM22 complex (Murcha *et al.*, 2005b). This extension is cleaved in the matrix in two steps. The first cleavage is catalysed by the MPP and the second by an unidentified endopeptidase (Fig. 12.1, B′) (Murcha *et al.*, 2004).

3.2.2. Signals for the import of β-barrel proteins into the OM

β-barrel proteins are found only in gram-negative bacteria and in the OM of mitochondria and chloroplasts. In mitochondria, three β-barrel proteins, Tom40, Sam50 and VDAC (voltage-dependent anion channel) are universally conserved. The mechanism of insertion of these proteins into the OM has been mainly studied in yeast and is still puzzling (Becker *et al.*, 2009; Dukanovic & Rapaport, 2011). At the mitochondrial surface, β-barrel precursors are recognized by Tom20 (Krimmer *et al.*, 2001; Rapaport & Neupert, 1999). It has been proposed that Tom20 recognizes structural elements with high β-sheet content but not a discrete linear signal (Walther & Rapaport, 2009). Once in the IMS, the β-signal targets precursors, associated with small Tims, to the SAM complex (Kutik *et al.*, 2008). This β-signal is located in the last β-strand at the C-terminal extremity of the proteins and is composed of eight amino acids: P_0xGxxHyxHy (P_0 is a polar residue, G a Gly residue, Hy a large hydrophobic residue and x any amino acid). More recently, Imai and colleagues have proposed a redefinition of this motif based on the analysis of the C-terminal end of 70 β-barrels proteins from different organisms (including plants): P_0HyGhyxHyxHy (hy is a hydrophobic residue) (Imai, Fujita, Gromiha, & Horton, 2011) (Fig. 12.3). This study has also shown a highly conserved position of β-signals among species but β-signal sequences sometime differ from the consensus. Functional analyses have demonstrated that β-signals are involved in the binding of precursors to the Sam35 subunit of the SAM complex in yeast (Kutik *et al.*, 2008). This interaction leads to a conformational change of Sam50 that switches to an opening state and drives the insertion of β-barrels into the OM (Kutik *et al.*, 2008). Despite the conservation of β-signals among organisms, the mechanism of β-barrels targeting to SAM complex is still elusive in animals and plants because no clear homologue of Sam35 has been identified (Fig. 12.2B). The involvement of metaxin has been suggested by functional studies in both mammals and plants (Kozjak-Pavlovic *et al.*, 2007; Lister *et al.*, 2007). In plants, mitochondria depleted of metaxin are impaired in the import of β-barrel proteins, such as Tom40 and VDAC, and some MTS-containing precursors (Lister *et al.*, 2007). Furthermore, plant metaxin is able to bind Tom40 and VDAC (Kozjak-Pavlovic *et al.*, 2007; Lister *et al.*, 2007). Further studied are needed to understand the mechanism of β-barrel protein insertion in plant OM.

3.2.3. Signals for import of tail-anchored proteins into the OM

Tail-anchored (TA) proteins are found in the mitochondrial OM and in the endoplasmic reticulum (ER), peroxisomes and plant plastids (Borgese & Fasana, 2011). In mitochondria, these proteins are composed of three domains: a large N-terminal cytosolic domain, a transmembrane domain (TMD) and a short C-terminal domain (CTD) present in the IMS (Walther & Rapaport, 2009). Examples of such proteins are the small Tom proteins (Tom5, 6 and 7) and cytochrome b5 (Cb5) (Dukanovic & Rapaport, 2011; Henderson, Hwang, Dyer, Mullen & Andrews, 2007; Lee *et al.*, 2011; Walther & Rapaport, 2009). Overall, targeting signals of TA proteins are conserved between yeast, mammals and plants (Dukanovic & Rapaport, 2011; Horie, Suzuki, Sakaguchi, & Mihara, 2002; Horie, Suzuki, Sakaguchi, & Mihara, 2003; Kaufmann *et al.*, 2003; Kuroda *et al.*, 1998). Targeting signals are present in the TMD and in the CTD. TA proteins targeted to mitochondria have a shorter TMD that is less hydrophobic compared with ER-targeted TA proteins. The CTD of mitochondrial TA proteins is short and enriched in basic residues. In plants, two contiguous basic residues are necessary to target the Cb5-D isoform to mitochondria (Henderson *et al.*, 2007; Hwang *et al.*, 2004).

It seems that different pathways are involved in the insertion of TA proteins into the mitochondrial OM (Borgese & Fasana, 2011; Dukanovic & Rapaport, 2011; Walther & Rapaport, 2009).

3.2.4. Signals for other OM proteins

Mitochondrial OM also contains signal-anchored (SA) proteins that are anchored by a single N-terminal membrane-spanning domain, and proteins with multiple membrane-spanning segments. The signals and pathways involved in the targeting and import of such proteins have been studied in yeast and mammals but not in plants (see Dukanovic & Rapaport, 2011; Walther & Rapaport, 2009 for review). Examples of SA proteins are Tom20 and Tom70. Targeting signals of SA proteins are close to those of TA proteins and are composed of the N-terminal TMD flanked by positively charged residues. The TMDs have moderate hydrophobicity and the flanking positively charged residues are essential for OM targeting in mammals but not in yeast. The insertion of SA proteins into the OM seems to be independent of the protein import machinery but the subsequent integration of Tom20 and Tom70 into the TOM complex involved Tom40 and Mim1. Examples of proteins with multiple membrane-spanning segments are Fzo1 in yeast and Mfn1/2 in mammals. Targeting of such proteins seems to depend mainly on TMD but the mechanism of insertion is still elusive.

3.2.5. Signals for import of Cys-rich proteins into IMS: internal targeting signal (ITS)

After translocation in the IMS, Cys-rich proteins are stabilized by intramolecular disulphide bonds formed by the Mia40/Erv1 disulphide relay system (Neupert & Herrmann, 2007). ITSs have been identified in yeast. In small Tims proteins, the ITS is a twin CX_3C motif and in Cox17, it is a twin CX_9C motif (C is a Cys residue and X is any amino acid) (Fig. 12.3) (Riemer, Fischer, & Herrmann, 2011). In the IMS, Mia40 binds the ITS-containing precursors and catalyses the formation of intramolecular disulphide bonds allowing the correct folding of proteins and preventing their subsequent escape from the IMS. Thus, Mia40 has a role as an intramitochondrial receptor (Riemer *et al.*, 2011). Erv1 is involved in the reoxidation of Mia40 (Stengel, Benz, Soll, & Bolter, 2010). ITS signals are conserved between plants and yeast but plant Mia40 seems to be involved to a lesser extent in the disulphide relay system compared with yeast (see Section II.B.5) (Carrie *et al.*, 2010a).

3.3. Roles of Cytosolic Factors in Mitochondria Targeting and Import

It is admitted that import into mitochondria can be a post-translational process and that precursors are translocated into mitochondria in an unfolding state (Neupert, 1997). Consequently, cytosolic chaperones have to interact with precursors to prevent their aggregation in the cytosol and to keep them in an import-competent state. The main factor involved in these processes is the well-conserved cytosolic chaperone, Hsp70. Hsp70 has been shown to interact with different mitochondrial precursors, to stimulate translocation of proteins into mitochondria and to prevent aggregation of precursors in the cytosol (Lithgow, 2000; Mihara & Omura, 1996; Sheffield,

Shore & Randall et al., 1990). Furthermore, a fraction of cytosolic Hsp70 has been found at the surface of the OM in rat and plant mitochondria reinforcing its role in mitochondrial import (Lithgow, Ryan, Anderson, Hoj, & Hoogenraad, 1993; Mooney & Harmey, 1996). In mammals, another cytosolic chaperon, Hsp90, in concert with Hsp70, is involved in the import of highly hydrophobic carrier proteins of the IM (Young, Hoogenraad, & Hartl, 2003; Zara, Ferramosca, Robitaille-Foucher, Palmieri, & Young, 2009). In plants, Hsp90 is involved in the import of chloroplastic but not mitochondrial precursors (Kriechbaumer, von Loffelholz, & Abell, 2011; Qbadou et al., 2006). Other cytosolic factors, such as mitochondrial import stimulating factor (MSF) or arylhydrocarbon receptor-interacting protein (AIP), have also been shown to stimulate protein import into mitochondria in yeast or in mammals but their roles in this process are still unclear (see Beddoe & Lithgow, 2002; Lithgow, 2000 for review). Cytosolic chaperons are also able to bind different organellar receptors containing a tetratricopeptide repeat (TPR) domain (Kriechbaumer *et al.*, 2011). This property could be involved in the function of chaperones in precursor targeting. Tom70 in yeast and mammals and mtOM64 in plants are OM proteins with a cytosolic TPR domain (Chan, Likic, Waller, Mulhern, & Lithgow, 2006; Chew *et al.*, 2004; Kriechbaumer *et al.*, 2011). In mammals, the Tom70 receptor is able to bind Hsp70 and Hsp90 in association with precursor proteins (Young *et al.*, 2003). Interaction of the chaperon complex with Tom70 is essential to import carrier proteins into mitochondria (Fan & Young, 2011; Young *et al.*, 2003). In plants, interaction of cytosolic chaperons with mtOM64 has never been demonstrated until now. However, Toc64, the chloroplastic homologue of mtOM64, is able to interact with the cytosolic chaperon Hsp90 (Qbadou *et al.*, 2006).

4. DUAL TARGETING OF PROTEINS

Dual-targeted proteins are proteins encoded by a single gene and localized in two (or more) subcellular compartments. The first example of dual targeting reported in higher plant mitochondria was glutathione reductase (GR) in pea, which is also imported into chloroplasts (Creissen, Reynolds, Xue, & Mullineaux et al., 1995). Since this discovery, many examples have been described in plants. Most of them are mitochondrial and chloroplastic dual-targeted proteins (Table 12.1). Here, we describe examples of proteins targeting to mitochondria and another compartments in plant cells.

Table 12.1 Dual-targeted proteins to mitochondria and another cellular compartment in plants

Name	Accession number/locus	Species	Localization	Mechanism	Reference
Ala-tRNA synthetase 1	AT1G50200	At	Mito/Cyto/Cp	Ambiguous (Mito/cp) + alternative translation initiation (Cyto)	(Duchene *et al.*, 2005; Mireau *et al.*, 1996)
Ala-tRNA synthetase 2	AT5G22800	At	Mito/Cp	Ambiguous	(Duchene *et al.*, 2005)
Alanine:glyoxylate aminotransferase 2	AT4G39660	At	Mito/Perox		(Carrie *et al.*, 2009b)
Amidotransferase GatA	AT3G25660	At	Mito/Cp	Ambiguous	(Pujol *et al.*, 2008)
Amidotransferase GatB	AT1G48520	At	Mito/Cp	Ambiguous	(Pujol *et al.*, 2008)
Amidotransferase GatC	AT4G32915	At	Mito/Cp	Ambiguous	(Pujol *et al.*, 2008)
Aminoimidazole Ribonucleotide synthetase (AIRS)	P52424	Vu	Mito/Cp		(Goggin *et al.*, 2003)
Ascorbate peroxidase	AT4G08390	At	Mito/Cp	Ambiguous	(Chew, Whelan, & Miller, 2003b)
Asn-tRNA synthetase	AT433760	At	Mito/Cp	Ambiguous	(Duchene *et al.*, 2005)
Asp-tRNA synthetase	AT4G17300	At	Mito/Cp	Ambiguous	(Peeters *et al.*, 2000)
Britle 1 AtBT1 (carrier)	AT4G32400	At	Mito/Cp	Ambiguous	(Bahaji *et al.*, 2011)
Britle 1 ZmBT1 (carrier)	NM_001112419	Zm	Mito/Cp	Ambiguous	(Bahaji *et al.*, 2011)
Carrier protein	AT3G55640	At	Mito/Perox		(Carrie *et al.*, 2009b)
Cryptochrome 3	AT5G24850	At	Mito/Cp	Ambiguous	(Kleine, Lockhart, & Batschauer, 2003)
Cys-tRNA synthetase	AT2G31170	At	Mito/Cp	Ambiguous	(Peeters *et al.*, 2000)
Cytochrome c1	S66866	St	Mito/Cp	Ambiguous	(Rodiger, Baudisch, Langner, & Klosgen, 2011)

(*Continued*)

Table 12.1 Dual-targeted proteins to mitochondria and another cellular compartment in plants—cont'd

Name	Accession number/locus	Species	Localization	Mechanism	Reference
Cytochrome B5	M87514	Bo	Mito/ER		(Zhao *et al.*, 2003)
DNA helicase	AT1G30680	At	Mito/Cp	Ambiguous	(Carrie *et al.*, 2009b)
DNA ligase I	AT1G08130	At	Mito/ nucleus	Alternative transcription initiation	(Sunderland *et al.*, 2004)
DNA polymerase 1	BAE45850	Nt	Mito/Cp	Ambiguous	(Ono *et al.*, 2007)
DNA polymerase 2	BAE45851	Nt	Mito/Cp	Ambiguous	(Ono *et al.*, 2007)
DNA polymerase γ1	AT3G20540	At	Mito/Cp	Ambiguous	(Christensen *et al.*, 2005b)
DNA polymerase γ2	AT1G50840	At	Mito/Cp	Alternative translation initiation	(Christensen *et al.*, 2005b)
DNA topoisomerase	AT4G31210	At	Mito/Cp	Ambiguous	(Carrie *et al.*, 2009b)
Farnesyl-diphosphate synthase 1	AT5G47770	At	Mito/Cyto	Alternative transcription initiation	(Cunillera *et al.*, 1997)
FISSION1A	AT3G57090	At	Mito/ Perox		(Zhang and Hu, 2008)
FISSION1B	AT5G12390	At	Mito/ Perox		(Zhang and Hu, 2008)
Glu-tRNA synthetase	AT5G64050	At	Mito/Cp	Ambiguous	(Duchene *et al.*, 2005)
Glutamine synthase 2	AT5G35630	At	Mito/Cp	Ambiguous	(Taira, Valtersson,, Burkhardt, & Ludwig, et al., 2004)
Glutathione reductase	AT3G54660	At	Mito/Cp	Ambiguous	(Chew *et al.*, 2003a)
Glutathione reductase	P27456	Ps	Mito/Cp	Ambiguous	(Creissen *et al.*, 1995)
Gly-tRNA synthetase 1	AT1G29880	At	Mito/Cyto	Alternative transcription initiation	(Duchene *et al.*, 2001)
Gly-tRNA synthetase 2	AT3G48110	At	Mito/Cp	Ambiguous	(Duchene *et al.*, 2001)

His-tRNA synthetase	AT3G46100	At	Mito/Cp	Ambiguous	(Akashi *et al.*, 1998)
Holocarboxylase synthetase 1	AT2G25710	At	Mito/Cp/Cyto	Alternative translation initiation + alternative splicing	(Puyaubert *et al.*, 2008)
Leu-tRNA synthetase 1	AT1G09620	At	Mito/Cyto		(Duchene *et al.*, 2005)
Lys-tRNA synthetase	AT3G13490	At	Mito/Cp	Ambiguous	(Duchene *et al.*, 2005)
Malonyl-CoA decarboxylase	AT4G04220	At	Mito/Perox		(Carrie *et al.*, 2009b)
Mercaptopyruvate sulphurtransferase	AT1G79230	At	Mito/Cp	Ambiguous	(Nakamura, Yamaguchi, & Sano, 2000)
Met-tRNA synthetase	AT3G55400	At	Mito/Cp	Ambiguous	(Menand, Marechal-Drouard, Sakamoto, Dietrich, & Wintz, 1998)
Methionine aminopeptidase MAP1C	AT3G25740	At	Mito/Cp	Ambiguous	(Giglione, Serero, Pierre, Boisson, & Meinnel, 2000)
Methionine aminopeptidase MAP1D	AT4G37040	At	Mito/Cp	Ambiguous	(Giglione *et al.*, 2000)
Mia40	AT5G23395	At	Mito/Perox		(Carrie *et al.*, 2009b, Carrie *et al.*, 2010a)
Mia40	Os04g44550	Os	Mito/Perox		(Carrie *et al.*, 2010a)
Monodehydroascorbate reductase 6 (MDAR6)	AT1G63940	At	Mito/Cp	Alternative transcription initiation	(Obara *et al.*, 2002)
Peptide deformylase PBF1B	AT5G14660	At	Mito/Cp	Ambiguous	(Giglione *et al.*, 2000)

(Continued)

Table 12.1 Dual-targeted proteins to mitochondria and another cellular compartment in plants—cont'd

Name	Accession number/locus	Species	Localization	Mechanism	Reference
Peptide deformylase PBF1B (OsPDF1B)	Os01g45070	Os	Mito/Cp	Ambiguous	(Moon *et al.*, 2008)
Phe-tRNA synthetase	AT3G58140	At	Mito/Cp	Ambiguous	(Duchene *et al.*, 2005)
Phosphatidylglycerol phosphate synthase	AT2G39290	At	Mito/Cp	Ambiguous	(Babiychuk *et al.*, 2003)
PMD1	?	At	Mito/Perox		(Aung *et al.*, 2010)
PNM1	AT5G60960	At	Mito/nucleus		(Hammani *et al.*, 2011)
Pro-tRNA synthetase	AT5G52520	At	Mito/Cp	Ambiguous	(Duchene *et al.*, 2005)
PRORP1	AT2G32230	At	Mito/Cp	Ambiguous	(Gobert *et al.*, 2010)
Protoporphyrinogen oxidase II	BAB60710	So	Mito/Cp	Alternative translation initiation	(Watanabe *et al.*, 2001)
pt/mtGuanylate kinase	AB267729	Os	Mito/Cp	Ambiguous	(Sugimoto *et al.*, 2007)
RCC1-like	AT5G08710	At	Mito/Cp	Ambiguous	(Carrie *et al.*, 2009b)
RNA polymerase 1	CAC95163	Pp	Mito/Cp	Alternative translation initiation	(Richter *et al.*, 2002)
RNA polymerase 2	AT5G15700	At	Mito/Cp	Ambiguous	(Hedtke, Borner, & Weihe, 2000)
RNA polymerase 2	Q8L6J3	Nt	Mito/Cp	Ambiguous	(Richter *et al.*, 2002)
RNA polymerase 2	CAC95164	Pp	Mito/Cp	Alternative translation initiation	(Richter *et al.*, 2002)
RNA polymerase B	BAB69431	Ns	Mito/Cp	Alternative translation initiation	(Kobayashi, Dokiya, & Sugita, 2001)

RPS16	AT5G56940	At	Mito/Cp	Ambiguous (non-cleavable)	(Ueda *et al.*, 2008)
RPS16	Os08g0517900	Os	Mito/Cp	Ambiguous (non-cleavable)	(Ueda *et al.*, 2008)
RPS16	AB365530	Pt	Mito/Cp	Ambiguous (non-cleavable)	(Ueda *et al.*, 2008)
RPS16	AB365527	Mt	Mito/Cp	Ambiguous (non-cleavable)	(Ueda *et al.*, 2008)
Ser-tRNA synthetase	AT1G11870	At	Mito/Cp	Ambiguous	(Duchene *et al.*, 2005)
Sigma 2B	AAD17855	Zm	Mito/Cp	Ambiguous	(Beardslee *et al.*, 2002)
Sulphur protein E (AtSufE)	AT4G26500	At	Mito/Cp	Ambiguous	(Xu & Moller, 2006)
Sulfiredoxin	GU223224	Ps	Mito/Cp	Ambiguous	(Iglesias-Baena, Barranco-Medina, Sevilla, & Lazaro, 2011)
Sulfiredoxin	AT1G31170	At	Mito/Cp	Ambiguous	(Iglesias-Baena *et al.*, 2011)
THI1	AT5G54770	At	Mito/Cp	Alternative translation initiation	(Chabregas *et al.*, 2001)
Thioredoxin 1	AM235208	Ps	Mito/nucleus	Two distinct signals	(Marti *et al.*, 2009)
Thr-tRNA synthetase 1	AT5G26830	At	Mito/Cyto	Alternative translation initiation	(Souciet *et al.*, 1999)
Thr-tRNA synthetase 2	AT2G04842	At	Mito/Cp	Ambiguous	(Duchene *et al.*, 2005)
tRNA nucleotidyltransferase	AT1G22660	At	Mito/Cp/Cyto	Alternative transcription and translation initiation	(von Braun *et al.*, 2007)
Trp-tRNA synthetase	AT2G25840	At	Mito/Cp	Ambiguous	(Duchene *et al.*, 2005)
	AT2G07180	At		Two distinct signals	(Carrie *et al.*, 2008)

(*Continued*)

Table 12.1 Dual-targeted proteins to mitochondria and another cellular compartment in plants—cont'd

Name	Accession number/locus	Species	Localization	Mechanism	Reference
Type II NAD(P)H dehydrogenase A1 (NDA1)			Mito/ Perox		
Type II NAD(P)H dehydrogenase A2 (NDA2)	AT2G2990	At	Mito/ Perox	Two distinct signals	(Carrie *et al.*, 2008)
Type II NAD(P)H dehydrogenase B1 (NDB1)	AT4G28220	At	Mito/ Perox	Two distinct signals	(Carrie *et al.*, 2008)
Type II NAD(P)H dehydrogenase C1 (NDC1)	AT5G08740	At	Mito/Cp	Ambiguous	(Carrie *et al.*, 2008)
Tyr-tRNA synthetase	AT3G02660	At	Mito/Cp	Ambiguous	(Duchene *et al.*, 2005)
Val-tRNA synthetase 1	AT3G14610	At	Mito/Cyto	Alternative translation initiation	(Souciet *et al.*, 1999)
Zinc metalloprotease AtPreP1	AT3G19170	At	Mito/Cp	Ambiguous	(Bhushan *et al.*, 2003)
Zinc metalloprotease AtPreP2	AT1G49630	At	Mito/Cp	Ambiguous	(Bhushan *et al.*, 2005)

At, Arabidopsis thaliana; Bo, Brassica oleracea; Mt, Medicago truncatula; Ns, Nicotiana sylvestris; Nt, Nicotiana tabacum; Os, Oryza sativa; Pp, Physcomitrella patens; Ps, Pisum sativum; So, Spinacia oleracea; St, Solanum tuberosum; Vu, Vigna unguiculata; Zm, Zea mays; Mito, mitochondrion; Cp, chloroplast; Cyto, cytosol; Perox, peroxisome.

4.1. Mechanisms of Dual Targeting

To be localized in two or more compartments, proteins have to retain all the signals necessary to be targeted to these compartments. Two major classes of signals can be involved in dual targeting: ambiguous signals and twin presequences (Fig. 12.4) (Peeters & Small, 2001). A signal can be defined as ambiguous when the same sequence is necessary and sufficient to target a protein in two or more compartments. On the contrary, when two separated targeting sequences can be clearly defined, each allowing the targeting of the protein in one or the other compartment, the signal can be defined as a twin presequence. In the case of ambiguous presequences, mainly described for dual targeting of proteins to mitochondria and chloroplasts (see Section IV.B.2.a), only one protein, with one targeting signal, is synthesized from a single mRNA (Fig. 12.4A). On the contrary, several mechanisms can be used to synthesize proteins with twin presequences (Fig. 12.4B, C, D and E) (see Karniely & Pines, 2005; Silva-Filho, 2003; Yogev & Pines, 2011 for review). Such mechanisms can be alternative transcription initiation or alternative splicing that lead to the synthesis of two mRNAs encoding two protein isoforms. Each isoform is targeted to a particular compartment (Fig. 12.4C and E). By a mechanism of alternative translation initiation, two proteins, with variations in the N-terminal domain, can be synthesized from a single mRNA. Each protein is targeted to one compartment (Fig. 12.4B). This mechanism involves the presence of two in-frame start codons on the mRNA. Selection of start codons depends on sequence context, which can be more or less favourable for translation initiation, and on other *cis*- and *trans*-acting factors (Fig. 12.4E). Furthermore, the same protein can retain separately two signals involved in the targeting of the protein in two different compartments (Fig. 12.4D and Section IV.D).

4.2. Dual Targeting to Mitochondria and Chloroplasts

Chloroplasts have emerged from a second process of endosymbiosis of a cyanobacterium by a eukaryotic cell (Gould, Waller, & McFadden, 2008). As for mitochondria, a massive transfer of genes to the nucleus has occurred during evolution and today most of the chloroplastic proteins are encoded by the nuclear genome. Chloroplasts have developed specific protein import machineries and pathways to fulfil their correct biogenesis in a context where mitochondria were already present in cells. Consequently, the protein import machineries of plant chloroplasts and mitochondria have

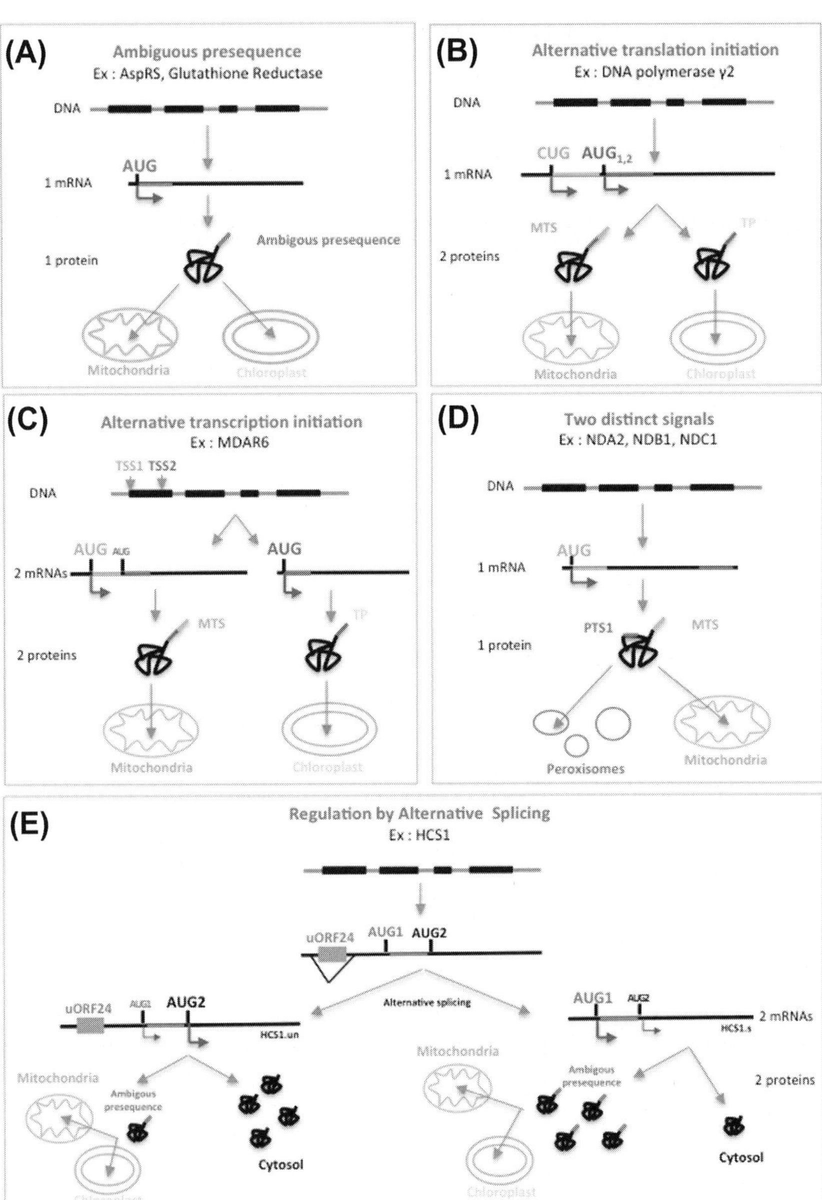

Figure 12.4 Examples of mechanisms involved in dual targeting of proteins to mitochondria and another compartments by ambiguous presequences (A) or twin presequences (B–E). (A) Dual targeting of proteins to mitochondria and chloroplasts by ambiguous presequences. Examples of proteins involved in this mechanism are aspartyl-tRNA synthetase (AspRS) or glutathione reductase. (B) Dual mitochondrial–chloroplastic targeting of DNA polymerase γ2 by a mechanism of alternative translation initiation.

coevolved in a way that avoids the mistargeting of chloroplastic proteins into mitochondria and vice versa. This coevolution can be seen at different levels. For example, plant mitochondria present noticeable variations in the OM receptors compared with mitochondria in other organisms (see Section II.B.1 and Fig. 12.2A) (Macasev *et al.*, 2000; Macasev *et al.*, 2004). Another result of this coevolution is the presence of protein dual targeting to mitochondria and chloroplasts. Despite a high functional specialization of both organelles, some fundamental processes, such as DNA replication, transcription, translation and proteolysis, still occur in mitochondria and chloroplasts (Mackenzie, 2005; Millar, Whelan, & Small, 2006). Thus, many of the enzymes involved in these processes can be shared between the two organelles.

4.2.1. Overview of protein import into chloroplasts and the main differences from mitochondria protein import

Most chloroplastic proteins have an N-terminal cleavable extension, called transit peptide (TP), necessary and sufficient to fulfil their targeting and import into chloroplasts. TPs are removed in the stroma by a metalloendonuclease, the stromal processing peptidase (SPP) (Balsera *et al.*, 2009; Li & Chiu, 2010). The overall amino acid composition of TP is similar to that of MTS. TPs are enriched in positively charged, hydroxylated and hydrophobic

Translation initiation from the CUG start codon leads to the synthesis of the mitochondrial-targeted protein with a mitochondrial targeting signal (MTS). Proteins synthesized from the two AUG codons ($AUG_{1,2}$) have a transit peptide (TP) and are targeted to chloroplasts. (C) Dual mitochondrial–chloroplastic targeting of monodehydroascorbate reductase 6 (MDAR6) by a mechanism of alternative transcription initiation. Two transcription start sites (TSS) are present on the gene. Translation of the mRNA transcribed from TSS1 leads to a protein with an MTS localized to mitochondria. Translation of the mRNA transcribed from TSS2 leads to a protein with only a TP localized to chloroplasts. (D) Dual targeting of type II NAD(P)H dehydrogenases (ND) to mitochondria and peroxisomes. The proteins have retained one MTS to be targeted to mitochondria and a peroxisomal targeting signal 1 (PTS1) to be targeted to peroxisomes. (E) Regulation of the localization of holocarboxylase synthase 1 (HCS1) to mitochondria, chloroplasts and cytosol by alternative splicing. Translation from the AUG1 codon allows the synthesis of a protein with an ambiguous presequence that is dual targeted to mitochondria and chloroplasts. Translation from the AUG2 gives a cytosolic protein without targeting signals. The main start codon used for translation initiation is regulated by the 5′ UTR of mRNAs. One splicing variant (HCS1.un) presents an upstream open reading frame (uORF24) that favours initiation from AUG2. Translation from the second variant (HCS1.s), lacking uORF24, is mainly initiated at AUG1. See the colour plate.

residues and contain few acidic residues (Pujol *et al.*, 2007; Zhang & Glaser, 2002; Zybailov *et al.*, 2008). However, TPs are longer than MTSs (5–10 amino acids longer) (Huang *et al.*, 2009; Zhang & Glaser, 2002; Zybailov *et al.*, 2008), they have less propensity to fold into secondary structures (von Heijne & Nishikawa, 1991; Zhang & Glaser, 2002) and they are slightly more hydrophobic than MTSs (Huang *et al.*, 2009). Another characteristic of TPs is the presence of an Ala residue in position 2 for half of them (Pujol *et al.*, 2007; Zybailov *et al.*, 2008).

TPs are recognized by receptors of the TOC complex and precursors reach the stroma after travelling through the TOC and the translocase of the inner envelop of chloroplast (TIC) complexes (Balsera *et al.*, 2009). The two main differences observed between the mechanisms of import into chloroplasts and mitochondria are in the precursor binding step and in the chaperones able to bind each precursor. TOC complexes contain two receptors that have a cytosolic GTPase domain (Toc34 and Toc159). Thus, precursors binding to the OM of chloroplasts are dependent on GTP contrary to precursors binding to mitochondria, which seems to be a more passive process (Bedard & Jarvis, 2005; Li & Chiu, 2010; Schleiff & Becker, 2011). Furthermore, TP can be phosphorylated in the cytosol (Martin *et al.*, 2006; Waegemann & Soll, 1996). This phosphorylation leads to the binding of the cytosolic 14-3-3 chaperons (May & Soll, 2000). 14-3-3 chaperons are then able to recruit Hsp70 to form the so-called guidance complex that stimulates import of precursors into chloroplasts (Kriechbaumer *et al.*, 2012; May & Soll, 2000). Such a mechanism has never been observed for mitochondrial precursors.

4.2.2. Signals of dual targeting to chloroplasts and mitochondria

Most of the proteins targeted to mitochondria and chloroplasts have an N-terminal ambiguous presequence (Table 12.1) but a few proteins have twin presequences (Peeters & Small, 2001; Yogev & Pines, 2011).

a. Ambiguous presequences

Ambiguous presequences are N-terminal cleavable sequences necessary and sufficient to fulfil protein targeting and import into mitochondria and chloroplasts (Fig. 12.4A). The amino acid composition of ambiguous presequences is overall very similar to MTS and TP as they are also enriched in positively charged hydroxylated residues and contain few acidic residues (Peeters & Small, 2001; Pujol *et al.*, 2007; Rudhe, Clifton, Whelan, & Glaser, 2002). However, they have a higher Phe and Leu content compared with MTS and TP (Berglund *et al.*, 2009b; Peeters & Small, 2001; Rudhe *et al.*, 2002). As for MTS and TP, the primary structure of ambiguous presequences is not conserved and no consensus motifs can be defined (Peeters & Small, 2001; Pujol *et al.*, 2007). Concerning the specific features of MTS or TP (Ala in position 2 in TP and enrichment of Arg residues in the N-terminal part of MTS), ambiguous presequences present an intermediary profile, meaning that they share features of both types of presequences (Pujol *et al.*, 2007). Secondary structure predictions show that ambiguous presequences fold into α-helices to a lesser extent compared with MTS (Peeters & Small, 2001; Pujol *et al.*, 2007). However, the presequence of the dual-targeted ThrRS is able to adopt helical structures in membrane mimetic media (Berglund *et al.*, 2009b).

Several mutagenesis studies have been done to understand the organization of ambiguous presequences and to identify regions or residues involved in targeting to either mitochondria or chloroplasts (Berglund, Pujol, Duchene, & Glaser, 2009a, Berglund *et al.*, 2009b, Chew, Rudhe, Glaser, & Whelan, 2003a, Pujol *et al.*, 2007; Rudhe *et al.*, 2002). Overall, these studies show that there are no rules to describe the organization of these presequences. Signals involved in targeting to either chloroplasts or mitochondria seem to be distinct but overlapping. The three main results highlighted by these studies are: (1) the importance of Arg and hydrophobic residues for efficient mitochondrial import of precursors (Chew *et al.*, 2003a, Pujol *et al.*, 2007). These observations are in accordance with studies on MTS showing the involvement of these residues in fulfilling efficient mitochondrial import of proteins. (2) The residue in position 2 can play an important role in targeting and import to either chloroplasts and/or mitochondria (Pujol *et al.*, 2007). (3) The N-terminal domain of ambiguous presequences is essential for import into mitochondria and/or chloroplasts (Berglund *et al.*, 2009a, Pujol *et al.*, 2007; Rudhe *et al.*, 2002).

Some studies have been performed to understand the mechanisms involved in the targeting and import of dual-targeting proteins into mitochondria and chloroplasts. The binding of precursors seems to be more efficient at the surface of isolated mitochondria compared with chloroplasts but the import and maturation of precursors are more efficient in chloroplasts (Pujol *et al.*, 2007). In mitochondria, only the metaxin protein has been shown to interfere with the import of GR suggesting that metaxin could be a receptor for dual-targeted proteins containing an ambiguous signal (Lister *et al.*, 2007). The receptors involved in the interaction between precursors and chloroplasts are unknown. However, most of the ambiguous presequences contain a predicted phosphorylation site that could be recognized by the chloroplastic guidance complex and targeted to the Toc34 receptor (Berglund *et al.*, 2009b; Carrie, Giraud, & Whelan, 2009a). Import of precursors into both organelles seems to involve the same import machinery as the specific mitochondrial or chloroplastic precursors (TOM/TIM and TOC/TIC complexes, respectively) (Berglund *et al.*, 2009b). Once in the matrix or in the stroma, ambiguous presequences are removed by MPP and SPP, respectively. The cleavage can be achieved at the same site in both compartments or at different sites depending on the precursors (Pujol *et al.*, 2007; Rudhe *et al.*, 2004). It has also been reported that the dual-targeted protein RPS16 has a non-cleavable ambiguous presequence (Ueda *et al.*, 2008).

b. Twin presequences

Some proteins have been reported to be dual targeted to mitochondria and chloroplasts by twin presequences generated by alternative transcription or translation initiation (Table 12.1, Fig. 12.4B and C). One of the best-studied examples is the DNA polymerase γ2 (Polγ2) in *Arabidopsis thaliana*, which is dual targeted to mitochondria and chloroplasts by a mechanism of alternative translation initiation (Fig. 12.4B) (Christensen *et al.*, 2005b; Wamboldt *et al.*, 2009). The 5′ untranslated region (UTR) of Polγ2 mRNA has three putative in-frame start codons (one non-AUG start codon (CUG) and two AUG separated by one Ala codon). Prediction analyses suggest that the sequence after the AUG codons is a putative TP and the sequence between the CUG and AUG codons is a putative MTS. In agreement with these predictions, proteins translated from the AUG codons are targeted to chloroplasts and proteins synthesized from the CUG codon are targeted to mitochondria (Fig. 12.4B) (Christensen *et al.*, 2005a; Wamboldt *et al.*, 2009). This is in agreement with experiments showing the predominance of the most N-terminal signal when an MTS and a TP are fused in tandem (de Castro Silva Filho, Chaumont, Leterme, & Boutry, 1996). *In planta*, Polγ2 seems to be mainly present in chloroplasts but the repartition of the protein between the two organelles seems to vary during development. The proportion of proteins produced from each start codon is regulated by a mechanism of leaky ribosome scanning but also seems to depend on *cis*- and *trans*-acting factors (Wamboldt *et al.*, 2009). Further studies are needed to understand the mechanisms involved in controlling translation initiation.

Monodehydroascorbate reductase 6 (MDAR6) in *A. thaliana* is dual targeted to mitochondria and chloroplasts by a mechanism of alternative transcription initiation (Obara, Sumi, & Fukuda, 2002). Two transcription start sites (TSS) are present in the MDAR6 gene. Thus, two mRNA isoforms with variation in the 5′ end are transcribed: a short form and a long form (Fig. 12.4C). The protein synthesized from the long mRNA (MDAR-L) has an N-terminal extension of seven amino acids compared with the protein synthesized from the short form (MDAR-S). *In vivo* studies have shown that the first 43 residues of MDAR-S and the first 50 residues of MDAR-L are sufficient to target green fluorescent protein fusions exclusively to chloroplasts and mitochondria, respectively (Obara *et al.*, 2002).

4.3. Dual Targeting to Mitochondria and ER

Until now, only the TA protein cytochrome b5 in cauliflower has been reported to be dual targeted to mitochondria and ER in plants (Zhao *et al.*, 2003). Dual targeting is mediated by a signal of 11 amino acids located in the C-terminal part of the protein (LVVRQYTKKE). Experimental analyses have shown that the two contiguous Lys residues are essential for the mitochondrial targeting as shown for the Cb5-D in *Vernicia fordii* (Henderson *et al.*, 2007; Hwang *et al.*, 2004). The residues essential for ER targeting have not been identified (Zhao *et al.*, 2003).

4.4. Dual Targeting to Mitochondria and Nucleus

In *A. thaliana*, ligase 1 (AtLig1) is dual targeted to mitochondria and the nucleus by a mechanism of alternative translation initiation (Sunderland, West, Waterworth, & Bray *et al.*, 2004; Sunderland, West, Waterworth, & Bray *et al.*, 2006). Two in-frame AUG start codons are used to synthesize the mitochondrial or the nuclear isoforms. Translation from the first AUG gives the AtLig1 isoform with an MTS that is exclusively localized in mitochondria despite the presence of an internal bipartite nuclear localization signal (NLS) in the protein. This result highlights the dominance of the MTS over the NLS in AtLig1. The isoform synthesized from the second downstream AUG has only the NLS signal and is exclusively targeted to the nucleus. This alternative translation initiation is mediated by a leaky ribosome scanning of the mRNA but also depends on the nature of the 5′ UTR. The sequence context around the first AUG is less favourable for initiation than the sequence around the second AUG, but the efficiency of translation initiation at the second AUG is affected by the length of the 5′ UTR (Sunderland *et al.*, 2006).

Another example of dual mitochondrial–nucleus localization is the pea thioredoxin 1 (PsTrxo1) protein. The same protein, containing an N-terminal MTS and an NLS, is targeted to both mitochondria and nucleus (Marti *et al.*, 2009). In this case, the MTS is not predominant over the NLS and mechanisms that regulate the partition of PsTrxo1 between mitochondria and nucleus are unknown. The MTS is still present at the N-terminal end of the nuclear protein but is cleaved after import of the mitochondrial isoform (Marti *et al.*, 2009).

PNM1, a pentatricopeptide repeat (PPR) protein, is also dual localized to mitochondria and nucleus. An N-terminal MTS and a C-terminal NLS are involved in PNM1 targeting to mitochondria and nucleus, respectively (Hammani *et al.*, 2011) (see Chapter 10).

4.5. Dual Targeting to Mitochondria and Peroxisomes

Dual targeting of proteins to mitochondria and peroxisomes was discovered a few years ago and the signals and mechanisms involved in such targeting are not well identified. In *A. thaliana*, FISSION1A (Fis1A) and FISSION1B (Fis1B) are TA proteins present in the mitochondrial OM and in the peroxisomal membrane where they play a role in the fission of both organelles (Zhang & Hu, 2008; Zhang & Hu, 2009). As for other TA proteins, targeting signals are located in the C-terminal part of the protein but the residues involved in the targeting to mitochondria and/or peroxisomes have not been defined (Zhang & Hu, 2009).

Dual targeting to mitochondria and peroxisomes has also been demonstrated for three type II NAD(P)H dehydrogenases (NDA2, NDB1 and NDC1, see Table 12.1) in *A. thaliana* (Fig. 12.4D) (Carrie *et al.*, 2008; Michalecka *et al.*, 2003). These three proteins contain two distinct targeting signals: an N-terminal MTS allowing the import into mitochondrial IM and a C-terminal peroxisomal targeting signal 1 (PTS1) that is needed for protein targeting to peroxisomes (Carrie *et al.*, 2008; Michalecka *et al.*, 2003).

4.6. Dual Targeting to Mitochondria and Plasma Membrane

Recently, three of the four isoforms of the β-barrel VDAC protein (AtVDAC1, 2 and 3) in *A. thaliana* have been reported to be dual localized in both mitochondria and plasma membranes (Robert *et al.*, 2012), but the signals involved in such dual targeting are not known. The mechanism of dual targeting to mitochondria and plasma membrane has been elucidated for one VDAC isoform in mice. It involves an alternative splicing of

VDAC1 mRNA (Buettner, Papoutsoglou, Scemes, Spray, & Dermietzel, 2000). The protein translated from one mRNA isoform (pl-VDAC1) presents an N-terminal extension of 12 amino acids corresponding to a signal peptide. This leader sequence triggers pl-VDAC1 to the secretory pathway (ER, then Golgi) before reaching the plasma membrane (Buettner *et al.*, 2000). The protein translated from the second mRNA isoform (mt-VDAC1) does not contain a signal peptide in the N-terminal and is subsequently targeted to mitochondria (Buettner *et al.*, 2000). However, signals involved in the mitochondrial targeting of VDAC proteins are still unknown (de Pinto, Messina, Lane, & Lawen, 2010).

4.7. Dual Targeting to Mitochondria and Cytosol

Dual targeting to mitochondria and cytosol often involves two different isoforms of the protein, one with an MTS and another without any targeting signals that will be retained in the cytosol by default. These two isoforms are synthesized by mechanisms of alternative translation or transcription initiation (Table 12.1). For the ValRS and ThrRS proteins in *A. thaliana*, one mRNA containing two AUG start codons in-frame is transcribed from a single gene. Translation initiation at the first AUG produces the isoform with an MTS that is targeted to mitochondria, and initiation from the second AUG produces the cytosolic isoform without any targeting signal (Souciet *et al.*, 1999).

In the cases of GlyRS1 and farnesyl-diphosphate synthase 1 (FPS1) in *A. thaliana*, two transcripts, with variations in the length of their 5′end, are synthesized from a single gene. The short mRNA contains only one AUG start codon and is translated into the cytosolic isoform without a targeting signal. The long transcript contains a second in-frame start codon upstream of the first AUG. The protein synthesized from this second AUG has an N-terminal MTS and is targeted to mitochondria (Cunillera, Boronat, & Ferrer, 1997; Duchene *et al.*, 2001). For GlyRS1, the cytosolic isoform can also be translated from the long transcript.

4.8. Targeting to Mitochondria, Chloroplasts and Cytosol

In *A. thaliana*, three proteins have been reported to localize in three different compartments (Table 12.1) (Duchene *et al.*, 2005; Mireau, Lancelin, & Small, 1996; Puyaubert, Denis, & Alban, 2008; von Braun *et al.*, 2007). For AlaRS, a mechanism of alternative translation initiation mediates the triple localization of the protein. One isoform, containing an N-terminal ambiguous presequence, is dual targeted to mitochondria and chloroplasts and a second isoform, without a targeting signal, is retained in the cytosol (Duchene *et al.*, 2005; Mireau *et al.*, 1996).

The mechanism involved in the regulation of the localization of holocarboxylase synthase 1 (HCS1) in *A. thaliana* is of particular interest (Puyaubert *et al.*, 2008). As for AlaRS, two protein isoforms, one with an ambiguous presequence targeted to mitochondria and chloroplasts and one cytosolic, are synthesized from two in-frame start codons (AUG1 and AUG2, respectively) (Fig. 12.4E). Selection of the AUG start codon is regulated by a mechanism of alternative splicing and two mRNA isoforms with variations in the 5′ end are generated in cells. The unspliced isoform (HCS1.un) contains an upstream ORF of 24 nucleotides (uORF24) in its 5′ leader sequence that is absent from the spliced isoform (HCS1.s) (Fig. 12.4E). The presence of uORF24 (HCS1.un) stimulates translation initiation from AUG2, so the cytosolic isoform is mainly synthesized (Puyaubert *et al.*, 2008). In contrast, translation initiation mainly occurs at AUG1 when uORF24 is absent (HCS1.s), so the mitochondrial/chloroplastic isoform is predominant. This is an elegant example showing the influence of the 5′ UTR on the regulation of translation initiation and on the subcellular localization of proteins.

5. CYTOSOLIC mRNA TARGETING TO THE VICINITY OF MITOCHONDRIA

In the 1970s, pioneer work from Butow's team proposed a mitochondrial import model with precursor synthesis at the mitochondrial surface (Kellems, Allison, & Butow, 1974; Kellems, Allison, & Butow, 1975; Kellems & Butow, 1972; Kellems & Butow, 1974; Suissa & Schatz, 1982). This model arises from different observations: (1) cytosolic polysomes are tightly associated with yeast mitochondria; (2) translation products from the mitochondria-bound cytosolic polysomes encode mitochondrial proteins; (3) yeast mitochondria are able to bind cytosolic ribosomes *in vitro*. Further studies also suggest that association of cytosolic polysomes with mitochondria could be involved in a cotranslational import of precursors into mitochondria (Fujiki & Verner, 1993; Verner, 1993). However, these models were not well accepted at first glance because most of the mitochondrial proteins studied were easily imported into isolated mitochondria in a post-translational manner. Further results supporting the presence of cytosolic mRNAs at the surface of mitochondria came from Jacq's team in the early 2000s (Corral-Debrinski, Blugeon, Jacq, 2000; Marc *et al.*, 2002; Margeot *et al.*, 2002). These studies clearly show a differential association of cytosolic mRNAs to yeast mitochondria, meaning that some mRNAs are enriched in the mitochondrial fraction compared with others and suggesting a specific targeting of some mRNAs to the mitochondrial surface. The relevance of this process was demonstrated by establishing a direct link between mRNA targeting and the functionality of mitochondria (Margeot *et al.*, 2002; Margeot *et al.*, 2005). Now the process of mRNA targeting to mitochondria is well accepted and has also been demonstrated in mammals and in plants (Michaud, Marechal-Drouard, & Duchene, 2010; Sylvestre *et al.*, 2003a). However, the cotranslational protein import model into mitochondria is still a matter of debate. In the last 10 years, large-scale mutagenesis studies, mainly undertaken in yeast, have provided a more global view of the mRNA targeting process in terms of extent, mechanisms and regulation.

5.1. Genome-Wide Studies of mRNA Targeting to Mitochondria in Yeast

Large-scale analyses of mRNAs copurified with mitochondria in yeast have allowed the mitochondrial localization ratio (MLR) to be determined for every mRNA detected in these fractions. This ratio represents the proportion of mRNA found in a mitochondrial fraction compared with the proportion found in a cytosolic fraction. A high MLR means that the mRNA is mainly associated with mitochondria (mMLR) and a low MLR implies a weak association of the mRNA with mitochondria (Garcia *et al.*, 2007; Marc *et al.*, 2002; Sylvestre, Vialette, Corral Debrinski, & Jacq, 2003b). Most mRNAs with a high MLR encode mitochondrial proteins, and half of mitochondrial proteins are translated from mitochondrial-targeted mRNA. A clear correlation has been detected between MLR of mRNAs and the origin of the encoded protein. The mitochondrial proteome of yeast is composed of proteins of both prokaryotic and eukaryotic origin (50.6% and 49.2%m respectively) (Karlberg, Canback, Kurland, & Andersson, 2000). Messenger RNAs encoding proteins of prokaryotic origin are mostly translated at the mitochondrial surface compared with mRNAs encoding eukaryotic proteins (Garcia *et al.*, 2007; Marc *et al.*, 2002; Sylvestre *et al.*, 2003b). It has also been reported that mRNAs encoding essential components for mitochondrial biogenesis are mainly associated with mitochondria (Garcia *et al.*, 2007) and that long mRNAs have a tendency to be more associated with mitochondria. At first glance, mRNA targeting to mitochondria was proposed to be a possible mechanism to address proteins without classic MTSs at the surface of mitochondria. However, there is no correlation between the MLR of mRNAs and the presence of MTSs at the N-terminal part of proteins meaning that the presence of targeting signals at both protein and mRNA levels are not exclusive (Karlberg & Andersson, 2003). There is also no correlation between mRNA targeting and the final mitochondrial localization (matrix, IMS, IM or OM) of the protein (Karlberg & Andersson, 2003; Zahedi *et al.*, 2006). It has also been proposed that translation of mRNAs at the mitochondrial surface could be an interesting process to prevent aggregation of highly hydrophobic proteins in the cytosol but there is no correlation between the MLR of mRNAs and the hydrophobicity of the protein (Karlberg & Andersson, 2003).

5.2. Mechanisms Involved in mRNA Targeting to Mitochondria in Yeast and Mammals

Messenger RNA targeting to mitochondria relies on a combination of *cis*- and *trans*-acting factors. Studies undertaken in yeast in the last 10 years show that (1) *cis* elements on the mRNAs are mainly present in the 3′ UTR, (2) the MTS can act as a *cis*-acting element to mediate mRNA binding to mitochondria, (3) mRNA translation can be important for mRNAs sorting to mitochondria, (4) at least two *trans*-acting factors are involved in mitochondrial targeting of several mRNAs in yeast: the Pumillio protein Puf3p and the Tom20 receptor (Table 12.2).

5.2.1. Cis- *and* trans-*acting factors*

a. Localization of *cis*-acting signals on mRNAs

Most of the studies done on yeast and mammals have shown that the 3′ UTR of mMLR is necessary for mRNA targeting to mitochondria (Corral-Debrinski *et al.*, 2000; Ginsberg, Feliciello, Jones, Avvedimento, & Gottesman, 2003; Marc *et al.*, 2002; Margeot *et al.*, 2002; Russo, Russo, Cuccurese, Garbi, & Pietropaolo, 2006; Sylvestre *et al.*, 2003a). When the 3′ UTR of different mMLRs, such as ATP2, ATM1 or OXA1, are replaced by the 3′ UTR of a non-MLR mRNA, chimeric mRNAs are less associated to mitochondria. Moreover, the 3′ UTRs of several mMLR are able to target reporter RNAs to mitochondria *in vivo* suggesting that targeting signals are mainly present in these regions (Corral-Debrinski *et al.*, 2000; Marc *et al.*, 2002; Margeot *et al.*, 2002; Sylvestre *et al.*, 2003a). *Cis*-acting

Table 12.2 Mechanisms involved in mitochondrial targeting of mRNAs in yeast. for each mRNA, the need for mRNA translation and the identification of *cis* elements and *trans* factors are indicated.

			Cis-acting factors				*Trans*-acting factors			
Gene	**Locus**	**Translation**	**MTS**	**3′ UTR**	**Puf3 binding site**	**Other**	**Tom20**	**Puf3p**	**Other**	**Reference**
ACO1	YLR304 W	+	+	nd	–		+	–	Ssa1	(Eliyahu *et al*., 2010; Eliyahu *et al*., 2012; Gadir *et al*., 2011)
ATM1	YMR301C	nd	+	+	–		nd	–		(Corral-Debrinski *et al*., 2000; Gadir *et al*., 2011)
ATP2	YJR121C	+	+	+	–	Elements in CDS	–	+	Tom70, Tom7, Ssa1	(Eliyahu *et al*., 2010; Eliyahu *et al*., 2012 Gadir *et al*., 2011; Garcia, Delaveau, Goussard, & Jacq, 2010; Margeot *et al*., 2002; Saint-Georges *et al*., 2008)
BCS1	YDR375C	nd	nr	+	+		nd	+		(Eliyahu *et al*., 2010; Saint-Georges *et al*., 2008)
COX17	YLL009C	+	nr	+	+		–	+		(Eliyahu *et al*., 2010; Saint-Georges *et al*., 2008)
MRLP9	YGR220C	nd	nd	nd	+		+	+		(Eliyahu *et al*., 2010)
OXA1	YER154 W	+	nd	+	+		–	+	Tom70, Tom6, Tom7	(Gadir *et al*., 2011)
TOM70	YNL121C	–	nr	nd	–		–	–		(Eliyahu *et al*., 2010)

CDS, coding sequence; nd, not determined; nr, not relevant.

signals are composed of motifs on primary sequences and/or structural elements on mRNAs. In yeast, the association of BCS1 mRNA with mitochondria is mediated by a short sequence of 10 nt and mutations of this sequence decrease the association of BCS1 mRNA with mitochondria (Saint-Georges *et al.*, 2008). In contrast, in mice, the localization at the mitochondrial surface of two mRNAs (manganese superoxide dismutase (MnSOD) and Fo-f) is mediated by a conserved structural element in their 3′ UTR composed of two successive stem loops (Ginsberg *et al.*, 2003). Association with the *trans*-acting factor AKAP121 (see Section V.C.2) involved in mitochondrial targeting of both mRNAs is abolished when the structure of this *cis* element is modified but not when the primary sequence is changed (Ginsberg *et al.*, 2003).

b. Involvement of Puf3p in mRNA targeting to yeast mitochondria

On a large scale, one motif, the Puf3p binding site (P3BS), has been identified as a putative signal involved in mRNA targeting in yeast (Gerber, Herschlag, & Brown, 2004; Saint-Georges *et al.*, 2008). Puf3p is an RNA-binding protein that belongs to the Pumillio family. Puf3p is a multifunctional protein involved in different cellular processes such as mRNA decay and mitochondria motility and inheritance (Garcia-Rodriguez, Gay, & Pon, 2007; Jackson, Houshmandi, Lopez Leban, & Olivas, 2004; Olivas & Parker, 2000; Quenault, Lithgow, & Traven, 2011). In yeast, mRNAs associated *in vivo* with each of the six Pumillio proteins were identified. This study has revealed that more than 80% of the mRNAs associated with Puf3p encode mitochondrial proteins (Gerber *et al.*, 2004). From this study, a consensus P3BS motif has been defined: [C/U]UGUA[A/U/C]AUA. This motif was found in nearly half of the mRNAs targeted to mitochondria (Saint-Georges *et al.*, 2008), and Puf3p has been shown to be localized at the mitochondrial surface (Garcia-Rodriguez *et al.*, 2007; Olivas & Parker, 2000). The involvement of Puf3p in mRNA targeting to mitochondria was supported by the cytosolic delocalization of 72% of mMLRs containing a P3BS in *ΔPuf3* yeast strains (Saint-Georges *et al.*, 2008). This delocalization is also observed in wild-type cells when the P3BS of the BCS1 mRNA is mutated (Saint-Georges *et al.*, 2008). However, further studies have revealed that the correlation between the presence of P3BS and Puf3p-dependence for mRNA sorting was more complicated than expected (Table 12.2): (1) in *ΔPuf3* strains, some mMLRs containing a P3BS are completely delocalized from mitochondria while others are just less associated (Saint-Georges *et al.*, 2008). These results suggest that Puf3p can act alone or in cooperation with other signals and/or factors to fulfil mRNA targeting to mitochondria. (2) Targeting of some mRNAs without identified P3BS can be sensitive to Puf3p depletion. This is the case for ATP2 mRNA, which is less associated with mitochondria in a *ΔPuf3* strain despite the absence of a canonical P3BS (Gadir, Haim-Vilmovsky, Kraut-Cohen, & Gerst, 2011). This observation suggests that either Puf3p could act indirectly on ATP2 mRNA targeting to mitochondria or ATP2 mRNA contains a P3BS that differs from the consensus motif and has not been identified yet. (3) P3BS is also found in the 3′ UTR of mRNAs encoding mitochondrial proteins not targeted to this organelle (Saint-Georges *et al.*, 2008). Overall, these data show that Puf3p is involved in the targeting of several mRNAs to mitochondria in yeast but different mechanisms seem to be involved. This is probably linked to the multifunctionality of Puf3p, and further studies are needed to understand what regulates the different functions of Puf3p.

c. Involvement of MTS and Tom20 in mRNA targeting to yeast mitochondria

Potential candidates to anchor mRNAs at the surface of mitochondria are proteins of the TOM complexes that are involved in precursor binding. In this case, the interaction of mRNA with mitochondria would be indirect and mediated by the interaction of the nascent chain with components of the protein import machinery. In such a model, an active translation of mRNAs is a prerequisite for their association with mitochondria. Eliyahu *et al.* (2010) have performed a large-scale analysis of mRNA association to mitochondria in yeast strains lacking the Tom20 receptor (*Δtom20*). In this strain, nearly 100 mRNAs are less associated to mitochondria suggesting that Tom20 has a role in mRNA targeting. ACO1, an mRNA encoding a matrix protein involved in the citrate cycle, is greatly delocalized from mitochondria in yeast *Δtom20* strains (Table 12.2). Association between ACO1 mRNA and mitochondria seems to be mainly mediated by the MTS itself, as association is greatly affected by mutations that decrease the hydrophobicity and charge of the MTS (Eliyahu *et al.*, 2010). These two features were shown to be important for MTS binding to Tom20 (Abe *et al.*, 2000; Rimmer *et al.*, 2011). Furthermore, active translation is necessary to achieve ACO1 mRNA localization to the mitochondrial surface (Eliyahu *et al.*, 2010). Tom70, Tom7 and/or Tom6 seem to be also involved in mitochondrial targeting of ATP2 and OXA1 mRNAs (Table 12.2) (Gadir *et al.*, 2011). For these two mRNAs, inhibition of mRNA translation decreases the association with mitochondria (Gadir *et al.*, 2011) supporting a link between the presence of MTS in the nascent precursors and mRNA targeting to mitochondria.

5.2.2. Models of mRNA targeting to mitochondria

The major information arising from studies in yeast is that there is no single mechanism involved in mRNA targeting to mitochondria, rather several *cis* and *trans* factors acting in different combination exist to fulfil the correct sorting of mRNAs to mitochondria and to regulate this process. As an example, the localization of ACO1 mRNA depends mainly on the interaction between the MTS and Tom20 (see Section V.B.1.c) (Eliyahu *et al.*, 2010). In this case, the localization of the mRNA is dependent on translation (Fig. 12.5A). The mitochondrial localization of ATP2 mRNA relies on both MTS and *cis*-acting signals on the mRNA. 3′ UTR deletion or translation inhibition leads to a decrease in mRNA targeting to mitochondria and the combination of both 3′ UTR deletion and translation inhibition completely abolishes its association with mitochondria (Table 12.2) (Gadir *et al.*, 2011). The localization of ATP2 at the mitochondrial surface involves Puf3p (see Section V.B.1.b) (Fig. 12.5B). Another mechanism of interaction with mitochondria has been described for COX17 mRNA, an mRNA encoding an IMS protein without MTS (Saint-Georges *et al.*, 2008). The association of COX17 mRNA, which contains two P3BS, with mitochondria is mainly mediated by Puf3p as total delocalization of this mRNA from mitochondria is observed when Puf3p is depleted (Fig. 12.5C, Table 12.2) (Saint-Georges *et al.*, 2008). However, translation is also needed for COX17 mRNA targeting to mitochondria (Saint-Georges *et al.*, 2008). Localization of Tom70 mRNA to mitochondria is not dependent on Tom20, Tom7 or Puf3p and it does not contain a P3BS. Moreover, association of Tom70 with mitochondria is insensitive to drugs that inhibit translation or lead to ribosome dissociation from mRNAs, such as puromycin, meaning that mRNA targeting to mitochondria can occur in a translation-independent way (Eliyahu *et al.*, 2010; Saint-Georges *et al.*, 2008). Elements involved in Tom70 mRNA association with mitochondria are still unknown (Fig. 12.5D, Table 12.2).

Recently, it was shown that interaction between nascent chain and Tom receptors could also be mediated by cytosolic chaperones. In yeast, Ssa1, a cytosolic Hsp70 chaperone, is involved, at least in part, in the association of some mRNAs with mitochondria (Eliyahu, Lesnik, & Arava, 2012). Ssa1 is able to bind MTS and the Tom70 receptor (see Section III.c) (Endo, Mitsui, Nakai, Roise, 1996; Young *et al.*, 2003). Thus, Ssa1 could act as an adapter

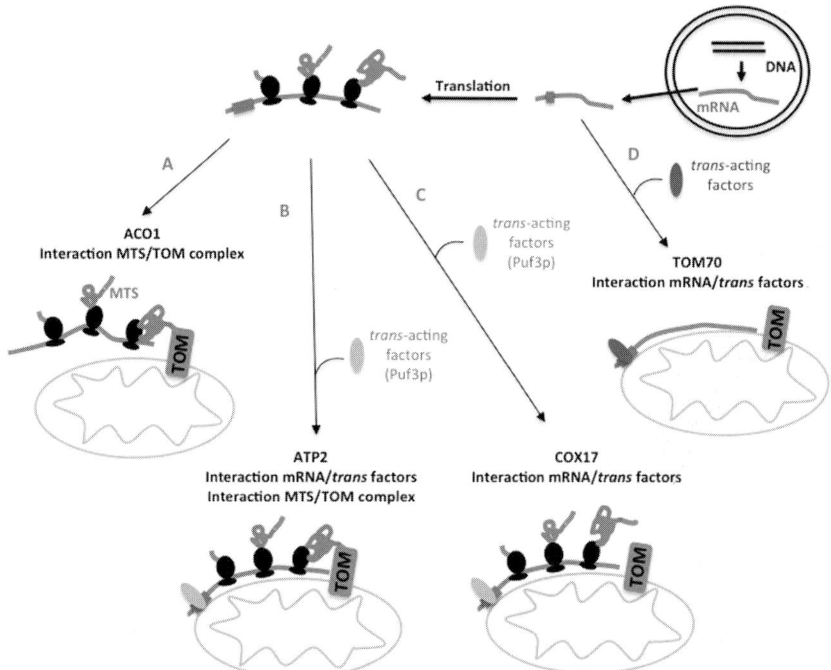

Figure 12.5 Mechanisms of targeting of different mRNAs to mitochondria in yeast. (A) Association of ACO1 mRNA with mitochondria is mainly mediated by interaction between MTS and Tom20. Translation is needed for mRNA localization. (B) Targeting of ATP2 mRNA to mitochondria involves interactions between MTS and proteins of the TOM complex and between mRNA and *trans*-acting factors like Puf3p. Translation is needed for mRNA localization. (C) Association of COX17 mRNA to mitochondria involves interactions between the two P3BS of COX17 and the *trans*-acting factor Puf3p and is dependent on translation. To date, no component of the TOM complex has been demonstrated to be involved in COX17 mRNA targeting. (D) Tom70 mRNA targeting to mitochondria is mediated by unidentified *trans*-acting factors and mRNA translation is not necessary. See the colour plate.

between Tom70 and the nascent precursor to trigger mRNAs at the surface of mitochondria.

Others interesting aspects of mRNA targeting to mitochondria are how mRNAs reach the OM and how mRNA metabolism (translation and decay) is regulated once at the surface of mitochondria. The movement of mRNAs in the cytosol can be an active process mediated by the cytoskeleton or mRNAs can diffuse in the cytosol until they reach a binding site at the surface of mitochondria. In mammals, the localization of the mRNA encoding the ribosomal protein mtRPS12 is mediated by microtubules (Russo *et al.*, 2008;

Russo *et al.*, 2006). Depolymerization of the microtubule network, but not of the actin network, leads to the delocalization of mtRPS12 mRNA from the mitochondria to the cytosol. The 3′ UTR of mtRPS12 is involved in both mRNA targeting to mitochondria and microtubule association (Russo *et al.*, 2008; Russo *et al.*, 2006). In mammals, YB-1 is a multifunctional DNA and RNA-binding protein involved in DNA replication, DNA repair, transcription, mRNA splicing and translation (see Eliseeva, Kim, Guryanov, Ovchinnikov & Lyabin, 2011 for review). Recently, YB-1 has been shown to regulate the translation state of several mRNAs encoding respiratory chain subunits (Matsumoto *et al.*, 2012b). Moreover, a proportion of YB-1 is associated with the OM of mitochondria in HeLa cells (Matsumoto *et al.*, 2012a). YB-1 is able to interact with four mRNAs encoding respiratory chain subunits that are targeted to mitochondria (Matsumoto *et al.*, 2012a). Thus, YB-1 was proposed to play a role in mRNA metabolism at the surface of mitochondria.

5.3. Regulation of mRNA Targeting to Mitochondria

Studies in yeast support the idea that mRNA targeting to mitochondria is essential for the correct import of a small subset of mitochondrial proteins and that this process is mostly involved in the regulation of mitochondria biogenesis and functions. These are illustrated by the fact that the *ΔPuf3* yeast strain does not present any growth defect despite the delocalization of nearly half the mMLRs in this strain, even in glycerol medium in which mitochondria are highly activated (Eliyahu *et al.*, 2010; Saint-Georges *et al.*, 2008).

5.3.1. mRNA targeting and mitochondrial biogenesis

In yeast grown in controlled conditions, O_2 consumption oscillates in a periodic manner that defines the yeast metabolic cycle (YMC) (Tu, Kudlicki, Rowicka & Mcknight, 2005). The YMC is characterized by a reductive phase, where O_2 consumption is very low, and by an oxidative phase, where O_2 is highly consumed (Tu *et al.*, 2005). Transcriptomic analyses of gene expression along YMC have revealed that genes involved in the biogenesis of mitochondria are highly expressed during the reductive phase of the cycle. Among them, expression of genes involved in the early phase of mitochondria biogenesis, such as those involved in translation, protein import and assembly factors (group A), precedes expression of genes encoding structural subunits of respiratory chain complexes or enzymes of the citrate cycle (group B) (Lelandais *et al.*, 2009). It was found that 80% of mMLRs containing a P3BS belong to group A, suggesting a common post-transcriptional regulation by Puf3p in this group. The best example to illustrate this regulatory model is the assembly of the cytochrome *c* oxidase complex (COX). COX complex is composed of 12 structural subunits (three encoded by the mitochondrial genome), and 18 assembly factors are required for the correct assembly of this complex (Fontanesi, Soto, Horn, & Barrientos, 2006). All the assembly factors are mMLRs and are highly expressed in the early reductive phase of the YMC (Garcia *et al.*, 2007; Lelandais *et al.*, 2009; Tu *et al.*, 2005). Furthermore, 17 of the 18 assembly factors have a P3BS and are less associated to mitochondria in *ΔPuf3* yeast strains (Saint-Georges *et al.*, 2008). In contrast, all the structural subunits of the COX complex are encoded by non-MLR mRNAs and are expressed later during the reductive phase. Thus, the proposed model arising from these data is that assembly factors are expressed before structural subunits and are coregulated by a post-transcriptional mechanism involving Puf3p (Devaux *et al.*, 2010; Gelly, Orguer, Jacq, & Lelandais, 2011; Lelandais *et al.*, 2009). This regulation is characterized by mRNA targeting to mitochondria, which probably increases protein import efficiency into mitochondria. Once assembly factors are imported, their genes are downregulated and structural subunits are synthesized and correctly assembled into complexes. On the one hand, downregulation of assembly factor expression could be mediated by Puf3p, which is involved in mRNA decay and has been shown to be directly involved in the mRNA degradation of the assembly factor COX17 (Jackson *et al.*, 2004; Olivas & Parker, 2000). On the other hand, the expression of the structural subunits seems regulated at the transcriptional level as common transcription factor binding sites can be found in the promoter of these genes (Lelandais *et al.*, 2009).

5.3.2. mRNA targeting and signal transduction

In mammals, one study has shown that mRNA targeting to mitochondria can be regulated by signal transduction mediated by cAMP (Ginsberg *et al.*, 2003). AKAP121 is an A-kinase anchor protein that is associated with the OM of mitochondria. This protein is involved in the recruitment of protein kinase A (PKA) at the surface of mitochondria. AKAP121 contains a KH RNA-binding domain that is able to interact with the 3′ UTR of mRNAs encoding MnSOD and the F0-f subunit of ATP synthase. Interaction between mRNAs and AKAP121 is significantly increased when the KH domain is phosphorylated. As described above (see Section V.B.1.a), a structural element in the 3′ UTR of both mRNAs is necessary for this interaction. Further studies with MnSOD have shown that, under physiological conditions, MnSOD mRNA is mainly translated in the cytosol. When cAMP concentration increases, PKA is activated and phosphorylates AKAP121 leading to an increased affinity of AKAP121 for MnSOD mRNA. MnSOD mRNA is thus anchored at the mitochondrial surface by interactions between its 3′ UTR and the phosphorylated KH domain of AKAP121. Translation of MnSOD at the surface of mitochondria greatly increases the proportion of MnSOD protein into mitochondria. This example shows that mRNA targeting to mitochondria could be increased in response to an intracellular cell signal (cAMP) to enhance protein import into mitochondria.

5.4. mRNA Targeting in Plant Cells

In plants, in addition to the roles that have been highlighted in yeast and mammals, mRNA targeting to mitochondria could be an interesting hypothesis to achieve specific sorting of proteins between mitochondria and chloroplasts. Some studies have reported the localization of cytosolic mRNAs encoding chloroplastic proteins at the surface of chloroplasts in higher plants and in *Chlamydomonas reinhardtii* but nothing is known about the extent of this process and the mechanisms that are involved (Marrison, Schunmann, Ougham, & Leech, 1996; Uniacke & Zerges, 2009). More recently, mRNA targeting to mitochondria has been demonstrated in potato (Michaud *et al.*, 2010). The differential association of 17 cytosolic mRNAs has been studied by quantitative reverse transcriptase-polymerase chain reaction in mitochondrial fractions prepared from potato tubers (Table 12.3). As plant mitochondria are easily contaminated by plastidial fractions, two cytosolic mRNAs encoding plastidial proteins (RPL12c and ATP5c, Table 12.3) have been used to evaluate the contamination background of mitochondrial preparations. Taking this background into account, six mRNAs are significantly enriched in mitochondrial fractions (Fig. 12.6). Four of them encode mitochondrial proteins involved in the citrate cycle or in acetyl-CoA synthesis. In yeast, all the enzymes of the citrate cycle are encoded by mMLR, suggesting a conserved function of mRNA targeting in mitochondria biogenesis and activities (Table 12.3) (Marc *et al.*, 2002). In potato, two mMLRs encode cytosolic proteins (glyceraldehyde-3-phosphate dehydrogenase (GAPDH) and the translation initiation factor, eIF4E1) (Fig. 12.6). GAPDH is one of the seven glycolytic

Table 12.3 Function, localization and mitochondrial association of 17 cytosolic mRNAs

Gene	Function	Localization of proteins	Metabolic pathway	Potato mRNA association to mitochondria (Michaud *et al.*, 2010)	Yeast mRNA association to mitochondria (Garcia *et al.*, 2007; Marc *et al.*, 2002)
CS	Citrate synthase	Matrix	Citrate cycle	+	+
IDH	Isocitrate dehydrogenase	Matrix	Citrate cycle	−	+
SUCLA2	Succinyl-CoA ligase, subunit β	Matrix	Citrate cycle	+	+
MDH	Malate dehydrogenase	Matrix	Citrate cycle	+	+
PDH-E1α	Pyruvate dehydrogenase, subunit E1α	Matrix	Acetyl-CoA synthesis	+	+
PDH-E2	Pyruvate dehydrogenase, subunit E2 matrix		Acetyl-CoA synthesis	−	+
RPL12 m	Mitochondrial ribosomal protein L12	Matrix	Translation	−/+	nd
ATP5 m	Mitochondrial ATP synthase, subunit 5	IMM	ATP synthesis	−	−
SYS	Seryl-tRNA synthetase	Matrix/stroma	Translation	−	−
SYE	Glutamyl-tRNA synthetase	Matrix/stroma	Translation	−/+	nr
SYP	Prolyl-tRNA synthetase	Matrix/stroma	Translation	−	nr
RPL12 p	Plastid ribosomal protein L12	Stroma	Translation	−	nr
ATP5 p	Plastid ATP synthase, subunit 5	IPM	ATP synthesis	−	nr

(*Continued*)

Table 12.3 Function, localization and mitochondrial association of 17 cytosolic mRNAs—cont'd

Gene	Function	Localization of proteins	Metabolic pathway	Potato mRNA association to mitochondria (Michaud *et al.*, 2010)	Yeast mRNA association to mitochondria (Garcia *et al.*, 2007; Marc *et al.*, 2002)
GAPDH	Glyceraldehyde-3-phodphate-dehydrogenase	Cytosol	Glycolysis	+	–
eEF1α	Elongation factor 1α	Cytosol	Translation	–	–
eIF4E1	Initiation factor 4E1	Cytosol	Translation	+	–
RPL12c	Cytosol ribosomal protein L12	Cytosol	Translation	–	–

nd, not determined; nr, not relevant (plastid proteins, or different origins between yeast and potato SYE and SYP); IMM, inner mitochondria membrane; IMP, inner plastid membrane.

enzymes that are associated with OM in *Arabidopsis* and in potato (Giege *et al.*, 2003; Graham *et al.*, 2007). Until now, nothing was known about the signals allowing targeting and anchoring of this protein to the OM. Association of GAPDH mRNA with the mitochondrial surface could be involved in the targeting of the protein. The meaning of eIF4E1 mRNA association with mitochondria is still elusive. This mRNA has also been shown to copurify with chloroplastic fractions in several plant species (Nicolai *et al.*, 2007) showing a potential link between eIF4E1 and organelles of endosymbiotic origin. The study on potato also includes three mRNAs encoding aminoacyl-tRNA synthetases that are dual targeted to mitochondria and chloroplasts (Fig. 12.6, Table III), but no significant association of these mRNAs with mitochondria has been detected (Michaud *et al.*, 2010).

Differential association of mRNAs with mitochondria seems to be conserved between plant species because this process has also been

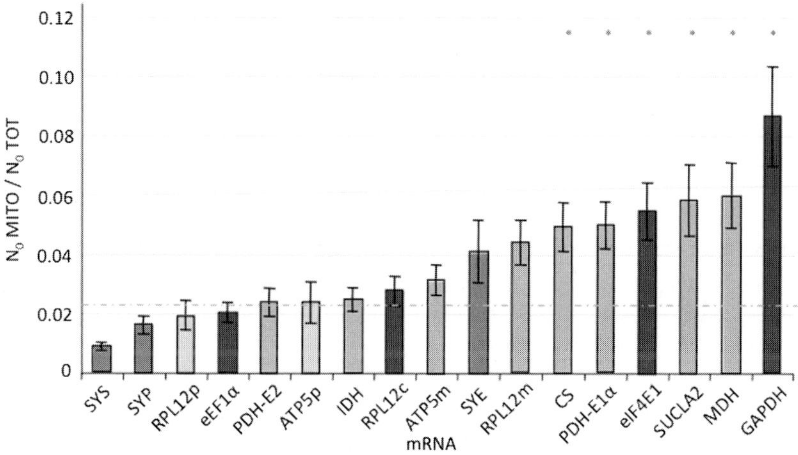

Figure 12.6 Differential association of 17 mRNAs at the surface of potato mitochondria. The histogram represents the amount of mRNA associated with mitochondria (N_0 MITO) compared with their level of expression in the total fraction (N_0 TOT). mRNAs encoding plastidial proteins were not expected to be found in mitochondrial fractions and were used as controls to estimate the background of contamination (dashed line). Statistical analyses showed that six mRNAs can be considered associated to mitochondria (*). Colours represent the final localization of the protein encoded by mRNAs: blue is for cytosolic proteins, orange for mitochondrial proteins, green for plastidial proteins and red for dual-targeted proteins to mitochondria and chloroplasts. See Table 12.3 for more information about the name and the function of the different candidates analysed. For interpretation of the references to colour in this figure legend, the reader is referred to the online version of this book.

demonstrated in *A. thaliana* cell cultures (unpublished data). Large-scale analysis of mRNAs enriched in plant mitochondrial fractions will give more information about the extent of this process and the nature of mRNAs associated to plant mitochondria. Until now, nothing has been known about the mechanisms involved in such targeting in plants. We can speculate that MTSs could also play a role in mRNA anchoring to mitochondria according to the similarities in yeast and plant MTS binding to the Tom20 receptor (Abe *et al.*, 2000; Rimmer *et al.*, 2011). It is also possible that RNA-binding proteins of the Puf family play a role in this process. The Puf family is poorly characterized in plants. The *A. thaliana* genome contains 25 genes encoding proteins from this family (only six Puf proteins in yeast) (Francischini & Quaggio, 2009; Tam *et al.*, 2010). The few studies done in *Arabidopsis* show that some Puf proteins are present in punctuate motile structures in the cytosol but until now, no mitochondrial localization has been clearly demonstrated (Tam *et al.*, 2010). Thus, mRNA targeting in plants is still largely unknown and much work is needed to elucidate the mechanisms and roles of this process in plant cells.

6. TRANSFER RNA IMPORT

In many organisms, mitochondria not only import proteins, but also tRNAs. Transfer RNA import was first observed in the protozoan *Tetrahymena* in 1967, then in yeast (1977), in plants (1988) and now in numerous other organisms (Schneider, 2011). Analyses of mitochondrial genomes show that some tRNA genes (*trn*) are missing in numerous organisms. Import of cytosolic tRNAs is expected to complete the mitochondrial tRNAs set. Transfer RNA import into mitochondria is thus a widespread phenomenon that is essential for mitochondrial translation. Transfer RNA gene loss and tRNA import acquisition have occurred early during evolution of plant mitochondria. Absence of genes corresponding to specific tRNAs is observed in green and red algae and in bryophytes (Duchene, El Farouk, Sieber, & Maréchal-Drouard, 2011), and tRNA import has been demonstrated in the green algae *C. reinhardtii* (Vinogradova, Salinas, Cognat, Remacle, & Marechal-Drouard, 2009) and in the bryophyte *Marchantia polymorpha* (Akashi, Sakurai, Hirayama, Fukuzama, & Ohyama, 1996). Moreover, import of other RNAs has been described. For example, 5S rRNA import in mitochondria has been shown in mammals (Smirnov *et al.*, 2010).

6.1. Transfer RNA Gene Content in Plant Mitochondrial Genomes

Complete mitochondrial genome sequences are now available for numerous plants, allowing the prediction of *trn*. Extensive variations in the number and origin of *trn* genes occur in plant mitochondria. In angiosperm, *trn* are either native or chloroplastic-like (cp-like) (Fig. 12.7). Native *trn* derive from genes of the ancestral bacterium at the origin of mitochondria. Cp-like *trn* correspond to chloroplastic DNA transfer into the mitochondrial genome during evolution. In addition, a new type of *trn*, a *trn*-C(GCA) of ambiguous origin, has recently been identified in *Beta* and other plant mitochondria (Kitazaki *et al.*, 2011). In addition, *trn* corresponding to 5–7 amino acids are missing in angiosperm mitochondria (Fig. 12.7). Import of cytosolic tRNAs is expected to compensate for the missing tRNAs (see Section VI.B) (Duchene *et al.*, 2011).

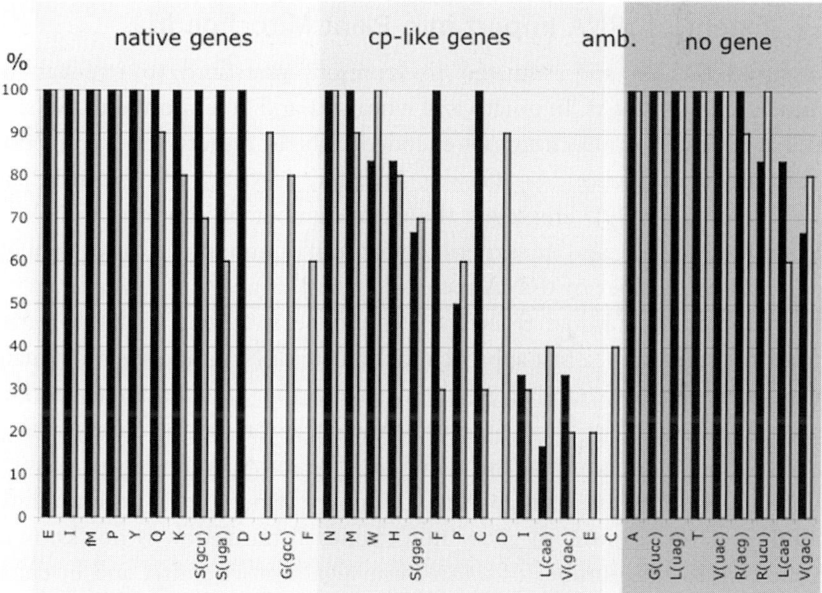

Figure 12.7 Origin of mitochondrial *trn* genes. *trn* genes have been predicted in the mitochondrial genome of numerous angiosperms. Analysis of six monocots (*Bambusa, Oryza, Triticum, Sorghum, Tripsacum, Zea mays*) and ten dicots (*Arabidopsis, Brassica, Carica, Cucumis, Citullus, Cucurbita, Vigna, Nicotiana, Beta, Silene*) is presented here (Alverson *et al.*, 2011; Sloan *et al.*, 2010). Mitochondrial *trn* genes were classified according to their type (native, cp-like, ambiguous) or absence, and to their frequency in monocots (in black) and dicots (in grey). *trn* are identified by the one-letter symbol of their cognate amino acid and eventually by their anticodon. I codes for $tRNA^{Ile}$(LAU). For interpretation of the references to colour in this figure legend, the reader is referred to the online version of this book.

Although great variations from one plant to another are observed, a few general features can be described. For instance, the native *trn*-E, -I, -K, -fM, -P, -Q and -Y, as well as the cp-like *trn*-N, -M, -W and -H are present in all or most angiosperm mitochondria, and *trn* genes for Ala, Leu, Thr, Val and Gly(UCC) are missing in most of these genomes (Fig. 12.7).

Divergences between monocots and dicots are also observed: *trn*-D is native in monocots and cp-like in dicots. *trn*-G(GCC) is native in dicots but tRNA^{Gly}(GCC) is imported in monocots. Transfer RNA^{Cys} and tRNA^{Phe} are cp-like in monocots but of various origins in dicots: mitochondrial-encoded (native, cp-like or ambiguous), or nuclear-encoded and imported. Changes can occur in individual species. For example, *trn*Q(UUG) cannot be found in *Silene* mitochondria (Sloan, Alverson, Storchova, Palmer & Taylor, 2010).

6.2. Extent of tRNA Import into Plant Mitochondria

Imported tRNAs are estimated to represent one-third to one-half of mitochondrial tRNAs. In potato and wheat, 11 and 16 imported tRNAs, as well as 20 and 14 mitochondrial-encoded tRNAs, respectively, have been identified in mitochondria (Glover, Spencer, & Gray, 2001; Maréchal-Drouard *et al.*, 1990). However, the exact number of imported tRNAs is difficult to evaluate, and the experimental proof of import was only obtained for a limited number of tRNA species (Table 12.4).

As the mitochondrial tRNA set should be sufficient to allow mitochondrial translation, a first approach to evaluate tRNA import is to evaluate what is needed for translation and what are the mitochondrial *trn*. For example, import of at least some tRNA isoacceptors can be clearly anticipated when mitochondrial *trn* genes such as *trn*-A or *trn*-T are missing (Table 12.4). However this approach is imperfect because: (1) the minimal number of isoacceptors to allow mitochondrial translation is not known. According to the number of *trn* in some algal mitochondria and in chloroplasts, it is evaluated to be about 25 in plants (Duchene *et al.*, 2011). (2) Some mitochondrial *trn* appear to be not expressed (Duchêne & Maréchal-Drouard, 2001). (3) The post-transcriptional modifications of nucleotides that should affect the decoding capacity of a tRNA are not known (Weber, Dietrich, Weil, & Marechal-Drouard, 1990).

A first strategy to determine the imported tRNAs is to evaluate the whole mitochondrial tRNA population. In potato and wheat, tRNAs were first extracted from highly purified mitochondria and separated on two-

Table 12.4 Loss of mitochondrial *trn* genes and import of nuclear-encoded tRNAs into mitochondria of a few angiosperms

	Missing *trn* genes			Other missing *trn* genes or loss of *trn* expression	Verification of tRNA import
Dicots	A,R,L,T,V and	Brassicales loss of trnF	*A. thaliana* (Chen, Viry-Moussaid,, Dietrich, & Wintz, 1997; Duchêne & Maréchal-Drouard, 2001)	W, M-e	W F
	1 trnG		*B. napus*	–	
		Solanales	*N. tabacum* (Delage *et al.*, 2003b; Salinas *et al.*, 2005)	–	V G
			S. tuberosum (Brubacher-Kauffmann *et al.*, 1999; Kumar, Maréchal-Drouard, Akama, & Small, 1996; Maréchal-Drouard *et al.*, 1990)	nd	A, R, L, T, I G V
Monocots	A,R,L,T,V and		*T. aestivum* (Glover *et al.*, 2001)	H	A, R, L, V, G, H, I(IAU)
	2 trnG		*Z. mays* strain NB (Kumar *et al.*, 1996)	–	A, R, L, T, V, G

The mitochondrial *trn*-A, R, L, T, V and one *trn*-G were lost in all angiosperms, the 2nd *trn*-G was lost in all monocots, and *trn*-F in Brassicales. Loss of other *trn* in particular plant species was also observed (*A. thaliana, T. aestivum*). Transfer RNA import can be predicted from analysis of mitochondrial genomes, but was also demonstrated by hybridization, sequencing or aminoacylation. nd, not determined.

dimensional gel (Glover *et al.*, 2001; Maréchal-Drouard *et al.*, 1990). Each spot was then identified by hybridization, aminoacylation or sequencing. However, such a strategy does not allow the detection of rare tRNAs. Moreover, it is not easy to differentiate low import and cytosolic contamination. Some tRNAs, such as $tRNA^{Thr}$ in wheat, can escape detection (Glover *et al.*, 2001).

Another approach is to systematically check the import status of every nuclear-encoded tRNA. This was performed in *C. reinhardtii* by Northern blot (Vinogradova *et al.*, 2009). Only three *trn* were found and expressed in the mitochondrial genome. So most mitochondrial tRNAs are nuclear encoded. Among the 49 nuclear-encoded isoacceptor families, 15 are considered to be cytosol specific, and 34 to be shared between the cytosol and mitochondria. Once again, the distinction between low import and cytosolic contamination is problematic. However, the association between a systematic analysis of mitochondrial *trn* genes and the evaluation of import of each cytosolic tRNA family gives the most accurate determination of the whole mitochondrial tRNA population.

6.3. Mechanism of tRNA Import into Plant Mitochondria

The steady-state level of a tRNA in mitochondria is the result of numerous processing and transport steps (Fig. 12.8).

6.3.1. *From the nucleus to the mitochondrial surface*

The nuclear *trn* is first transcribed and its RNA product is end processed, spliced and modified. Once matured, the tRNA is exported from the nucleus into the cytosol. Transfer RNAs are not free in the cell, rather they are thought to be always associated with components of the translation machinery and recycled during protein synthesis. To be targeted towards the surface of mitochondria, a fraction of the nuclear-encoded tRNAs needs to escape the cytosolic translational machinery by interacting with protein factors that will direct them to mitochondria. In yeast, the $tRNA^{Lys}$ transporter is enolase2p (Entelis *et al.*, 2006). In *Trypanosoma brucei*, binding to eEF1α is a prerequisite for a tRNA to be imported into mitochondria (Bouzaidi-Tiali, Aeby, Charriere, Pusnik, & Schneider, 2007). In plants, although tRNA sequences are highly conserved, both the number and the identity of imported tRNAs vary dramatically from one plant species to another, even between closely related species. It is thus quite difficult to understand how different plants regulate the import of identical tRNAs

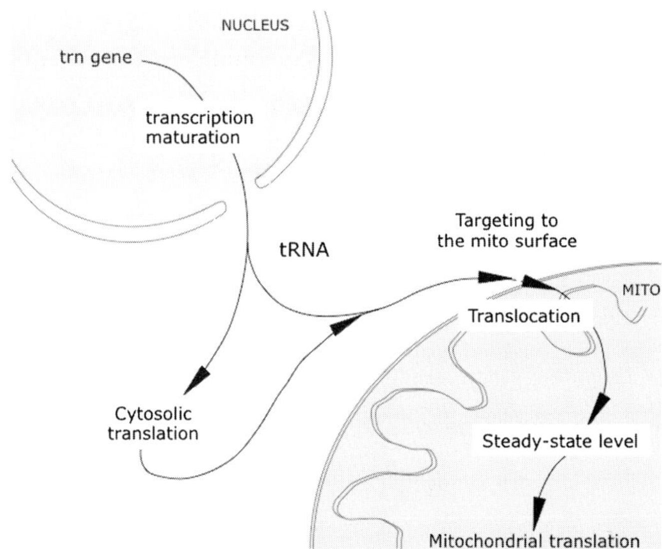

Figure 12.8 Transport of tRNAs from the nucleus to mitochondria. After transcription and maturation, tRNAs are exported from the nucleus to the cytosol. A fraction of tRNAs are targeted to the mitochondrial surface and translocate through the double mitochondrial membrane. It is not clear if tRNAs interact directly with the mitochondrial surface or if they escape the cytosolic translational machinery to reach mitochondria. Once in the matrix, the nuclear-encoded tRNAs are used in the mitochondrial translation. For colour version of this figure, the reader is referred to the online version of this book.

(Brubacher-Kauffmann, Marechal-Drouard, Cosset, Dietrich, & Duchene, 1999; Salinas, Schaeffer, Marechal-Drouard, & Duchene, 2005). Using *in vivo* approaches, different nucleotides or domains were shown to be essential for import of a few cytosolic tRNAs: position 73 in $tRNA^{Ala}$, the anticodon and the D-arm in $tRNA^{Val}$ and $tRNA^{Gly}$ (Table 12.5) (Delage, Duchêne, Zaepfel, & Maréchal-Drouard, 2003b; Dietrichm, Maréchal-Drouard, Carneiro, Cosset, & Small, 1996; Salinas *et al.*, 2005). Two couples of imported versus non-imported tRNAs have been extensively studied in BY2 cells: $tRNA^{Val}$(AAC)/$tRNA^{Met\text{-}e}$(CAU) (Delage *et al.*, 2003b; Laforest, Delage, & Marechal-Drouard, 2005) and $tRNA^{Gly}$(UCC)/ $tRNA^{Gly}$(GCC) (Salinas *et al.*, 2005). Sequence exchanges between tRNAs in each couple were performed in the anticodon stem, D- or T-domains, and identity elements (Table 12.5). All mutant tRNAs are highly expressed. Overall, the results show that: (1) anticodon stem mutations are silent for import and do not affect the imported or non-imported status of tRNAs. (2) Sequence exchanges in the D- or T-domains inhibit import of the naturally imported tRNA but do not allow import of the naturally non-

Table 12.5 *In Vivo* studies of tRNA import into plant mitochondria

tRNA region	tRNA	Mutation	Aminoacylation	Import of mutant tRNA
Anticodon	$tRNA^{Val}$(AAC)	U27-A43 + C28-G42 → C27-G43+U28-A42	Yes	Yes
Stem	$tRNA^{Gly}$(UCC)	U29-A41 → A29-U41	Yes	Yes
	$tRNA^{Gly}$(GCC)	C30-G40 → G30-C40	Yes	No
D-domain	$tRNA^{Val}$(AAC)	D-domain (Val → Met)	Yes	No
	$tRNA^{Gly}$(UCC)	D-domain (UCC → GCC)	Yes	No
	$tRNA^{Gly}$(GCC)	D-domain (GCC → UCC)	Yes	No
T-stem	$tRNA^{Val}$(AAC)	T-domain (Val → Met)	Yes	No
Anticodon	$tRNA^{Val}$(AAC)	Anticodon AAC(Val) → CAU(Met)	No by ValRS Yes by MetRS	No
	$tRNA^{Met-e}$(CAU)	Anticodon CAU(Met) → CAC(Val)	No	No
	$tRNA^{Gly}$(UCC)	Anticodon UCC → GCC	Yes	No
	$tRNA^{Gly}$(GCC)	Anticodon GCC → UCC	Yes	No
	$tRNA^{Leu}$(CAA)	Insertion of four nucleotides in the anticodon	Yes	Yes
	$tRNA^{Ala}$(UGC)	Insertion of four nucleotides in the anticodon	Yes	Yes
Multiple	$tRNA^{Met-e}$(CAU)	Anticodon CAU(Met) → CAC(Val) + D-domain (Met → Val)	No	No
Multiple	$tRNA^{Ala}$(UGC)	Insertion of four nucleotides in the anticodon + U70 → C70 (identity element)	No	No

Transgenic mutated tRNAs were expressed in tobacco (all except $tRNA^{Leu}$) or potato (for $tRNA^{Leu}$). $tRNA^{Val}$ is imported in all angiosperms, $tRNA^{Met-e}$ is not imported in tobacco (Delage *et al.*, 2003b; Laforest *et al.*, 2005). $tRNA^{Gly}$(UCC) is imported in all angiosperms, $tRNA^{Gly}$(GCC) is imported in monocots but not in dicots (Salinas *et al.*, 2005). $tRNA^{Leu}$(CAA) (Small *et al.*, 1992) and $tRNA^{Ala}$(UGC) (Dietrich *et al.*, 1996) are imported in all angiosperms.

imported tRNA. These mutations in the D- or T-domain probably affect the tertiary structure of the tRNA suggesting that the three-dimensional structure or the tRNA flexibility could be important for efficient import into plant mitochondria. (3) Concerning mutations in the anticodon, no general rules can be observed. GCC mutation in $tRNA^{Gly}$(UCC) background inhibits import of the tRNA, but UCC mutation in $tRNA^{Gly}$(GCC) background does not allow import. By contrast, insertion of four nucleotides in the imported $tRNA^{Leu}$(CAA) and $tRNA^{Ala}$ (UGC) anticodons do not affect their import (Dietrich *et al.*, 1996; Small *et al.*, 1992). (4) Aminoacylation by the cognate aminoacyl-tRNA synthetase (aaRS) appears to be essential for import, and this is in accordance with a previous study on $tRNA^{Ala}$(UGC) (Dietrich *et al.*, 1996). However, recognition of a tRNA by an imported aaRS is not sufficient to permit its import into the organelle (Brubacher-Kauffmann *et al.*, 1999; Mireau *et al.*, 2000). These results show that import determinants carried by cytosolic tRNAs can be distinct from the identity elements required for recognition by the cognate aaRS.

In conclusion, the control of tRNA import selectivity appears complex. Some features (three-dimensional structure of the tRNA, aminoacylation) appear essential for import and conserved from one tRNA model to the other, but particular species or tRNA characteristics also influence import (e.g. for $tRNA^{Gly}$). Moreover, the import determinants could act on the targeting of tRNAs to the mitochondrial surface and on the next steps, that is, translocation through the mitochondrial membranes and establishment of a steady-state level in mitochondria (see Sections VI.C.2 and VI.C.3).

6.3.2. The translocation step across mitochondrial membranes

By contrast to the well-known and versatile set of machineries and mechanisms involved in protein mitochondrial import, it is only during the last decade that tRNA import pathways have begun to be elucidated. As tRNAs represent large hydrophilic and negatively charged macromolecules (25 kDa), their import into the matrix is not a passive process. What is common to all systems studied so far (yeast, protozoans and plants) is the requirement of protease-sensitive receptor(s) at the surface of mitochondria and of exogenous ATP. In plants, an *in vitro* system was set up with isolated mitochondria (Delage, Dietrich, Cosset, & Marechal-Drouard, 2003a; Salinas *et al.*, 2006). Radioactive tRNAs can be imported into isolated potato mitochondria in the presence of ATP and a mitochondrial membrane potential but without any cytosolic factors (Fig. 12.9A) (Delage *et al.*, 2003a). Moreover, *in vitro* tRNA import can occur when the protein import

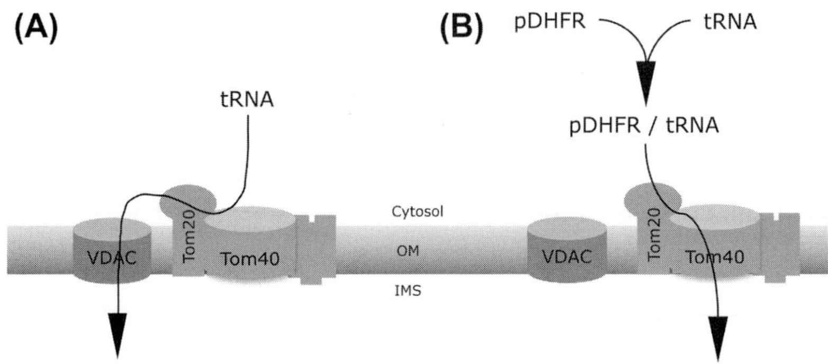

Figure 12.9 tRNA translocation through the plant mitochondrial outer membrane. Two *in vitro* systems have been set up with plant mitochondria. Direct import into isolated mitochondria was demonstrated (A) (Salinas *et al.*, 2006), but the use of a protein carrier greatly improves import (B) (Sieber *et al.*, 2011). Components of the TOM complex are involved in both systems. In (A), VDAC is also implicated. In (B), a protein carrier (pDHFR) is used. This last system is artificial because pDHFR is of mammalian origin. Whether both pathways (direct or with an unknown protein carrier) exist *in vivo* is an open question. For colour version of this figure, the reader is referred to the online version of this book.

channel is blocked. However, two major components of the TOM complex (TOM20 and TOM40) are involved in tRNA binding at the surface of mitochondria (Salinas *et al.*, 2006). Under these *in vitro* conditions, the VDAC, which plays a key role in metabolite transport, is implicated in the tRNA translocation step through the outer mitochondrial membrane. Both ruthenium red, which induces the closure of VDAC, and antibodies against VDAC are able to inhibit *in vitro* import of tRNA.

However, selectivity of import was partially lost in the *in vitro* import system (Delage *et al.*, 2003a; Salinas *et al.*, 2005) and VDAC was shown to interact with different RNA and DNA substrates (Salinas *et al.*, 2006). For example, cytosol-restricted tRNA^{Gly}(GCC) was efficiently imported into isolated mitochondria, as efficiently as the imported tRNA^{Gly}(UCC) (Salinas *et al.*, 2005). Among the different hypotheses to explain this discrepancy, one is that still unknown cytosolic specificity-regulating factors are missing in the *in vitro* system.

Very recently, a new strategy was developed to import tRNAs and other RNAs into mitochondria isolated from potato (Fig. 12.9B) (Sieber, Placido, El Farouk-Ameqrane, Duchene & Marechal-Drouard, 2011). This strategy is based on the use of a protein as a vector. *In vitro*, the mammalian dihydrofolate reductase (DHFR) is able to bind RNAs in a non-specific manner, and DHFR fused to a mitochondrial targeting sequence (pDHFR) can act as a carrier to introduce foreign RNAs into isolated plant mitochondria. It was previously shown that tRNAs could enter plant mitochondria at a low level without any added protein factors (see above). However, when pDHFR is added to the import medium, tRNA mitochondrial import efficiency is considerably increased (10–20 times more) (Sieber *et al.*, 2011). These new data support the idea that, *in vitro* and as in yeast and humans, plants can import tRNA thanks to a protein carrier and components of the protein import machinery. Although this system is artificial, it raises interesting questions on the natural mitochondrial tRNA import mechanisms. Understanding whether plant mitochondria use carrier proteins and whether the two options (with and without a carrier protein) occur *in vivo* will be the next challenge in tRNA import studies in plants.

6.3.3. The steady-state level of imported tRNAs in mitochondria

Once tRNAs have reached the mitochondrial matrix, the translation machinery is expected to recruit them. We cannot exclude the possibility that tRNAs, once translocated, are differentially degraded inside the mitochondrion and that their steady-state level results from both their import efficiency and their stability within the organelle.

6.4. Imported tRNAs and Mitochondrial Translation

Transfer RNAs, as adapter molecules between mRNA and proteins, represent key elements of all translation machineries.

6.4.1. Complementarity between imported and mitochondrial-encoded tRNAs

If the minimal number of tRNAs for translation is not expressed from a mitochondrial genome, then the imported cytosolic tRNA species will be useful to compensate for this incomplete set. In that case, the import of nuclear-encoded tRNAs into mitochondria will be a prerequisite for mitochondrial protein synthesis. From an evolutionary point of view and to keep the mitochondrial translation machinery active, it would make sense if the acquisition of a new cytosolic tRNA in mitochondria precedes the loss of the corresponding mitochondrial *trn*. Thus, in theory, all intermediary steps must be visible; in particular, in plant mitochondria that are able to import a nuclear-encoded tRNA even though the mitochondrial counterpart is still expressed.

Analysis of mitochondrial tRNA populations shows a quite strong correlation between the missing mitochondria-encoded tRNAs and the imported nuclear-encoded tRNAs. For example, in wheat, mitochondrial *trn* corresponding to six amino acids have been lost during evolution and the corresponding cytosolic tRNAs are now found in mitochondria (Glover *et al.*, 2001). Only a few cases of redundancy between imported and mitochondrial-encoded tRNAs have been described in higher plants and bryophytes (Akashi *et al.*, 1996; Duchene *et al.*, 2011). However, an exhaustive approach was only performed in *Chlamydomonas* (Vinogradova *et al.*, 2009), and similar in depth analyses are needed in higher plants to evaluate tRNA import and complementarity between nuclear and mitochondrial-encoded tRNAs in mitochondria.

6.4.2. Imported tRNAs and codon usage

In *Escherichia coli*, *Saccharomyces cerevisiae* and chloroplasts, it has been shown that the tRNA population is adapted to the codon usage (Pfitzinger, Guillemaut, Weil, & Pillay, 1987). An analysis of the expression and sub-localization of the nuclear-encoded tRNA species was performed in *C. reinhardtii* (Vinogradova *et al.*, 2009). The mitochondrion of this green alga has only three mitochondrial-encoded tRNAs, and mitochondrial localization for nuclear-encoded tRNAs varies from 0 to 98% (Fig. 12.10). The work of Vinogradova and colleagues has shown that the intracellular

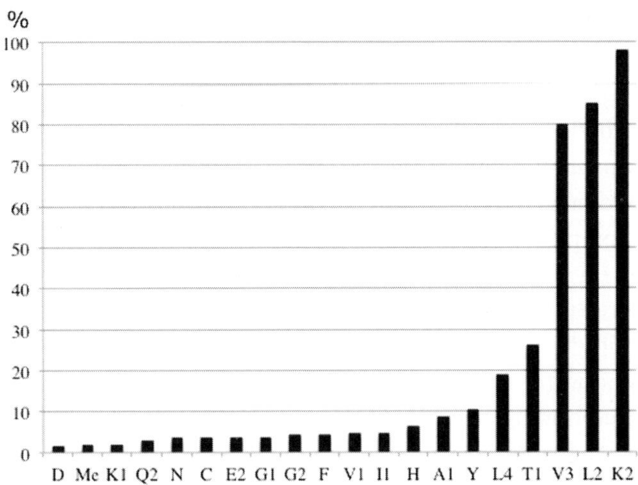

Figure 12.10 Percentage of steady-state levels of imported tRNAs into *C. reinhardtii* mitochondria relative to their abundance in the whole-cell fraction. Only percentages above 1% are indicated. tRNAs are identified by the one-letter symbol of their cognate amino acid; the numbers refer to isoacceptors as in Vinogradova *et al.* (2009).

localization of tRNA isoacceptors is tightly linked to both the nuclear and the mitochondrial codon usage (Table 12.6). In higher plants, cytosolic and mitochondrial codon usage is quite similar (http://www.kazusa.or.jp/codon/) and finding such correlation is more difficult, if there is any.

6.4.3. Aminoacyl-tRNA synthetases (aaRS)

In plants, protein synthesis occurs in three cellular compartments: the cytosol, the mitochondrion and the chloroplast. Thus, three sets of aaRS activities are expected. In *A. thaliana*, only 45 aaRS expressed genes have been identified in the nuclear genome and none in the mitochondrial and

Table 12.6 Codon usages and steady-state levels of imported tRNAs in *C. reinhardtii* mitochondria for five amino acids (Gly, Lys, Leu, Thr, Val)

Amino acid	Anticodon	Codons	% codon in cytosol	% codon in mitochondrion	% import in mitochondrion
Gly	CCC	GGG	11	1	0.2
	GCC	GGU-GGC	83	93	3.6
	UCC	GGA	6	5	4.2
		All	100	100	
Lys	CUU	AAG	95	16	2.1
	UUU	AAA	5	84	98.0
		All	100	100	
Leu	AAG	CUU-CUC	19	9	0.2
	CAG	CUG	73	6	0.3
	UAA	UUA	1	0	0.6
	UAG	CUA	3	31	19.0
	CAA	UUG	4	55	85.0
		All	100	100	
Thr	CGU	ACG	29	0	0.1
	UGU	ACA	8	1	0.7
	AGU	ACU-ACC	63	99	26.0
		All	100	100	
Val	CAC	GUG	67	4	0.2
	AAC	GUU-GUC	30	49	4.5
	UAC	GUA	3	46	80.0
		All	100	100	

The codon frequency is calculated per amino acid.
From Vinogradova *et al.* (2009).

chloroplastic genomes (Duchene *et al.*, 2005; Small and Wendel, 1999). These 45 proteins are expected to catalyse the formation of the 20 sets of aminoacylated tRNAs in the three cellular compartments. One, two or three aaRSs are found for each amino acid, but the situation with one cytosolic and one mitochondrial chloroplastic enzyme is the most frequent (Table 12.7).

Half of the 45 *A. thaliana* aaRSs subunits are localized in a unique compartment, and half of them are in at least two compartments. Twenty-one aaRS proteins are only cytosolic, two are only chloroplastic and none is only mitochondrial. Four are cytosolic and mitochondrial and 17 are mitochondrial and chloroplastic (Duchene *et al.*, 2005). The three subunits of the amidotransferase (AdT), which catalyses the formation of Gln-$tRNA^{Gln}$ in organelles, are also mitochondrial and chloroplastic (Pujol *et al.*, 2008). AlaRS is found in the cytosol, the mitochondria and the chloroplasts.

6.4.4. tRNA/aaRS coevolution

It is generally admitted that a strong coevolution must exist between aaRS and their cognate tRNAs. In plants, cytosolic tRNAs are in most cases aminoacylated by typical cytosolic aaRSs. In photosynthetic plants, all chloroplastic tRNAs are encoded by the plastidial genome. These chloroplastic tRNAs are highly similar to bacterial tRNAs and are mostly aminoacylated by bacterial-like aaRSs. So a good correlation is observed between the origin of aaRSs and of tRNAs in plant cytosol and chloroplasts.

The tRNA/aaRS coevolution appears far less evident in plant mitochondria. Some tRNAs are encoded by the mitochondrial genome and are highly similar to bacterial tRNAs. Most of these tRNAs are aminoacylated by bacterial-like aaRSs. The other tRNAs are encoded by the nucleus. In *A. thaliana*, cytosolic tRNAs corresponding to nine amino acids are expected to be imported into mitochondria (Table 12.8). All their cognate aaRSs were shown to be dual targeted to another compartment, either the cytosol or the chloroplasts (Table 12.8). The cytosolic mitochondrial aaRSs are cytosolic enzymes that have acquired an MTS during evolution. The mitochondrial chloroplastic enzymes present no similarities with mitochondrial aaRS in other organisms (except for PheRS) but have strong similarities with bacterial aaRSs. In contrast to the reduction in aaRS gene number and sharing of enzymes, four examples of two aaRSs with the same specificity are found in *A. thaliana* mitochondria (for Ala, Gly, Thr, Val). One enzyme is cytosolic and mitochondrial and the other is chloroplastic and mitochondrial. The role of these enzymes in mitochondria is an open

Table 12.7 Localization of aaRSs in *A. thaliana* cells

Amino acid	1st aaRS	2nd aaRS	3rd aaRS
Glu, His, Ile, Lys, Met, Phe, Pro, Ser, Trp	Cytosol	Mitochondrion-chloroplast	
Gly, Val	Cytosol-mitochondrion	Mitochondrion-chloroplast	
Ala	Cytosol-mitochondrion-chloroplast	Mitochondrion-chloroplast	
Leu	Cytosol-mitochondrion	Chloroplast	
Arg	Cytosol	Cytosol-chloroplast-(mitochondrion?)	
Thr	Cytosol	Cytosol-mitochondrion	Mitochondrion-chloroplast
Asp, Cys, Asn, Tyr	Cytosol	Cytosol	Mitochondrion-chloroplast
Gln	Cytosol	AdT mitochondrion-chloroplast	

Table 12.8 Imported tRNAs and their aaRS in *A. thaliana* Mitochondria

Imported tRNAs	Cytosol-mitochondrion aaRS	Mitochondrion-chloroplast aaRS
$tRNA^{Ala}$	Cytosolic-like Also chloroplast	Bacterial-like
$tRNA^{Gly}$(UCC and CCC) + mitochondrion-encoded $tRNA^{Gly}$(GCC)	Cytosolic-like Inactive in mitochondrion	Bacterial-like Active in mitochondrion
$tRNA^{Val}$	Cytosolic-like Inactive in mitochondrion	Bacterial-like
$tRNA^{Thr}$	Cytosolic-like	Bacterial-like
$tRNA^{Arg}$	Cytosolic-like, cytosol-chloroplast Indirect proof for mitochondrion	
$tRNA^{Leu}$	Cytosolic-like	
$tRNA^{Ile}$(IAU) + mitochondrion-encoded $tRNA^{Ile}$(LAU)		Bacterial-like
$tRNA^{Phe}$		Mitochondrial-like
$tRNA^{Trp}$		Bacterial-like

question. Specialization is suggested in the case of GlyRSs, because the cytosolic mitochondrial GlyRS has no apparent aminoacylation activity within the organelle, this activity being assured by the chloroplastic mitochondrial GlyRS (Duchene *et al.*, 2001). Two imported tRNAs coexist with native tRNA isoacceptors ($tRNAs^{Gly}$ and $tRNAs^{Ile}$), and all are aminoacylated by a mitochondrial chloroplastic bacterial-type aaRS (Table 12.8). Therefore, the tRNA/aaRS coevolution is not evident for imported tRNAs into plant mitochondria.

6.5. Conclusion

tRNA import into mitochondria is a widespread phenomenon in plants and in numerous other organisms. In some protozoans, it represents the only source of tRNAs in mitochondria. In plants, some cytosolic tRNAs are found shared between the cytosol and mitochondria of all species studied; others are imported into mitochondria of only a few species. How the specificity of import is achieved is a major question to be solved in plants. The other main question is the identification of translocation pathways(s) through the mitochondrial membranes. Information is already available for translocation through the OM but nothing is known at the level of the IM.

7. DNA UPTAKE INTO MITOCHONDRIA

DNA is another macromolecule involved in mitochondrial transport. Plant mitochondria are able to take up double-strand linear DNA (Koulintchenko, Konstantinov, & Dietrich, 2003). The mechanism appears to be an active process that requires the IM electrochemical potential. It involves the VDAC and the ANT (Koulintchenko *et al.*, 2003). DNA import appears to be sequence dependent for large fragments (more than 7 kb) (Ibrahim *et al.*, 2011). The incorporated DNA can be transcribed and processed in the organelles and serves as a template for repair mechanisms (Boesch *et al.*, 2009; Koulintchenko *et al.*, 2003; Placido, Gagliardi, Gallerani, Grienenberger, & Marechal-Drouard, 2005). *In organello* homologous recombination of the imported DNA with the resident DNA has been demonstrated (Mileshina, Koulintchenko, Konstantinov, & Dietrich, 2011). DNA uptake has also been shown with yeast and mammalian mitochondria (Koulintchenko, Temperley, Mason, Dietrich, & Lightowlers, 2006; Weber-Lotfi *et al.*, 2009).

REFERENCES

Abe, Y., Shodai, T., Muto, T., Mihara, K., Torii, H., Nishikawa, S., Endo, T., & Kohda, D. (2000). Structural basis of presequence recognition by the mitochondrial protein import receptor Tom20. *Cell, 100*, 551–560.

Akashi, K., Grandjean, O., & Small, I. (1998). Potential dual targeting of an *Arabidopsis* archaebacterial-like histidyl-tRNA synthetase to mitochondria and chloroplasts. *FEBS Letters, 431*, 39–44.

Akashi, K., Sakurai, K., Hirayama, J., Fukuzama, H., & Ohyama, K. (1996). Occurrence of nuclear-encoded $tRNA^{Ile}$ in mitochondria of the liverwort *Marchantia polymorpha*. *Current Genetics, 30*, 181–185.

Alverson, A. J., Rice, D. W., Dickinson, S., Barry, K., & Palmer, J. D. (2011). Origins and recombination of the bacterial-sized multichromosomal mitochondrial genome of cucumber. *The Plant Cell, 23*, 2499–2513.

Armstrong, L. C., Komiya, T., Bergman, B. E., Mihara, K., & Bornstein, P. (1997). Metaxin is a component of a preprotein import complex in the outer membrane of the mammalian mitochondrion. *Journal of Biological Chemistry, 272*, 6510–6518.

Aung, K., Zhang, X., & Hu, J. (2010). Peroxisome division and proliferation in plants. *Biochemical Society Transactions, 38*, 817–822.

Babiychuk, E., Muller, F., Eubel, H., Braun, H. P., Frentzen, M., & Kushnir, S. (2003). *Arabidopsis* phosphatidylglycerophosphate synthase 1 is essential for chloroplast differentiation, but is dispensable for mitochondrial function. *The Plant Journal, 33*, 899–909.

Bahaji, A., Ovecka, M., Barany, I., Risueno, M. C., Munoz, F. J., Baroja-Fernandez, E., et al. (2011). Dual targeting to mitochondria and plastids of AtBT1 and ZmBT1, two members of the mitochondrial carrier family. *Plant and Cell Physiology, 52*, 597–609.

Balsera, M., Soll, J., & Bolter, B. (2009). Protein import machineries in endosymbiotic organelles. *Cellular and Molecular Life Sciences, 66*, 1903–1923.

Beardslee, T. A., Roy-Chowdhury, S., Jaiswal, P., Buhot, L., Lerbs-Mache, S., Stern, D. B., & Allison, L. A. (2002). A nucleus-encoded maize protein with sigma factor activity accumulates in mitochondria and chloroplasts. *The Plant Journal, 31*, 199–209.

Becker, T., Gebert, M., Pfanner, N., & Van Der Laan, M. (2009). Biogenesis of mitochondrial membrane proteins. *Current Opinion in Cell Biology, 21*, 484–493.

Becker, T., Pfannschmidt, S., Guiard, B., Stojanovski, D., Milenkovic, D., Kutik, S., et al. (2008). Biogenesis of the mitochondrial TOM complex: Mim1 promotes insertion and assembly of signal-anchored receptors. *Journal of Biological Chemistry, 283*, 120–127.

Bedard, J., & Jarvis, P. (2005). Recognition and envelope translocation of chloroplast preproteins. *Journal of Experimental Botany, 56*, 2287–2320.

Beddoe, T., & Lithgow, T. (2002). Delivery of nascent polypeptides to the mitochondrial surface. *Biochimica et Biophysica Acta, 1592*, 35–39.

Berglund, A. K., Pujol, C., Duchene, A. M., & Glaser, E. (2009a). Defining the determinants for dual targeting of amino acyl-tRNA synthetases to mitochondria and chloroplasts. *Journal of Molecular Biology, 393*, 803–814.

Berglund, A. K., Spanning, E., Biverstahl, H., Maddalo, G., Tellgren-Roth, C., Maler, L., & Glaser, E. (2009b). Dual targeting to mitochondria and chloroplasts: characterization of Thr-tRNA synthetase targeting peptide. *Molecular Plant, 2*, 1298–1309.

Bhushan, S., Kuhn, C., Berglund, A. K., Roth, C., & Glaser, E. (2006). The role of the N-terminal domain of chloroplast targeting peptides in organellar protein import and miss-sorting. *FEBS Letters, 580*, 3966–3972.

Bhushan, S., Lefebvre, B., Stahl, A., Wright, S. J., Bruce, B. D., Boutry, M., & Glaser, E. (2003). Dual targeting and function of a protease in mitochondria and chloroplasts. *EMBO Reports, 4*, 1073–1078.

Bhushan, S., Stahl, A., Nilsson, S., Lefebvre, B., Seki, M., Roth, C., et al. (2005). Catalysis, subcellular localization, expression and evolution of the targeting peptides degrading protease, AtPreP2. *Plant and Cell Physiology, 46*, 985–996.

Boesch, P., Ibrahim, N., Paulus, F., Cosset, A., Tarasenko, V., & Dietrich, A. (2009). Plant mitochondria possess a short-patch base excision DNA repair pathway. *Nucleic Acids Research, 37*, 5690–5700.

Bolender, N., Sickmann, A., Wagner, R., Meisinger, C., & Pfanner, N. (2008). Multiple pathways for sorting mitochondrial precursor proteins. *EMBO Reports, 9*, 42–49.

Borgese, N., & Fasana, E. (2011). Targeting pathways of C-tail-anchored proteins. *Biochimica et Biophysica Acta, 1808*, 937–946.

Bouzaidi-Tiali, N., Aeby, E., Charriere, F., Pusnik, M., & Schneider, A. (2007). Elongation factor 1a mediates the specificity of mitochondrial tRNA import in T. brucei. *EMBO Journal, 26*, 4302–4312.

Brandner, K., Rehling, P., & Truscott, K. N. (2005). The carboxyl-terminal third of the dicarboxylate carrier is crucial for productive association with the inner membrane twin-pore translocase. *Journal of Biological Chemistry, 280*, 6215–6221.

Braun, H. P., Emmermann, M., Kruft, V., & Schmitz, U. K. (1992). Cytochrome c1 from potato: a protein with a presequence for targeting to the mitochondrial intermembrane space. *Molecular and General Genetics, 231*, 217–225.

Brubacher-Kauffmann, S., Marechal-Drouard, L., Cosset, A., Dietrich, A., & Duchene, A. M. (1999). Differential import of nuclear-encoded tRNAGly isoacceptors into *Solanum tuberosum* mitochondria. *Nucleic Acids Research, 27*, 2037–2042.

Buettner, R., Papoutsoglou, G., Scemes, E., Spray, D. C., & Dermietzel, R. (2000). Evidence for secretory pathway localization of a voltage-dependent anion channel isoform. *Proceedings of the National Academy of Sciences of the United States of America, 97*, 3201–3206.

Carrie, C., Giraud, E., Duncan, O., Xu, L., Wang, Y., Huang, S., et al. (2010a). Conserved and novel functions for *Arabidopsis thaliana* MIA40 in assembly of proteins in mitochondria and peroxisomes. *Journal of Biological Chemistry, 285*, 36138–36148.

Carrie, C., Giraud, E., & Whelan, J. (2009a). Protein transport in organelles: dual targeting of proteins to mitochondria and chloroplasts. *FEBS Journal, 276*, 1187–1195.

Carrie, C., Kuhn, K., Murcha, M. W., Duncan, O., Small, I. D., O'Toole, N., et al. (2009b). Approaches to defining dual-targeted proteins in *Arabidopsis*. *The Plant Journal, 57*, 1128–1139.

Carrie, C., Murcha, M. W., Kuehn, K., Duncan, O., Barthet, M., Smith, P. M., et al. (2008). Type II NAD(P)H dehydrogenases are targeted to mitochondria and chloroplasts or peroxisomes in *Arabidopsis thaliana*. *FEBS Letters, 582*, 3073–3079.

Carrie, C., Murcha, M. W., & Whelan, J. (2010b). An in silico analysis of the mitochondrial protein import apparatus of plants. *BMC Plant Biology, 10*, 249.

de Castro Silva Filho, M., Chaumont, F., Leterme, S., & Boutry, M. (1996). Mitochondrial and chloroplast targeting sequences in tandem modify protein import specificity in plant organelles. *Plant Molecular Biology, 30*, 769–780.

Chabregas, S. M., Luche, D. D., Farias, L. P., Ribeiro, A. F., Van Sluys, M. A., Menck, C. F., et al. (2001). Dual targeting properties of the N-terminal signal sequence of *Arabidopsis thaliana* THI1 protein to mitochondria and chloroplasts. *Plant Molecular Biology, 46*, 639–650.

Chacinska, A., Koehler, C. M., Milenkovic, D., Lithgow, T., & Pfanner, N. (2009). Importing mitochondrial proteins: machineries and mechanisms. *Cell, 138*, 628–644.

Chacinska, A., Lind, M., Frazier, A. E., Dudek, J., Meisinger, C., Geissler, A., et al. (2005). Mitochondrial presequence translocase: switching between TOM tethering and motor recruitment involves Tim21 and Tim17. *Cell, 120*, 817–829.

Chan, N. C., Likic, V. A., Waller, R. F., Mulhern, T. D., & Lithgow, T. (2006). The C-terminal TPR domain of Tom70 defines a family of mitochondrial protein import receptors found only in animals and fungi. *Journal of Molecular Biology, 358*, 1010–1022.

Chan, N. C., & Lithgow, T. (2008). The peripheral membrane subunits of the SAM complex function codependently in mitochondrial outer membrane biogenesis. *Molecular Biology of the Cell, 19*, 126–136.

Chen, H. C., Viry-Moussaid, M., Dietrich, A., & Wintz, H. (1997). Evolution of a mitochondrial tRNA PHE gene in *A. thaliana*: import of cytosolic tRNA PHE into mitochondria. *Biochemical and Biophysical Research Communications, 237*, 432–437.

Chew, O., Lister, R., Qbadou, S., Heazlewood, J. L., Soll, J., Schleiff, E., Millar, A. H., & Whelan, J. (2004). A plant outer mitochondrial membrane protein with high amino acid sequence identity to a chloroplast protein import receptor. *FEBS Letters, 557*, 109–114.

Chew, O., Rudhe, C., Glaser, E., & Whelan, J. (2003a). Characterization of the targeting signal of dual-targeted pea glutathione reductase. *Plant Molecular Biology, 53*, 341–356.

Chew, O., Whelan, J., & Millar, A. H. (2003b). Molecular definition of the ascorbate-glutathione cycle in *Arabidopsis* mitochondria reveals dual targeting of antioxidant defenses in plants. *Journal of Biological Chemistry, 278*, 46869–46877.

Christensen, A. C., Lyznik, A., Mohammed, S., Elowsky, C. G., Elo, A., Yule, R., et al. (2005a). Dual-domain, dual-targeting organellar protein presequences in *Arabidopsis* can use non-AUG start codons. *Plant Cell, 17*, 2805–2816.

Christensen, A. C., Lyznik, A., Mohammed, S., Elowsky, C. G., Elo, A., et al. (2005b). Dual-domain, dual-targeting organellar protein presequences in *Arabidopsis* can use non-AUG start codons. *The Plant Cell, 17*, 2805–2816.

Corral-Debrinski, M., Blugeon, C., & Jacq, C. (2000). In yeast, the 3′ untranslated region or the presequence of ATM1 is required for the exclusive localization of its mRNA to the vicinity of mitochondria. *Molecular and Cellular Biology, 20*, 7881–7892.

Creissen, G., Reynolds, H., Xue, Y., & Mullineaux, P. (1995). Simultaneous targeting of pea glutathione reductase and of a bacterial fusion protein to chloroplasts and mitochondria in transgenic tobacco. *The Plant Journal, 8*, 167–175.

Cunillera, N., Boronat, A., & Ferrer, A. (1997). The *Arabidopsis thaliana* FPS1 gene generates a novel mRNA that encodes a mitochondrial farnesyl-diphosphate synthase isoform. *Journal of Biological Chemistry, 272*, 15381–15388.

Delage, L., Dietrich, A., Cosset, A., & Marechal-Drouard, L. (2003a). *In vitro* import of a nuclearly encoded tRNA into mitochondria of *Solanum tuberosum*. *Molecular and Cellular Biology, 23*, 4000–4012.

Delage, L., Duchêne, A. M., Zaepfel, M., & Maréchal-Drouard, L. (2003b). The anticodon and the D-domain sequences are essential determinants for plant cytosolic tRNA(Val) import into mitochondria. *The Plant Journal, 34*, 623–633.

Devaux, F., Lelandais, G., Garcia, M., Goussard, S., & Jacq, C. (2010). Posttranscriptional control of mitochondrial biogenesis: spatio-temporal regulation of the protein import process. *FEBS Letters, 584*, 4273–4279.

Dietrich, A., Maréchal-Drouard, L., Carneiro, V., Cosset, A., & Small, I. (1996). A single base change prevents import of cytosolic tRNA-Ala into mitochondia in transgenic plants. *The Plant Journal, 1996*, 913–918.

Duchene, A. M., El Farouk, S., Sieber, F., & Maréchal-Drouard, L. (2011). Import of RNAs into plant mitochondria. In F. Kempken (Ed.), *Plant Mitochondria. Advances in Plant Biology*, Vol. 1 (pp. 241–260). Springer.

Duchene, A. M., Giritch, A., Hoffmann, B., Cognat, V., Lancelin, D., Peeters, N. M., Zaepfel, M., Marechal-Drouard, L., & Small, I. D. (2005). Dual targeting is the rule for organellar aminoacyl-tRNA synthetases in *Arabidopsis thaliana*. *Proceedings of the National Academy of Sciences of the United States of America, 102*, 16484–16489.

Duchêne, A. M., & Maréchal-Drouard, L. (2001). The chloroplast-derived *trnW* and *trnM-e* genes are not expressed in *Arabidopsis* mitochondria. *Biochemical and Biophysical Research Communications, 285*, 1213–1216.

Duchene, A. M., Peeters, N., Dietrich, A., Cosset, A., Small, I. D., & Wintz, H. (2001). Overlapping destinations for two dual targeted glycyl-tRNA synthetases in *Arabidopsis thaliana* and *Phaseolus vulgaris*. *Journal of Biological Chemistry, 276*, 15275–15283.

Dukanovic, J., & Rapaport, D. (2011). Multiple pathways in the integration of proteins into the mitochondrial outer membrane. *Biochimica et Biophysica Acta, 1808*, 971–980.

Eliseeva, I. A., Kim, E. R., Guryanov, S. G., Ovchinnikov, L. P., & Lyabin, D. N. (2011). Y-box-binding protein 1 (YB-1) and its functions. *Biochemistry (Moscow), 76*, 1402–1433.

Eliyahu, E., Lesnik, C., & Arava, Y. (2012). The protein chaperone Ssa1 affects mRNA localization to the mitochondria. *FEBS Letters, 586*, 64–69.

Eliyahu, E., Pnueli, L., Melamed, D., Scherrer, T., Gerber, A. P., Pines, O., et al. (2010). Tom20 mediates localization of mRNAs to mitochondria in a translation-dependent manner. *Molecular and Cellular Biology, 30*, 284–294.

Endo, T., Mitsui, S., Nakai, M., & Roise, D. (1996). Binding of mitochondrial presequences to yeast cytosolic heat shock protein 70 depends on the amphiphilicity of the presequence. *Journal of Biological Chemistry, 271*, 4161–4167.

Entelis, N., Brandina, I., Kamenski, P., Krasheninnikov, I. A., Martin, R. P., & Tarassov, I. (2006). A glycolytic enzyme, enolase, is recruited as a cofactor of tRNA targeting toward mitochondria in *Saccharomyces cerevisiae*. *Genes & Development, 20*, 1609–1620.

Eriksson, A. C., Sjoling, S., & Glaser, E. (1996). Characterization of the bifunctional mitochondrial processing peptidase (MPP)/bc1 complex in *Spinacia oleracea*. *Journal of Bioenergetics and Biomembranes, 28*, 285–292.

Fan, A. C., & Young, J. C. (2011). Function of cytosolic chaperones in Tom70-mediated mitochondrial import. *Protein and Peptide Letters, 18*, 122–131.

Fontanesi, F., Soto, I. C., Horn, D., & Barrientos, A. (2006). Assembly of mitochondrial cytochrome *c*-oxidase, a complicated and highly regulated cellular process. *American Journal of Physiology - Cell Physiology, 291*, C1129–C1147.

Francischini, C. W., & Quaggio, R. B. (2009). Molecular characterization of *Arabidopsis thaliana* PUF proteins–binding specificity and target candidates. *FEBS Journal, 276*, 5456–5470.

Fujiki, M., & Verner, K. (1993). Coupling of cytosolic protein synthesis and mitochondrial protein import in yeast. Evidence for cotranslational import in vivo. *Journal of Biological Chemistry, 268*, 1914–1920.

Gadir, N., Haim-Vilmovsky, L., Kraut-Cohen, J., & Gerst, J. E. (2011). Localization of mRNAs coding for mitochondrial proteins in the yeast *Saccharomyces cerevisiae*. *RNA, 17*, 1551–1565.

Gakh, O., Cavadini, P., & Isaya, G. (2002). Mitochondrial processing peptidases. *Biochimica et Biophysica Acta, 1592*, 63–77.

Garcia, M., Darzacq, X., Delaveau, T., Jourdren, L., Singer, R. H., & Jacq, C. (2007). Mitochondria-associated yeast mRNAs and the biogenesis of molecular complexes. *Molecular Biology of the Cell, 18*, 362–368.

Garcia, M., Delaveau, T., Goussard, S., & Jacq, C. (2010). Mitochondrial presequence and open reading frame mediate asymmetric localization of messenger RNA. *EMBO Reports, 11*, 285–291.

Garcia-Rodriguez, L. J., Gay, A. C., & Pon, L. A. (2007). Puf3p, a Pumilio family RNA binding protein, localizes to mitochondria and regulates mitochondrial biogenesis and motility in budding yeast. *Journal of Cell Biology, 176*, 197–207.

Gelly, J. C., Orgeur, M., Jacq, C., & Lelandais, G. (2011). MitoGenesisDB: an expression data mining tool to explore spatio-temporal dynamics of mitochondrial biogenesis. *Nucleic Acids Research, 39*, D1079–D1084.

Gerber, A. P., Herschlag, D., & Brown, P. O. (2004). Extensive association of functionally and cytotopically related mRNAs with Puf family RNA-binding proteins in yeast. *PLoS Biology, 2*. E79.

Giege, P., Heazlewood, J. L., Roessner-Tunali, U., Millar, A. H., Fernie, A. R., Leaver, C. J., et al. (2003). Enzymes of glycolysis are functionally associated with the mitochondrion in *Arabidopsis* cells. *The Plant Cell, 15*, 2140–2151.

Giglione, C., Serero, A., Pierre, M., Boisson, B., & Meinnel, T. (2000). Identification of eukaryotic peptide deformylases reveals universality of N-terminal protein processing mechanisms. *EMBO Journal, 19*, 5916–5929.

Ginsberg, M. D., Feliciello, A., Jones, J. K., Avvedimento, E. V., & Gottesman, M. E. (2003). PKA-dependent binding of mRNA to the mitochondrial AKAP121 protein. *Journal of Molecular Biology, 327*, 885–897.

Glaser, E., Eriksson, A., & Sjoling, S. (1994). Bifunctional role of the bc1 complex in plants. Mitochondrial bc1 complex catalyses both electron transport and protein processing. *FEBS Letters, 346*, 83–87.

Glaser, E., Sjoling, S., Tanudji, M., & Whelan, J. (1998). Mitochondrial protein import in plants. Signals, sorting, targeting, processing and regulation. *Plant Molecular Biology, 38*, 311–338.

Glover, K. E., Spencer, D. F., & Gray, M. W. (2001). Identification and structural characterization of nucleus-encoded transfer RNAs imported into wheat mitochondria. *Journal of Biological Chemistry, 276*, 639–648.

Gobert, A., Gutmann, B., Taschner, A., Gossringer, M., Holzmann, J., Hartmann, R. K., et al. (2010). A single *Arabidopsis* organellar protein has RNase P activity. *Nature Structural & Molecular Biology, 17*, 740–744.

Goggin, D. E., Lipscombe, R., Fedorova, E., Millar, A. H., Mann, A., Atkins, C. A., et al. (2003). Dual intracellular localization and targeting of aminoimidazole ribonucleotide synthetase in cowpea. *Plant Physiology, 131*, 1033–1041.

Gould, S. B., Waller, R. F., & McFadden, G. I. (2008). Plastid evolution. *Annual Review of Plant Biology, 59*, 491–517.

Graham, J. W., Williams, T. C., Morgan, M., Fernie, A. R., Ratcliffe, R. G., & Sweetlove, L. J. (2007). Glycolytic enzymes associate dynamically with mitochondria in response to respiratory demand and support substrate channeling. *The Plant Cell, 19*, 3723–3738.

Hammani, K., Gobert, A., Hleibieh, K., Choulier, L., Small, I., & Giege, P. (2011). An *Arabidopsis* dual-localized pentatricopeptide repeat protein interacts with nuclear proteins involved in gene expression regulation. *The Plant Cell, 23*, 730–740.

Hedtke, B., Borner, T., & Weihe, A. (2000). One RNA polymerase serving two genomes. *EMBO Reports, 1*, 435–440.

Henderson, M. P., Hwang, Y. T., Dyer, J. M., Mullen, R. T., & Andrews, D. W. (2007). The C-terminus of cytochrome b5 confers endoplasmic reticulum specificity by preventing spontaneous insertion into membranes. *Biochemical Journal, 401*, 701–709.

Horie, C., Suzuki, H., Sakaguchi, M., & Mihara, K. (2002). Characterization of signal that directs C-tail-anchored proteins to mammalian mitochondrial outer membrane. *Molecular Biology of the Cell, 13*, 1615–1625.

Horie, C., Suzuki, H., Sakaguchi, M., & Mihara, K. (2003). Targeting and assembly of mitochondrial tail-anchored protein Tom5 to the TOM complex depend on a signal distinct from that of tail-anchored proteins dispersed in the membrane. *Journal of Biological Chemistry, 278*, 41462–41471.

Huang, S., Taylor, N. L., Whelan, J., & Millar, A. H. (2009). Refining the definition of plant mitochondrial presequences through analysis of sorting signals, N-terminal modifications, and cleavage motifs. *Plant Physiology, 150*, 1272–1285.

Hwang, Y. T., Pelitire, S. M., Henderson, M. P., Andrews, D. W., Dyer, J. M., & Mullen, R. T. (2004). Novel targeting signals mediate the sorting of different isoforms of the tail-anchored membrane protein cytochrome b5 to either endoplasmic reticulum or mitochondria. *The Plant Cell, 16*, 3002–3019.

Ibrahim, N., Handa, H., Cosset, A., Koulintchenko, M., Konstantinov, Y., Lightowlers, R. N., et al. (2011). DNA delivery to mitochondria: sequence specificity and energy enhancement. *Pharmaceutical Research, 28*, 2871–2882.

Iglesias-Baena, I., Barranco-Medina, S., Sevilla, F., & Lazaro, J. J. (2011). The dual-targeted plant sulfiredoxin retroreduces the sulfinic form of atypical mitochondrial peroxiredoxin. *Plant Physiology, 155*, 944–955.

Imai, K., Fujita, N., Gromiha, M. M., & Horton, P. (2011). Eukaryote-wide sequence analysis of mitochondrial beta-barrel outer membrane proteins. *BMC Genomics, 12*, 79.

Jackson, J. S., Jr., Houshmandi, S. S., Lopez Leban, F., & Olivas, W. M. (2004). Recruitment of the Puf3protein to its mRNA target for regulation of mRNA decay in yeast. *RNA, 10*, 1625–1636.

Karlberg, E. O., & Andersson, S. G. (2003). Mitochondrial gene history and mRNA localization: is there a correlation? *Nature Reviews Genetics, 4*, 391–397.

Karlberg, O., Canback, B., Kurland, C. G., & Andersson, S. G. (2000). The dual origin of the yeast mitochondrial proteome. *Yeast, 17*, 170–187.

Karniely, S., & Pines, O. (2005). Single translation–dual destination: mechanisms of dual protein targeting in eukaryotes. *EMBO Reports, 6*, 420–425.

Kaufmann, T., Schlipf, S., Sanz, J., Neubert, K., Stein, R., & Borner, C. (2003). Characterization of the signal that directs Bcl-x(L), but not Bcl-2, to the mitochondrial outer membrane. *Journal of Cell Biology, 160*, 53–64.

Kellems, R. E., Allison, V. F., & Butow, R. A. (1974). Cytoplasmic type 80 S ribosomes associated with yeast mitochondria. II. Evidence for the association of cytoplasmic ribosomes with the outer mitochondrial membrane in situ. *Journal of Biological Chemistry, 249*, 3297–3303.

Kellems, R. E., Allison, V. F., & Butow, R. A. (1975). Cytoplasmic type 80S ribosomes associated with yeast mitochondria. IV. Attachment of ribosomes to the outer membrane of isolated mitochondria. *Journal of Cell Biology, 65*, 1–14.

Kellems, R. E., & Butow, R. A. (1972). Cytoplasmic-type 80 S ribosomes associated with yeast mitochondria. I. Evidence for ribosome binding sites on yeast mitochondria. *Journal of Biological Chemistry, 247*, 8043–8050.

Kellems, R. E., & Butow, R. A. (1974). Cytoplasmic type 80 S ribosomes associated with yeast mitochondria. 3. Changes in the amount of bound ribosomes in response to changes in metabolic state. *Journal of Biological Chemistry, 249*, 3304–3310.

Kimura, T., Takeda, S., Kyozuka, J., Asahi, T., Shimamoto, K., & Nakamura, K. (1993). The presequence of a precursor to the delta-subunit of sweet potato mitochondrial F1ATPase is not sufficient for the transport of beta-glucuronidase (GUS) into mitochondria of tobacco, rice and yeast cells. *Plant and Cell Physiology, 34*, 345–355.

Kitazaki, K., Kubo, T., Kagami, H., Matsumoto, T., Fujita, A., Matsuhira, H., et al. (2011). A horizontally transferred tRNA(Cys) gene in the sugar beet mitochondrial genome: evidence that the gene is present in diverse angiosperms and its transcript is aminoacylated. *The Plant Journal, 68*, 262–272.

Kleine, T., Lockhart, P., & Batschauer, A. (2003). An *Arabidopsis* protein closely related to *Synechocystis* cryptochrome is targeted to organelles. *The Plant Journal, 35*, 93–103.

Kobayashi, Y., Dokiya, Y., & Sugita, M. (2001). Dual targeting of phage-type RNA polymerase to both mitochondria and plastids is due to alternative translation initiation in single transcripts. *Biochemical and Biophysical Research Communications, 289*, 1106–1113.

Komiya, T., Rospert, S., Koehler, C., Looser, R., Schatz, G., & Mihara, K. (1998). Interaction of mitochondrial targeting signals with acidic receptor domains along the protein import pathway: evidence for the 'acid chain' hypothesis. *EMBO Journal, 17*, 3886–3898.

Koulintchenko, M., Konstantinov, Y., & Dietrich, A. (2003). Plant mitochondria actively import DNA via the permeability transition pore complex. *EMBO Journal, 22*, 1245–1254.

Koulintchenko, M., Temperley, R. J., Mason, P. A., Dietrich, A., & Lightowlers, R. N. (2006). Natural competence of mammalian mitochondria allows the molecular investigation of mitochondrial gene expression. *Human Molecular Genetics, 15*, 143–154.

Kozjak-Pavlovic, V., Ross, K., Benlasfer, N., Kimmig, S., Karlas, A., & Rudel, T. (2007). Conserved roles of Sam50 and metaxins in VDAC biogenesis. *EMBO Reports, 8*, 576–582.

Kriechbaumer, V., von Loffelholz, O., & Abell, B. M. (2011). Chaperone receptors: guiding proteins to intracellular compartments. *Protoplasma, 249*, 21–30.

Krimmer, T., Rapaport, D., Ryan, M. T., Meisinger, C., Kassenbrock, C. K., Blachly-Dyson, E., et al. (2001). Biogenesis of porin of the outer mitochondrial membrane involves an import pathway via receptors and the general import pore of the TOM complex. *Journal of Cell Biology, 152*, 289–300.

Kumar, R., Maréchal-Drouard, L., Akama, K., & Small, I. (1996). Striking differences in mitochondrial tRNA import between different plant species. *Molecular and General Genetics, 252*, 404–411.

Kuroda, R., Ikenoue, T., Honsho, M., Tsujimoto, S., Mitoma, J. Y., & Ito, A. (1998). Charged amino acids at the carboxyl-terminal portions determine the intracellular locations of two isoforms of cytochrome b5. *Journal of Biological Chemistry, 273*, 31097–31102.

Kurz, M., Martin, H., Rassow, J., Pfanner, N., & Ryan, M. T. (1999). Biogenesis of Tim proteins of the mitochondrial carrier import pathway: differential targeting mechanisms and crossing over with the main import pathway. *Molecular Biology of the Cell, 10*, 2461–2474.

Kutik, S., Stojanovski, D., Becker, L., Becker, T., Meinecke, M., Kruger, V., et al. (2008). Dissecting membrane insertion of mitochondrial beta-barrel proteins. *Cell, 132*, 1011–1024.

Laforest, M. J., Delage, L., & Marechal-Drouard, L. (2005). The T-domain of cytosolic tRNA(Val), an essential determinant for mitochondrial import. *FEBS Letters, 579*, 1072–1078.

Lee, J., Lee, H., Kim, J., Lee, S., Kim, D. H., Kim, S., & Hwang, I. (2011). Both the hydrophobicity and a positively charged region flanking the C-terminal region of the transmembrane domain of signal-anchored proteins play critical roles in determining their targeting specificity to the endoplasmic reticulum or endosymbiotic organelles in *Arabidopsis* cells. *The Plant Cell, 23*, 1588–1607.

Lee, M. N., & Whelan, J. (2004). Identification of signals required for import of the soybean F(A)d subunit of ATP synthase into mitochondria. *Plant Molecular Biology, 54*, 193–203.

Lelandais, G., Saint-Georges, Y., Geneix, C., Al-Shikhley, L., Dujardin, G., & Jacq, C. (2009). Spatio-temporal dynamics of yeast mitochondrial biogenesis: transcriptional and post-transcriptional mRNA oscillatory modules. *PLoS Computational Biology, 5*. e1000409.

Li, H. M., & Chiu, C. C. (2010). Protein transport into chloroplasts. *Annual Review of Plant Biology, 61*, 157–180.

Lister, R., Carrie, C., Duncan, O., Ho, L. H., Howell, K. A., Murcha, M. W., & Whelan, J. (2007). Functional definition of outer membrane proteins involved in preprotein import into mitochondria. *The Plant Cell, 19*, 3739–3759.

Lister, R., Chew, O., Lee, M. N., Heazlewood, J. L., Clifton, R., Parker, K. L., et al. (2004). A transcriptomic and proteomic characterization of the *Arabidopsis* mitochondrial protein import apparatus and its response to mitochondrial dysfunction. *Plant Physiology, 134*, 777–789.

Lister, R., Murcha, M. W., & Whelan, J. (2003). The Mitochondrial Protein Import Machinery of Plants (MPIMP) database. *Nucleic Acids Research, 31*, 325–327.

Lister, R., & Whelan, J. (2006). Mitochondrial protein import: convergent solutions for receptor structure. *Current Biology, 16*, R197–R199.

Lithgow, T. (2000). Targeting of proteins to mitochondria. *FEBS Letters, 476*, 22–26.

Lithgow, T., Ryan, M., Anderson, R. L., Hoj, P. B., & Hoogenraad, N. J. (1993). A constitutive form of heat-shock protein 70 is located in the outer membranes of mitochondria from rat liver. *FEBS Letters, 332*, 277–281.

Macasev, D., Newbigin, E., Whelan, J., & Lithgow, T. (2000). How do plant mitochondria avoid importing chloroplast proteins? Components of the import apparatus Tom20 and Tom22 from *Arabidopsis* differ from their fungal counterparts. *Plant Physiology, 123*, 811–816.

Macasev, D., Whelan, J., Newbigin, E., Silva-Filho, M. C., Mulhern, T. D., & Lithgow, T. (2004). Tom22′, an 8-kDa trans-site receptor in plants and protozoans, is a conserved feature of the TOM complex that appeared early in the evolution of eukaryotes. *Molecular Biology and Evolution, 21*, 1557–1564.

Mackenzie, S. A. (2005). Plant organellar protein targeting: a traffic plan still under construction. *Trends in Cell Biology, 15*, 548–554.

Marc, P., Margeot, A., Devaux, F., Blugeon, C., Corral-Debrinski, M., & Jacq, C. (2002). Genome-wide analysis of mRNAs targeted to yeast mitochondria. *EMBO Reports, 3*, 159–164.

Maréchal-Drouard, L., Guillemaut, P., Cosset, A., Arbogast, M., Weber, F., Weil, J. H., et al. (1990). Transfer RNAs of potato (*Solanum tuberosum*) mitochondria have different genetic origins. *Nucleic Acids Research, 18*, 3689–3696.

Margeot, A., Blugeon, C., Sylvestre, J., Vialette, S., Jacq, C., & Corral-Debrinski, M. (2002). *Saccharomyces cerevisiae*, ATP2 mRNA sorting to the vicinity of mitochondria is essential for respiratory function. *EMBO Journal, 21*, 6893–6904.

Margeot, A., Garcia, M., Wang, W., Tetaud, E., Di Rago, J. P., & Jacq, C. (2005). Why are many mRNAs translated to the vicinity of mitochondria: a role in protein complex assembly? *Gene, 354*, 64–71.

Marrison, J. L., Schunmann, P., Ougham, H. J., & Leech, R. M. (1996). Subcellular visualization of gene transcripts encoding key proteins of the chlorophyll accumulation process in developing chloroplasts. *Plant Physiology, 110*, 1089–1096.

Marti, M. C., Olmos, E., Calvete, J. J., Diaz, I., Barranco-Medina, S., Whelan, J., Lazaro, J. J., Sevilla, F., & Jimenez, A. (2009). Mitochondrial and nuclear localization of a novel pea thioredoxin: identification of its mitochondrial target proteins. *Plant Physiology, 150*, 646–657.

Martin, T., Sharma, R., Sippel, C., Waegemann, K., Soll, J., & Vothknecht, U. C. (2006). A protein kinase family in *Arabidopsis* phosphorylates chloroplast precursor proteins. *Journal of Biological Chemistry, 281*, 40216–40223.

Matsumoto, S., Uchiumi, T., Saito, T., Yagi, M., Takazaki, S., Kanki, T., & Kang, D. (2012a). Localization of mRNAs encoding human mitochondrial oxidative phosphorylation proteins. *Mitochondrion*.

Matsumoto, S., Uchiumi, T., Tanamachi, H., Saito, T., Yagi, M., Takazaki, S., Kanki, T., & Kang, D. (2012b). Ribonucleoprotein Y-box binding protein-1 regulates mitochondrial oxidative phosphorylation (OXPHOS) protein expression after serum stimulation through binding to OXPHOS mRNA. *Biochemical Journal*.

May, T., & Soll, J. (2000). 14-3-3 proteins form a guidance complex with chloroplast precursor proteins in plants. *The Plant Cell, 12*, 53–64.

Meisinger, C., Rissler, M., Chacinska, A., Szklarz, L. K., Milenkovic, D., Kozjak, V., et al. (2004). The mitochondrial morphology protein Mdm10 functions in assembly of the preprotein translocase of the outer membrane. *Developmental Cell, 7*, 61–71.

Menand, B., Marechal-Drouard, L., Sakamoto, W., Dietrich, A., & Wintz, H. (1998). A single gene of chloroplast origin codes for mitochondrial and chloroplastic methionyl-tRNA synthetase in *Arabidopsis thaliana. Proceedings of the National Academy of Sciences of the United States of America, 95*, 11014–11019.

Michalecka, A. M., Svensson, A. S., Johansson, F. I., Agius, S. C., Johanson, U., Brennicke, A., et al. (2003). *Arabidopsis* genes encoding mitochondrial type II NAD(P) H dehydrogenases have different evolutionary origin and show distinct responses to light. *Plant Physiology, 133*, 642–652.

Michaud, M., Marechal-Drouard, L., & Duchene, A. M. (2010). RNA trafficking in plant cells: targeting of cytosolic mRNAs to the mitochondrial surface. *Plant Molecular Biology, 73*, 697–704.

Mihara, K., & Omura, T. (1996). Cytosolic factors in mitochondrial protein import. *Experientia, 52*, 1063–1068.

Mileshina, D., Koulintchenko, M., Konstantinov, Y., & Dietrich, A. (2011). Transfection of plant mitochondria and in organello gene integration. *Nucleic Acids Research, 39*, e115.

Millar, A. H., Whelan, J., & Small, I. (2006). Recent surprises in protein targeting to mitochondria and plastids. *Current Opinion in Plant Biology, 9*, 610–615.

Mireau, H., Cosset, A., Marechal-Drouard, L., Fox, T. D., Small, I. D., & Dietrich, A. (2000). Expression of *Arabidopsis thaliana* mitochondrial alanyl-tRNA synthetase is not sufficient to trigger mitochondrial import of tRNAAla in yeast. *Journal of Biological Chemistry, 275*, 13291–13296.

Mireau, H., Lancelin, D., & Small, I. D. (1996). The same *Arabidopsis* gene encodes both cytosolic and mitochondrial alanyl-tRNA synthetases. *The Plant Cell, 8*, 1027–1039.

Moberg, P., Nilsson, S., Stahl, A., Eriksson, A. C., Glaser, E., & Maler, L. (2004). NMR solution structure of the mitochondrial F1beta presequence from *Nicotiana plumbaginifolia*. *Journal of Molecular Biology, 336*, 1129–1140.

Moczko, M., Bomer, U., Kubrich, M., Zufall, N., Honlinger, A., & Pfanner, N. (1997). The intermembrane space domain of mitochondrial Tom22 functions as a trans binding site for preproteins with N-terminal targeting sequences. *Molecular and Cellular Biology, 17*, 6574–6584.

Mokranjac, D., & Neupert, W. (2009). Thirty years of protein translocation into mitochondria: unexpectedly complex and still puzzling. *Biochimica et Biophysica Acta, 1793*, 33–41.

Mokranjac, D., Sichting, M., Popov-Celeketic, D., Mapa, K., Gevorkyan-Airapetov, L., Zohary, K., et al. (2009). Role of Tim50 in the transfer of precursor proteins from the outer to the inner membrane of mitochondria. *Molecular Biology of the Cell, 20*, 1400–1407.

Moon, S., Giglione, C., Lee, D. Y., An, S., Jeong, D. H., Meinnel, T., et al. (2008). Rice peptide deformylase PDF1B is crucial for development of chloroplasts. *Plant and Cell Physiology, 49*, 1536–1546.

Mooney, B., & Harmey, M. A. (1996). The occurrence of hsp70 in the outer membrane of plant mitochondria. *Biochemical and Biophysical Research Communications, 218*, 309–313.

Mozo, T., Fischer, K., Flugge, U. I., & Schmitz, U. K. (1995). The N-terminal extension of the ADP/ATP translocator is not involved in targeting to plant mitochondria in vivo. *The Plant Journal, 7*, 1015–1020.

Murcha, M. W., Elhafez, D., Millar, A. H., & Whelan, J. (2004). The N-terminal extension of plant mitochondrial carrier proteins is removed by two-step processing: the first cleavage is by the mitochondrial processing peptidase. *Journal of Molecular Biology, 344*, 443–454.

Murcha, M. W., Elhafez, D., Millar, A. H., & Whelan, J. (2005a). The C-terminal region of TIM17 links the outer and inner mitochondrial membranes in *Arabidopsis* and is essential for protein import. *Journal of Biological Chemistry, 280*, 16476–16483.

Murcha, M. W., Lister, R., Ho, A. Y., & Whelan, J. (2003). Identification, expression, and import of components 17 and 23 of the inner mitochondrial membrane translocase from *Arabidopsis*. *Plant Physiology, 131*, 1737–1747.

Murcha, M. W., Millar, A. H., & Whelan, J. (2005b). The N-terminal cleavable extension of plant carrier proteins is responsible for efficient insertion into the inner mitochondrial membrane. *Journal of Molecular Biology, 351*, 16–25.

Nakamura, T., Yamaguchi, Y., & Sano, H. (2000). Plant mercaptopyruvate sulfurtransferases: molecular cloning, subcellular localization and enzymatic activities. *European Journal of Biochemistry, 267*, 5621–5630.

Neupert, W. (1997). Protein import into mitochondria. *Annual Review of Biochemistry, 66*, 863–917.

Neupert, W., & Herrmann, J. M. (2007). Translocation of proteins into mitochondria. *Annual Review of Biochemistry, 76*, 723–749.

Nicolai, M., Duprat, A., Sormani, R., Rodriguez, C., Roncato, M. A., Rolland, N., et al. (2007). Higher plant chloroplasts import the mRNA coding for the eucaryotic translation initiation factor 4E. *FEBS letters, 581*, 3921–3926.

Obara, K., Sumi, K., & Fukuda, H. (2002). The use of multiple transcription starts causes the dual targeting of *Arabidopsis* putative monodehydroascorbate reductase to both mitochondria and chloroplasts. *Plant and Cell Physiology, 43*, 697–705.

Olivas, W., & Parker, R. (2000). The Puf3protein is a transcript-specific regulator of mRNA degradation in yeast. *EMBO Journal, 19*, 6602–6611.

Ono, Y., Sakai, A., Takechi, K., Takio, S., Takusagawa, M., & Takano, H. (2007). NtPolI-like1 and NtPolI-like2, bacterial DNA polymerase I homologs isolated from BY-2 cultured tobacco cells, encode DNA polymerases engaged in DNA replication in both plastids and mitochondria. *Plant and Cell Physiology, 48*, 1679–1692.

Peeters, N., & Small, I. (2001). Dual targeting to mitochondria and chloroplasts. *Biochimica et Biophysica Acta, 1541*, 54–63.

Peeters, N. M., Chapron, A., Giritch, A., Grandjean, O., Lancelin, D., Lhomme, T., et al. (2000). Duplication and quadruplication of *Arabidopsis thaliana* cysteinyl- and asparaginyl-tRNA synthetase genes of organellar origin. *Journal of Molecular Evolution, 50*, 413–423.

Perry, A. J., Hulett, J. M., Likic, V. A., Lithgow, T., & Gooley, P. R. (2006). Convergent evolution of receptors for protein import into mitochondria. *Current Biology, 16*, 221–229.

Perry, A. J., Rimmer, K. A., Mertens, H. D., Waller, R. F., Mulhern, T. D., Lithgow, T., et al. (2008). Structure, topology and function of the translocase of the outer membrane of mitochondria. *Plant Physiology and Biochemistry, 46*, 265–274.

Pfanner, N., & Geissler, A. (2001). Versatility of the mitochondrial protein import machinery. *Nature Reviews Molecular Cell Biology, 2*, 339–349.

Pfitzinger, H., Guillemaut, P., Weil, J. H., & Pillay, D. T. (1987). Adjustment of the tRNA population to the codon usage in chloroplasts. *Nucleic Acids Research, 15*, 1377–1386.

de Pinto, V., Messina, A., Lane, D. J., & Lawen, A. (2010). Voltage-dependent anion-selective channel (VDAC) in the plasma membrane. *FEBS Letters, 584*, 1793–1799.

Placido, A., Gagliardi, D., Gallerani, R., Grienenberger, J. M., & Marechal-Drouard, L. (2005). Fate of a larch unedited tRNA precursor expressed in potato mitochondria. *Journal of Biological Chemistry, 280*, 33573–33579.

Pujol, C., Bailly, M., Kern, D., Marechal-Drouard, L., Becker, H., & Duchene, A. M. (2008). Dual-targeted tRNA-dependent amidotransferase ensures both mitochondrial and chloroplastic Gln-tRNAGln synthesis in plants. *Proceedings of the National Academy of Sciences of the United States of America, 105*, 6481–6485.

Pujol, C., Marechal-Drouard, L., & Duchene, A. M. (2007). How can organellar protein N-terminal sequences be dual targeting signals? In silico analysis and mutagenesis approach. *Journal of Molecular Biology, 369*, 356–367.

Puyaubert, J., Denis, L., & Alban, C. (2008). Dual targeting of *Arabidopsis* holocarboxylase synthetase1: a small upstream open reading frame regulates translation initiation and protein targeting. *Plant Physiology, 146*, 478–491.

Qbadou, S., Becker, T., Mirus, O., Tews, I., Soll, J., & Schleiff, E. (2006). The molecular chaperone Hsp90 delivers precursor proteins to the chloroplast import receptor Toc64. *EMBO Journal, 25*, 1836–1847.

Quenault, T., Lithgow, T., & Traven, A. (2011). PUF proteins: repression, activation and mRNA localization. *Trends in Cell Biology, 21*, 104–112.

Rapaport, D., & Neupert, W. (1999). Biogenesis of Tom40, core component of the TOM complex of mitochondria. *Journal of Cell Biology, 146*, 321–331.

Richter, U., Kiessling, J., Hedtke, B., Decker, E., Reski, R., Borner, T., & Weihe, A. (2002). Two RpoT genes of *Physcomitrella patens* encode phage-type RNA polymerases with dual targeting to mitochondria and plastids. *Gene, 290*, 95–105.

Riemer, J., Fischer, M., & Herrmann, J. M. (2011). Oxidation-driven protein import into mitochondria: insights and blind spots. *Biochimica et Biophysica Acta, 1808*, 981–989.

Rimmer, K. A., Foo, J. H., Ng, A., Petrie, E. J., Shilling, P. J., Perry, A. J., Mertens, H. D., Lithgow, T., Mulhern, T. D., & Gooley, P. R. (2011). Recognition of mitochondrial targeting sequences by the import receptors Tom20 and Tom22. *Journal of Molecular Biology, 405*, 804–818.

Robert, N., D'erfurth, I., Marmagne, A., Erhardt, M., Allot, M., Boivin, K., Gissot, L., Monachello, D., Michaud, M., Duchene, A. M., Barbier-Brygoo, H., Marechal-Drouard, L., Ephritikhine, G., & Filleur, S. (2012). Voltage-dependent anion channels (VDACs) in *Arabidopsis* have a dual localization in the cell but show a distinct role in mitochondria. *Plant Molecular Biology*.

Rodiger, A., Baudisch, B., Langner, U., & Klosgen, R. B. (2011). Dual targeting of a mitochondrial protein: the case study of cytochrome c1. *Molecular Plant, 4*, 679–687.

Roise, D., Theiler, F., Horvath, S. J., Tomich, J. M., Richards, J. H., Allison, D. S., & Schatz, G. (1988). Amphiphilicity is essential for mitochondrial presequence function. *EMBO Journal, 7*, 649–653.

Rudhe, C., Clifton, R., Chew, O., Zemam, K., Richter, S., Lamppa, G., et al. (2004). Processing of the dual targeted precursor protein of glutathione reductase in mitochondria and chloroplasts. *Journal of Molecular Biology, 343*, 639–647.

Rudhe, C., Clifton, R., Whelan, J., & Glaser, E. (2002). N-terminal domain of the dual-targeted pea glutathione reductase signal peptide controls organellar targeting efficiency. *Journal of Molecular Biology, 324*, 577–585.

Russo, A., Cirulli, C., Amoresano, A., Pucci, P., Pietropaolo, C., & Russo, G. (2008). cis-acting sequences and trans-acting factors in the localization of mRNA for mitochondrial ribosomal proteins. *Biochimica et Biophysica Acta, 1779*, 820–829.

Russo, A., Russo, G., Cuccurese, M., Garbi, C., & Pietropaolo, C. (2006). The 3′-untranslated region directs ribosomal protein-encoding mRNAs to specific cytoplasmic regions. *Biochimica et Biophysica Acta, 1763*, 833–843.

Saint-Georges, Y., Garcia, M., Delaveau, T., Jourdren, L., Le Crom, S., Lemoine, S., et al. (2008). Yeast mitochondrial biogenesis: a role for the PUF RNA-binding protein Puf3p in mRNA localization. *PloS ONE, 3*. e2293.

Salinas, T., Duchene, A. M., Delage, L., Nilsson, S., Glaser, E., Zaepfel, M., & Marechal-Drouard, L. (2006). The voltage-dependent anion channel, a major component of the tRNA import machinery in plant mitochondria. *Proceedings of the National Academy of Sciences of the United States of America, 103*, 18362–18367.

Salinas, T., Schaeffer, C., Marechal-Drouard, L., & Duchene, A. M. (2005). Sequence dependence of tRNA(Gly) import into tobacco mitochondria. *Biochimie, 87*, 863–872.

Schleiff, E., & Becker, T. (2011). Common ground for protein translocation: access control for mitochondria and chloroplasts. *Nature Reviews Molecular Cell Biology, 12*, 48–59.

Schneider, A. (2011). Mitochondrial tRNA import and its consequences for mitochondrial translation. *Annual Review of Biochemistry, 80*, 1033–1053.

Sheffield, W. P., Shore, G. C., & Randall, S. K. (1990). Mitochondrial precursor protein. Effects of 70-kilodalton heat shock protein on polypeptide folding, aggregation, and import competence. *Journal of Biological Chemistry, 265*, 11069–11076.

Sieber, F., Placido, A., El Farouk-Ameqrane, S., Duchene, A. M., & Marechal-Drouard, L. (2011). A protein shuttle system to target RNA into mitochondria. *Nucleic Acids Research*.

Silva-Filho, M. C. (2003). One ticket for multiple destinations: dual targeting of proteins to distinct subcellular locations. *Current Opinion in Plant Biology, 6*, 589–595.

Sjoling, S., & Glaser, E. (1998). Mitochondrial targeting peptides in plants. *Trends in Plant Science, 3*, 136–140.

Sloan, D. B., Alverson, A. J., Storchova, H., Palmer, J. D., & Taylor, D. R. (2010). Extensive loss of translational genes in the structurally dynamic mitochondrial genome of the angiosperm *Silene latifolia. BMC Evolutionary Biology, 10*, 274.

Small, I., Marechal-Drouard, L., Masson, J., Pelletier, G., Cosset, A., Weil, J. H., & Dietrich, A. (1992). In vivo import of a normal or mutagenized heterologous transfer RNA into the mitochondria of transgenic plants: towards novel ways of influencing mitochondrial gene expression? *EMBO Journal, 11*, 1291–1296.

Small, R. L., & Wendel, J. F. (1999). The mitochondrial genome of allotetraploid cotton (*Gossypium* L.). *Journal of Heredity, 90*, 251–253.

Smirnov, A., Comte, C., Mager-Heckel, A. M., Addis, V., Krasheninnikov, I. A., Martin, R. P., et al. (2010). Mitochondrial enzyme rhodanese is essential for 5 S ribosomal RNA import into human mitochondria. *Journal of Biological Chemistry, 285*, 30792–30803.

Souciet, G., Menand, B., Ovesna, J., Cosset, A., Dietrich, A., & Wintz, H. (1999). Characterization of two bifunctional *Arabidopsis thaliana* genes coding for mitochondrial and cytosolic forms of valyl-tRNA synthetase and threonyl-tRNA synthetase by alternative use of two in-frame AUGs. *European Journal of Biochemistry, 266*, 848–854.

Stengel, A., Benz, J. P., Soll, J., & Bolter, B. (2010). Redox-regulation of protein import into chloroplasts and mitochondria: similarities and differences. *Plant Signaling & Behavior, 5*, 105–109.

Sugimoto, H., Kusumi, K., Noguchi, K., Yano, M., Yoshimura, A., & Iba, K. (2007). The rice nuclear gene, VIRESCENT 2, is essential for chloroplast development and encodes a novel type of guanylate kinase targeted to plastids and mitochondria. *The Plant Journal, 52*, 512–527.

Suissa, M., & Schatz, G. (1982). Import of proteins into mitochondria. Translatable mRNAs for imported mitochondrial proteins are present in free as well as mitochondria-bound cytoplasmic polysomes. *Journal of Biological Chemistry, 257*, 13048–13055.

Sunderland, P. A., West, C. E., Waterworth, W. M., & Bray, C. M. (2004). Choice of a start codon in a single transcript determines DNA ligase 1 isoform production and intracellular targeting in *Arabidopsis thaliana. Biochemical Society Transactions, 32*, 614–616.

Sunderland, P. A., West, C. E., Waterworth, W. M., & Bray, C. M. (2006). An evolutionarily conserved translation initiation mechanism regulates nuclear or mitochondrial targeting of DNA ligase 1 in *Arabidopsis thaliana. The Plant Journal, 47*, 356–367.

Sylvestre, J., Margeot, A., Jacq, C., Dujardin, G., & Corral-Debrinski, M. (2003a). The role of the 3′ untranslated region in mRNA sorting to the vicinity of mitochondria is conserved from yeast to human cells. *Molecular Biology of the Cell, 14*, 3848–3856.

Sylvestre, J., Vialette, S., Corral Debrinski, M., & Jacq, C. (2003b). Long mRNAs coding for yeast mitochondrial proteins of prokaryotic origin preferentially localize to the vicinity of mitochondria. *Genome Biology, 4*, R44.

Taira, M., Valtersson, U., Burkhardt, B., & Ludwig, R. A. (2004). *Arabidopsis thaliana* GLN2-encoded glutamine synthetase is dual targeted to leaf mitochondria and chloroplasts. *The Plant Cell, 16*, 2048–2058.

Tam, P. P., Barrette-Ng, I. H., Simon, D. M., Tam, M. W., Ang, A. L., & Muench, D. G. (2010). The Puf family of RNA-binding proteins in plants: phylogeny, structural modeling, activity and subcellular localization. *BMC Plant Biology, 10*, 44.

Tu, B. P., Kudlicki, A., Rowicka, M., & Mcknight, S. L. (2005). Logic of the yeast metabolic cycle: temporal compartmentalization of cellular processes. *Science, 310*, 1152–1158.

Ueda, M., Nishikawa, T., Fujimoto, M., Takanashi, H., Arimura, S., Tsutsumi, N., & Kadowaki, K. (2008). Substitution of the gene for chloroplast RPS16 was assisted by generation of a dual targeting signal. *Molecular Biology and Evolution, 25*, 1566–1575.

Uniacke, J., & Zerges, W. (2009). Chloroplast protein targeting involves localized translation in *Chlamydomonas*. *Proceedings of the National Academy of Sciences of the United States of America, 106*, 1439–1444.

Vasiljev, A., Ahting, U., Nargang, F. E., Go, N. E., Habib, S. J., Kozany, C., et al. (2004). Reconstituted TOM core complex and Tim9/Tim10 complex of mitochondria are sufficient for translocation of the ADP/ATP carrier across membranes. *Molecular Biology of the Cell, 15*, 1445–1458.

Verner, K. (1993). Co-translational protein import into mitochondria: an alternative view. *Trends in Biochemical Sciences, 18*, 366–371.

Vinogradova, E., Salinas, T., Cognat, V., Remacle, C., & Marechal-Drouard, L. (2009). Steady-state levels of imported tRNAs in *Chlamydomonas* mitochondria are correlated with both cytosolic and mitochondrial codon usages. *Nucleic Acids Research, 37*, 1521–1528.

von Braun, S. S., Sabetti, A., Hanic-Joyce, P. J., Gu, J., Schleiff, E., & Joyce, P. B. (2007). Dual targeting of the tRNA nucleotidyltransferase in plants: not just the signal. *Journal of Experimental Botany, 58*, 4083–4093.

von Heijne, G. (1986). Mitochondrial targeting sequences may form amphiphilic helices. *EMBO Journal, 5*, 1335–1342.

von Heijne, G., & Nishikawa, K. (1991). Chloroplast transit peptides. The perfect random coil? *FEBS Letters, 278*, 1–3.

von Heijne, G., Steppuhn, J., & Herrmann, R. G. (1989). Domain structure of mitochondrial and chloroplast targeting peptides. *European Journal of Biochemistry, 180*, 535–545.

Waegemann, K., & Soll, J. (1996). Phosphorylation of the transit sequence of chloroplast precursor proteins. *Journal of Biological Chemistry, 271*, 6545–6554.

Wagner, K., Gebert, N., Guiard, B., Brandner, K., Truscott, K. N., Wiedemann, N., Pfanner, N., & Rehling, P. (2008). The assembly pathway of the mitochondrial carrier translocase involves four preprotein translocases. *Molecular and Cellular Biology, 28*, 4251–4260.

Waizenegger, T., Schmitt, S., Zivkovic, J., Neupert, W., & Rapaport, D. (2005). Mim1, a protein required for the assembly of the TOM complex of mitochondria. *EMBO Reports, 6*, 57–62.

Walther, D. M., & Rapaport, D. (2009). Biogenesis of mitochondrial outer membrane proteins. *Biochimica et Biophysica Acta, 1793*, 42–51.

Wamboldt, Y., Mohammed, S., Elowsky, C., Wittgren, C., de Paula, W. B., & Mackenzie, S. A. (2009). Participation of leaky ribosome scanning in protein dual targeting by alternative translation initiation in higher plants. *The Plant Cell, 21*, 157–167.

Watanabe, N., Che, F. S., Iwano, M., Takayama, S., Yoshida, S., & Isogai, A. (2001). Dual targeting of spinach protoporphyrinogen oxidase II to mitochondria and chloroplasts by alternative use of two in-frame initiation codons. *Journal of Biological Chemistry, 276*, 20474–20481.

Weber, F., Dietrich, A., Weil, J. H., & Marechal-Drouard, L. (1990). A potato mitochondrial isoleucine tRNA is coded for by a mitochondrial gene possessing a methionine anticodon. *Nucleic Acids Research, 18*, 5027–5030.

Weber-Lotfi, F., Ibrahim, N., Boesch, P., Cosset, A., Konstantinov, Y., Lightowlers, R. N., & Dietrich, A. (2009). Developing a genetic approach to investigate the mechanism of mitochondrial competence for DNA import. *Biochimica et Biophysica Acta, 1787*, 320–327.

Winning, B. M., Sarah, C. J., Purdue, P. E., Day, C. D., & Leaver, C. J. (1992). The adenine nucleotide translocator of higher plants is synthesized as a large precursor that is processed upon import into mitochondria. *The Plant Journal, 2*, 763–773.

Xu, X. M., & Moller, S. G. (2006). AtSufE is an essential activator of plastidic and mitochondrial desulfurases in *Arabidopsis*. *EMBO Journal, 25*, 900–909.

Yamano, K., Tanaka-Yamano, S., & Endo, T. (2010). Mdm10 as a dynamic constituent of the TOB/SAM complex directs coordinated assembly of Tom40. *EMBO Reports, 11*, 187–193.

Yogev, O., & Pines, O. (2011). Dual targeting of mitochondrial proteins: mechanism, regulation and function. *Biochimica et Biophysica Acta, 1808*, 1012–1020.

Young, J. C., Hoogenraad, N. J., & Hartl, F. U. (2003). Molecular chaperones Hsp90 and Hsp70 deliver preproteins to the mitochondrial import receptor Tom70. *Cell, 112*, 41–50.

Zahedi, R. P., Sickmann, A., Boehm, A. M., Winkler, C., Zufall, N., Schonfisch, B., et al. (2006). Proteomic analysis of the yeast mitochondrial outer membrane reveals accumulation of a subclass of preproteins. *Molecular Biology of the Cell, 17*, 1436–1450.

Zara, V., Ferramosca, A., Palmisano, I., Palmieri, F., & Rassow, J. (2003). Biogenesis of rat mitochondrial citrate carrier (CIC): the N-terminal presequence facilitates the solubility of the preprotein but does not act as a targeting signal. *Journal of Molecular Biology, 325*, 399–408.

Zara, V., Ferramosca, A., Robitaille-Foucher, P., Palmieri, F., & Young, J. C. (2009). Mitochondrial carrier protein biogenesis: role of the chaperones Hsc70 and Hsp90. *Biochemical Journal, 419*, 369–375.

Zara, V., Palmieri, F., Mahlke, K., & Pfanner, N. (1992). The cleavable presequence is not essential for import and assembly of the phosphate carrier of mammalian mitochondria but enhances the specificity and efficiency of import. *Journal of Biological Chemistry, 267*, 12077–12081.

Zhang, X. P., & Glaser, E. (2002). Interaction of plant mitochondrial and chloroplast signal peptides with the Hsp70 molecular chaperone. *Trends in Plant Science, 7*, 14–21.

Zhang, X., & Hu, J. (2009). Two small protein families, DYNAMIN-RELATED PROTEIN3 and FISSION1, are required for peroxisome fission in *Arabidopsis*. *The Plant Journal, 57*, 146–159.

Zhang, X. C., & Hu, J. P. (2008). FISSION1A and FISSION1B proteins mediate the fission of peroxisomes and mitochondria in *Arabidopsis*. *Molecular Plant, 1*, 1036–1047.

Zhang, X. P., Sjoling, S., Tanudji, M., Somogyi, L., Andreu, D., Eriksson, L. E., et al. (2001). Mutagenesis and computer modelling approach to study determinants for recognition of signal peptides by the mitochondrial processing peptidase. *The Plant Journal, 27*, 427–438.

Zhao, J., Onduka, T., Kinoshita, J. Y., Honsho, M., Kinoshita, T., Shimazaki, K., et al. (2003). Dual subcellular distribution of cytochrome b5 in plant, cauliflower, cells. *Journal of Biochemistry, 133*, 115–121.

Zybailov, B., Rutschow, H., Friso, G., Rudella, A., Emanuelsson, O., Sun, Q., et al. (2008). Sorting signals, N-terminal modifications and abundance of the chloroplast proteome. *PloS ONE, 3*. e1994.

INDEX

Note: Page numbers with "f" denote figures; "t" tables.

A

D

K

L

M

N

R

U

V

W

Y

Z

Color Plates

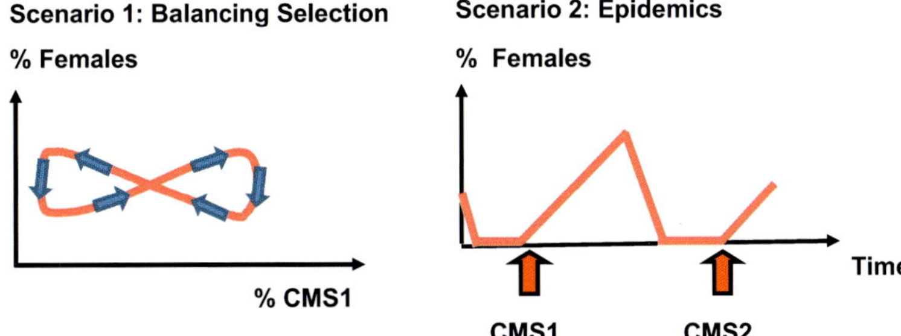

Fig. 4.1 Two evolutionary scenarios of gynodioecy. Scenario 1: Balancing selection. Under the hypothesis of two CMSs (CMS1 and CMS2) with their corresponding dominant restorer alleles, R1 and R2, on two independent loci occurring in the same large population. Restorer alleles are assumed to be costly when useless (R1 on CMS2 or R2 on CMS1). Female frequency varies over time (arrows indicate the evolution over time), as the two CMSs and restorer alleles vary in frequency under balancing selection. When CMS1 is rare, R1 is rare since it is costly on CMS2, which is the common cytoplasm. Most individuals on CMS1 are females. Due to female advantage, CMS1 spreads in the population until R1 is selected. With R1 increasing, plants on CMS1 are restored more and more, losing female advantage. Then CMS2 is mostly found on females and thus increases in frequency at the expense of CMS1. These dynamics can induce cycles of female frequency in the population over time (Gouyon *et al.*, 1991). Scenario 2: Epidemics. The variation in female frequency is due to recurrent arrival of new CMSs. In a cycle, CMS1 spreads in the population due to female advantage until R1 is selected. CMS1 and R1 go to fixation. The population is thus hermaphroditic until the arrival of a new CMS (CMS2). *(Frank, 1989)*

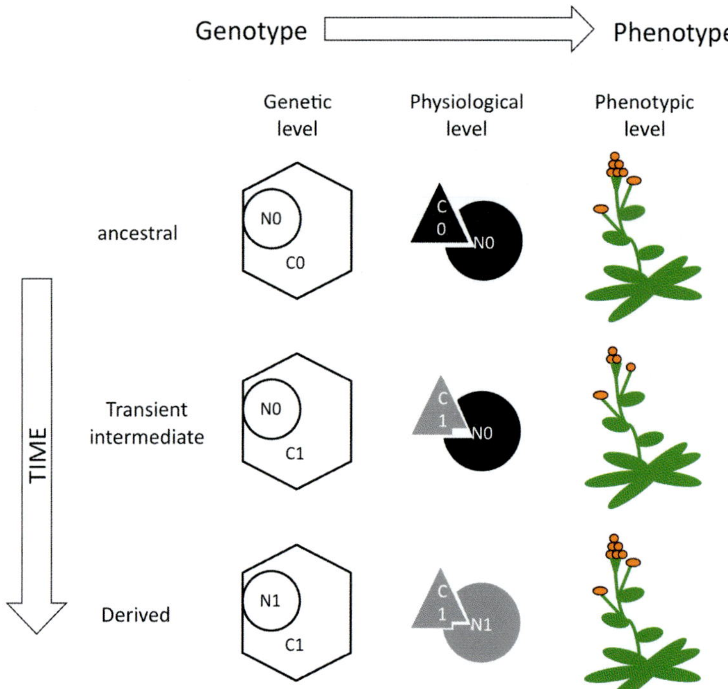

Figure 5.1 Cooperative cytonuclear coadaptation. Each row of the scheme represents one stage of a coevolutionary history, at different integration levels, from the genotype to the plant phenotype. In this figure, we consider the interaction between partners (N and C) that are encoded in the nucleus and an organelle (mt or pt). The nuclear genotype is represented in the circle and is designated by N0 (ancestral allele) or N1 (derived allele). For simplicity, only homozygous individuals are depicted. The cytotype is designated by C0 (ancestral form) and C1 (derived form). The products of these genes are designated with the allele names. At the physiological level, the correct interaction between N and C partners participate in the global phenotype and the fitness of the plant. Transiently, a cytoplasmic variant (C1) may be fixed in a population. The variation may affect the interaction with the N0 partner at the physiological level, and this may affect the global fitness of the plant. It is also possible that, although the N0/C1 interaction is less efficient than the N0/C0 one, the global fitness of the plant in the intermediate stage is not lower than that of the ancestral step, due to new environmental conditions, for example (see text). The N1 new allele for the nuclear-encoded partner restores an optimal interaction with C1 at the physiological level and therefore can be selected because the fitness of plants with the N1/C1 combination will be higher than that of plants with the N0/C1 combination.

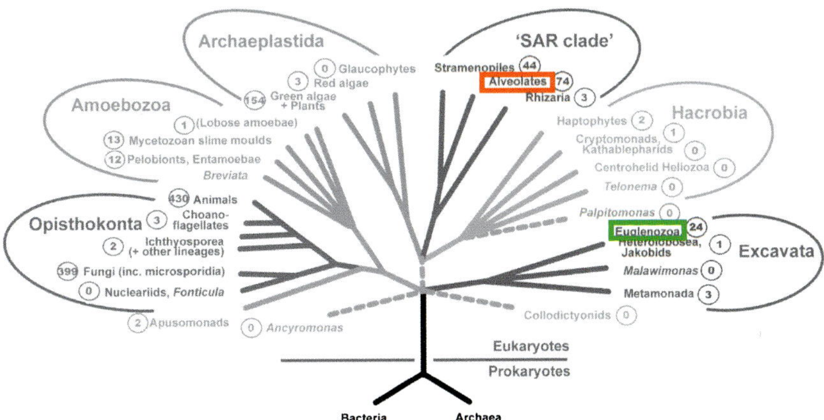

Figure 6.1 Current view of eukaryotic diversity; both protist groups dealt within this chapter are highlighted in colour. The tree is based on Roger and Simpson (2009). The numbers of whole-genome sequencing projects for a given group are shown in circles.

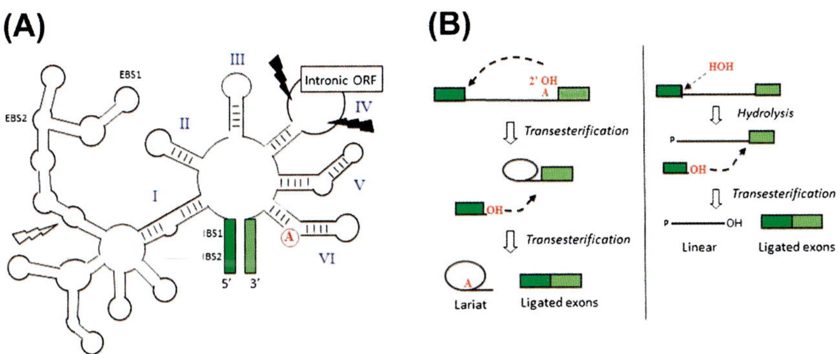

Figure 7.2 (A) Secondary structure model of a classic group II intron, showing exons as green blocks and base pairing within helical domains I–VI. Intronic ORFs (with reverse transcriptase-maturase domains) are located within domain IV, but the reading frame can also extend through domains I–III. Black lightning rod symbols indicate positions of disruption within dIV observed in plant/algal mitochondria that convert the intron from a *cis*-splicing to *trans*-splicing form, and the grey symbol is for an additional break seen in the tripartite *Oenothera berteriana nad2* intron 3 (Knoop *et al.*, 1997). (B) Biochemistry of group II splicing showing the two transesterification steps that generate spliced exons and either a classic lariat excised intron or a linear form if water serves as the initiating nucleophile. Other unusual excised intronic forms have also been observed for certain flowering plant mitochondrial introns that lack conventional group II features (Li-Pook-Than & Bonen, 2006).

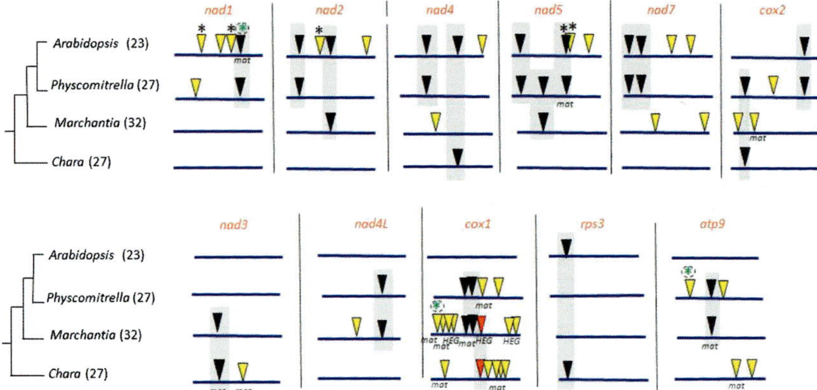

Figure 7.4 Mitochondrial introns that are located at homologous sites in the mitochondrial genomes of selected plants/algae. Representatives of flowering plants (*A. thaliana*, Unseld, Marienfeld, Brandt, & Brennicke, 1997), mosses (*P. patens*, Terasawa *et al.*, 2007), liverworts (*M. polymorpha,* Oda *et al.*, 1992) and green algae (*C. vulgaris*, Turmel *et al.*, 2003). The total number of mitochondrial introns present is shown in parentheses. Only those genes that have at least one shared intron among two or more of these four organisms are shown. Black arrowheads (and shading) indicate shared intron positions, yellow arrows represent non-shared sites and red arrows depict the strong homing site in *cox1*, which is also occupied by an intron in certain flowering plants. Black asterisks indicated *trans*-splicing introns in the plants shown and circled green asterisks indicate that this intron is *trans*-splicing in certain species not shown in the figure, namely, *nad1* in many flowering plants; *atp9* and *cox1* in *S. moellendorfii* (see text). HEG and *matR* depict intronic ORFs of the group I or group II intron type, respectively. *Data are compiled from various sources including Hecht* et al. *(2011) and Turmel* et al. *(2003), and are depicted as in Terasawa* et al. *(2007)*

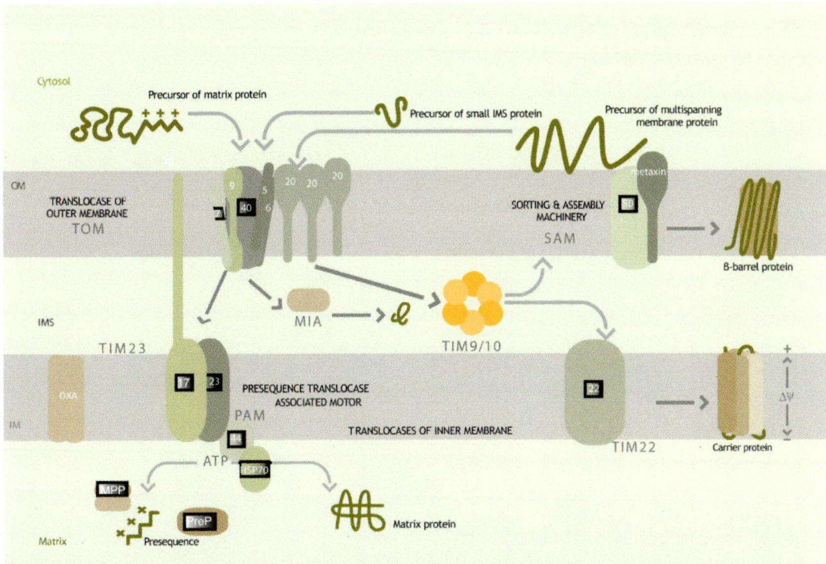

Figure 11.1 ***An overview of the protein import pathways into mitochondria***. The import of precursor proteins via the general, carrier or mitochondrial intermembrane space assembly (MIA) pathways are shown. The numbers relate to the designation of the protein in each complex, e.g. Tom9 retc. The numbers within boxes indicate proteins in which orthologous protein in yeast can be identified.

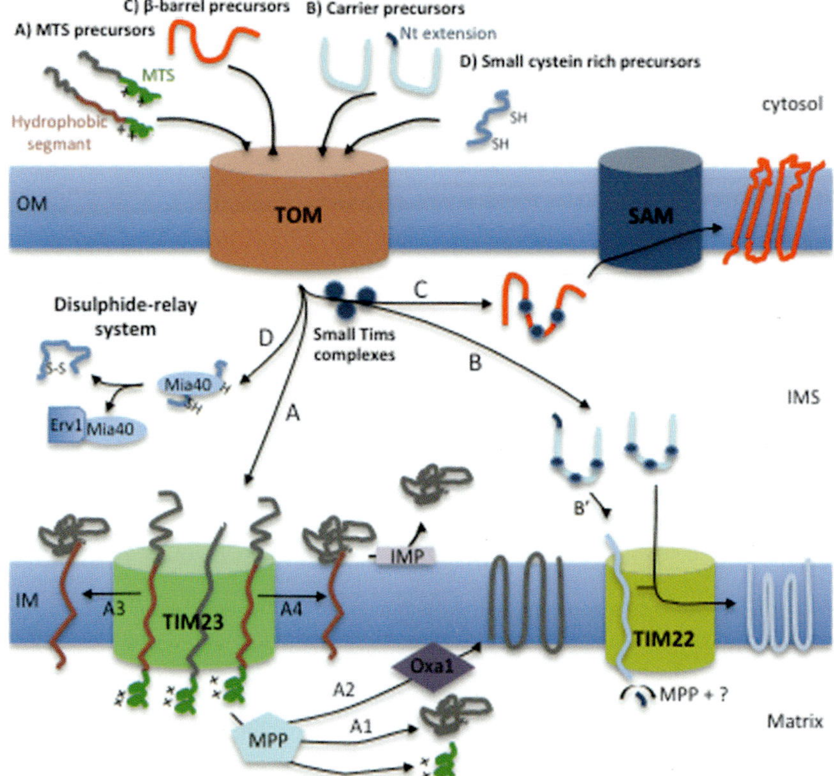

Figure 12.1 Protein import pathways into mitochondria. (A) Import pathways for MTS-containing precursors. After their translocation through the TOM complex, proteins with a cleavable MTS reach the TIM23 complex. MTS are cleaved by the MPP (present in the matrix or the MPP/bc1 complex in yeast and plants, respectively). Matrix proteins and multiple membrane-spanning proteins of the IM are completely translocated into the matrix (A1, A2) and membrane proteins are inserted into the IM with the help of Oxa1 (A2). Transfer of precursors with a hydrophobic segment is stopped in the Tim23 channel and proteins are released into the IM (A3, A4). IMS proteins are released in the IMS by a cleavage catalysed by the IMP (A4). No homologues of IMP have been found in plants and the presence of the A4 pathway is still elusive. (B) Import pathways for carrier proteins, with or without N-terminal extension. N-terminal extension has been found only in plants and mammals. After crossing the TOM complex, carrier precursors bind to the small Tims protein complexes in the IMS. Then they reach the TIM22 complex involved in their insertion into the IM. The N-terminal extension present in some precursors is cleaved by MPP and an unknown endopeptidase (B′). (C) Import pathway of β-barrel precursors. Precursors interact with the small Tims complexes in the IMS after their translocation through the TOM complex. Proteins are inserted into the OM by the SAM complex. (D) Import pathway of small cystein-rich proteins. After translocation in the IMS by the TOM complex, the disulphide relay system catalyses the formation of intramolecular disulphide bonds allowing the folding of the proteins. MTS, mitochondrial targeting signal; TOM, translocase of outer membrane of mitochondria; TIM, translocase of inner membrane of mitochondria; SAM, sorting and assembly machinery; IMS, intermembrane space; MPP, mitochondrial processing peptidase; IM, inner membrane; OM, outer membrane; IMP, intermediate peptidase.

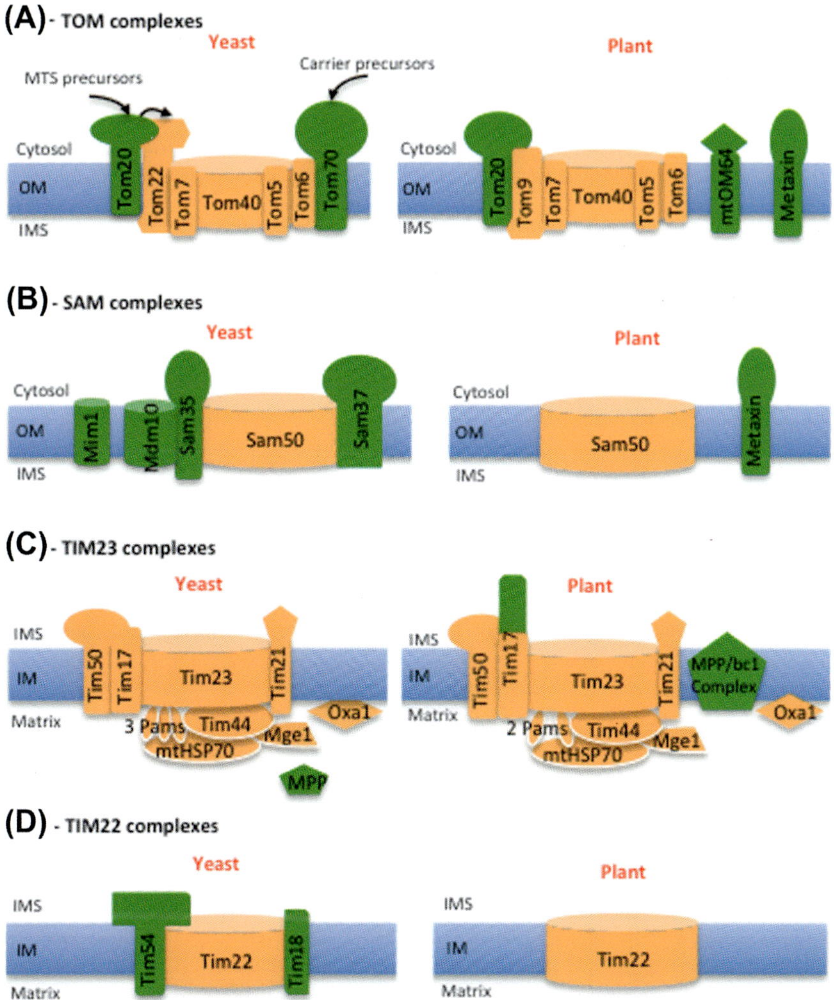

Figure 12.2 Composition of the four main complexes involved in protein import into yeast and plant mitochondria. The evolution-related subunits between yeast and plant are in orange and the non-conserved subunits are in green. (A) Translocase of the outer membrane of mitochondria (TOM) complexes. (B) Sorting and assembly machinery (SAM). Note that the association of metaxin with TOM and/or SAM complexes is still elusive in plants. (C) Translocase of the inner membrane of mitochondria 23 (TIM23) complexes. The mitochondrial processing peptidase (MPP) and Oxa1 are not part of the TIM23 complexes but are involved in the processing of mitochondrial targeting signals (MTS) and the insertion of IM proteins, respectively. Note that in plants, MPP is part of the bc1 complex. (D) Translocase of the inner membrane of mitochondria 22 (TIM22) complexes. IMS, intermembrane space; IM, inner membrane; OM, outer membrane.

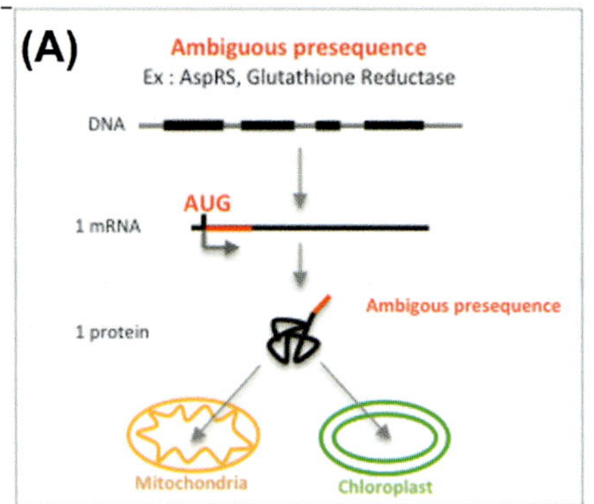
(A)
Ambiguous presequence
Ex : AspRS, Glutathione Reductase
DNA
AUG
1 mRNA
Ambigous presequence
1 protein
Mitochondria
Chloroplast

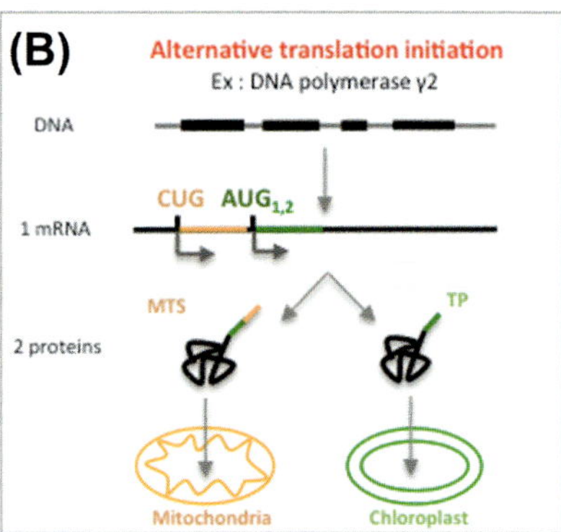
(B)
Alternative translation initiation
Ex : DNA polymerase γ2
DNA
CUG
AUG1,2
1 mRNA
MTS
TP
2 proteins
Mitochondria
Chloroplast

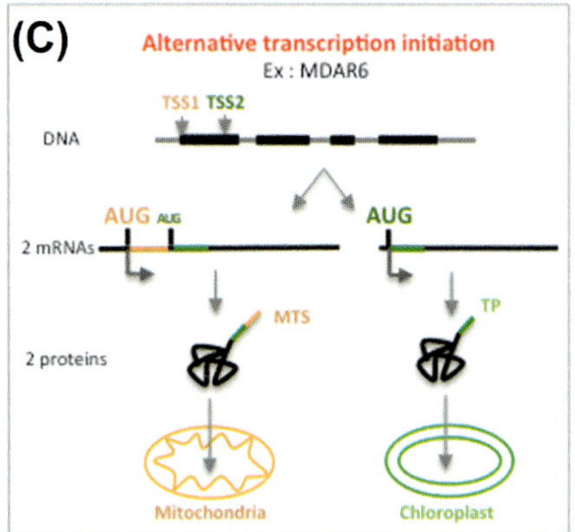
(C)
Alternative transcription initiation
Ex : MDAR6
TSS1
TSS2
DNA
AUG
AUG
AUG
2 mRNAs
MTS
TP
2 proteins
Mitochondria
Chloroplast

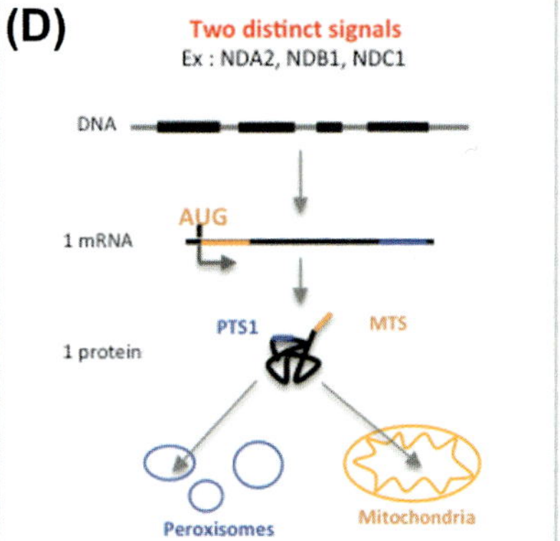
(D)
Two distinct signals
Ex : NDA2, NDB1, NDC1
DNA
AUG
1 mRNA
PTS1
MTS
1 protein
Peroxisomes
Mitochondria

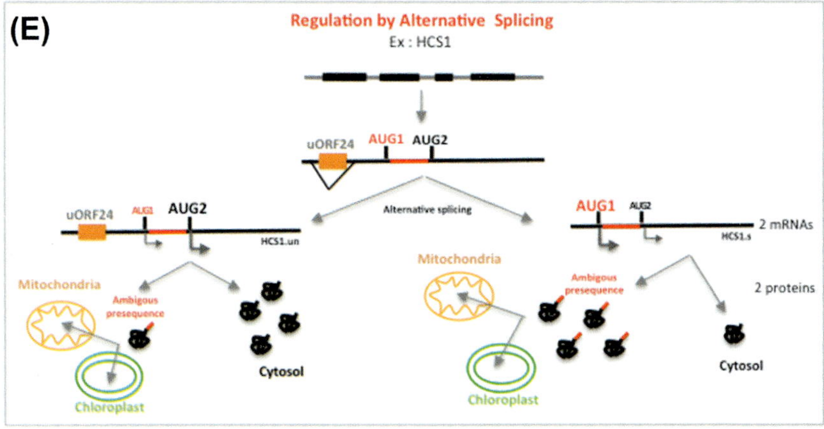
(E)
Regulation by Alternative Splicing
Ex : HCS1
uORF24
AUG1
AUG2
Alternative splicing
uORF24
AUG1
AUG2
HCS1.un
AUG1
AUG2
2 mRNAs
HCS1.s
Mitochondria
Ambigous presequence
Cytosol
Chloroplast
Mitochondria
Ambigous presequence
2 proteins
Cytosol
Chloroplast

Figure 12.4 Examples of mechanisms involved in dual targeting of proteins to mitochondria and another compartments by ambiguous presequences (A) or twin presequences (B–E). (A) Dual targeting of proteins to mitochondria and chloroplasts by ambiguous presequences. Examples of proteins involved in this mechanism are aspartyl-tRNA synthetase (AspRS) or glutathione reductase. (B) Dual mitochondrial–chloroplastic targeting of DNA polymerase Y2 by a mechanism of alternative translation initiation. Translation initiation from the CUG start codon leads to the synthesis of the mitochondrial-targeted protein with a mitochondrial targeting signal (MTS). Proteins synthesized from the two AUG codons ($AUG_{1,2}$) have a transit peptide (TP) and are targeted to chloroplasts. (C) Dual mitochondrial–chloroplastic targeting of monodehydroascorbate reductase 6 (MDAR6) by a mechanism of alternative transcription initiation. Two transcription start sites (TSS) are present on the gene. Translation of the mRNA transcribed from TSS1 leads to a protein with an MTS localized to mitochondria. Translation of the mRNA transcribed from TSS2 leads to a protein with only a TP localized to chloroplasts. (D) Dual targeting of type II NAD(P)H dehydrogenases (ND) to mitochondria and peroxisomes. The proteins have retained one MTS to be targeted to mitochondria and a peroxisomal targeting signal 1 (PTS1) to be targeted to peroxisomes. (E) Regulation of the localization of holocarboxylase synthase 1 (HCS1) to mitochondria, chloroplasts and cytosol by alternative splicing. Translation from the AUG1 codon allows the synthesis of a protein with an ambiguous presequence that is dual targeted to mitochondria and chloroplasts. Translation from the AUG2 gives a cytosolic protein without targeting signals. The main start codon used for translation initiation is regulated by the 5′ UTR of mRNAs. One splicing variant (HCS1.un) presents an upstream open reading frame (uORF24) that favours initiation from AUG2. Translation from the second variant (HCS1.s), lacking uORF24, is mainly initiated at AUG1.

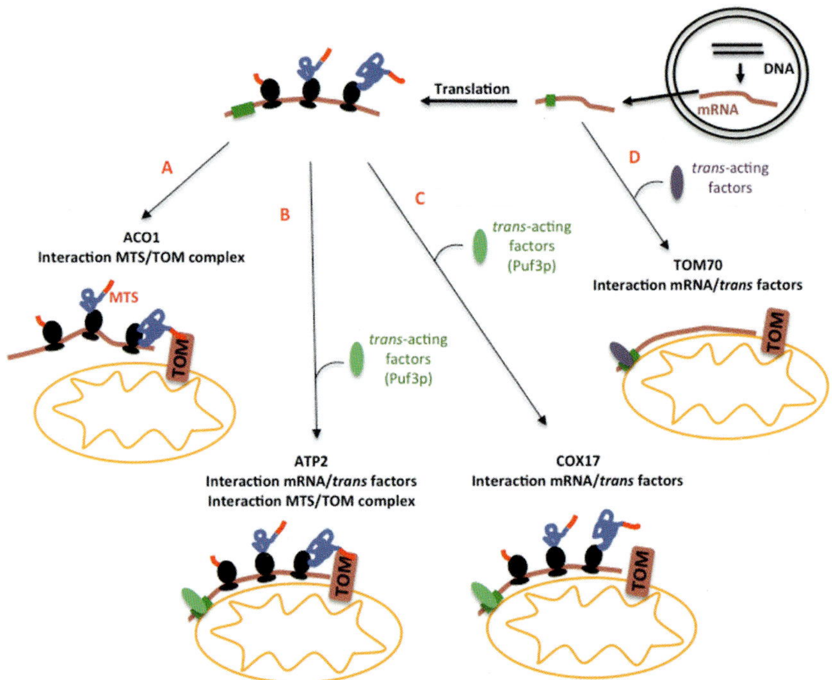

Figure 12.5 Mechanisms of targeting of different mRNAs to mitochondria in yeast. (A) Association of ACO1 mRNA with mitochondria is mainly mediated by interaction between MTS and Tom20. Translation is needed for mRNA localization. (B) Targeting of ATP2 mRNA to mitochondria involves interactions between MTS and proteins of the TOM complex and between mRNA and *trans*-acting factors like Puf3p. Translation is needed for mRNA localization. (C) Association of COX17 mRNA to mitochondria involves interactions between the two P3BS of COX17 and the *trans*-acting factor Puf3p and is dependent on translation. To date, no component of the TOM complex has been demonstrated to be involved in COX17 mRNA targeting. (D) Tom70 mRNA targeting to mitochondria is mediated by unidentified *trans*-acting factors and mRNA translation is not necessary.